Sustainable Crop Production

Sustainable Crop Production

Edited by Alabaster Jenkins

New York

Published by Syrawood Publishing House,
750 Third Avenue, 9th Floor,
New York, NY 10017, USA
www.syrawoodpublishinghouse.com

Sustainable Crop Production
Edited by Alabaster Jenkins

International Standard Book Number: 978-1-68286-374-9 (Hardback)

Cataloging-in-publication Data

Sustainable crop production / edited by Alabaster Jenkins.
p. cm.
Includes bibliographical references and index.
ISBN 978-1-68286-374-9
1. Crop science. 2. Sustainable agriculture. 3. Field crops. I. Jenkins, Alabaster.
SB91 .S87 2017
630--dc23

Printed in the United States of America.

TABLE OF CONTENTS

PREFACE

Sustainable crop production came as a revolution in the industry of crop production and agriculture. This field has emerged as a way of questioning the current practices of crop production which are proving hazardous to the ecology and detrimental to human health. The goal of this text is to strive for environmentally healthy and economically feasible alternative practices. This book contains some path-breaking studies in the field of sustainable crop production which will not only offer a critical insight into theory but will also highlight the recent developments in the field with its applications. In this book, using case studies and examples, constant effort has been made to make the understanding of the difficult concepts of sustainable crop production as easy and informative as possible, for the readers.

This book has been the outcome of endless efforts put in by authors and researchers on various issues and topics within the field. The book is a comprehensive collection of significant researches that are addressed in a variety of chapters. It will surely enhance the knowledge of the field among readers across the globe.

It gives us an immense pleasure to thank our researchers and authors for their efforts to submit their piece of writing before the deadlines. Finally in the end, I would like to thank my family and colleagues who have been a great source of inspiration and support.

Editor

1

The Phytotoxin Coronatine Induces Abscission-Related Gene Expression and Boll Ripening during Defoliation of Cotton

Mingwei Du[1]ⁱ, Yi Li[1], Xiaoli Tian[1], Liusheng Duan[1], Mingcai Zhang[1], Weiming Tan[1], Dongyong Xu[2], Zhaohu Li[1]*

1 State Key Laboratory of Plant Physiology and Biochemistry, Engineering Research Center of Plant Growth Regulator, Ministry of Education, College of Agronomy and Biotechnology, China Agricultural University, Beijing, China, 2 Hebei Provincial Engineering Research Center of Cotton Seed, Hejian, Hebei, China

Abstract

Defoliants can increase machine harvest efficiency of cotton (*Gossypium hirusutum* L.), prevent lodging and reduce the time from defoliation to harvest. Coronatine (COR) is a chlorosis-inducing non-host-specific phytotoxin that induces leaf and/or fruit abscission in some crops. The present study investigates how COR might induce cotton leaf abscission by modulating genes involved in cell wall hydrolases and ACC (ethylene precursor) in various cotton tissues. The effects of COR on cotton boll ripening, seedcotton yield, and seed development were also studied. After 14 d of treatment with COR, cells within the leaf abscission zone (AZ) showed marked differentiation. Elevated transcripts of *GhCEL1*, *GhPG* and *GhACS* were observed in the AZs treated with COR and Thidiazuron (TDZ). The relative expression of *GhCEL1* and *GhACS* in TDZ treated plants was approximately twice that in plants treated with COR for 12 h. However, only *GhACS* expression increased in leaf blade and petiole. There was a continuous increase in the activity of hydrolytic enzymes such as cellulase (CEL) and polygalacturonase (PG), and ACC accumulation in AZs following COR and TDZ treatments, but there was greater increase in ACC activity of COR treated boll crust, indicating that COR had greater ripening effect than TDZ. Coronatine significantly enhanced boll opening without affecting boll weight, lint percentage and seed quality. Therefore, COR can be a potential cotton defoliant with different physiological mechanism of action from the currently used TDZ.

Editor: Jinfa Zhang, New Mexico State University, United States of America

Funding: This work was supported by the National Natural Science Foundation of China (31301257), the Special Fund for Agro-scientific Research in the Public Interest (201203057-02), China Agriculture Research System (CARS-18-18), and China Postdoctoral Science Foundation (2013M530077). The funders had no role in study design, data collection and analysis, decision to publish, or preparation of the manuscript.

Competing Interests: The authors have declared that no competing interests exist.

* E-mail: lizhaohu@cau.edu.cn

These authors contributed equally to this work.

Introduction

Cotton is an important commercial crop worldwide, and serves as a significant source of fiber, feed, foodstuff, oil and biofuel [1]. Defoliation or leaf abscission is induced in cotton as a natural physiological process which usually is inadequate or not timely enough for a complete mechanical harvest of cotton. Therefore, defoliation before harvest is often induced by managing the plants so that senescence, abscission (separation) layer development and leaf drop are encouraged [2,3]. The ultimate goal of defoliants is to facilitate mechanical harvest, reduce trash and protect fiber and seed quality from weathering and staining by allowing earlier harvest [4]. Another benefit is the reduced moisture content in the raw fibers and seed which is essential for storage of seedcotton.

Selection of appropriate abscission chemicals is one of the critical decisions in cotton production. Herbicidal or hormonal defoliants, such as dimethipin and thidiazuron, are widely used in many cotton producing areas [5]. Dimethipin is a plant growth regulator used as a harvest aid on a variety of crops [6]. It causes leaf cells to slowly lose water and generates ethylene within plants. Dimethipin is considered a contact-type defoliant, whereas thidiazuron has growth-regulator properties and moves through the plant [7]. Thidiazuron increases the concentration of ethylene relative to auxin in leaf petioles and results in the activation of the leaf abscission layer [8,9]. However, these types of defoliants induce drastic leaf abscission which inhibits timely transport of nutrients from leaves to cotton bolls. Also, these defoliants do not directly influence boll ripening and must be applied in combination with ethephon, a boll opener, to provide satisfactory defoliation and boll opening [5]. An abscission chemical with improved defoliation and boll opening properties is needed for cotton harvest practices.

Coronatine (COR) is a chlorosis-inducing non-host-specific phytotoxin produced by several members in the *Pseudomonas syringae* group of pathovars [10,11]. It induces inhibition of root elongation, senescence, production of defense-related protease inhibitors, and resistance to abiotic stresses [12–16]. COR also induces growth regulator-like effects such as hypertrophy and stimulation of ethylene production and tendril coiling [17–20]. In addition, COR has been reported to be a structural and functional analog of jasmonic acid and methyl jasmonate, which are important plant growth substances in octadecanoid signaling

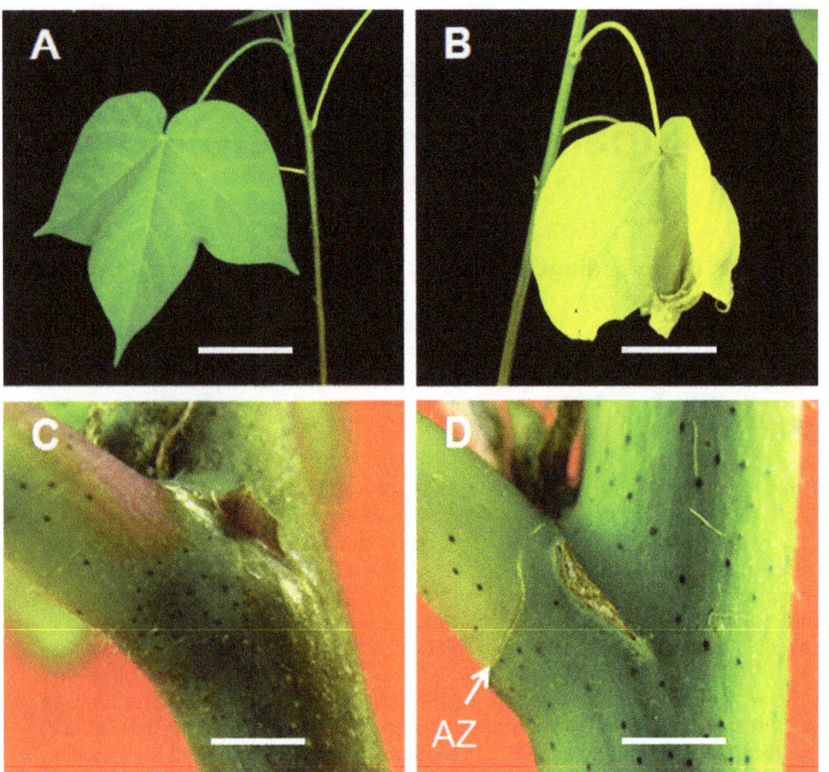

Figure 1. Phenotypic changes in the leaf blade and petiole of coronatine (COR) and water (Control) treated plants. **A** and **C** are phenotypes for 14 d distilled water treated materials, **B** and **D** are phenotypes for 14 d COR treated materials. AZ: leaf abscission zone. *Bar:* **A, B** 4 cm; **C, D** 3 mm.

[21–23]. Components of the octadecanoid pathway have been shown to affect the regulation of wounding [24], fruit ripening [25], and abscission [26]. External application of methyl jasmonate and COR likely induced abscission by stimulating levels of ethylene when applied to the entire citrus (*Citrus sinensis*) tree canopy [27,28]. However, the ability of COR to cause leaf abscission in cotton is unclear.

Abscission is the main process that involves structural, biochemical, and molecular changes resulting in the detachment of plant organs, including leaves, flowers and fruits [29,30]. Abscission occurs at predetermined sites referred to as abscission zones, which consist of a few layers of small, densely packed cells that respond in different ways from neighbouring cells to the same hormonal or environmental cues [31,32]. Knowledge of mechanisms involved in abscission of leaves or other organs is essential to develop strategies to control them and improve harvesting practices or unwanted crop loss in fruit crops [33]. Once abscission is initiated, cells in the abscission zone begin to enlarge, followed by increased expression of genes and the activities of cell wall-degrading enzymes such as β-1, 4-glucanase or cellulase (CEL) and polygalacturonase (PG) [32,34–36]. As a result, the middle lamellae of abscission zone cells dissolve and, ultimately, the organ abscises.

Ethylene plays a primary role in accelerating leaf abscission and fruit ripening [37–40]. The conversions of S-AdoMet (SAM) to 1-aminocyclopropane-1-carboxylic acid (ACC, metabolic precursor of ethylene) is the rate-limiting step in ethylene biosynthesis, and is catalysed by ACC synthese (ACS) [39,41]. The observations that expression of the ACS genes is highly regulated by a variety of signals and that active ACC synthase is labile and present at low levels suggest that ethylene biosynthesis is tightly controlled [41]. Both positive and negative feedback regulation of ethylene biosynthesis have been reported in different plant species [30,31,42–44]. Most studies addressing ACS regulation have focused on ACS gene expression in response to various endogenous cues and environmental stimuli [31,41,42]. In an attempt to understand how responses to COR operate, some physiological- and transcriptional- level responses of cotton to the application of COR need further study.

The purpose of this study was first to investigate the possible roles of COR during cotton leaf abscission compared with using TDZ or water (control). In the present work, the phenotypic and anatomical changes in leaves, leaf detachment force (break strength), activity of abscission-related enzymes, and expression of genes encoding the enzymes in different cotton tissues were determined under greenhouse and/or field conditions. We also estimated the transcript levels of two hydrolytic enzyme genes (*GhCEL1* and *GhPG*) and one ethylene biosynthesis enzyme gene (*GhACS*) in leaf, petiole and leaf abscission zone as well as during leaf abscission. Finally, we determined boll opening, seedcotton yield and seed quality to elucidate whether and how COR affects cotton boll ripening and seed development.

Materials and Methods

Plant Material and Coronatine Preparation

The cotton cultivar, Guoxin 3 (GX 3), was selected for the experiment. Seeds of GX 3 were provided by Guoxin Corporation, China. Standard coronatine was provided by Carol L.

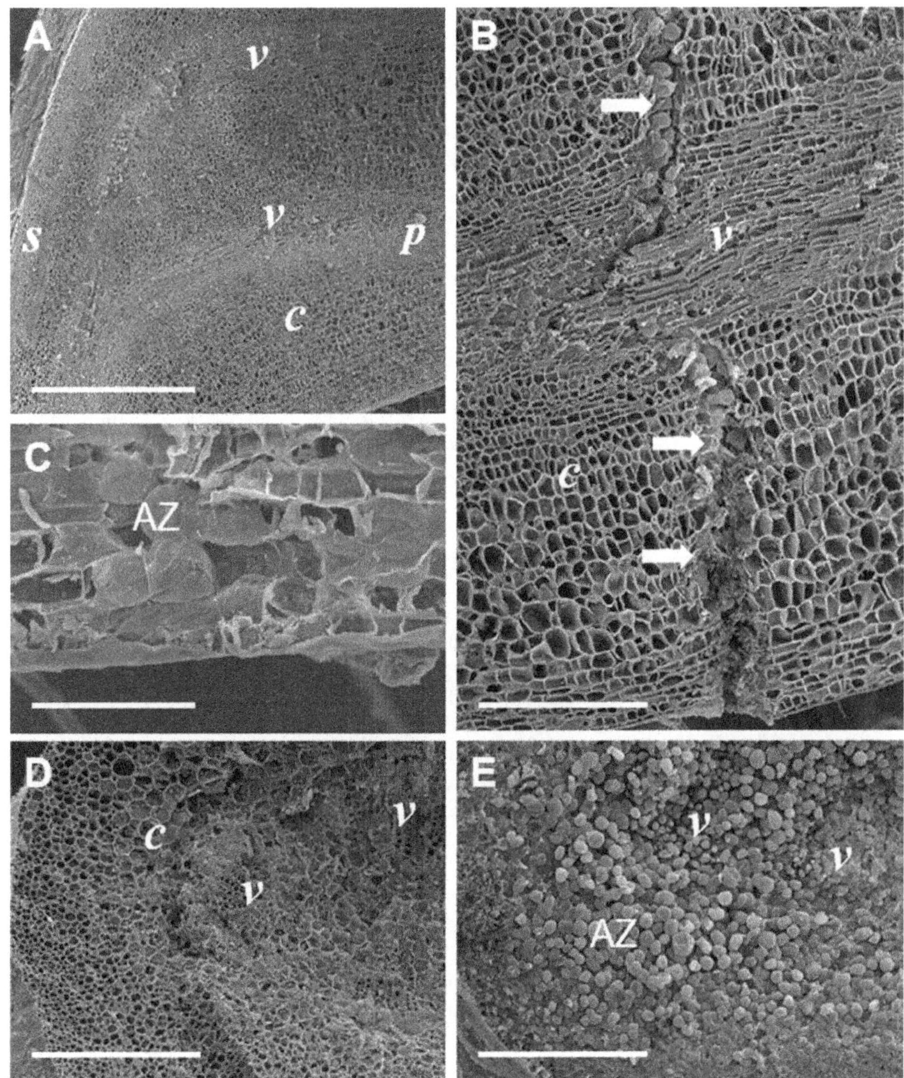

Figure 2. Scanning electron micrograph of cells at the petiole and stem juncture, abscission zone (A, B, C), and fracture plane (D, E) of the cotton leaf abscission zone. **A** and **D** are micrographs of 14 d distilled water treated materials, **B**, **C** and **E** are micrographs of 14 d COR treated materials. *s*: stem, *v*: vascular bundles, *c*: cortex, *p*: petiole, AZ: leaf abscission zone. *Bar*: **A** 1 mm; **B** 400 μm; **C** 100 μm; **D**, **E** 500 μm.

Bender, Oklahoma State University, Stillwater, OK, USA. The coronatine was prepared as described in Palmer and Bender [45].

Experiment 1

Seeds of GX 3 were sown in 28 cm diameter pots maintained in a glasshouse under controlled temperature (30±3°C) for about 2 months until the 7th true leaf stage which was approximately 35 days after sowing. At this growth stage, 300 mg L^{-1} COR and TDZ solution were applied evenly to the 7th leaves of ten randomly selected plants at a rate of 1 ml per leaf. Distilled water was similarly applied to the 7th leaves of another ten randomly selected plants as a control. The leaf abscission zone (AZ) was sampled after COR treatment for observation under the electron microscope. Break strength and abscission-related gene expression were determined.

Experiment 2

Seeds of GX 3 were sown in the field on 29 April 2010 and 27 April 2011. The experimental unit consisted of four rows, 6.5 m long and 0.9 m apart. A randomized complete block design with three replications was used each year. The thidiazuron (TDZ) and coronatine (COR) concentration was 300 mg L^{-1}, each applied at 225 L ha^{-1}. All treatments were applied during 45–50% boll opening in late September. Break strength, defoliation and ripening effects, cotton yield, and seed quality were examined. Leaf abscission zones (1–2 mm on either side of fracture plane) and other tissues, including leaf blade, petiole and boll crust were harvested, frozen in liquid nitrogen and stored at −80°C for the analysis of hydrolytic enzymes and ACC activities in 2011.

Scanning Electron Microscopy of Leaf Abscission Zone (AZ)

For the electron microscopy observations, the abscission zones of the 7th leaf were fixed in 4% (v: v) glutaraldehyde in 0.5 mol L^{-1} potassium phosphate buffer (pH 7.4) for 4 h at 25°C, rinsed four times in buffer, and then dehydrated in ethanol through a series of increasing concentrations. Sputter coated sections were

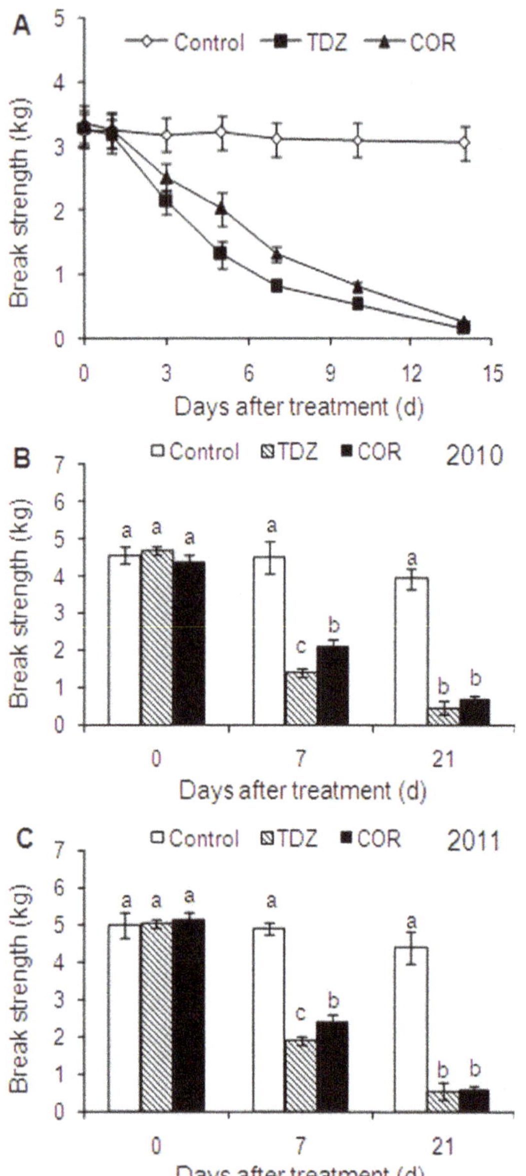

Figure 3. Changes in break strength in abscission zone and adjacent cells treated with water (Control), thidiazuron (TDZ) and coronatine (COR) under glasshouse (A) and field (B, C) conditions. Each value represents the mean ± SE of three replicates. Bars with the same letters are not significantly different.

then examined at different magnifications with Hitachi S-3400N scanning electron microscope (Hitachi, Japan).

Determination of Break Strength

Break strength was measured as the force necessary to cause the petiole to separate from the stem across the abscission zone according to Malladi and Burns [33]. This measurement was used as an indicator of the progressive weakening of the tissue in this area at different time points. Petioles from uppermost 3 or 4 nodes were clamped to a digital force gauge (Tayasaf Corporation, China) and force was applied mechanically to the stem. The force at which separation occurred was recorded as the break strength. Petioles from ten randomly selected plants per replicate were used for each treatment. Break strength values obtained were recorded to compute the average break strength per node. Each treatment was repeated three times.

Hydrolytic Enzyme and ACC Activities

Tissues were pulverized in a mortar under liquid nitrogen and the powder resuspended in extraction buffer (100 mM Tris–HCl, 0.5% PVPP, 10 mM $MgCl_2$,10 mM $NaHCO_3$, 10 mM DTT, 0.15 mM PMSF, 0.3% (w: v) X-100 Triton and 0.03% sodium azide). The homogenised liquid was filtered through a nylon mesh and centrifuged at 20 000 g for 20 min. The supernatant was dialysed for 16 h at 4°C in extraction buffer diluted 1:9 (v: v) in water. The samples were then frozen until used.

The extracts were assayed to determine cellulase (CEL) and polygalacturonase (PG) activities, by the viscosity method [46]. A unit of enzyme was expressed as specific activity (U mg^{-1} protein), being the reciprocal of the time in hours to obtain the 50% viscosity loss $\times 10^3$.

ACC contents were analysed as previously described in Yuan et al. [31]. Tissues were ground to a fine powder in liquid nitrogen using a prechilled mortar and pestle. The powdered tissue was transferred to a centrifuge tube and 10 ml of 80% ethanol was added. The homogenate was centrifuged at 10 000 g for 30 min after incubating the powdered tissue in ethanol at 65°C for 15 min. The residue was reextracted in 10 ml of 80% ethanol at 65°C for 15 min. The supernatants were combined and dried under vacuum. The dry pellet was dissolved in 1 ml of water and extracted once with an equal volume of chloroform. The aqueous phase was collected by centrifugation, dried under vacuum, and redissolved in 0.7 ml of water.

Quantitative Real-time Polymerase Chain Reaction (qRT-PCR) Analysis

Total RNA was extracted from the leaf abscission zone, leaf blade and petiole using Trizol according to the supplier's recommendation. Residual DNA was removed with a purifying column. One microgram of total RNA was reverse transcribed using 0.5 μg of oligo(dT) 18 (Invitrogen) and 200 units of Superscript II (Invitrogen) following the supplier's recommendation. On the basis of expressed sequence tag (EST) sequences, the gene-specific primers were designed and used for amplification [47].

The PCR amplification was performed with gene-specific primers. Primer sequences were designed as follows: *GhCEL1*, forward primer 5′-TTATGGAGAGGTGGGCGATGGT-3′ and reverse primer 5′-CGGATTGCTTGGGTCTTTCTTGT-3′; *GhPG*, forward primer 5′-CACTGCGGCATATGTGTCTAA-3′ and reverse primer 5′-CCTCCCTGCCATGTTTTTATT-3′; *GhACS*, forward primer 5′-GGACTTGTGGCGAGTGAT-TATC-3′ and reverse primer 5′-AAGCAAACCCTGAAC-CAACC-3′. The *GhUBQ* gene was used as an internal control to normalize small differences in template amounts with the forward primer 5′-AAGAAGAAGACCTACACCAAGCC-3′ and the reverse primer 5′-GCCCACACTTACCGCAATA-3′.

An Applied Biosystems 7500 Fast Real-Time PCR System (Applied Biosystems, USA) was used for quantitative real-time PCR analyses. Analysis was performed on 1 μl of diluted cDNA in a final reaction volume of 20 μl using the SYBR® Green PCR Master Mix (Applied Biosystems). The PCR conditions consisted of denaturation at 95°C for 3 min, followed by 40 cycles of denaturation at 95°C for 30 s, annealing at 62°C for 30 s, extension at 72°C for 30 s, and a final elongation step of 7 min at 72°C. Primer concentration was optimized and primer validation

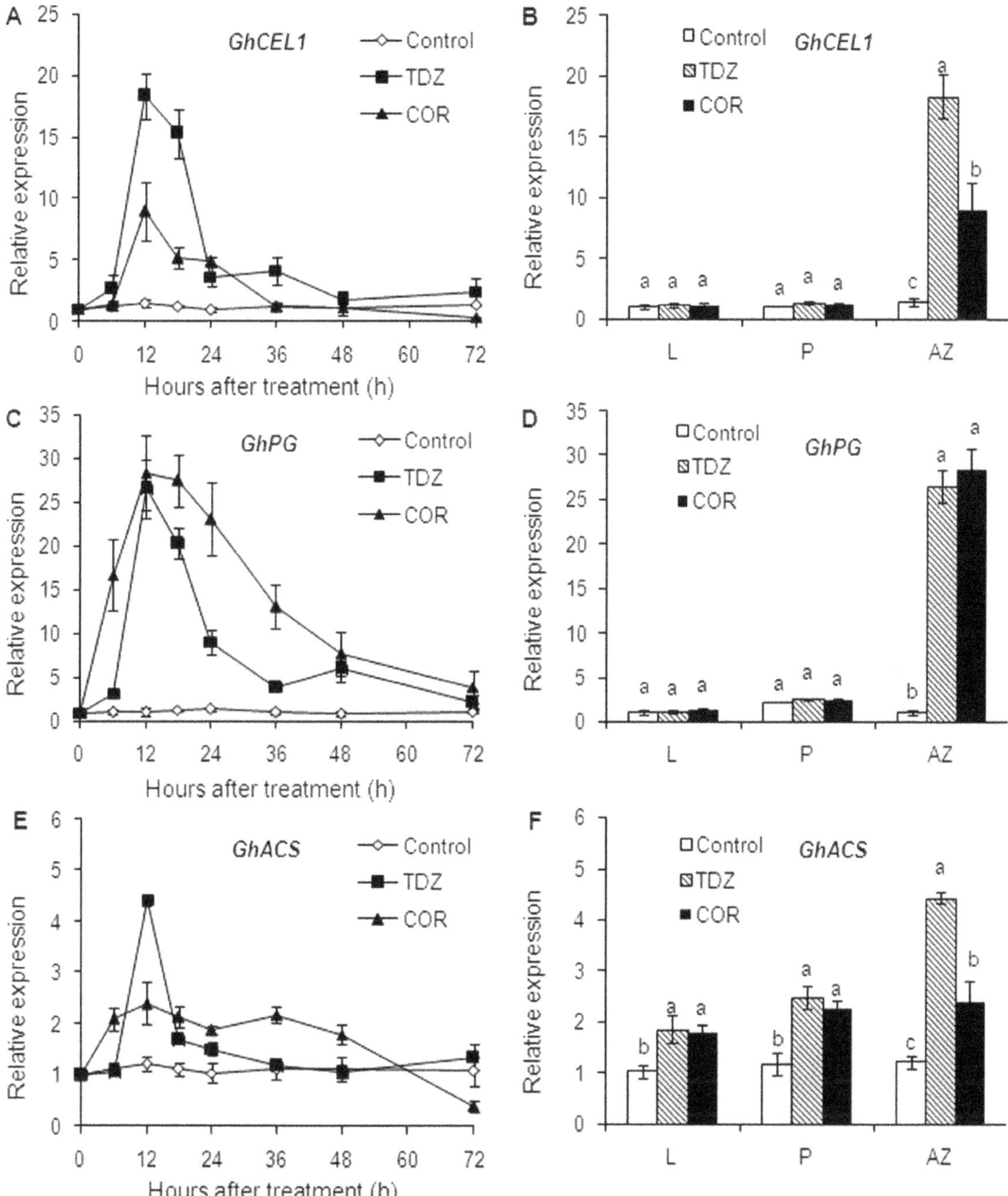

Figure 4. Temporal changes in the relative expression of *GhCEL1*, *GhPG* and *GhACS* in AZ treated with water (Control), thidiazuron (TDZ) and coronatine (COR), and signalling pathways in various cotton tissues at 12 hours after treatment. L: leaf; P: petiole; AZ: leaf abscission zone. Each value represents the mean ± SE of three replicates. Bars with the same letters are not significantly different.

was performed to enable relative gene expression analysis using the ΔΔCt method [48].

Defoliation and Boll Opening

Prior to treatment application, 10 plants were randomly tagged from two rows at the center of each plot to count the number of leaves on each plant. The number of leaves was counted again 21 days after treatment (DAT) on the same tagged plants. Defoliation percentage was calculated by equation (1). Opened bolls were determined 21 DAT on the same 10 plants tagged. Bolls on each plant were examined and recorded as either opened or closed and boll opening percentage was calculated by equation (2).

$$DP = \frac{Lb - La}{Lb} \times 100\% \qquad (1)$$

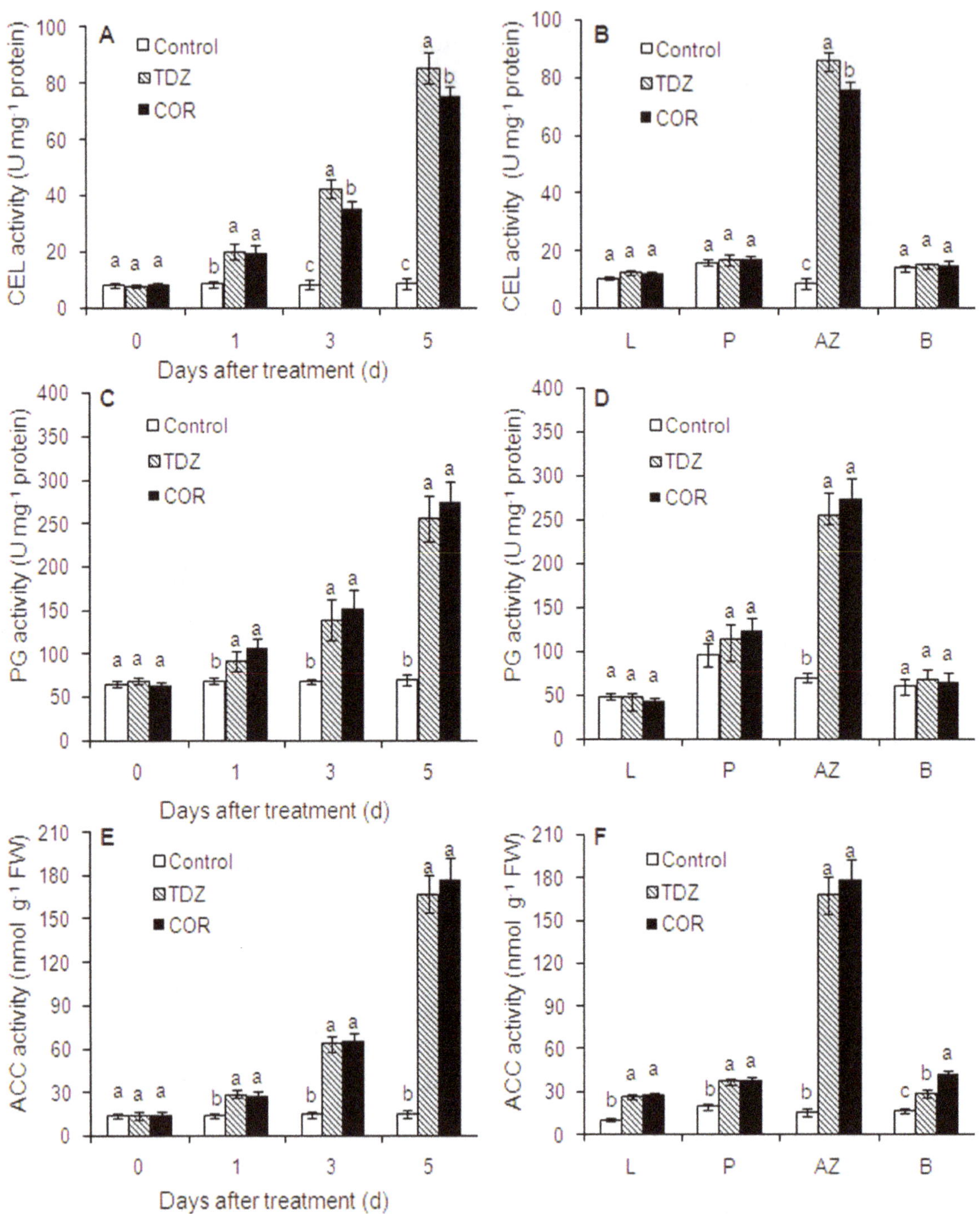

Figure 5. Temporal changes in the activity of cellulase (CEL), polygalacturonase (PG) and ACC in AZ treated with water (Control), thidiazuron (TDZ) and coronatine (COR). Different tissues from Control, TDZ or COR treated plants for 5 d were collected and activities measured. L: leaf; P: petiole; AZ: leaf abscission zone; B: boll crust. Each value represents the mean ± SE of three replicates. Bars with the same letters are not significantly different.

$$BO = \frac{Ob}{Tb} \times 100\% \quad (2)$$

Where DP = Defoliation percentage; Lb = Number of leaves before treatment; La = Number of leaves at 21 days after treatment; BO = Boll opening percentage; Ob = Open bolls; Tb = Total number of bolls.

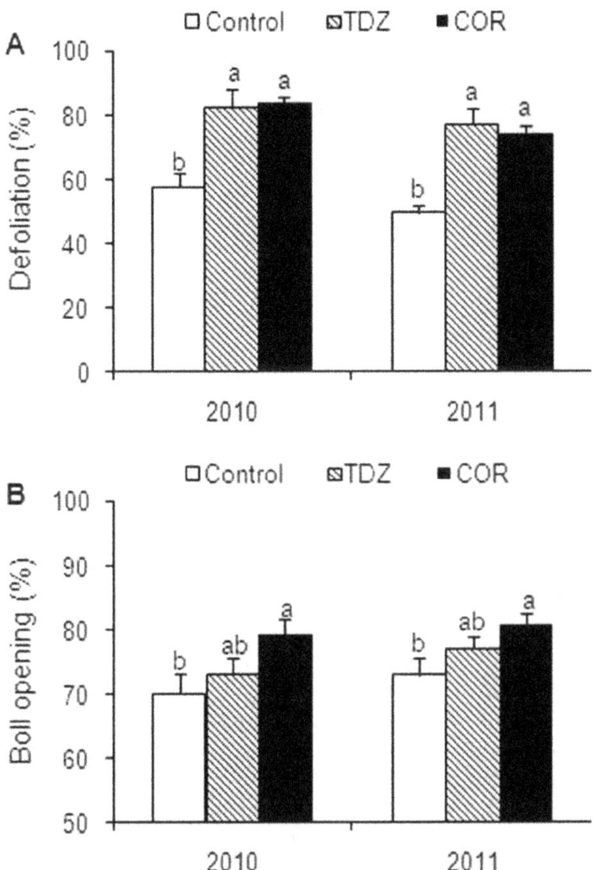

Figure 6. Effect of Thidiazuron (TDZ) and coronatine (COR) on defoliation and boll opening at 21 days after treatment in 2010 and 2011. Each value represents the mean ± SE of three replicates. Bars with the same letters are not significantly different.

Yield and Seed Quality

Each year, plants from the central two rows in each plot were harvested by hand two times. The first harvest was 21 DAT. The final harvest occurred on 28 October in 2010 and 2 November 2011, about 2 weeks after the first harvest. Seed cotton from each plot was weighed, and subsamples (~1 kg) were collected, air-dried, and ginned on a 10-saw, hand-fed laboratory gin. Ginning percentage was determined after ginning. Boll weights were determined from 20 randomly selected plants in the central two rows from each plot at first harvest. Seed quality parameters such as seed index (fresh weight per 100 seeds (g)) and germination percentage were determined.

Statistical Analysis

The experimental data were subjected to an analysis of variance and treatment means were compared using the least significant difference (LSD) at the 5% probability level.

Results

Changes in Phenotypic and Anatomical Features of Leaves during Abscission Induced by Coronatine

Abscission was accelerated when 300 mg L^{-1} coronatine (COR) solution was administered to cotton leaves (Fig. 1). Disassembly of cell walls in the leaf abscission zone (AZ) should lead to altered anatomical features in this separation layer. The

Table 1. Effect of Thidiazuron (TDZ) and coronatine (COR) on seedcotton yield and seed quality in 2010 and 2011.

Year	Treatment	1st harvest yield (kg ha^{-1})	Total yield (kg ha^{-1})	1st harvest percentage (%)	Boll weight (g)	Ginning percentage (%)	Seed index (g)	Germination percentage (%)
2010	Control	2599.8^{b*}	3650.8^{a}	71.2^{b}	5.67^{a}	39.1^{a}	10.9^{a}	91.3^{a}
	TDZ	2877.1^{a}	3681.4^{a}	78.2^{ab}	5.40^{a}	38.7^{a}	10.1^{a}	90.2^{a}
	COR	3049.9^{a}	3703.8^{a}	82.6^{a}	5.53^{a}	39.4^{a}	10.8^{a}	91.8^{a}
2011	Control	3011.9^{b}	3925.5^{a}	76.7^{b}	5.86^{a}	38.3^{a}	11.3^{a}	89.8^{a}
	TDZ	3289.1^{ab}	3909.4^{a}	84.0^{ab}	5.83^{a}	39.2^{a}	10.7^{a}	88.7^{a}
	COR	3461.8^{a}	3978.5^{a}	87.1^{a}	5.96^{a}	38.7^{a}	10.2^{a}	90.3^{a}

*For each factor, means within the same column followed by different letters differ significantly (P≤0.05).

AZs of plants treated with COR and their control were examined under scanning microscopy in order to elucidate the anatomical alterations in AZs (Fig. 2). Compact, well-organized and pentagonal cells were observed on the petiole and stem junction (Fig. 2 A, D) of control plants 14 d after treatment with water; in the COR treated plants, cells of the abscission zone became differentiated and formed (Fig. 2 B, C, E). The treated cells appeared to be elongated and disorganized with a thin cell wall compared to the control.

Changes in Break Strength of AZ during Leaf Abscission Induced by COR and TDZ

A significant decrease in break strength was observed in TDZ- and COR-treated plants (Fig. 3A). Although break strength in COR treatment was higher than that in TDZ treatment at 7 DAT, no difference was observed between both treatments at 21 DAT under field conditions (Fig. 3B, C). The break strength in TDZ and COR treatments decreased by approximately 87% at 21 DAT in both 2010 and 2011.

Changes in Relative Expression of *GhCEL1*, *GhPG* and *GhACS* during Leaf Abscission Induced by COR and TDZ

To determine the mechanism of COR induced leaf abscission, we analyzed the expression patterns of several abscission-related genes. Elevated transcripts of *GhCEL1*, *GhPG* and *GhACS* were observed in AZs treated with COR and TDZ (Fig. 4A, C, E). The relative expression of *GhCEL1* and *GhACS* in TDZ treated plants was approximately twice as much as in plants treated with COR for 12 h. However, prolonged expression of *GhPG* and *GhACS* was detected in COR treatment in comparison to TDZ treatment.

Expressions of *GhCEL1*, *GhPG* and *GhACS* were also observed in other tissues such as leaf and petiole at 12 h (Fig. 4B, D, F). No significant effects of TDZ and COR treatments were observed for *GhCEL1* and *GhPG* expression in any tissues other than the leaf abscission zone. A substantial increase in *GhACS* expression was observed in leaf and petiole following TDZ and COR treatment.

Changes in Activities of Cellulase (CEL), Polygalacturonase (PG) and ACC during Leaf Abscission Induced by COR and TDZ

The activities of CEL, PG and ACC in different tissues and AZ during TDZ and COR induced abscission are shown in Fig. 5. There was a continuous increase in the activities of the three enzymes in AZs under TDZ and COR treatments (Fig. 5A, C, E). A 4.9- and 9.7-fold increase in CEL activity was observed in the AZs of TDZ treated plants at 3 and 5 DAT. Similarly a continuous increase (4.1- and 8.6-fold) in cellulase activity was observed in AZs of COR treated plants at 3 and 5 DAT, respectively. A substantial increase in ACC and PG activities were observed after TDZ and COR treatment although no difference was observed between these two treatments in each enzyme.

The CEL, PG and ACC activities were also observed in other tissues such as the leaf, petiole, and boll crust at 5 DAT (Fig. 5B, D, F). No significant effects of TDZ and COR treatments were observed on CEL and PG activities in any tissue other than the leaf abscission zone. However, a substantial increase in ACC activity was observed in petiole, leaf abscission zone, and boll crust after TDZ and COR treatment. In addition, a 50.1% increase in ACC activity of COR treated boll crust relative to the treatment of TDZ was observed.

Changes in Defoliation and Boll Opening of Cotton Treated with COR and TDZ

Defoliation was increased by TDZ and COR treatments at 21 DAT in both experiments of 2010 and 2011 (Fig. 6). Whereas the defoliation percentage for the control plants averaged 54.2% in 2010 and 2011, it averaged above 80.0% for the TDZ and COR treatments. Boll opening increased by about 8.3% in the COR treatment but not was significantly increased in the TDZ treatment.

Changes in Seedcotton Yield and Seed Quality following Treatment with COR and TDZ

First harvest yield and first harvest percentage significantly increased in the COR treatments, but not in TDZ treatment except the first harvest yield in 2010 (Table 1). Although the difference between COR and TDZ treatments was not significant, a trend was noticed that COR treatment was more effective in increasing the first harvest yield. For the controls, the first harvest yield ranged from 70.8 to 77.1% of total yield. This percentage increased to about 83.4 to 87.3% of the total yield in the COR treatment. Boll weight, ginning percentage, seed index, and germination percentage were not influenced by COR treatment.

Discussion

Appropriate and safe abscission chemicals will improve timing and facilitate harvest of cotton. In this study, we demonstrated that the phytotoxin, coronatine induced leaf abscission during cotton defoliation. Abscission occurs in an anatomically distinct cell layer known as the abscission zone (AZ) [49]. The abscission zone is defined as the region at base of abscising organs through which abscission eventually occurs. The anatomy of abscission is important for understanding the biology of a given plant species since form and structure comprise an appropriate starting point for potential functional comparisons between botanically distinct organs [50,51]. Our data showed that abscission was accelerated when COR solution was applied to cotton leaves at 300 mg L^{-1}. Disassembly of cell walls in the AZ should lead to alteration in anatomical structures in this separation layer. Leaf abscission zone cells were examined by scanning microscopy to elucidate the anatomic mechanisms of COR induced abscission in cotton leaves. After 14 d treatment with COR, the cells of AZ became elongated and disorganized, and the cell wall became thinner than that of control plants. It was also observed that COR alone could initiate the abscission process. The enlarged cells of the abscission zone seemed to have undergone a programmed cell death or physical dissolution in which the cells lost integrity. These results are consistent with a previous argument that while the abscission zone consists of several layers of cells across the petiole, the vascular bundles remain intact, allowing transportation of water and nutrients in and out of leaves [52].

The COR treated leaf abscission zone showed a greater decrease in break strength than the control, suggested that the COR effect was over and above the wounding effect. Similar observations have been made in citrus fruit abscission zones in which the break strength decreased after COR treatment [28]. The break strength under COR treatment was higher than that under TDZ treatment at 7 DAT, but not at 21 DAT. This suggests that leaf abscission induced by COR is relatively moderate, and could allow timely nutrient transport from cotton leaves to bolls.

High synthesis and activities of cell wall hydrolases, including β-1, 4-glucanase or cellulase (CEL) and polygalacturonase (PG), were observed in most abscising events which could be responsible for the degradation of middle lamella and the loosening of primary cell wall in separation layers [29,36]. Mishra et al. analyzed the effects of some phytohormones such as ABA and IAA on cellulase and PG activities of cotton leaf explants. The increase in cellulase and PG activity in the LAZ of the ABA treated explants relative to control explants suggested the roles of ABA in this increment. The process of leaf abscission in cotton was associated with higher biosynthesis of ethylene in abscission zones along with elevated levels of cellulase activity [3]. In the current study, both COR and TDZ induced elevated transcripts of *GhCEL1* and *GhPG* in AZs, but not in leaf blades and petioles (Fig. 4). No differences were observed in the maximum expression of *GhPG* and PG activity between COR and TDZ treatments. *GhCEL1* maximum expression and CEL activity in AZ treated with TDZ were higher than those in plants treated with COR. This resulted to a smaller reduction in break strength under COR treatment than under TDZ treatment (Fig. 3). Nevertheless, for the final levels of break strength and defoliation, there were no differences between COR and TDZ treatments (Fig. 3 and 6).

Ethylene production and ACC accumulation increased in abscised tissues or organs such as leaves and fruits treated with ethephon or other exogenous chemicals [31,53]. Increased ethylene biosynthesis through over-expression of ACS leads to premature flower abscission, while a block in ethylene perception in the never ripe (nr) mutant delays petal abscission in tomato [37,54]. ACS1 is mainly involved in system II-like ethylene biosynthesis in citrus. Increased expression of ACS1 in mature fruit and leaf abscission zones was associated with ethephon-induced abscission [31]. In this work, *GhACS* in AZs was upregulated by TDZ and COR treatment whether in leaf blades or petioles (Fig. 4). A substantial increase in ACC activity was observed in petiole, leaf abscission zone, and boll crust after TDZ and COR treatment. Although *GhACS* expression in AZ treated with TDZ was approximately 2.0-fold higher than that in plants treated with COR for 12 h, prolonged expression time of *GhACS* was observed in COR treatment compared with TDZ treatment. Thus, no difference in ACC activity was noted between COR and TDZ treatment. The application of ethylene induced cotton defoliation and increased the percentage of open bolls [55,56]. The increased ACC activity of COR treated boll crust relative to that of TDZ indicated that COR can induce more ethylene in boll crust. Thus, it is beneficial to increasing the percentage of open bolls.

Abscission-inducing chemicals can increase machine harvest efficiency, improving lodging, and reducing the time from defoliation to harvest. Numerous studies have focused on effects of exogenous chemicals on defoliation and boll opening [4,57,58]. Defoliants such as TDZ had greater defoliation effects but did not directly influence boll opening [5,59]. In this study, it was found that COR induced both defoliation and boll opening. The higher boll opening under COR treatment might have been associated with increased ACC activity in boll crust. Although approximately 85% abscission and 80% boll opening were observed for COR treatment, defoliation and boll opening were lower than those reported in Gwathmey and Hayes [59]. Further studies on both dosage and application timing of COR are necessary in cotton to optimize its use as a harvest aid chemical.

Defoliation allows producers to harvest earlier than allowing crops to mature naturally. However, the practice with defoliation may reduce yield and alter fiber quality if the application of harvest aids is premature (e.g., prior to 60% open bolls) [2]. Defoliation may increase the total harvest yield only if defoliant or boll opener increases the number of open bolls at harvest. On the other hand, it may reduce boll weight by opening small bolls prematurely and further decrease yield [60]. Recent evidence suggests that defoliation could be initiated before 60% open bolls if fruiting is compact (i.e., fruit set over eight to ten nodes); however, a crop with extended fruiting may require delayed defoliation to achieve maximum yields [61]. Although our study was conducted with a relatively early application of abscission chemicals (45–50% open bolls), the total seedcotton yield, boll weight, lint percentage, seed quality, and fiber quality (data not shown) were unaffected by either COR or TDZ treatment. In addition, the first harvest yield and first harvest percentage were significantly increased by COR. Although the difference between COR and TDZ treatments was not significant, COR was more effective in increasing the first harvest yield than TDZ.

In conclusion, this work provides structural, biochemical and molecular evidence that the phytotoxin, coronatine affects cotton abscission by increasing *GhCEL1*, *GhPG* and *GhACS* expression, and activity of hydrolytic enzymes such as CEL and PG as well as ACC accumulation in AZ through mechanisms dissimilar to those of TDZ. In particular, the greater increase in ACC activity of COR treated boll crust suggests that COR has better ripening effect than TDZ. It is possible that COR can induce both defoliation and boll ripening in cotton without adverse effects on yield and seed development.

Acknowledgments

We thank Dr. Edward Deckard (Professor of Plant Science, North Dakota State University, Fargo) and Dr. Eneji A. Egrinya (Professor of Soil Science, University of Calabar, Nigeria) for technical improvement of the manuscript.

Author Contributions

Conceived and designed the experiments: MWD YL ZHL. Performed the experiments: MWD YL XLT MCZ WMT. Analyzed the data: MWD XLT LSD. Contributed reagents/materials/analysis tools: MWD YL LSD WMT DYX. Wrote the paper: MWD MCZ ZHL.

References

1. Sunilkumar G, Campbell LM, Puckhaber L, Stipanovic RD, Rathore KS (2006) Engineering cottonseed for use in human nutrition by tissue-specific reduction of toxic gossypol. Proceedings of the National Academy of Sciences 103: 18054–18059.
2. Snipes CE, Baskin CC (1994) Influence of early defoliation on cotton yield, seed quality, and fiber properties. Field Crops Research 37: 137–143.
3. Mishra A, Khare S, Trivedi PK, Nath P (2008) Effect of ethylene, 1-MCP, ABA and IAA on break strength, cellulase and polygalacturonase activities during cotton leaf abscission. South African Journal of Botany 74: 282–287.
4. Siebert JD, Stewart AM (2006) Correlation of defoliation timing methods to optimize cotton yield, quality and revenue. Journal of Cotton Science 10: 146–154.
5. Gwathmey CO, Craig Jr CC (2006) Defoliants for cotton. Encyclopedia of Pest Management 1: 1–3.
6. Metzger JD, Keng J (1984) Effects of dimethipin, a defoliant and desiccant, on stomatal behavior and protein synthesis. Journal of Plant Growth Regulation 3: 141–156.
7. Snipes CE, Wills GD (1994) Influence of temperature and adjuvants on thidiazuron activity in cotton leaves. Weed Science 42: 13–17.
8. Suttle JC (1985) Involvement of ethylene in the action of the cotton defoliant thidiazuron. Plant Physiology 78: 272–276.
9. Suttle JC (1988) Disruption of the polar auxin transport system in cotton seedlings following treatment with the defoliant thidiazuron. Plant physiology 86: 241–245.
10. Bender CL, Alarcón-Chaidez F, Gross DC (1999) Pseudomonas syringae phytotoxins: mode of action, regulation, and biosynthesis by peptide and polyketide synthetases. Microbiology and Molecular Biology Reviews 63: 266–292.

11. Cintas NA, Koike ST, Bull CT (2002) A new pathovar, *Pseudomonas syringae* pv. *alisalensis* pv. nov., proposed for the causal agent of bacterial blight of broccoli and broccoli raab. Plant disease 86: 992–998.
12. Schüler G, Mithöfer A, Baldwin IT, Berger S, Ebel J, et al. (2004) Coronalon: a powerful tool in plant stress physiology. FEBS letters 563: 17–22.
13. Uppalapati SR, Ayoubi P, Weng H, Palmer DA, Mitchell RE, et al. (2005) The phytotoxin coronatine and methyl jasmonate impact multiple phytohormone pathways in tomato. The Plant Journal 42: 201–217.
14. Braun Y, Smirnova AV, Weingart H, Schenk A, Ullrich MS (2009) Coronatine gene expression In Vitro and In Planta, and protein accumulation during temperature downshift in *Pseudomonas syringae*. Sensors 9: 4272–4285.
15. Xie Z, Duan L, Tian X, Wang B, Egrinya Eneji A, et al. (2008) Coronatine alleviates salinity stress in cotton by improving the antioxidative defense system and radical-scavenging activity. Journal of Plant physiology 165: 375–384.
16. Wu H, Wu X, Li Z, Duan L, Zhang M, et al. (2012) Physiological evaluation of drought stress tolerance and recovery in cauliflower (*Brassica oleracea* L.) seedlings treated with methyl jasmonate and coronatine. Journal of Plant Growth Regulation 31: 113–123.
17. Ferguson IB, Mitchell RE (1985) Stimulation of ethylene production in bean leaf discs by the pseudomonad phytotoxin coronatine. Plant Physiology 77: 969–973.
18. Kenyon JS, Turner JG (1992) The stimulation of ethylene synthesis in Nicotiana tabacum leaves by the phytotoxin coronatine. Plant Physiology 100: 219–224.
19. Perner B, Schmauder HP, Mueller J, Greulich F, Bublitz F (1994) Effect of coronatine on ethylene release and ATPase activity of tomato cell cultures. Journal of Phytopathology 142: 27–36.
20. Stelmach BA, Müller A, Weiler EW (1999) 12-Oxo-phytodienoic acid and indole-3-acetic acid in jasmonic acid-treated tendrils of *Bryonia dioica*. Phytochemistry 51: 187–192.
21. Koda Y, Takahashi K, Kikuta Y, Greulich F, Toshima H, et al. (1996) Similarities of the biological activities of coronatine and coronafacic acid to those of jasmonic acid. Phytochemistry 41: 93–96.
22. Koch T, Krumm T, Jung V, Engelberth J, Boland W (1999) Differential induction of plant volatile biosynthesis in the lima bean by early and late intermediates of the octadecanoid-signaling pathway. Plant Physiology 121: 153–162.
23. Haider G, Schrader T, Füßlein M, Blechert S, Kutchan TM (2000) Structure-activity relationships of synthetic analogs of jasmonic acid and coronatine on induction of benz [*c*] phenanthridine alkaloid accumulation in *Eschscholzia californica* cell cultures. Biological Chemistry 381: 741–748.
24. Benedetti CE, Costa CL, Turcinelli SR, Arruda P (1998) Differential expression of a novel gene in response to coronatine, methyl jasmonate, and wounding in the *Coi1* mutant of Arabidopsis. Plant Physiology 116: 1037–1042.
25. Fan X, Mattheis JP, Fellman JK (1998) A role for jasmonates in climacteric fruit ripening. Planta 204: 444–449.
26. Miyamoto K, Oka M, Ueda J (1997) Update on the possible mode of action of the jasmonates: Focus on the metabolism of cell wall polysaccharides in relation to growth and development. Physiologia Plantarum 100: 631–638.
27. Hartmond U, Yuan R, Burns JK, Grant A, Kender WJ (2000) Citrus fruit abscission induced by methyl-jasmonate. Journal of the American Society for Horticultural Science 125: 547–552.
28. Burns JK, Pozo LV, Arias CR, Hockema B, Rangaswamy V, et al. (2003) Coronatine and abscission in citrus. Journal of the American Society for Horticultural Science 128: 309–315.
29. Sakamoto M, Munemura I, Tomita R, Kobayashi K (2008) Involvement of hydrogen peroxide in leaf abscission signaling, revealed by analysis with an in vitro abscission system in Capsicum plants. The Plant Journal 56: 13–27.
30. Parra-Lobato MC, Gomez-Jimenez MC (2011) Polyamine-induced modulation of genes involved in ethylene biosynthesis and signalling pathways and nitric oxide production during olive mature fruit abscission. Journal of Experimental Botany 62: 4447–4465.
31. Yuan R, Wu Z, Kostenyuk IA, Burns JK (2005) G-protein-coupled α2A-adrenoreceptor agonists differentially alter citrus leaf and fruit abscission by affecting expression of ACC synthase and ACC oxidase. Journal of Experimental Botany 56: 1867–1875.
32. Roberts JA, Elliott KA, Gonzalez-Carranza ZH (2002) Abscission, dehiscence, and other cell separation processes. Annual Review of Plant Biology 53: 131–158.
33. Malladi A, Burns JK (2008) *CsPLDα1* and *CsPLDγ1* are differentially induced during leaf and fruit abscission and diurnally regulated in *Citrus sinensis*. Journal of Experimental Botany 59: 3729–3739.
34. David A, Brummell I, Bradford D (1999) Antisense suppression of tomato endo-1, 4-glucanase *Cel2* mRNA accumulation increases the force required to break fruit abscission zones but does not affect fruit softening. Plant Molecular Biology 40: 615–622.
35. González-Carranza ZH, Whitelaw CA, Swarup R, Roberts JA (2002) Temporal and spatial expression of a polygalacturonase during leaf and flower abscission in oilseed rape and Arabidopsis. Plant Physiology 128: 534–543.
36. González-Carranza ZH, Elliott KA, Roberts JA (2007) Expression of polygalacturonases and evidence to support their role during cell separation processes in Arabidopsis thaliana. Journal of Experimental Botany 58: 3719–3730.
37. Wilkinson JQ, Lanahan MB, Yen H, Giovannoni JJ, Klee HJ (1995) An ethylene-inducible component of signal transduction encoded by Never-ripe. Science 270: 1807–1809.
38. Brown KM (1997) Ethylene and abscission. Physiologia Plantarum 100: 567–576.
39. Alexander L, Grierson D (2002) Ethylene biosynthesis and action in tomato: a model for climacteric fruit ripening. Journal of Experimental Botany 53: 2039–2055.
40. Taylor JE, Whitelaw CA (2001) Signals in abscission. New Phytologist 151: 323–340.
41. Wang KL, Li H, Ecker JR (2002) Ethylene biosynthesis and signaling networks. The Plant Cell 14: S131-S151.
42. Nakatsuka A, Murachi S, Okunishi H, Shiomi S, Nakano R, et al. (1998) Differential expression and internal feedback regulation of 1-aminocyclopropane-1-carboxylate synthase, 1-aminocyclopropane-1-carboxylate oxidase, and ethylene receptor genes in tomato fruit during development and ripening. Plant Physiology 118: 1295–1305.
43. Kumar A, Taylor MA, Arif SA, Davies HV (1996) Potato plants expressing antisense and sense S-adenosylmethionine decarboxylase (SAMDC) transgenes show altered levels of polyamines and ethylene: antisense plants display abnormal phenotypes. The Plant Journal 9: 147–158.
44. Barry CS, Llop-Tous MI, Grierson D (2000) The regulation of 1-aminocyclopropane-1-carboxylic acid synthase gene expression during the transition from system-1 to system-2 ethylene synthesis in tomato. Plant Physiology 123: 979–986.
45. Palmer DA, Bender CL (1993) Effects of environmental and nutritional factors on production of the polyketide phytotoxin coronatine by Pseudomonas syringae pv. glycinea. Applied and Environmental Microbiology 59: 1619–1626.
46. Garcia Garrido JM, Tribak M, Rejon Palomares A, Ocampo JA, Garcia Romera I (2000) Hydrolytic enzymes and ability of arbuscular mycorrhizal fungi to colonize roots. Journal of Experimental Botany 51: 1443–1448.
47. Xia XJ, Zhou YH, Ding J, Shi K, Asami T, et al. (2011) Induction of systemic stress tolerance by brassinosteroid in Cucumis sativus. New Phytologist 191: 706–720.
48. Livak KJ, Schmittgen TD (2001) Analysis of Relative Gene Expression Data Using Real-Time Quantitative PCR and the $2^{-\Delta\Delta CT}$ Method. Methods 25: 402–408.
49. Patterson SE (2001) Cutting loose. Abscission and dehiscence in Arabidopsis. Plant Physiology 126: 494–500.
50. van Nocker S (2009) Development of the abscission zone. Stewart Postharvest Review 5: 1–6.
51. Wang H, Friedman CMR, Shi J, Zheng Z (2010) Anatomy of leaf abscission in the Amur honeysuckle (Lonicera maackii, Caprifoliaceae): a scanning electron microscopy study. Protoplasma 247: 111–116.
52. Ayala F, Silvertooth JC (2001) Physiology of cotton defoliation. University of Arizona Publication AZ 1240.
53. Kende H (1993) Ethylene biosynthesis. Annual Review of Plant Physiology and Plant Molecular Biology 44: 283–307.
54. Lanahan MB, Yen H, Giovannoni JJ, Klee HJ (1994) The never ripe mutation blocks ethylene perception in tomato. The Plant Cell 6: 521–530.
55. Stewart AM, Edmisten KL, Wells R (2000) Boll openers in cotton: Effectiveness and environmental influences. Field Crops Research 67: 83–90.
56. Bange MP, Long RL (2011) Optimizing timing of chemical harvest aid application in cotton by predicting its influence on fiber quality. Agronomy Journal 103: 390–395.
57. Faircloth JC, Edmisten KL, Wells R, Stewart AM (2004) The influence of defoliation timing on yields and quality of two cotton cultivars. Crop Science 44: 165–172.
58. Snipes CE, Cathey GW (1992) Evaluation of defoliant mixtures in cotton. Field Crops Research 28: 327–334.
59. Gwathmey CO, Hayes RM (1997) Harvest-aid interactions under different temperature regimes in field-grown cotton. Journal of Cotton Science 1: 1–9.
60. Smith CW, Cothren JT, Varvil JJ (1986) Yield and fiber quality of cotton following application of 2-chloroethyl phosphonic acid. Agronomy Journal 78: 814–818.
61. Collins GD, Edmisten KL, Jordan DL, Wells R, Lanier JE, et al. (2007) Defining optimal defoliation timing and harvest timing for compact, normal, and extended fruiting patterns of cotton (Gossypium hirsutum L.) Achieved by Cultivar Maturity Groups. The World Cotton Research Conference.

2

Comparison of the Transcriptome between Two Cotton Lines of Different Fiber Color and Quality

Wenfang Gong[1¶], **Shoupu He**[1¶], **Jiahuan Tian**[1], **Junling Sun**[1], **Zhaoe Pan**[1], **Yinhua Jia**[1], **Gaofei Sun**[1,2], **Xiongming Du**[1]*

1 State Key Laboratory of Cotton Biology, Institute of Cotton Research, Chinese Academy of Agricultural Sciences, Anyang, China, **2** Department of Computer Science and Information Engineering, Anyang Institute of Technology, Anyang, China

Abstract

To understand the mechanism of fiber development and pigmentation formation, the mRNAs of two cotton lines were sequenced: line Z128 (light brown fiber) was a selected mutant from line Z263 (dark brown fiber). The primary walls of the fiber cell in both Z263 and Z128 contain pigments; more pigments were laid in the lumen of the fiber cell in Z263 compared with that in Z128. However, Z263 contained less cellulose than Z128. A total of 71,895 unigenes were generated: 13,278 (20.26%) unigenes were defined as differentially expressed genes (DEGs) by comparing the library of Z128 with that of Z263; 5,345 (8.16%) unigenes were up-regulated and 7,933 (12.10%) unigenes were down-regulated. qRT-PCR and comparative transcriptional analysis demonstrated that the pigmentation formation in brown cotton fiber was possibly the consequence of an interaction between oxidized tannins and glycosylated anthocyanins. Furthermore, our results showed the pigmentation related genes not only regulated the fiber color but also influenced the fiber quality at the fiber elongation stage (10 DPA). The highly expressed flavonoid gene in the fiber elongation stage could be related to the fiber quality. DEGs analyses also revealed that transcript levels of some fiber development genes (Ca^{2+}/CaM, reactive oxygen, ethylene and sucrose phosphate synthase) varied dramatically between these two cotton lines.

Editor: Xianlong Zhang, National Key Laboratory of Crop Genetic Improvement, China

Funding: This study was supported by the "Twelfth Five-Year Plan" of the National Science and Technology Support Project (2011BAD35B05 and 2013BAD01B03) and Basic Scientific Research funds in National Nonprofit Institutes (SJA0608). The funders had no role in study design, data collection and analysis, decision to publish, or preparation of the manuscript.

Competing Interests: The authors have declared that no competing interests exist.

* Email: dujeffrey8848@hotmail.com

¶ These authors contributed equally to this work.

Introduction

Upland cotton (*Gossypium hirsutum* L.) is the largest natural fiber producer of the plants. In recent years, interest in naturally colored cotton has grown because it may reduce pollution, making it preferable to white fiber which requires a dyeing process [1–3]. However, its commercial application is very limited due to the lack of fiber color diversity and low fiber quality [2]. Limited brown (different color depth) and green fiber lines, among other varieties, have been used in the textile industry. A previous study demonstrated that there was a significant negative correlation between the degree of fiber color and lint percentage and fiber quality traits in cotton [4]. Therefore, subsequent studies should focus on improving the fiber quality and revealing the underlying mechanisms for pigmentation formation in naturally colored cotton.

Early genetic analysis suggested that the brown color of cotton fiber was controlled by one incompletely dominant major gene [5]. Furthermore, gene expression analysis and dimethylaminocinnaldehyde staining showed that tannins could be the key chemical responsible for the brown color in cotton fiber [6]. Subsequent chemical research indicated that the brown pigment in cotton fiber might be the chinone compound oxidated from condensed tannins, and the accumulation period of condensed tannins was from 10 DPA-25 DPA [7,8]. However, the molecular mechanism that underlies pigmentation in colored cotton fiber is still unknown.

With the development of next generation sequencing technology, RNA-seq provides a powerful tool to rebuild our knowledge of transcriptomics. By directly sequencing and assembling the mRNA, the whole transcriptome could be de novo reconstructed precisely and efficiently [9], aligned with public databases for function annotation and the critical genes could be assessed using pathway classification. In addition, gene expression can be measured and the number of transcripts can be obtained if the appropriate level of sequencing was performed. The application of RNA-seq technology has been used successfully for various species [10–14].

Lines Z263 and Z128, two brown fiber inbred lines with dark and light brown fiber respectively, both derived from a cross between white and brown fiber cotton. To reveal the whole transcriptome landscape of the natural colored cotton fiber, and to understand the molecular mechanism of pigment formation, the transcriptomes of these two lines at the whole early developmental stage (0 dpa–20 dpa) were sequenced using RNA-seq technology and analyzed.

Materials and Methods

1. Plant material

Line Z263 (*Gossypium hirsutum* L.), with dark brown fiber, was selected from a cross between white cotton (Zhong 6331) and brown fiber cotton (Crd). Line Z128 (*G. hirsutum* L.) was a selected mutant from Z263. They were planted in an experimental field at the Institute of Cotton Research, Chinese Academy of Agricultural Sciences (ICR, CAAS) under normal agronomic management conditions. The bolls at the day of anthesis (0 day post anthesis/DPA), 5 DPA, 15 DPA and 20 DPA were harvested and stored in an ice box, then the ovules of 0 DPA and fibers of other stages were dissected and separated on ice as fast as possible, and then stored at −80°C immediately.

2. Fiber quality measurement, fiber microstructure detection and cellulose test

To test the fiber quality of Z263 and Z128, the following measurements were included: upper half mean length (mm), uniformity index (%), micronaire, fiber elongation (%), fiber strength, 15 g lint samples of each line were analyzed using USTER HVI 1000 (USTER Technologies, Inc., Uster, Switzerland). The fiber microstructure detection and cellulose test were performed according to Ru et al [15].

3. RNA extraction and cDNA library construction

Total RNAs were extracted from each sample using the CTAB method described by Wan and Wilkins [16], with minor modifications to increase the yield. All RNA quality and quantity were measured by 1.0% agarose gel and an ultraviolet spectrophotometer. Four stages of RNAs (0 DPA, 5 DPA, 15 DPA and 20 DPA) with the same concentration and quality from each line were mixed together as one mixed library. The two libraries of Z263 and Z128 were constructed using the method described by Xia et al. [17].

4. RNA-seq and sequence de novo assembly

The sequencing of two libraries of Z263 and Z128 were performed on HiSeq 2000 (illumina) by the Beijing Genomics Institute (BGI) (Shenzhen, Guangdong, China). The raw reads, transformed from images, were first processed by removing adaptors and redundant fragments to generate clean reads. The clean reads de novo assembly was carried out using the short reads assembling program-SOAPdenovo [18]. Clean reads with a certain length of overlap were first combined to form "contigs". Then the clean reads were mapped back to the contigs; it was possible to detect contigs from the same transcript, as well as the distances between them, by paired-end reads. Next, SOAPdenovo connected the contigs into "scaffolds"; "N" was used to present unknown sequences. Paired-end reads were used again to fill the gaps between scaffolds, longer sequences which could not extend at either end were assembled and defined as "unigenes". In this study, the unigenes from the two libraries were further processed by sequence splicing and redundancy removal with the sequence cluster program-TGICL [19] to acquire non-redundant unigenes that were as long as possible.

5. Unigene functional annotation and functional categorization

All-unigenes were first searched using the blastx tool against public protein databases such as Non-redundant protein sequences (nr, http://www.ncbi.nlm.nih.gov), Swiss-Prot (http://www.expasy.org/sprot/), Kyoto Encyclopedia of Genes and Genomes (KEGG, http://www.genome.jp/kegg/pathway.html) and Cluster of orthologous groups for eukaryotic complete genomes (KOG,

Table 1. The primers used in this study.

Gene name	NCBI accession no.	Direction	Primer Sequence (5′→3′)	Product length
CHS	EF643507	F	GGTGTGGACATGCCTGGGGC	265
		R	CAGCTGCGGCACCATCACCA	
CHI	EF187439	F	ATCCGTTGAGTTTTTCAGAG	127
		R	CCAAATAGCAACGCAAT	
F3′H		F	CGAGGAGATGGATAAGGTGATTG	128
		R	GCAAGTTCAAGGGAGTAGATGGA	
F3H	EF187440	F	GCTTCTTGAGGTGTTGTCAGAGG	116
		R	CAGGTTGAGGGCATTTAGGATAG	
DFR	FJ713480	F	CGCGACCCTGGCAACTCGAA	417
		R	CCAGGCTGCTTGCTCTGCCA	
ANS	EU921264	F	AAGAGAAGTATGCCAACGAC	102
		R	AGAAGTAGTCCTCCCACTCA	
ANR	FJ713479	F	TCCTCAACAAAAGATACCCTGACTT	147
		R	CGGTTTGGTCGTAGATTTCCTC	
LAR		F	AAAGTAGCCAAAGCCCTTCA	260
		R	TAACAGTGCCGACAGAGTGAA	
Gh18S	L24145	F	TGACGGAGAATTAGGGTTCGA	100
		R	CCGTGTCAGGATTGGGTAATTT	

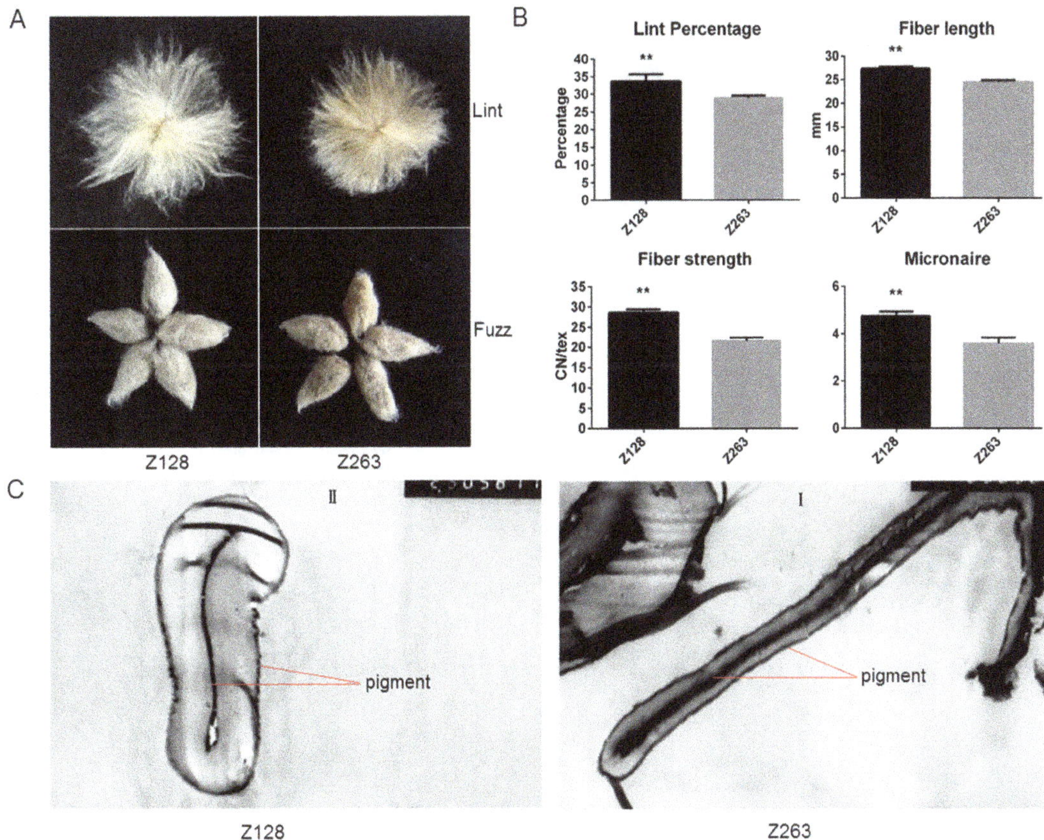

Figure 1. The fiber color, yield, quality and microstructure of Z128 and Z263. Statistical analyses were performed at 95% condence with IBM SPSS Statistics 11.0 (SPSS Inc., Chicago, USA). Values with an asterisk represented a significant differe- nce at P<0.05. The fiber microstructure of Z263 (C I) and Z128 (C II) was observed under a microscope (×3000 and ×2500, respectively).

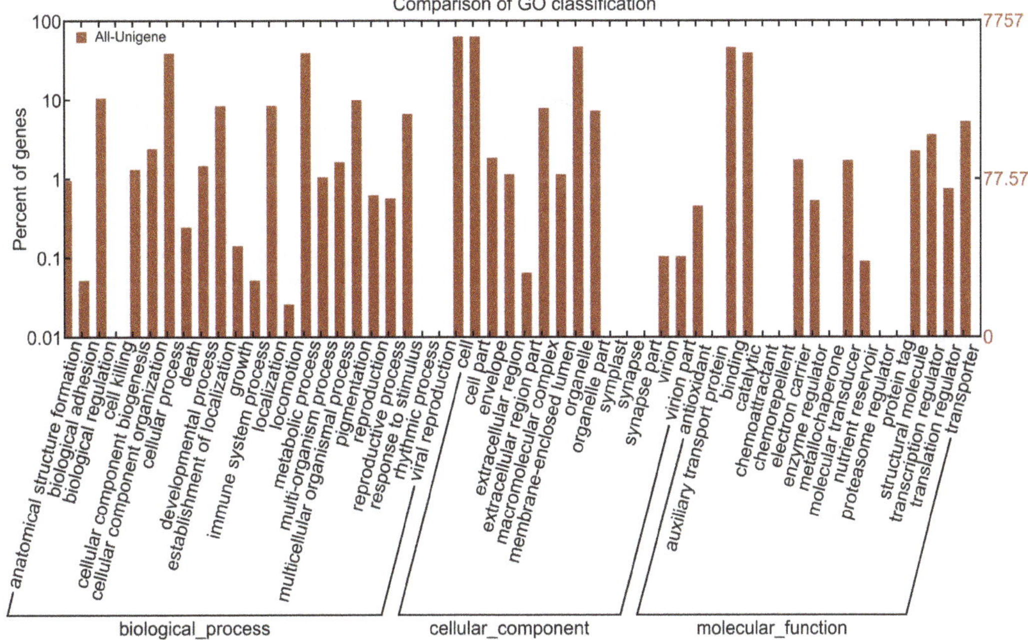

Figure 2. All-unigenes classified by GO analysis.

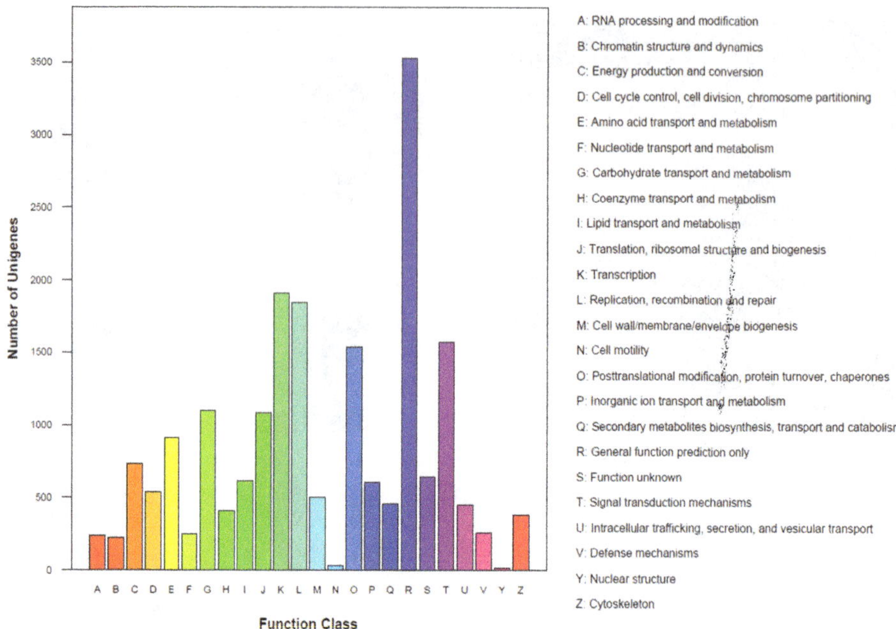

Figure 3. COG classification of all-unigenes. A–Z represented different functions classified by GO analysis, respectively.

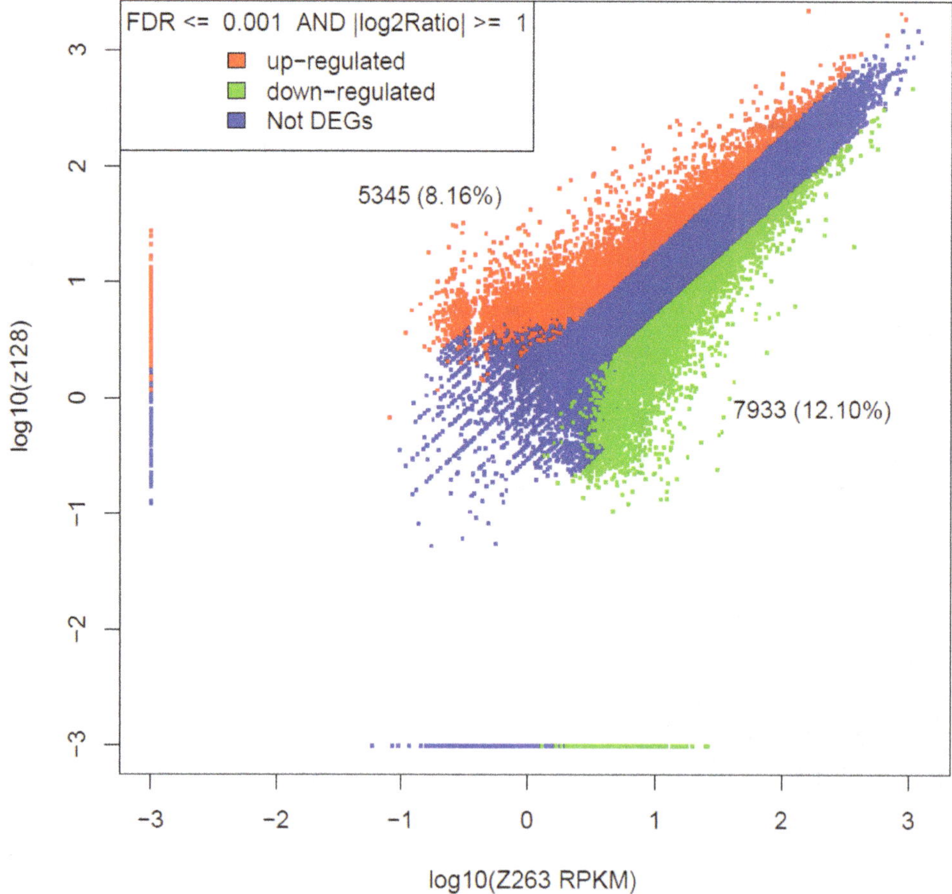

Figure 4. Mapping of all expressed unigenes. This figure was created by comparing gene expression levels of Z128 to Z263. FDR≤0.001, $|log_2Ratio| \geq 1$ was used as threshold, red dots represented up-regulated genes, green dots represented the down-regulated genes, and blue dots represented gene expression with no significant difference.

Table 2. The DEGs enriched terms in GO analysis (P-value<1).

Ontology	Gene Ontology term	Cluster frequency	Genome Frequency of use	Corrected P-value
Biological process	RNA-dependent DNA replication	15/597, 2.5%	47/4110, 1.1%	0.82163
Cellular component	cytoplasmic vesicle	137/696, 19.7%	750/4911, 15.3%	0.06344
	cytoplasmic membrane-bounded vesicle	135/696, 19.4%	745/4911, 15.2%	0.10093
	spliceosome	7/696, 1.0%	14/4911, 0.3%	0.23912
Molecular function	oxidoreductase activity, acting on paired donors, with incorporation or reduction of molecular oxygen, 2-oxoglutarate as one donor, and incorporation of one atom each of oxygen into both donors	7/769, 0.9%	13/5273, 0.2%	0.32203
	RNA-directed DNA polymerase activity	15/769, 2.0%	48/5273, 0.9%	0.76482
	DNA polymerase activity	15/769, 2.0%	49/5273, 0.9%	0.95719

http://genome.jgi.d oe.gov/Tutorial/tutorial/kog.html). An e-value$<10^{-5}$ was used as the threshold. To further understand the distribution of gene function, the protein functional classification and pathway were annotated by Gene Ontology (http://www.geneontology.org/), KOGs and KEGG. With NR annotation, GO functional annotation and classification were obtained using the Blast2GO program [20] and WEGO software [21], respectively.

6. Differential expressed genes (DEGs) identification and enrichment analysis

Referring to the method described by Audic and Claverie [22], the Beijing Genomics Institute (BGI) developed a rigorous

Table 3. The top 10 DEGs enriched pathways in KEGG analysis.

No.	Pathway	DEGs with pathway annotation (3,213)	All genes with pathway annotation (20,242)	P-value	Pathway ID
1	Zeatin biosynthesis	16 (0.5%)	41 (0.2%)	0.000299	ko00908
2	Anthocyanin biosynthesis	4 (0.12%)	5 (0.02%)	0.002767	ko00942
3	ABC transporters	40 (1.24%)	164 (0.81%)	0.003	ko02010
4	Diterpenoid biosynthesis	15 (0.47%)	47 (0.23%)	0.004717	ko00904
5	Phenylpropanoid biosynthesis	56 (1.74%)	255 (1.26%)	0.006239	ko00940
6	3-Chloroacrylic acid degradation	18 (0.56%)	63 (0.31%)	0.007641	ko00641
7	Flavone and flavonol biosynthesis	14 (0.44%)	45 (0.22%)	0.007987	ko00944
8	Ubiquitin mediated proteolysis	108 (3.36%)	550 (2.72%)	0.00975	ko04120
9	Taurine and hypotaurine metabolism	5 (0.16%)	10 (0.05%)	0.012555	ko00430
10	Base excision repair	36 (1.12%)	158 (0.78%)	0.014168	ko03410

Table 4. The involved unigenes in Flavone and flavonol biosynthesis and Anthocyanin biosynthesis pathways.

	Orthology	Entry of enzyme	Unigenes	Ratio[1]	Status[2]	Encoded protein
Flavonoid biosynthesis	K00660	2.3.1.74	Unigene56280_All	−1.9	D	chalcone synthase (CHS)
	K00588	2.1.1.104	Unigene34142_All	−1.4	D	caffeoyl-CoA O-methyltransferase
			Unigene42690_All	−3.6	D	
			Unigene10308_All	−1.1	D	
	K01859	5.5.1.6	Unigene69406_All	−1.6	D	chalcone isomerase (CHI)
	K05277	1.14.11.19	Unigene56780_All	−2.3	D	anthocyanidin synthase (ANS)
			Unigene57093_All	−3.0	D	
			Unigene5167_All	−1.6	D	
	K08695	1.3.1.77	Unigene43073_All	−1.1	D	anthocyanidin reductase (ANR)
	K05280	1.14.13.21	Unigene71334_All	2.8	U	flavonoid 3′-hydroxylase (F3′H)
	K00475	1.14.11.9	Unigene70267_All	3.9	U	flavanone 3-hydroxylase (F3H)
			Unigene71550_All	4.2	U	
	K05278	1.14.11.23	Unigene70581_All	4.1	U	flavonol synthase (FLS)
			Unigene6472_All	1.0	U	
Flavone and flavonol biosynthesis	K05280	1.14.13.21	Unigene71334_All	2.8	U	flavonoid 3′-hydroxylase (F3′H)
	K05279	2.1.1.76	Unigene13368_All	−2.2	D	flavonol 3-O-methyltransferase (FOMT)
			Unigene21288_All	−1.4	D	
			Unigene38018_All	−1.2	D	
			Unigene47446_All	−2.4	D	
			Unigene47563_All	−3.7	D	
			Unigene50636_All	−2.7	D	
			Unigene57809_All	−1.7	D	
			Unigene58594_All	−1.9	D	
			Unigene70908_All	−1.7	D	
			Unigene9272_All	−1.1	D	
	K10757	2.4.1.91	Unigene50839_All	1.0	U	flavonol 3-O-glucosyltransferase (FOGT)
			Unigene8998_All	2.1	U	
			Unigene54635_All	−4.3	D	
Anthocyanin biosynthesis	K12338	2.4.1.298	Unigene18122_All	2.2	U	anthocyanin 5-O-glucosyltransferase (5GT)
			Unigene59143_All	12.0	U	
			Unigene67316_All	12.5	U	
			Unigene71660_All	12.9	U	

1: ratio indicated log2(z128_RPKM/Z263_RPKM).
2: "U" indicated that this unigene was up-regulated; "D" indicated down-regulated.

algorithm to identify DEGs from two samples of RNA-seq data. Because the expression of each gene occupies only a small part of the library, we denote the number of unambiguous clean tags from gene A as x, and the p(x) is in the Poisson distribution. The formula is as follows:

$$p(x) = \frac{e^{-\lambda}\lambda^x}{x!} \text{ (λ is the real transcripts of the gene)}$$

The total clean tag number of sample 1 is N^1, and total clean tag number of sample 2 is N^2; gene A holds x tags in sample 1 and y tags in sample 2. The probability of gene A expressed equally between two samples can be calculated with the following formula:

$$2\sum_{i=0}^{i=y} p(i|x) \text{ or } 2\times(1-\sum_{i=0}^{i=y} p(i|x)) \text{ (if } \sum_{i=0}^{i=y} p(i|x) > 0.5)$$

$$p(y|x) = \left(\frac{N^2}{N^1}\right)^y \frac{(x+y)!}{x!y!\left(1+\frac{N2}{N1}\right)^{(x+y+1)}}$$

The P value corresponds to the differential gene expression test.

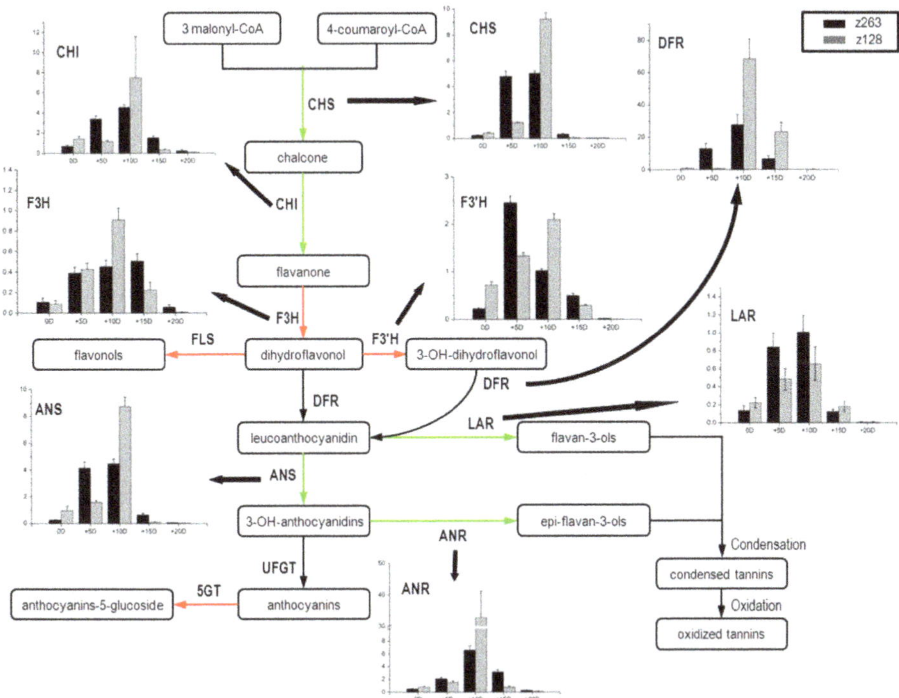

Figure 5. The schematic of pigment formation in cotton fiber. The red and green arrows indicated the up- and down-regulated status detected by comparing gene expression in Z128 with that in Z263. chalcone synthase (CHS); chalcone isomerase (CHI); flavanone 3-hydroxylase (F3H); flavonol synthase (FLS); dihydroflavonol 4-reductase (DFR); leucoanthocyanidin reductase (LAR); anthocyanidin synthase (ANS); UDP-favonoid glucosyl transferase (UFGT); anthocyanidin reductase (ANR); anthocyanin 5-O-glucosyltransferase (5GT).

The false discovery rate (FDR) is a method used to determine the threshold of the P value in multiple tests. In our study, the reads per kilobase of exon model per million mapped reads (RPKM) value, was calculated referring to the formula described by Mortazavi et al. [23] and was used to quantify the transcript level of Z128 versus Z263. FDR$\leq 10^{-3}$ and the absolute value of the log2Ratio (Z128_RPKM/Z263_RPKM) ≥ 1 as the threshold, were used to judge significant differences in gene expression. DEGs were then subjected to GO functional enrichment analysis and KEGG pathway enrichment analysis.

GO functional enrichment analysis provides GO terms, which significantly enrich in DEGs compared to the genome background, indicating that the DEGs are connected to interesting biological functions. All DEGs are firstly mapped to GO terms in the database, calculating gene numbers for every term, then the ultra-geometric test is used to find significantly enriched GO terms in DEGs compared to the genome background. The formula is:

$$P = 1 - \sum_{i=0}^{m-1} \frac{\binom{M}{i}\binom{N-M}{n-i}}{\binom{N}{n}}$$

Where N is the number of genes with GO annotation; n is the number of DEGs in N; M is the number of genes that are annotated according to the GO terms; m is the number of DEGs in M. The calculated P value$\leq$0.05 was taken as a threshold. GO terms fulfilling this condition are defined as significantly enriched GO terms in DEGs. This analysis recognizes the main biological functions that DEGs participate in.

Pathway enrichment analysis identifies significantly enriched metabolic pathways or signal transduction pathways in DEGs compared with the whole genome background. The formula is the same as that in GO analysis. Here, N is the number of genes with KEGG annotation, n is the number of DEGs in N, M is the number of genes related to specific pathways, and m is number of DEGs in M (Qvalue$\leq$0.05). All DEGs are further mapped on to each pathway; up-regulated genes are marked with red borders while down-regulated genes are marked with green borders.

7. qRT-PCR analysis for genes related to flavonoid synthesis

The genes and primers used for the gene expression analysis related to flavonoid synthesis were listed in Table 1. qRT-PCR was performed in a total volume of 20 μL with 10 μL SYBR Premix Ex Taq(2×) (Takara, Japan), PCR forward primer (10 μM) 0.4 μL,PCR reversed primer 0.4 μL (10 μM),ROX Reference Dye II (50×) 0.4 μL,cDNA template 2.0 μL and ddH_2O 6.8 μl on a 7500 real-time PCR machine (Applied Biosystems) according to the manufacturer's instructions. PCR amplification employed a 10 s denaturing step at 95°C, followed by 5 s at 95°C and 40 s at 60°C with 40 cycles. Relative mRNA levels were calculated by the $2^{-\Delta\Delta CT}$ method with *Gh18S* (accession number: L24145) as an internal control.

8. Statistical analysis

All of the experiments concerning data comparisons were performed three times. Statistical analyses were performed using the S-N-K method of independent-samples t-test at 95% confidence with IBM SPSS Statistics 11.0 (SPSS Inc., Chicago, USA). Values with different lowercases represent a significant difference at P<0.05.

Table 5. Fiber development related DEGs of brown and white fibers.

genes	Z263 (RPKM)	Z128 (RPKM)	p-value	FDR	Ratio[1]	Status[2]	Homologous proteins
Unigene31293_All	40.4792	19.2683	5.06E-15	2.40E-16	−1.1	D	extracellular Cu/Zn superoxide dismutase
Unigene15009_All	36.5823	3.0274	1.54E-52	2.92E-54	−3.6	D	class III peroxidase
Unigene22859_All	1.8516	5.7002	4.97E-04	1.53E-04	1.6	U	glutathione peroxidase
Unigene25776_All	2.1838	9.8675	1.78E-09	2.04E-10	2.2	U	peroxisomal membrane ABC transporter family
Unigene40537_All	80.13	230.84	2.93E-14	1.51E-15	1.5	U	fiber quinone-oxidoreductase
Unigene50168_All	2.78	1.20	3.83E-02	2.08E-02	−1.2	D	calcium-transporting ATPase(ACA9)
Unigene1176_All	16.85	45.17	2.03E-13	1.24E-14	1.4	U	calcium ion transmembrane transporter(ACA2)
Unigene55005_All	3.59	0.50	1.55E-03	5.46E-04	−2.8	D	calcium ion binding/transporter(ATNRT1:2)
Unigene15884_All	3.32	14.7	2.71E-14	1.38E-15	2.1	U	autoinhibited calcium ATPase
Unigene61435_All	1.1851	5.9315	3.37E-04	1.00E-04	2.3	U	ethylene receptor
Unigene40770_All	5.06	14.98	2.02E-07	3.10E-08	1.5	U	ethylene responsive element binding protein
Unigene26099_All	4.86	61.24	0.00	0.00	3.7	U	ethylene responsive transcription factor 2b
Unigene37372_All	22.708	90.2862	0.00	0.00	1.9	U	ethylene-responsive element binding protein ERF2
Unigene62869_All	0.86	9.38	2.65E-05	6.01E-06	3.4	U	Ethylene-responsive transcription factor 1B
Unigene30365_All	6.59	21.61	6.11E-14	3.33E-15	1.7	U	sucrose phosphate synthase

1: ratio indicated log2(z128_RPKM/Z263_RPKM);
2: "U" indicated that this unigene was up-regulated; "D" indicated down-regulated.

Results

1. The differences in fiber color and quality between Z263 and Z128

Line Z128 was a selected mutant from Z263 and both lines have similar genetic backgrounds. However, their fibers were different. As shown in Fig. 1, the fiber of Z263 was dark brown while that of Z128 was light. Though the primary walls of the fiber cell in both Z263 and Z128 contain pigments (Fig. 1), more pigments were laid in the lumen of the fiber cell in Z263 compared with that Z128. This resulted in a darker color in the Z263 fiber. However, the fiber yield and quality of Z128 was better than that of Z263. The lint percentages (%), upper half mean length (mm), micronaire and fiber strength in Z128 were also better than those in Z263 (Fig. 1). Furthermore, Z128 contained more cellulose (98.5%) than Z263 (94.5%).

2. The de novo assembled transcriptome of the fibers in Z263 and Z128

By removing useless sequences, a total of 38,114,054 (2,858,554,050 nucleotides) and 39,355,642 (2,951,673,150 nucleotides) 75 bp-length clean reads were obtained from the Z128 and Z263 mRNA libraries, respectively. A total of 170,201 and 182,404 contigs, 143,588 and 151,478 scaffolds and 59,926 and 71,895 unigenes were assembled in the Z128 and Z263 library, respectively. Because both of the samples for library construction were collected from the same tissue (ovule and fiber), two unigene libraries were taken forward for sequence clustering and redundancy removal to generate a new unigene library (All-unigene library) to make the non-redundant unigenes as long as possible. The All-unigene library contained 71,895 unigenes with an average length of 533 bp, which was obviously greater and longer than the other two libraries. The length of most of the unigenes in the three libraries was in the range of 100–1,000 bp, accounting for 93.08%, 92.44% and 88.21% in Z128-, Z263- and all-unigene library, respectively.

Since the all-unigene library contained the most complete and longest sequences, it was used to run batch alignment with a cut-off E-value of 10^{-5} on online public databases. A total of 49,941 (69.46%) and 31,714 (44.11%) unigenes received annotations from the NCBI non-redundant (nr) and Swiss-Prot databases, respectively. KEGG, KOGs and GO similarity analyses indicated that 20,241 (28.15%), 14,333 (19.94%) and 7,757 (10.79%) unigenes matched these databases, respectively. To investigate the genomic similarity between *Gossypium* and other species, we estimated all annotations of unigenes from nr. The result showed that the most abundant unigenes were annotated as "*Vitis vinifera*", "*Ricinus communis*" and "*Populus trichocarpa*", which accounted for 29.71%, 29.59% and 24.63%, respectively. Furthermore, we annotated 68.2% and 80.2% unigenes on the recently r- eleased A and D genome of diploid cotton (which were considered to be two donor g- enomes of the tetroploid cotton subgenome), respectively.

GO classification analysis showed that 7,757 all-unigenes were categorized into three main ontologies: biological process, cellular component and molecular function, which were further categorized as 54 terms, and all-unigenes were classified into different terms. One unigene might be repeatedly classified in different terms, therefore, a total of 32,935 all-unigenes (including 75 unigenes repeated in different categories) were distributed over 42 terms, 14,867 (45.14%) of them were categorized in cellular component ontology, 10,229 (31.06%) in biological process and 7,839 (23.80%) in molecular function (Fig. 2). The detailed classification demonstrated that the term "cell" (4,876) and "cell part" (4,876) in "cellular component" ontology, "metabolic process" (3,026) and "cellular process" (3,002) in "biological process" ontology, "binding" (3,538) and "catalytic" (3,030) in "molecular function" contained the most unigenes, respectively. Furthermore, 764 unigenes under the term "pigmentation" indicated that an abundance of pigment-related biological processes were involved in the development of colored cotton fiber.

The all-unigenes library was aligned to the KOG database to predict and classify possible functions. A total of 14,333 unigenes were matched and categorized into 24 classes (Fig. 3). The function class defined as "general function prediction" (code: R) had the most unigenes (24.61%), followed by "transcription" (K: 13.33%), "replication, recombination and repair" (L: 12.86%), "signal transduction mechanisms" (T: 10.95%) and "posttranslational modification, protein turnover, chaperones" (O: 10.70%).

The all-unigene library was aligned with the KEGG pathway database, and the result showed a total of 38,645 unigenes which were classified into six categories, mostly concentrated in three of them: metabolism (11,371; 29.42%), protein families (10,638; 27.53%) and cellular processes (6,959; 18.01%). Furthermore, 1,315 unigenes were clustered in the "biosynthesis of secondary metabolites" category (Table S1).

3. Differentially expressed genes (DEGs) analysis

The unigene expression level in the Z128 and Z263 libraries were compared. A total of 13,278 (20.26%) unigenes were significantly differentially expressed when these two non-redundant libraries were compared (FDR≤0.001, $|\log_2\text{Ratio}|\geq 1$). A total of 5,345 (8.16%) of them were up-regulated, while 7,933 (12.10%) of them were down-regulated; the others were not DEGs (Fig. 4).

The GO analysis results showed that when corrected (P-value≤ 1), seven enriched terms belonged to three ontologies, one of them was categorized in "biological process", three were in "cellular component" and three were in "molecular function" (Table 2).

Table 3 showed the top 10 DEGs enriched pathways, seven of which were categorized as a "metabolism" pathway, two were categorized as "genetic information processing" pathways and the other one was categorized as "environmental information processing". Further identification indicated that five of the "metabolism" terms belonged to the "biosynthesis of secondary metabolites" sub-category.

4. Expression of related genes for color and fiber development in two cotton lines of different fiber color and quality

The pigmentation deposits in the brown colored cotton fiber were closely related to the flavonoid and proanthocyanidins biosynthesis. In this study, the "anthocyanin biosynthesis" and "flavone and flavonol biosynthesis" appeared on the top 10 list of the KEGG enrichment analysis (Table 3). Further analysis using the KEGG database indicated that a total of 14 unigenes were involved in the "flavone and flavonol biosynthesis" pathway (Table 4), which could be further classified as three orthology categories. "K05280" contained one up-regulated unigene which encoded flavonoid 3′-hydroxylase (F3′H). In addition, 10 down-regulated unigenes that encoded flavonol 3-O-methyltransferase (OMT) belonged to "K05279"; in "K10757", all unigenes encoded flavonol 3-O-glucosyltransferase (FOGT), two of them were up-regulated, and one was down-regulated. A total of four unigenes were mapped in the "anthocyanin biosynthesis" pathway, all of them were up-regulated. All four unigenes were involved in anthocyanin 5-O-glucosyltransferase (5GT) encoding.

Another important pigmentation pathway in plants is the "flavonoid biosynthesis pathway", which is also considered to be a key pathway for pigment formation in brown fiber cotton. Although it was not shown in Table 3, eight important genes were involved in this pathway. Chalcone synthase (*CHS*), chalcone isomerase (*CHI*), leucoanthocyanidin reductase (*LAR*), anthocyanidin reductase (*ANR*) and anthocyanidin synthase (*ANS*) were down-regulated in this pathway, while flavonoid 3′-hydroxylase (*F3′H*), flavanone 3-hydroxylase (*F3H*) and flavonol synthase (*FLS*) were all up-regulated. According to the distribution of DEGs in the entire pathway, most of the down-regulated DEGs (*CHS, CHI, LAR, ANR, ANS*) were enriched in upstream and downstream pathways, and the up-regulated DEGs were in the middle of the flavonoid biosynthetic pathway (*F3H, F3′H, FLS*). However, another gene in the middle of the pathway, the dihydroflavonol 4-reductase (*DFR*), was unchanged (Fig. 5). Furthermore, a gene that encodes 5-O-glucosyltransferase in the anthocyanin biosynthesis pathway was up-regulated. The down-regulated genes in the flavonoid biosynthesis pathway suggested that there were less pigments in Z128 compared with that in Z263. This was confirmed by the lighter brown fiber color in Z128 compared to the dark brown fiber in Z263.

To test the reliability of comparative transcriptional data, qRT-PCR analysis was performed for *CHS, CHI, LAR, DFR, F3H, F3′H, ANR* and *ANS*. Samples were selected from the flavonoid biosynthesis pathway across five developmental time points from 0 DPA to 20 DPA. Overall, the results of the qRT-PCR analysis were consistent with the results from the transcriptome for the mixed mRNAs of five developmental time points of the eight selected genes. However, when Z128 was compared to Z263 at 10 DPA, the selected eight genes had significantly higher transcript levels in Z128 (Fig. 5). This suggests that the genes involved in the flavonoid biosynthesis pathway at 10 DPA are related to better fiber quality formation. Moreover, the genes involved in flavonoid biosynthesis affected the fiber color and fiber quality of the brown fiber cotton. Some other genes such as the ethylene related factors also regulated fiber development [24]. According to Table 5, all of the ethylene related factors had higher expression levels in Z128 than in Z263. The DEGs analysis also revealed that transcript levels of different fiber development related factors varied dramatically in cotton fibers, such as reactive oxygen [25], ethylene [24], Ca^{2+}/CaM [26], and sucrose phosphate synthase [27] (Table 5).

Discussion

The RNA-seq approach based on next generation sequencing technology provided us with a new method to study the transcriptome of developing cotton fibers. It was not dependent on existing genome information and was an efficient way to quantify the expression level of a single gene without high background noise [28]. In recent years, this technology has been successfully applied in transcriptome studies for many non-model organisms [13,17,29–30].

In this study, we mixed the RNA from four important fibers at various developmental stages (5, 10, 15 and 20 DPA) to de novo assemble the transcriptome of developmental colored cotton fiber. A total of 125,014 unigenes were generated in two sequencing libraries, which were further assembled into 71,895 all-unigenes with an average length of 533 bp, 69.46% of which could be matched in the NCBI database (E-value≤10^{-5}). This volume of data was greater than that reported in previous studies on other species [17,30]. Approximately 20,000 unigenes identified from *G. hirsutum* were recorded in the NCBI database (Nov 2011). In this study, approximately 50,000 unigenes (EST) were matched with nr database records. Therefore, we believe that our unigene library contained almost all of the known unigenes from *G. hirsutum*. In conclusion, we acquired a high quality and well-assembled transcriptome library for developing colored cotton fibers.

In all nr-annotated unigenes, only 1,537 (3.08%) were directly annotated with the field of "*Gossypium hirsutum*"; most other unigenes were assigned to other species, which mainly included "*Vitis vinifera*", "*Ricinus communis*" and "*Populus trichocarpa*". This implies that the genome of cotton may be very similar to these species and this could be a reference for a prospective cotton sequencing project.

Z263 was the offspring which derived from a cross between white fiber cotton and dark brown cotton, and Z128 was an inbred line selected from Z263 with a lighter color and better fiber quality. Therefore, the similar genetic backgrounds provided a fine model with which to study the mechanism underlying the formation of brown color in cotton fiber. We compared the whole transcriptome for each and, unexpectedly, an abundance of unigenes (20.26%) revealed significant differential expression between Z263 and Z128 (FDR≤0.001, $|\log_2\text{Ratio}|\geq 1$). This evidence demonstrates that all DEGs are relatively evenly distributed in most of the relevant metabolism pathways. Namely, the divergence of cotton fiber color and quality was the result of complex processes generated by multiple metabolic processes.

There is limited information in the literature on the molecular mechanism that underlies the formation of fiber color in cotton. Xiao et al [6] cloned five flavonoid structure genes from brown cotton fiber and found that all the cloned genes could be involved in pigmentation metabolism for brown fiber. Several studies focused on chemicals also suggested that proanthocyanidins (condensed tannins) might be the precursor of pigmentation in natural colored cotton fiber [7–8]. Here, almost of all the unigenes which encoded the key enzymes (CHS, CHI, LAR, ANS and ANR) of the flavonoid biosynthesis pathway were down-regulated in Z128 compared with that in Z263, implying that accumulation of proanthocyanidins in Z263 might be more than that in Z128. Zhan et al. [7–8] suggested that the depth of brown color might be closely related to the accumulated quantity of condensed tannins. Another unexpected finding in this study was related to the "anthocyanin biosynthesis" pathway. As a downstream metabolic pathway, it is one of the most important elements of pigment biosynthesis in plants [31–33]. We found that all unigenes involved in this pathway were significantly up-regulated (read in Z263 = 0) in Z128 and homologous with anthocyanin 5-O-glucosyltransferase (5GT), which could make anthocyanin more stable by modification [34]. Another recent study demonstrated that the lack of glucose at the 5 position of anthocyanin could lead to color variation in carnations [35]. This result implied that the depth of brown cotton fiber color variation might be the consequence of an interaction between oxidized tannins and glycosylated anthocyanin.

F3H has been thoroughly studied for decades and it is predominantly expressed during the fiber elongation stage in *G. barbadense*, a process that has no relation to pigment formation [36]. Furthermore, when the *F3H–RNAi* segment was transferred into brown fiber plants, they yielded more stunted fibers. Transgenic analysis showed that the suppression of *F3H* not only inhibited fiber elongation but also retarded fiber maturation [37]. *F3H* was suppressed in Z263 but up-regulated in Z128. This evidence suggests that *F3H* is important in fiber development. Like *F3H, F3′H* and *FLS* were very important in the pigments synthesis. Therefore, it is possible that the genes in the middle were up-regulated. According to Xiao et al [6] and Zhan et al [7], tannins could be the key chemical responsible for the brown color in cotton fiber. As shown in Fig. 5, *LAR*, *ANR*, and *ANS* were the key enzymes to accumulate the tannins. The down-regulated *CHS*, *CHI*, *LAR*, *ANR*, and *ANS* genes in the main pathway of pigment formation should be related to lighter fiber color of Z128 than that of Z263.

The pigmentation related genes not only regulated the fiber color but also influenced the fiber quality. The flavonoids are abundant and widely distributed plant secondary metabolites, and known to be an active participant in fiber development [38]. Previous studies showed that in fiber cells, the flavonoid genes were dominantly expressed in the fiber elongation stage [39–41]. In our study, *CHS*, *CHI*, *LAR*, *DFR*, *F3H*, *F3'H*, *ANR* and *ANS* showed higher expression levels at 10 DPA in Z128, thus highlighting that flavonoid metabolism represents a novel pathway with the potential for cotton fiber improvement. Our GWAS analysis of SNP in cotton germplasm indicated that the genes involved in flavonoid biosynthesis were also associated with fiber quality traits (unpublished data). Therefore, the highly expressed flavonoid gene in the fiber elongation stage in Z128 should be related to better fiber quality.

Cotton fibers are single-celled trichomes that differentiate from the ovule epidermis, including fiber initiation, elongation, secondary cell wall biosynthesis and maturation, leading to mature fibers. Ca^{2+} and ROS are two important factors involved in fiber cell growth [42]. Ca^{2+}/Calmodulin (CaM) is involved in plant growth and development through interaction with ROS signaling [43]. Based on gene expression profile analysis, Ca^{2+}/CaM is implicated in cotton fiber elongation. However, currently, there remains little direct evidence of the mechanism of Ca^{2+}/CaM on cotton fiber development. In our study, Ca^{2+} related genes were either down-regulated or up-regulated in Z128 compared with that in Z263.

Recent literature indicates that ethylene acts as a positive regulator of root hair, apical hook, and hypocotyl development [44–46]. Furthermore, Shi et al. [24] found that ethylene biosynthesis was one of the most significantly up-regulated biochemical pathways during fiber elongation. Exogenously applied ethylene promoted robust fiber cell expansion, whereas its biosynthetic inhibitor L-(2-aminoeth oxyvinyl)-glycine (AVG) specifically suppressed fiber growth. The ethylene biosynthesis pathways in our data were not shown in Table 3 as the "top 10 DEGs enriched pathways in KEGG analysis", however, a number of ethylene related genes were up-regulated in Z128 compared with that in Z263, such as Unigene61435_All, Unigene40770_All, and Unigene62869_All. This suggests that ethylene related genes may contribute to better fiber quality in Z128.

Author Contributions

Conceived and designed the experiments: WG SH XD. Performed the experiments: WG SH JT. Analyzed the data: WG SH GS XF. Contributed reagents/materials/analysis tools: JS ZP YJ. Contributed to the writing of the manuscript: WG.

References

1. Vreeland JM (1999) The revival of colored cotton. Sci Am 280: 90–96.
2. Dickerson DK, Lane EF, Rodriguez DF (1999) Naturally colored cotton: resistance to changes in color and durability when refurbished with selected laundry aids. California Agric Tech Inst, California State University, Fresno, California. 42p.
3. Murthy MSS (2001) Never say dye: the story of coloured cotton. Resonance 6, 29–35.
4. Feng HJ, Sun JL, Wang J, Jia YH, Zhang XY, et al. (2011) Genetic effects and heterosis of the fibre colour and quality of brown cotton (*Gossypium hirsutum*). Plant Breeding 130: 450–456.
5. Kohel R (1985) Genetic analysis of fiber color variants in cotton. Crop Sci 25: 793–797.
6. Xiao YH, Zhang ZS, Yin MH, Luo M, Li XB, et al. (2007) Cotton flavonoid structural genes related to the pigmentation in brown fibers. Biochem Biophys Res Commun 35: 873–78.
7. Zhan SH, Lin Y, Cai YP, Li ZP (2007) Preliminary deductions of the chemical structure of the pigment brown in cotton fiber. Chin Bull Bot 24: 99–104 (Chinese with english abstract).
8. Zhan SH, Lin Y, Cai YP, Li ZP (2007) Relationship between the pigment in natural brown cotton fiber and the condensed tannin. Cotton Sci 19: 183–188 (Chinese with english abstract).
9. Haas BJ, Zody MC (2010) Advancing RNA-seq analysis. Nat Biotechnol 28: 421–423.
10. Mortazavi A, Williams BA, McCue K, Schaeffer L, Wold B (2008) Mapping and quantifying mammalian transcriptomes by RNA-SNat Methods 5: 621–628.
11. Nagalakshmi U, Wang Z, Waern K, Shou C, Raha D, et al. (2008) The transcriptional landscape of the yeast genome defined by RNA sequencing. Science 320: 1344.
12. Hittinger CT, Johnston M, Tossberg JT, Rokas A (2010) Leveraging skewed transcript abundance by RNA-Seq to increase the genomic depth of the tree of life. Proc Natl Acad Sci USA, 107: 1476.
13. Guo S, Zheng Y, Joung JG, Liu S, Zhang Z, et al. (2010) Transcriptome sequencing and comparative analysis of cucumber flowers with different sex types. BMC Genomics 11: 384.
14. Tisserant E, Da Silva C, Kohler A, Morin E, Wincker P, et al. (2011) Deep RNA sequencing improved the structural annotation of the *Tuber melanosporum* transcriptome. New Phytol. 189: 883–891.
15. Ru Z, Wang G, He S, Du X (2013) The difference of fiber quality and fiber ultrastructure in different natural colored cotton. Cotton Sci 23: 184–188.
16. Wan CY, Wilkins TA (1994) A Modified Hot Borate Method Significantly Enhances the Yield of High-Quality RNA from Cotton (*Gossypium hirsutum* L.). Anal Biochem 223: 7–12.
17. Z Xia, H Xu, J Zhai, D Li, H Luo, C He, X Huang (2011) RNA-Seq analysis and de novo transcriptome assembly of *Hevea brasiliensis*. Plant Mol Biol 77: 1–10.
18. Li R, Zhu H, Ruan J, Qian W, Fang X, et al. (2010) De novo assembly of human genomes with massively parallel short read sequencing. Genome Res 20: 265–272.
19. Pertea G, Huang X, Liang F, Antonescu V, Sultana R, et al. (2003) TIGR Gene Indices clustering tools (TGICL): a software system for fast clustering of large EST datasets. Bioinformatics 19: 651–652.
20. Conesa A, Götz S, García-Gómez JM, Terol J, Talón M, et al. (2005) Blast2GO: a universal tool for annotation, visualization and analysis in functional genomics research. Bioinformatics 21: 3674–3676.
21. Ye J, Fang L, Zheng H, Zhang Y, Chen J (2006) WEGO: a web tool for plotting GO annotations. Nucleic Acids Res 34: 293–297.
22. Audic S, Claverie JM (1997) The significance of digital gene expression profiles. Genome Res 7: 986–995.
23. Mortazavi A, Williams BA, McCue K, Schaeffer L, Wold B (2008) Mapping and quantifying mammalian transcriptomes by RNA-SNat Methods 5: 621–628.
24. YH Shi, Zhu SW, Mao XZ, Feng JX, Qin YM, et al. (2006) Transcriptome profiling, molecular biological, and physiological studies reveal a major role for ethylene in cotton fiber cell elongation. Plant Cell 18(3): 651–664.
25. Chaudhary B, Hovav R, Flagel L, Mittler R, Wendel J (2009) Parallel expression evolution of oxidative stress-related genes in fiber from wild and domesticated diploid and polyploid cotton (*Gossypium*). BMC Genomics 10: 378.
26. Tang W, Tu L, Yang X, Tan J, Deng F, et al (2014) The calcium sensor GhCaM7 promotes cotton fiber elongation by modulating reactive oxygen species (ROS) production. New Phytol 202: 509–520.
27. Haigler CH, Singh B, Zhang D, Hwang S, Wu C, et al. (2007) Transgenic cotton over-producing spinach sucrose phosphate synthase showed enhanced leaf sucrose synthesis and improved fiber quality under controlled environmental conditions. Plant Mol Biol 63: 815–832.
28. Wang Z, Gerstein M, Snyder M. (2009) RNA-Seq: a revolutionary tool for transcriptomics. Nat Rev Genet 10(1): 57–63.
29. Zenoni S, Ferrarini A, Giacomelli E, Xumerle L, Fasoli M, et al. (2010) Characterization of transcriptional complexity during berry development in *Vitis vinifera* using RNA-SPlant Physiol 152: 1787–1795.

30. Wei W, Qi X, Wang L, Zhang Y, Hua W, et al. (2011) Characterization of the sesame (*Sesamum indicum* L.) global transcriptome using Illumina paired-end sequencing and development of EST-SSR markers. BMC Genomics 12: 451.
31. Dooner HK, Robbins TP, Jorgensen RA (1991) Genetic and developmental control of anthocyanin biosynthesis. Ann Rev Genet 25: 173–199.
32. Halloin JM (1982) Localization and changes in catechin and tannins during development and ripening of cottonseed. New Phytol 90: 651–657.
33. Grotewold E (2006) The genetics and biochemistry of floral pigments. Annu Rev Plant Biol 57: 761–780.
34. Yamazaki M, Gong Z, Fukuchi-Mizutani M, Fukui Y, Tanaka Y, et al. (1999) Molecular cloning and biochemical characterization of a novel anthocyanin 5-O-glucosyltransferase by mRNA differential display for plant forms regarding anthocyanin. J Biol Chem 274: 7405–7411.
35. Nishizaki Y, Matsuba Y, Okamoto E, Okamura M, Ozeki Y, et al. (2011) Structure of the acyl-glucose-dependent anthocyanin 5-O-glucosyltransferase gene in carnations and its disruption by transposable elements in some varieties. Mol Genet Genomics 286: 383–394.
36. Tu LL, Zhang XL, Liang SG, Liu DQ, LF Zhu, et al. (2007) Gene expression analyses of sea-island cotton (*Gossypium barbadense* L.) during fiber development. Plant Cell Rep 26: 1309–1320.
37. Tan J, Tu L, Deng F, Hu H, Nie Y, et al. (2013) A genetic and metabolic analysis revealed that cotton fiber cell development was retarded by flavonoid naringenin. Plant Physiol 162: 86–95.
38. Owens DK, Crosby KC, Runac J, Howard BA, Winkel B (2008) Biochemical and genetic characterization of Arabidopsis flavanone 3beta-hydroxylase. Plant Physiol Biochem 46: 833–843.
39. Gou JY, Wang LJ, Chen SP, Hu WL, Chen XY, et al. (2007) Gene expression and metabolite profiles of cotton fiber during cell elongation and secondary cell wall synthesis. Cell Res 17: 422–434.
40. Hovav R, Udall JA, Hovav E, Rapp R, Flagel L, et al. (2008) A majority of cotton genes are expressed in single-celled fiber. Planta 227: 319–329.
41. Rapp R, Haigler C, Flagel L, Hovav R, Udall J, et al. (2010) Gene expression in developing fibres of Upland cotton (*Gossypium hirsutum* L.) was massively altered by domestication. BMC Biol. 8: 139.
42. Qin YM, Zhu YX. (2011) How cotton fibers elongate: a tale of linear cell-growth mode. Curr. Opin. Plant Biol. 14: 106–111.
43. Lee SH, Seo HY, Kim JC, WD Heo, Chung WS, et al. (1997) Differential activation of NAD kinase by plant calmodulin isoforms. The critical role of domain I. J Biol Chem 272: 9252–9259.
44. Achard P, Vriezen WH, Van Der Straeten D, Harberd NP (2003) Ethylene regulates *Arabidopsis* development via the modulation of DELLA protein growth repressor function. Plant Cell 15: 2816–2825.
45. Seifert GJ, Barber C, Wells B, Roberts K (2004) Growth regulators and the control of nucleotide sugar flux. Plant Cell 16: 723–730.
46. Grauwe LD, Vandenbussche F, Tietz O, Palme K, Straeten DVD (2005) Auxin, ethylene and brassinosteroids: Tripartite control of growth in the *Arabidopsis* hypocotyl. Plant Cell Physiol 46: 827–836.

3

Evaluation of Cotton Leaf Curl Virus Resistance in BC_1, BC_2, and BC_3 Progenies from an Interspecific Cross between *Gossypium arboreum* and *Gossypium hirsutum*

Wajad Nazeer[1,2], Abdul Latif Tipu[2], Saghir Ahmad[2], Khalid Mahmood[2], Abid Mahmood[3], Baoliang Zhou[1]*

1 State Key Laboratory of Crop Genetics and Germpalsm Enhancement, MOE Hybrid Cotton R&D Engineering Research Center, Nanjing Agricultural University, Nanjing, Jiangsu Province, China, **2** Cotton Research Station, Multan, Ayub Agricultural Research Institute, Faisalabad, Punjab, Pakistan, **3** Cotton Research Institute, Ayub Agricultural Research Institute, Faisalabad, Punjab, Pakistan

Abstract

Cotton leaf curl virus disease (CLCuD) is an important constraint to cotton production. The resistance of *G. arboreum* to this devastating disease is well documented. In the present investigation, we explored the possibility of transferring genes for resistance to CLCuD from *G. arboreum* (2n = 26) cv 15-Mollisoni into *G. hirsutum* (2n = 52) cv CRSM-38 through conventional breeding. We investigated the cytology of the BC_1 to BC_3 progenies of direct and reciprocal crosses of *G. arboreum* and *G. hirsutum* and evaluated their resistance to CLCuD. The F_1 progenies were completely resistant to this disease, while a decrease in resistance was observed in all backcross generations. As backcrossing progressed, the disease incidence increased in BC_1 (1.7–2.0%), BC_2 (1.8–4.0%), and BC_3 (4.2–7.0%). However, the disease incidence was much lower than that of the check variety CIM-496, with a CLCuD incidence of 96%. Additionally, the disease incidence percentage was lower in the direct cross 2(*G. arboreum*)×*G. hirsutum* than in that of *G. hirsutum*×*G. arboreum*. Phenotypic resemblance of BC_1 ~BC_3 progenies to *G. arboreum* confirmed the success of cross between the two species. Cytological studies of CLCuD-resistant plants revealed that the frequency of univalents and multivalents was high in BC_1, with sterile or partially fertile plants, but low in BC_2 (in both combinations), with shy bearing plants. In BC_3, most of the plants exhibited normal bearing ability due to the high frequency of chromosome associations (bivalents). The assessment of CLCuD through grafting showed that the BC_1 to BC_3 progenies were highly resistant to this disease. Thus, this study successfully demonstrates the possibility of introgressing CLCuD resistance genes from *G. arboreum* to *G. hirsutum*.

Editor: David D. Fang, USDA-ARS-SRRC, United States of America

Funding: This work was supported by the following. Baoliang Zhou: Research was supported partially by National Key Technology Support Program of China during the twelfth Five-year Plan Period (2013BAD01B03-04), the Independent Innovation Funds for Agricultural Technology of Jiangsu Province, China [CX (14) 2065] and the Priority Academic Program Development of Jiangsu Higher Education Institutions. Saghir Ahmad: Punjab Agriculture Research Board (PARB) for providing financial support for these studies under PARB Project no. 27. The funders had no role in study design, data collection and analysis, decision to publish, or preparation of the manuscript.

Competing Interests: The authors have declared that no competing interests exist.

* Email: baoliangzhou@njau.edu.cn

Introduction

Cotton production is biotically constrained by various diseases, which lead to yield instability and reduced seed quality. Cotton leaf curl disease (CLCuD) is a debilitating disease of cotton in Africa, Pakistan, and Northwestern India [1–3]. CLCuD is caused by a pathogen complex of a virus and a DNA beta satellite (DNA-β) molecule [4]. There are seven such virus species, all belonging to the *Begomovirus* genus, and DNA-β satellites are associated with CLCuD in these regions [5–8].

CLCuD was first recorded in 1967 in the Multan district, Pakistan, on scattered *Gossypium hirsutum* plants [9–11], and it has spread rapidly to all cotton growing areas of Pakistan and throughout the Indian subcontinent. Two epidemics of this disease have been observed during the past three decades due to a loss of host-plant resistance in existing cotton varieties [12–13].

In Pakistan, an outbreak of CLCuD occurred in the early 1990s. This disease devastated the Pakistani cotton industry, where it caused an estimated yield reduction of 30–35%. Between 1992 and 1997, the economic losses due to CLCuD in Pakistan amounted to approximately 5 billion dollars (US) [14]. Similarly, in the Indian state of Punjab, this disease reduced cotton production by almost 70% in 1998 [15]. Singh et al. [16] observed a reduction of 52.7% in the number of bolls and a reduction of 54.2% in boll weight due to CLCuD, whereas the differences in yield loss between resistant and susceptible cultivars were almost 50% and 85–90%, respectively.

In the late 1990s, several resistant cotton varieties were gradually introduced into the Indo-Pak region, and losses due to the disease diminished [17–18]. However, resistance subsequently broke in 2001–2002 [3,12] due to new strains of CLCuD emerged, and all of the cotton varieties that were previously known resistant to CLCuD, such as LRA-5166, CP-15/2, and Cedex, have become susceptible to CLCuD [6–7,19–23]. Symptoms of this disease were also reported in China [24], which is located far from the hot spots of India and Pakistan, and there is great concern that CLCuD could spread from its origin to other cotton growing areas of the world where the disease is not currently present. Plant biologists have attempted to understand the molecular biology of this disease complex to control CLCuD [25], but the tricky nature of the pathogen and the rapid evolution/recombination of these genes have hindered the progress of this research [26–28].

In plant breeding, wild relatives have long been studied due to the presence of novel genes [29–31], and these wild species have been exploited most often as sources for biotic and abiotic stress resistance [32]. Among the wild species of cotton, especially, desi cotton (*G. arboreum* L.) has built in desirable resistant genes for all kind of *Begomoviruses* associated with CLCuD [33]. Additionally, *G. arboreum* is known to combat various stresses like drought [34–35], heat [36], root rot, cotton leaf curl virus [37] and insect pests (bollworms and aphids) [12]. Interspecific hybridization of cotton has been performed with varying degrees of success [21,38–40]. For example, Sacks and Robinson [41] transferred nematode (*Rotylenchulus reniformis*) resistance into tetraploid *G. hirsutum*. Chen et al. [42] and Nazeer et al. [2] employed *Gossypium australe* and *Gossypium stocksii* to introgress some novel genes for drought and CLCuD resistance into *G. hirsutum,* respectively. The interspecific hybridization is quite difficult, especially, between *G. arboreum* and *G. hirsutum*, and some scientists explored bridge lines for introgression of genes form wild species [43].

At present, no single variety of *G. hirsutum* is resistant to CLCuD; however, *G. arboreum* is documented to have resistance against CLCuD [31]. Due to the importance of this disease and significant features of this species, we initiated a project to explore the possibility of successful transferring CLCuD resistance genes from Desi cotton (*G. arboreum*, 2n = 26) into cultivated upland cotton (*G. hirsutum*, 2n = 52) genotypes through conventional hybridization and backcrossing without developing bridging line. In this way maximum desirable donor genes of *G. arboreum* can be transferred into *G. hirsutum* to improve the resistance to CLCuD of the cultivated *G. hirsutum*.

Materials and Methods

Plant materials

The plant materials used in this study include *G. hirsutum* cv CRSM-38 (2n = 4x = AADD = 52), *G. arboreum* cv 15-Mollisoni (2n = 2x = AA = 26), and an artificial autotetraploid of *G. arboreum* cv 15-Mollisoni (2n = 4x = 52; Figure 1). The F_1 CLCuD-resistant progeny involving these parents, which was developed by Ahmad et al. [44], comprising direct cross [2(*G. arboreum*)×G. *hirsutum*] and its reciprocal cross (*G. hirsutum*×G. *arboreum*), was utilized to directly backcross with *G. hirsutum*. The number of F_1 progenies was increased by cuttings. Thus total number of F_1 plant progenies for direct and reciprocal cross was 10 and 15, respectively, to generate BC_1 to BC_2 generations. CIM-496, a cotton variety highly susceptible to CLCuD, was employed as a standard/control in order to obtain a natural virus inoculum.

Development of backcross progenies

The F_1 CLCuD-resistant progenies consisting of two cross combinations, 2(*G. arboreum*)×G. *hirsutum* and *G. hirsutum*×G. *arboreum*, were backcrossed with *G. hirsutum* to produce the BC_1 progenies in 2011. The BC_1 progenies were planted in the field at Cotton Research Station in Multan, Pakistan in May 2012 and backcrossed with *G. hirsutum* to generate the BC_2 progenies. These BC_2 progenies were again backcrossed with *G. hirsutum* cv CRSM-38 to produce the BC_3 progenies in 2013. One thing should be noticed here that only those normal morphological plant progenies that produce more fruits but no symptoms of CLCuD were selected for backcrossing. The plant progenies that showed even minor spots of CLCuD were rejected to utilize for backcrossing. The scheme for the development of the backcross progenies are shown in Figure 2. Emasculation was carried out in the evening, and emasculated flowers were manually pollinated the next morning.

Use of plant growth hormones for hybridization

Normally, embryos fail to develop in hybridizations between *G. arboreum* and *G. hirsutum*. This obstacle was overcome by the application of plant hormones such as gibberellic acid (GA_3) and naphthalene acetic acid. Specifically, 50 mgL^{-1} GA_3 and 100 mg L^{-1} naphthalene acetic acid were applied to the bases of pedicles 24 hours after pollination for 3 consecutive days to reduce embryo and boll shedding. The number of cross boll sets was counted, and the bolls were picked at harvest time.

Cross fertility studies

Fertility studies for BC_1 to BC_3 progenies of 2(*G. arboreum*)×G. *hirsutum* and *G. hirsutum*×G. *arboreum* were measured in term of cross boll setting and their germination percentage by given formula:

$$\text{Boll setting}(\%) = \frac{\text{Total number of cross boll picked}}{\text{Total number of pollinations}} \times 100;$$

$$\text{Seed germination}(\%) = \frac{\text{Number of seed geminated}}{\text{Total number of seed tested}} \times 100$$

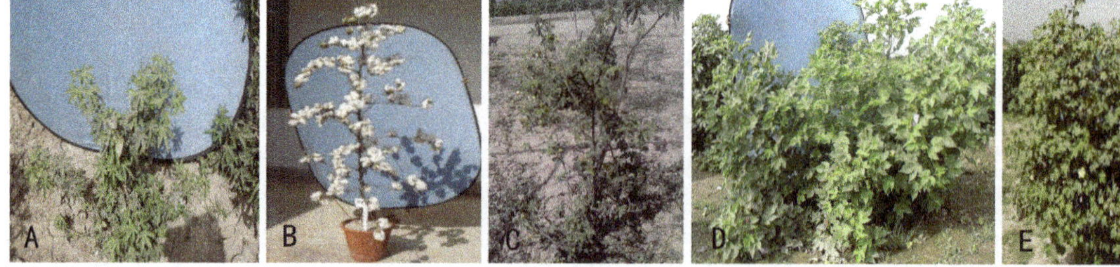

Figure 1. Parents of interspecific hybridization. A. *G. arboreum* cv 15-Mollisoni (2n = 2x = 26); B. *G. hirsutum* cv CRSM-38 (2n = 4x = 52); C. 2(*G. arboreum*) (2n = 4x = 52); D. [2(*G. arboreum*)×*G. hirsutum*] F_1 (2n = 4x = 52); E. (*G. hirsutum*×*G. arboreum*) F_1 (2n = 3x = 39).

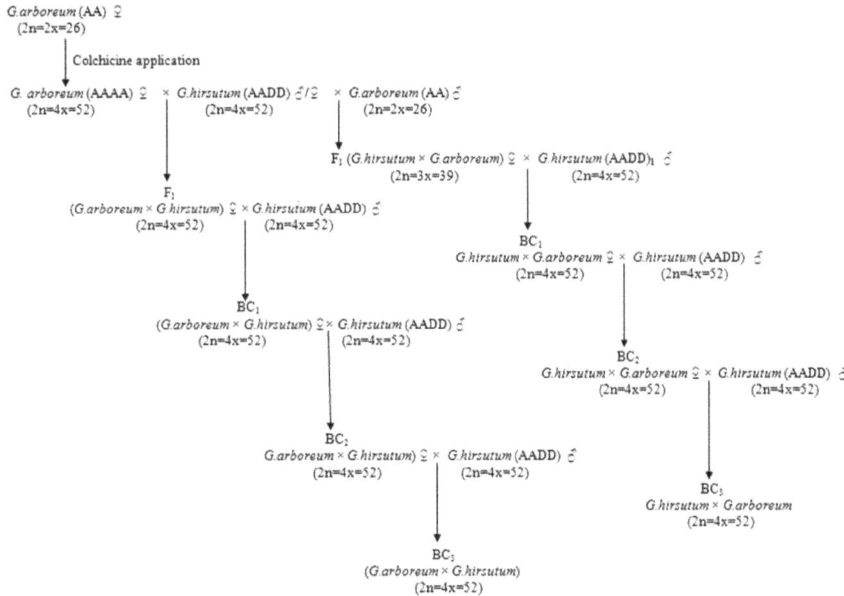

Figure 2. Scheme for the development of the BC_1 to BC_3 progenies for interspecific cross 2(*G. arboreum*)×G. *hirsutum* and *G. hirsutum*×*G. arboretum*.

Morphological characteristics

Observations of growth habit, stem color, leaf texture, leaf shape, leaf hairiness, bracteole size, corolla color, petal spots, the position of the staminal column, anther color, and dehiscence in the parents, as well as in BC_1 to BC_3, were recorded. The phenotypic resemblance of BC_1 to BC_3 progenies to *G. arboreum* having desirable traits with good resistance to CLCuD will be helpful for selection of introgression progenies.

Cytological studies

Morphological normal plants producing more fruits were selected from BC_1 to BC_3 progenies for cytological studies. Young buds of BC_1 to BC_3 plants, along with those of the parents, were collected and fixed in Carnoy's solution at 8 to 9 am and preserved in 70% ethanol after 24 h. Three to four anthers were squashed on a slide with a drop of 2.5% acetocarmine solution to examine the pollen mother cells (PMCs). Chromosomal configurations such as univalent (I's), bivalents (II's), trivalents (III's), quadrivalents (IV's), and division stage were examined under a Labomed microscope, and photographs were also taken using a camera mounted on a Labomed microscope.

Maintenance of virus inoculum and screening for CLCuD

Artificial inoculation techniques is not available for CLCuD, therefore, the only way to study the response of cotton germplasm is to expose the introgression progenies to high inoculum pressure by planting in natural hot spots [45], so sick plot technique was used to arrange spreader plants among BC introgression lines. In this sick plot technique, we planted susceptible variety CIM-496 after each two rows of CLCuD resistant lines to encourage uniform spread of the disease. Planting of BC_1~BC_3 progenies was done after 3rd week of May for the three seasons i.e. 2011–2013. Sowing was done manually and row to row (75 cm) and plant to plant (30 cm) distance was maintained. Row length for each genotype was 450cm and plot size was variable depending upon the seed availability.

Phenotypic assessment of BC_1 to BC_3 progenies against CLCuD

The resistance of the BC_1 to BC_3 progenies against CLCuD was assessed under natural field conditions using an inoculum of CIM-496 at Cotton Research Station in Multan, Pakistan which is hot spot of CLCuD. Data for CLCuD were recorded following the rating system described in Table 1 to calculate the severity index

Table 1. Disease rating (symptom rating) scale for evaluation of cotton leaf curl virus disease.

Disease index (%)	Severity grade	Symptoms	Remarks
0	0	No Symptoms	Resistant
1–20	1	Thickening of only secondary and tertiary veins.	Highly tolerant
21–30	2	Thickening of secondary and primary (mid rib) veins.	Tolerant
31–50	3	Vein thickening (V.T), leaf curling (L.C) or enation (E) or both.	Susceptible
>50	4	Stunting along with vein thickening leaf curling/enation.	Highly susceptible

Table 2. Fertility studies of interspecific hybrid between *G. arboreum* and *G. hirsutum* to produce the BC_1 to BC_3 progenies.

Parentage	Year	No. of plants*	No. of pollinations	No. of bolls picked	Boll setting (%)	No. of seed obtained	No. of seed germinated	Germination (%)
BC_1 [2(*G. arboreum*)×*G. hirsutum*]	2009–11	15	12890	338	2.6	57	20	35.1
BC_1 (*G. hirsutum*×*G. arboreum*)	2009–11	12	8144	265	3.3	48	25	52.1
BC_2 [2(*G. arboreum*)×*G. hirsutum*]	2012	14	1495	19	1.3	22	11	45.5
BC_2 (*G. hirsutum*×*G. arboreum*)	2012	155	299	117	39.1	519	340	65.5
BC_3 [2(*G. arboreum*)×*G. hirsutum*]	2013	12	980	15	1.5	52	24	46.1
BC_3 (*G. hirsutum*×*G. arboreum*)	2013	225	7263	3123	42.9	412	276	67.0

*Number of plants used for pollination and recording data.

(SI), percent disease index (%, DI), and disease reaction. Individual plant ratings for each genotype were added and means were calculated to generate the corresponding SI. The DI was calculated using the following formula:

$$\text{Percent disease index} = \frac{\text{Sum of all disease ratings}}{\text{Total plants}} \times \frac{100}{\text{Maximum grade}}$$

The percent disease tolerance (PDT) was calculated by selecting a minimum of 100 plants on a diagonal from one corner to the other, and diseased plants were counted to determine the PDT using the formula:

$$\text{Percent disease tolerance} = \frac{\text{Total plants - diseased plants}}{\text{Total plants}} \times 100$$

Data regarding the latent period, number of virus-infected plants, disease incidence percentage, disease severity index, infection type, and disease reaction were recorded.

Inoculation of CLCuD through grafting

A petiole and rootstock from CIM-496 were used to transfer virus inoculum into healthy plants. Two grafting techniques, i.e., approach grafting and petiole grafting, were employed to confirm the resistance against CLCuD in BC_1 to BC_3 plants. For approach grafting, the resistant plants of the BC_1, BC_2, and BC_3 progenies were used as scions, whereas virus-susceptible *G. hirsutum* plants were used as stock. For petiole grafting, young petioles from CLCuD-infected plants were selected and inserted into the test plants. Two infected petioles were also grafted onto the same plant to introduce additional virus inoculum. The following data were recorded: grafting success, infectivity, latent period, infection type, disease severity index, and disease incidence percentage at 40 and 70 days after grafting (DAG).

Results

Cross fertility studies

Examination of the cross ability of BC_1 to BC_3 of the combination 2(*G. arboreum*)×*G. hirsutum* and *G. hirsutum*×*G. arboreum* revealed that the maximum percentage of boll set (42.9%) and germination (67.0%) were observed in the BC_3 (*G. hirsutum*×*G. arboreum*) progenies (Table 2). Minimum boll setting (1.3%) was recorded in the cross BC_2 [2(*G. arboreum*)×*G. hirsutum*]. A minimum percentage of viable seeds (35.1%) were obtained in BC_1 [2(*G. arboreum*)×*G. hirsutum*]. The boll setting and germination (%) gradually increased from BC_1 to BC_3 (Figure 3).

Morphological studies

Examination of the morphological characteristics of the parents and BC_1 to BC_3 of 2(*G. arboreum*)×*G. hirsutum* revealed that in BC_1 to BC_3, leaf hairiness, flower size, corolla color, petal spots, and pollen color were segregated for the male and female parents. Stem color, leaf lobation, flower size, corolla color, petal number, petal size, anther dehiscence, and pollen color of BC_1 to BC_3 were similar to those of the female parents. Stem hairiness, gossypol glands, leaf size, leaf hairiness, leaf texture, petiole length, and bracteole number and size were dominant characters of the male parents. Bracteole dentation, petiole size, petal spots, and position

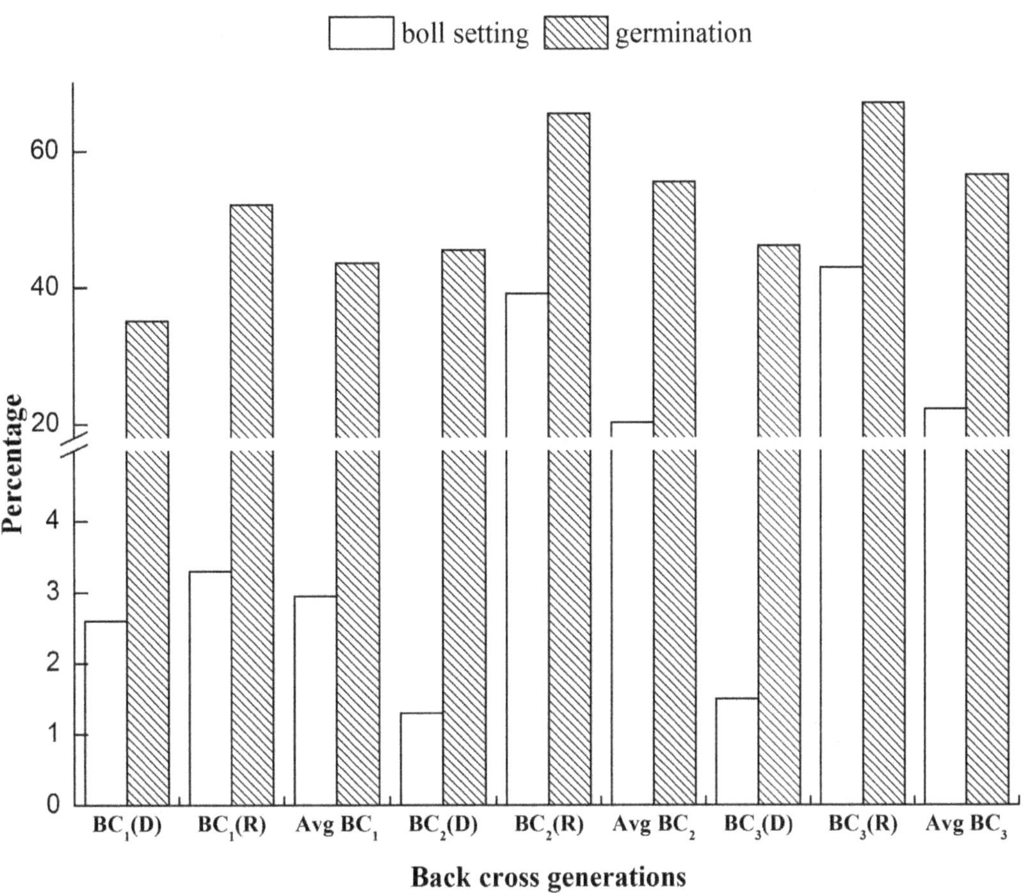

Figure 3. Advancement of boll setting and germination (%) across different generations (BC_1 to BC_3). D = Direct cross [2(*G. arboreum*) × *G. hirsutum*]; R = Reciprocal cross [*G. hirsutum* × *G. arboreum*].

of the staminal column of BC_1 to BC_3 were intermediate between those of both parents (Table 3). The hybrid plant progenies of BC_1 to BC_3 of 2(*G. arboreum*) × *G. hirsutum* are shown in Figure 4.

An analysis of the morphological characteristics of the parents and BC_1 to BC_3 of *G. hirsutum* × *G. arboreum* revealed that in BC_1 to BC_3, gossypol, bracteole number and size, and pistil size was dominant characters of the female parents. Leaf size, leaf lobation, corolla color, petal spots, and pollen color were the dominant characteristics of the male parents. Stem color and hairiness, leaf texture, bracteole number and size, and position of staminal column of BC_1 to BC_3 were intermediate between those of both parents, while leaf hairiness was segregated (Table 4). The hybrid plant progenies of BC_1 to BC_3 of *G. hirsutum* × *G. arboreum* are shown in Figure 5. Morphological characteristics particularly leaf texture, leaf size, bracteole size, corolla color and petal spots from *G. arboreum* into BC_1 to BC_3 progenies of both crosses helped for selection of plant progenies that have some resemblance of *G. arboreum* and also showed CLCuD resistance.

Cytological studies

Meiosis in parents. The course of meiosis was examined in the *G. hirsutum* and *G. arboreum* parents. In these species, the reduction division was normal, with regular pairing of chromosomes. The number of bivalents in *G. hirsutum* and *G. arboreum* at Metaphase-I was 26 and 13, respectively (Figures 6A and 6B). The disjunction of the chromosomes was normal at Anaphase-I. The meiotic behavior in the artificial autotetraploid of *G. arboreum* parent showed two I's, 23 II's, and one IV (Figure 6C).

Meiosis in BC_1. [2(*G. arboreum*) × *G. hirsutum*]. The progenies of this combination comprised 15 plants; only seven normal morphological plants with better boll setting were studied cytologically. Cytological studies at Metaphase-I revealed that there were six I's, 21 II's, and one IV's (Figure 6D). The number of I's, II's, III's, and IV's for 76 PMCs ranged from 5–12, 18–22, 0–1, and 0–1, respectively, for a total of 52 chromosomes (Table 5), while the average number of I's, II's, III's, and IV's was 8.2, 20.4, 0.2, and 0.7, respectively. A few lagging chromosomes were also observed at Anaphase-I. The high frequency of univalents (5–12) and multivalents (0–1) caused meiotic disturbance; the plants were partially fertile/sterile.

***G. hirsutum* × *G. arboretum*.** The plant progenies of this combination comprised 12 plants; cytological studies were conducted on six normal morphological plants with better boll setting. The cytological configuration at Metaphase-I of the BC_1 plants revealed two I's, 23 II's, and one IV's (Figure 6E). In the 53 PMCs of these hybrid plants, there were 1–5 I's, 20–25 II's, and 0–1 III's and IV's, for a total of 52 chromosomes (Table 5), while the average number of I's, II's, III's, and IV's for 53 PMCs was 2.2, 23.2, 0.4, and 0.6, respectively. Although multivalent association was observed, the high frequency of bivalents (20–25) caused these plants to be fertile or partially fertile.

Table 3. Morphological characteristics of parents and the BC_1 to BC_3 progenies from the cross 2(*G. arboreum*)×*G. hirsutum.*

Morphological characteristic	2(*G. arboreum*)	*G. hirsutum*	BC_1	BC_2	BC_3
Stem characteristics					
Stem color	Greenish brown	Green	Brown	Brown	Brown
Stem hairiness	Profusely hairy	Hairy	Hairy	Hairy	Hairy
Black glands	Dense	Sparse	Sparse	Sparse	Sparse
Leaf characteristics					
Leaf color	Dark green	Green	Light/dark green	Light/dark green	Light/dark green
Leaf size(cm)	Medium (7.3×9.9)	Large (10×14)	Small/Large (7.0×8.0 cm)/(9.0×11.0 cm)	Small/Large (7.1×8.3)/(9.2×10.8)	Medium/Large (7.6×8.9)/(9.6×11.4)
Leaf hairiness	Profusely hairy	Hairy	Hairy/profusely hairy	Hairy/profusely hairy	Hairy/profusely hairy
Leaf lobation	3–5 narrrow, deep lobed	3–5 broad, shallow lobed	3–5 broad lobed	3–5 broad lobed	3–5 broad lobed
Leaf texture	Thick, Leathery	Herbaceous	Herbaceous	Herbaceous	Herbaceous
Petiole length (cm)	Medium (4.4)	Long (8.8)	Long (7.3)	Long (7.2)	Long (7.5.0)
Boll characteristics					
Bracteole number and size (cm)	2–3, large (3.0×2.6), united at base	3 large (3.3×1.8)	3 Large (3.0×2.3)	3 Large (3.3×2.2)	3 Large (3.1×2.4)
Bracteole dentation	Entire	5–11, deep narrow	4–9 medium	3–10 medium	3–11 medium
Flower characteristics					
Flower size	Medium	Large	Medium	Medium	Medium
Pedicel size (cm)	Long (1.7)	Long (1.2)	Long (1.3)	medium(1.0)	medium(0.9)
Calyx	5 sepal forming a cup with wavy margins	5 sepals forming a cup with teeth	5 sepal forming a cup with wavy margins	5 sepal forming a cup with wavymargins	5 sepal forming a cup with wavymargins
Corolla color	Light yellow	Creamy	Creamy/light yellow	Creamy/ligt yellow	Creamy/light yellow
Petal number and size (cm)	5, medium, (3.0×2.6)	5, large (4.6×4.5)	5, medium (3.5×4.1)	5, medium (3.4×4.2)	5, medium (3.3×3.5)
Petal spot	Dark pink	Absent	Present/absent	Present/absent	Present/absent
Position of staminalcolum and size (cm)	Short (0.4)	Long (2.0)	Medium (1.5)	Medium (1.7)	Medium (1.6)
Anther dehiscence	Partial	Normal	Partial	Normal	Normal
Pollen color	Light yellow	Creamy	Yellow	Creamy/light yellow	Creamy/light yellow
Pistil size (cm)	Long (2.5)	Long (2.9)	Long (2.6)	Long (2.8)	Long (2.9)

Meiosis in BC_2. **[2(*G. arboreum*)×*G. hirsutum*].** These plant progenies consisted of 14 plants; only five normal morphological plants with better boll setting were studied cytologically. The chromosomal conformation at Metaphase-I was two I's+23 II's+1 IV's (Figure 6F). A study of 60 PMCs revealed 2–4 I's, 23–25 II's, and 0–1 IV's, for a total of 52 chromosomes (Table 5), while the average number of I's, II's, and IV's for 60 PMCs was 3.0, 24.1, and 0.2, respectively. Trivalents were not observed in these plants. The low frequency of uni- and multi-valents, as well as the high frequency of chromosome

Figure 4. Hybrid Progenies of [2(*G. arboreum*)×*G. hirsutum*]. A = BC_1; B = BC_2; C = BC_3.

association (23–25 II's), caused the plants to be fertile but shy bearing.

***G. hirsutum*×*G. arboretum*.** The plant progenies of this combination comprised 161 plants; only 10 normal morphological plants with better boll setting were studied cytologically. The chromosomal constitution at Metaphase-I revealed 3 I's+21 II's+1 III's+1 IV's (Figure 6G). However, in 110 PMCs, there were 1–4 I's, 21–25 II's, and 0–1 III's and IV's, for a total of 52 chromosomes (Table 5), and the average number of I's, II's, III's, and IV's was 2.4, 23.2, 0.4, and 0.5, respectively. Low frequencies of univalents (1–4) and multivalents (0–1), as well as high frequencies of bivalents (21–25), were observed. The plants were fertile. A few shy bearing plants were also observed.

Meiosis in BC_3. **[2(*G. arboreum*)×*G. hirsutum*].** The plant progenies consisted of 12 plants. A total of 35 PMCs were sampled from two plants for microscopic studies. Metaphase-I of these PMCs showed 2 I's+25 II's (Figure 6H). The average range of these PMCs revealed that there were 2 I's and 25 II's, for a total of 52 chromosomes (Table 5); the plants were fertile.

***G. hirsutum*×*G. arboretum*.** The plant progenies consisted of 225 plants. The chromosome pairing was normal (26 II's) in most of the PMCs (Figure 6I). The average number of chromosomes among 40 PMCs exhibited normal disjunction (Table 5); the plants were fertile.

Testing of BC_1 to BC_3 progenies against CLCuD through grafting

The resistance/susceptibility of the plants was confirmed through petiole and approach grafting, as indicated in Figure 7, and only resistant plants were used for backcrossing to produce the next generation. Grafting for BC_1 to BC_3 hybrid plants of 2(*G. arboreum*)×*G. hirsutum* and *G. hirsutum*×*G. arboreum* was carried out under greenhouse conditions as well as in the natural field. All plants from both crosses [2(*G. arboreum*)×*G. hirsutum* and *G. hirsutum*×*G. arboreum*] showed 100% infectivity and grafting success (Table 6). Plants of susceptible variety CIM-496 showed symptoms of CLCuD at 11–14 days after germination. Grafts of BC_1 from cross 2(*G. arboreum*)×*G. hirsutum* remained asymptomatic to this disease throughout their lifecycles, whereas only two grafts from BC_1 of *G. hirsutum*×*G. arboreum* showed minor spots (3–5) of vein thickening at 40 DAG, which appeared on a few leaves. These minor spots become quite small at 70 DAG and were only detected after careful observation. Therefore, the BC_1 hybrid plants of 2(*G. arboreum*)×*G. hirsutum* and *G. hirsutum*×*G. arboreum* were resistant to CLCuD, with good plant growth. The BC_2 hybrid plants of 2(*G. arboreum*)×*G. hirsutum* and *G. hirsutum*×*G. arboreum* developed disease symptoms at 30–35 and 28–30 DAG, respectively. The infection type range for 2(*G. arboreum*)×*G. hirsutum* and *G. hirsutum*×*G. arboreum* was 0–1 and 0–2, respectively, and the same trend for the first appearance of disease symptoms was observed for BC_3 plants from both crosses. All BC_3 progenies from 2(*G. arboreum*)×*G. hirsutum* and *G. hirsutum*×*G. arboreum* were highly tolerant to CLCuD, with good fruit bearing and normal growth compared with susceptible variety CIM-496. By and large, the hybrid plants of cross 2(*G. arboreum*)×*G. hirsutum* showed better resistance/tolerance to CLCuD than those of cross *G. hirsutum*×*G. arboreum*.

Testing of BC_1 to BC_3 progenies against CLCuD under natural field conditions

The BC_1 to BC_3 hybrid plants of 2(*G. arboreum*)×*G. hirsutum* and *G. hirsutum*×*G. arboreum* were tested under natural field conditions. Nineteen plants of BC_1 of the combination [2(*G. arboreum*)×*G. hirsutum*] and 15 plants of reciprocal cross *G. hirsutum*×*G. arboreum* revealed disease indices of 1.3% and 1.6%, respectively, whereas the average severity index was 0.05 and 0.06 at 40 DAS, respectively (Table 7). However, the disease index and severity index were zero after 70 DAS because the minor spots of vein thickening that were observed on a single plant of each cross disappeared after 70 DAS. CIM 496, the control variety used in this trial, had a disease index of 94.3%, and enation was also observed at 70 DAS. Fourteen plants of BC_2 of the combination 2(*G. arboreum*)×*G. hirsutum* and 161 plants of the combination *G. hirsutum*×*G. arboreum* raised through backcrossing of BC_1 with *G. hirsutum* had disease indices of 1.8% and 4.0%, respectively, at 40 DAS, and the disease index increased to 3.5% and 6.8%, respectively, at 70 DAS. The grade of disease severity in 2(*G. arboreum*)×*G. hirsutum* was 0.07 (40 DAS) to 0.1 (70 DAS), whereas it was 0.17 (40 DAS) to 0.2 (70 DAS) for *G. hirsutum*×*G. arboreum*. The susceptible cotton variety CIM 496 in this trial had a disease index of 97.7% with a disease severity grade of 3.9.

Twelve plants of BC_3 of the combination 2(*G. arboreum*)×*G. hirsutum* and 225 plants of the combination *G. hirsutum*×*G. arboreum* raised through backcrossing with *G. hirsutum* had a 4.2% and 7.0% disease index, respectively, at 40 DAS. And the

Table 4. Morphological characteristics of parents and the BC_1 to BC_3 progenies from the cross *G. hirsutum*×*G. arboretum.*

Morphological characteristics	*G. hirsutum*	*G. arboreum*	BC_1	BC_2	BC_3
Stem characteristcs					
Stem color	Green	Green	Green	Green	Green
Stem hairiness	Hairy	Hairy	Hairy	Hairy	Hairy
Black glands	Sparse	Sparse/dense	Sparse	Sparse	Sparse
Leaf characteristcs					
Leaf color	Green	Green	Green	Green/dark green	Green/dark green
Leaf size(cm)	large (10.0×14.0)	Small/medium (6.0×8.3)	Medium (7.1×7.9)	medium (7.0×8.4)	medium (7.3×8.3)
Leaf lobation	3–5 broad, shallow lobed	3–5 narrrow, deep lobed	3–5 medium lobed	3–5 broad lobed	3–5 broad lobed
Leaf texture	Herbaceous	Herbaceous	Herbaceous	Herbaceous	Herbaceous
Leaf hairiness	Hairy	Hairy/profusely hairy	Hairy	Hairy/profusely hairy	Hairy/profusely hairy
Petiole length (cm)	Long (8.8)	Medium (4.4)	Long (7.2)	Long (7.5)	Long (7.4)
Boll characteristcs					
Bracteole number and size (cm)	3 large (3.3×1.8)	3, small (2.7×2.1), united at base	3, large (3.0×2.6)	3, large (3.0×2.0)	3, large (3.2×2.2)
Bracteole dentation	3–7, superficial	5–11, deep narrow	4–9, superficial	3–11, superficial	3–9, superficial
Flower characteristcs					
Flower size	Large	Small	Medium	Medium	Medium/large
Pedicel size (cm)	Long (1.2)	Long (1.2)	Long (1.1)	Long (1.3)	Long (1.2)
Calyx	5 sepal forming a cup with teeth	5 sepals forming a cup with wavy margins	5 sepals forming a cup with wavy margins	5 sepals forming a cup with wavy margins	5 sepals forming a cup with wavy margins
Corolla color	Creamy	Yellow	Creamy/Light Yellow	Creamy/light yellow	Creamy/light yellow
Petal number and size (cm)	5, large, (4.6×4.5)	5, small (2.6×2.5)	5, large, (4.5×4.4)	5, large, (4.4×4.6)	5, large, (4.6×4.4)
Petal spot	Absent	Light pink	Present/absent	Present/absent	Present/absent
Position of staminalcolum and size (cm)	long (2.0)	small (1.0)	Medium (1.5)	Medium (1.5)	Medium (1.6)
Anther dehiscence	Normal	Normal	Partial/normal	Normal	Normal
Pollen color	Creamy	Yellow	Creamy/light Yellow	Creamy/light Yellow	Creamy/light Yellow
Pistil size (cm)	Long (2.9)	Small (2.1)	Long (3.1)	Long (2.9)	Long (3.2)

disease index increased to 8.3% and 12.0%, respectively, at 70 DAS. The grade of disease severity in 2(*G. arboreum*)×*G. hirsutum* was 0.17 (40 DAS) to 0.3 (70 DAS), whereas it was 0.28 (40 DAS) to 0.5 (70 DAS) for *G. hirsutum*×*G. arboreum*. CIM 496 had a disease index of 95.0% with a disease severity grade of 3.8 (Table 7).

As the backcross progressed from BC_1 to BC_3, the PDT gradually decreased (Figure 8). However, the PDT was fairly high in 2(*G. arboreum*)×*G. hirsutum* compared to the combination *G. hirsutum*×*G. arboreum*.

Discussion

G. hirsutum has low genetic diversity and lacks resistance against CLCuD. In general, wild diploid species of *Gossypium* possess resistance against many challenges, such as insects, pests, diseases, and many abiotic factors [37,46]. Hence, there is a great need to exploit this resource to develop resistance against CLCuD in cultivated tetraploid species [2]. Cotton breeders have long tried to obtain hybrids between diploid and tetraploid species [47]. However, several incompatibility factors hinder the development

Figure 5. Hybrid Progenies of (*G. hirsutum*×*G. arboreum*). A = BC_1; B, C = BC_2; D = BC_3.

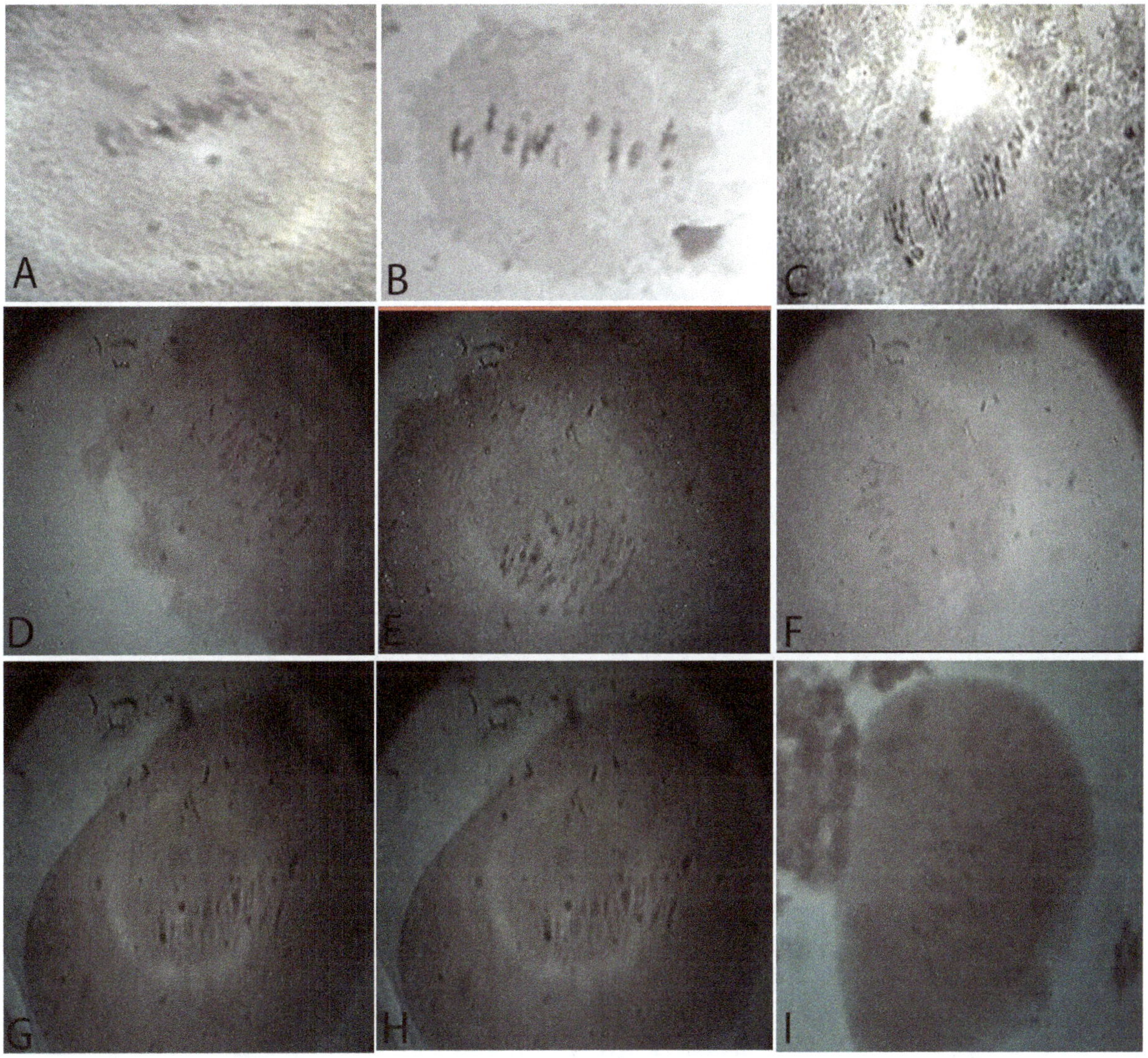

Figure 6. Chromosome configurations in PMCs at Metaphase-I of meiosis. A. *G. hirsutum*, 26 II's; B. *G. arboreum*, 13 II's; C. 2(*G. arboreum*), 2 I's+23 II's+1 IV; D. [2(*G. arboreum*)×*G. hirsutum*] BC_1, 6 I's+21 II's+1 IV; E. [*G. hirsutum*×*G. arboreum*] BC_1, 2 I's+23 II's+1 IV; F. [2(*G. arboreum*)×*G. hirsutum*] BC_2, 2 I's+23 II's+1 IV; G. [*G. hirsutum*×*G. arboreum*] BC_2, 3 I's+21 II's+1 III+1 IV; H. [2(*G. arboreum*)×*G. hirsutum*] BC_3, 2 I's+25 II's; I. [*G. hirsutum*×*G. arboreum*] BC_3, 26 II's.

Table 5. Cytological comparison of BC_1 to BC_3 plants from an interspecific cross between *G. arboreum* and *G. hirsutum*.

Cross congifuration	Plant number	PMC	I's	II's	III's	IV's	Total
Chromosomal configuarion for BC_1							
2(*G. arboreum*)×*G. hirsutum*	P2	12	5	20	1	1	52
//	P3	10	6	21	0	1	52
//	P4	8	6	21	0	1	52
//	P9	10	8	20	0	1	52
//	P11	15	8	22	0	0	52
//	P13	11	12	20	0	0	52
//	P4	10	12	18	0	1	52
	Range		5–12	18–22	0–1	0–1	
	Average of 76 cells		8.2	20.4	0.2	0.7	
G. hirsutum×*G. arboreum*	P1	5	2	25	0	0	52
//	P2	12	1	22	1	1	52
//	P5	8	5	20	1	1	52
//	P9	10	2	23	0	1	52
//	P10	6	2	25	0	0	52
//	P11	12	2	25	0	0	52
	Range		1–5	20–25	0–1	0–1	
	Average of 53 cells		2.2	23.2	0.4	0.6	
Chromosomal configuarion for BC_2							
2(*G. arboreum*)×*G. hirsutum*	P3	10	2	25	0	0	52
//	P4	15	4	24	0	0	52
//	P7	8	2	25	0	0	52
//	P10	12	2	23	0	1	52
//	P11-(1)	15	4	24	0	0	52
	Range		2–4	23–25	0	0–1	
	Average of 60 cells		3.0	24.1	0	0.2	
(*G. hirsutum*×*G. arboreum*)	P1(16)	15	1	24	1	0	52
	P2	10	3	23	1	0	52
		12	2	25	0	0	52
	P4 (1)	10	2	23	0	1	52
//	P5(3)	10	2	25	0	0	52
//		8	4	22	0	1	52
//	P5(14)	5	3	21	1	1	52
//	P7(15)	5	3	23	1	0	52
		10	2	23	0	1	52
//		10	2	23	0	1	52

Table 5. Cont.

Cross congifuration	Plant number	PMC	I's	II's	III's	IV's	Total
	P9(16)	8	4	22	0	1	52
//	P13(17)	7	3	21	1	1	52
	Range		1–4	21–25	0–1	0–1	
	Average of 110 cells		2.4	23.2	0.4	0.5	
Chromosomal configuarion for BC_3							
2(*G. arboreum*)×*G. hirsutum*	P1	15	2	25	0	0	52
	P2	20	2	25	0	0	52
	Range		2	25	0	0	
	Average of 35 cells		2	25	0	0	
G. hirsutum×*G. arboreum*	P1	20	0	26	0	0	52
//	P14	20	0	26	0	0	52
	Range		0	26	0	0	
	Average of 40 cells		0	26	0	0	

of hybrids under *in situ* conditions [48,49]. Abortion of the embryo after fertilization and the lack of retention of cross bolls [50,51] is a common stumble in interspecific crosses. Some species like *G. barbadense* can be hybridize easily with *G. hirsutum* and produce fertile F_1 progeny [52] without hormones application. These two species i.e. *G. hirsutum* and *G. barbadense* have chromosome homology and the tetraploid genomes, are not separated by any large scale chromosomal rearrangement [53]. However, crosses between *G. hirsutum* and *G. arboreum* L. are rarely successful without hormone application [44,54]. Plant hormones are known to control pollen tube growth [55]. Exogenous application of growth hormones has been used to overcome the crossing barrier and to facilitate interspecific crosses in many crops, i.e., cotton [56], wheat [57], and tomato [58]. Altman [56] compared exogenous application with *in vitro* techniques, i.e., ovule and embryo culture, and found that exogenous hormone application in conjunction with standard hybridization methods is superior to *in vitro* methods. Interspecific hybridization of cotton is enhanced by the application of exogenous hormones after pollination. Exogenous hormone application alone may be used to overcome certain crossing barriers within *Gossypium* [59–60]. The extract of garlic acid has been used as a growth regulator to obtain interspecific hybrids between tetraploid *G. hirsutum* and diploid *G. arboreum* species of cotton [56,61]. The *in situ* development of BC_1 to BC_3 plants using exogenous hormones in the current study was superior to that using *in vitro* methods, which is in agreement with an earlier report [55]. The average number of seeds per boll varied from immature seeds to 1.5 seeds per boll. In the absence of exogenous hormones, pollinated flowers produce 0.1% seed development [56].

The boll setting and seed germiantion is very low in interspecific crosses and fertility of interspecific crosses can be measured in terms of boll setting percentage [52,62]. The cross fertility of BC_1 to BC_3 between [2(*G. arboreum*)×*G. hirsutum*] and *G. hirsutum*×*G. arboreum* showed that the boll set was maximum (42.9%) in cross BC_3, *G. hirsutum*×*G. arboreum*, but minimum (1.3%) in cross BC_2, 2(*G. arboreum*)×*G. hirsutum* (Table 2). Viable seeds were obtained in both combinations. From BC_1 to BC_3, an increasing trend of boll setting and germination (%) was observed. Seed setting improvement was also recorded in *Brassica* by backcrossing with the recurrent parent [63]. The factor responsible for the semi-sterile condition are transmitted rarely through the pollen but readily through the egg cell. Boll setting and germination (%) was higher in reciprocal cross (*G. hirsutum*× *G.arboreum*) as compared to direct cross 2(*G. arboreum*)×*G. hirsutum* [64].

In general, the BC_1 to BC_3 hybrid plants of both cross combinations [2(*G. arboreum*)×*G. hirsutum* and *G. hirsutum*×*G. arboreum*] were intermediate in several traits between the two parents. The prevalence of yellow pollen in both crosses (direct and reciprocal) in most of the plants validated the inheritance of this character from *G. arboreum*, because this color is more common in *G. arboreum* species [65,66], and it revealed the dominance in inheritance [67]. By contrast, in BC_1 to BC_3, leaf hairiness, flower size, corolla color, petal spots, pollen color, and so on were segregated in both parents. Moreover, morphological characteristics particularly leaf texture, leaf size, bracteole size, corolla color and petal spots from *G. arboreum* into BC_1 to BC_3 progenies [68] of both crosses were helpful for selection of plant progenies that have resemblance to *G. arboreum* and also showed CLCuD resistance. The frequency of plant progenies that showed

Figure 7. Testing of CLCuD through grafting. A and B. Cleft grafts; C = Single petiole grafts; D = Double petiole grafts.

good plant architecture were higher in reciprocal crosses as compared to direct cross.

When developing interspecific hybrids for resistance, a thorough knowledge of the chromosomal behavior in hybrids and backcross progenies is essential. In the present study, in hybrid 2(*G. arboreum*)×*G. hirsutum*, the 'AD' genome was introgressed into the 'AA' genome of *G. arboreum*, producing an 'AAAD' genomic constitution. In hybrid *G. hirsutum*×*G. arboreum*, the A-genome of *G. arboreum* was introgressed into the 'AD' genome of *G. hirsutum,* producing the genomic constitution 'AAD'. In *G. arboreum* and *G. hirsutum*, normal orientation, association, and disjunction of chromosomes were observed, while in F_1 hybrids of the above genomic constitution, quadrivalents and a low frequency of chromosome association (bivalents) were observed. The univalents observed in this study can be attributed to asynapsis due to the lack of homology between the different sets of chromosomes. The presence of laggards demonstrates the occurrence of meiotic disturbances, leading to an imbalance in the daughter cells. In BC_1 hybrid plants of both combinations, the frequency of univalents and multivalents was high, and the plants were sterile/partially fertile. In BC_2 hybrids of both combinations, the frequency of univalents and multivalents was low, and the plants were shy bearing. In BC_3 hybrids of both combinations, the frequency of chromosome association (bivalents) was 25–26; hence, the plants were fertile. The average of univalent (I's) chromosomes was higher in 2(*G. arboreum*)×*G. hirsutum* in comparison with *G. hirsutum*×*G.arboreum*. However, the average of bivalents (II's) chromosomes was higher in *G. hirsutum*×*G.arboreum*. Thus *G. hirsutum*×*G.arboreum* was more fertile and more adaptive to the environment than 2(*G.arboreum*)×*G. hirsutum*.

Studies of resistance/susceptibility are rather difficult and laborious due to the involvement of vectors, the efficiency of transmission, and the persistent nature of the virus/CLCuD. Grafting may successfully lead to the transmission of the virus when other methods fail, as it involves the union of cambial layers of the root sock and scion [69–70]. Thus, to screen CLCuD-resistant germplasm, transmission by grafting is the best alternative to natural transmission by vector, as most viruses of a persistent nature, such as CLCuD, cannot be transmitted through mechanical inoculation [71]. Ahmad et al. [72] used sick plot techniques to screen the exotic and local germplasm against CLCuD.

The BC_1 to BC_3 of 2(*G. arboreum*)×*G. hirsutum* and *G. hirsutum*×*G. arboreum* were tested through grafting under natural field/greenhouse conditions. These hybrids remained resistant to CLCuD [39]. The results of evaluation of the BC_1 to BC_3 progenies revealed a high degree of variability for CLCuD in the field and through grafting. All plants from both crosses [2(*G. arboreum*)×*G. hirsutum* and *G. hirsutum*×*G. arboreum*] showed 100% infectivity and grafting success. However, latent period and infection type range for BC_1–BC_2 was better in 2(*G.arboreum)*×*G. hirsutum* cross than *G. hirsutum*×*G.arboreum*. The grafts of BC_1 from cross 2(*G. arboreum*)×*G. hirsutum* remained asymptomatic to this disease. However, BC_1 of *G. hirsutum*×*G. arboreum* showed minor vein thickening, but the vein thickening was highly reduced after 70 days of grafting [46], whereas CIM-496 showed symptoms of CLCuD within 11–14 days after germination. Although minor symptoms of CLCuD appeared in BC_1 of *G. hirsutum*×*G. arboreum*, this disease did not affect the growth of the plants. Therefore, we can conclude that these plants were also resistant to CLCuD. The BC_2 and BC_3 hybrid plants of both cross combinations 2(*G. arboreum*)×*G. hirsutum* and *G. hirsutum*×*G. arboreum* developed disease symptoms after 28–35 DAG, and the average disease severity was grade 1.0 (70 DAG). Additionally, these plants showed good tolerance to CLCuD, with no symptoms of stunted growth. Therefore, these BC_2 and BC_3 plants were highly tolerant to CLCuD compared with susceptible variety CIM-496, which showed CLCuD symptoms after 11 DAS with no boll setting. Ullah et al. [46] also observed mild symptoms of CLCuD on the introgressed material following grafting, but the amount of viral DNA was significantly lower than the levels found in *G. hirsutum*. The same trend/response for latent period to acquire CLCuD was observed in the field for BC_1 to BC_3 hybrid plants. The average severity index for 2(*G. arboreum*)×*G. hirsutum* and *G. hirsutum*×*G. arboreum* was 0.05 and 0.06 (40 DAS), respectively. However, the disease index and severity index were zero after 70 DAS. Thus, the resistant hybrid plants of both crosses 2(*G. arboreum*)×*G. hirsutum* and *G. hirsutum*×*G. arboreum* showed better tolerance to CLCuD, with not deleterious effects on yield or growth. However, 2(*G. arboreum*)×*G. hirsutum* plants were more tolerant regarding number of virus infected plants, disease index (%), severity index and infection type range than those of cross *G. hirsutum*×*G. arboreum.* Collectively, plants from these crosses had better tolerance to CLCuD than CIM-496. The PDT was higher in 2(*G. arboreum*)×*G. hirsutum* than in *G. hirsutum*×*G. arboreum*. The frequency of ideotype plants was higher in *G. hirsutum*×*G. arboreum* compared with 2(*G. arboreum)*×*G. hirsutum*.

Conclusions

The results indicate that the BC_1 to BC_3 progenies were highly tolerant to CLCuD, indicating the possibility of transferring

Table 6. Evaluation of plants from an interspecific cross between *G. arboreum* and *G. hirsutum* against cotton leaf curl virus disease through grafting.

	Progeny	Year	No. of plants tested	Grafting success (%)	Infectivity (%)	Latent period (days)	Infection type range[A]	Av disease severity after 70 (DAG)	Disease reaction
BC_1	[2(*G. arboreum*)×*G. hirsutum*]	2011	15	100	100	Symptomless	0	0	Resistant
	G. hirsutum×*G. arboreum*	2011	12	100	100	39–41	0–1	1	Highly tolerant
	CIM-496 (Std.)	2011	20	100	100	14	3–4E*	4E	Highly susceptible
BC_2	[2(*G. arboreum*)×*G. hirsutum*]	2012	14	100	100	30–35	0–1	1	Highly tolerant
	G. hirsutum×*G. arboreum*	2012	20	100	100	28–30	0–2	1	Highly tolerant
	CIM-496 (Std.)	2012	10	100	100	11	3–4E	4E	Highly susceptible
BC_3	[2(*G. arboreum*)×*G. hirsutum*]	2013	12	100	100	28–30	0–2	1	Highly tolerant
	G. hirsutum×*G. arboreum*	2013	16	100	100	25–30	0–3	1	Highly tolerant
	CIM-496 (Std.)	2013	10	100	100	11	3–4E	4E	Highly susceptible

[A]Infection type range is based on the 0–4 scale described in Table 1,
*Enation where observed.

Table 7. Evaluation of plants from an interspecific cross between *G. arboreum* and *G. hirsutum* against cotton leaf curl virus disease under natural field conditions.

Parentage		No. of plants tested	Latent period (days)	No. of virus infected plants		Disease index (%)		Severity index		Infection type range[A]		Disease reaction[#]
				40 DAS	70 DAS	40 DAS	70 DAS	40 DAS	70 DAS	40 DAS	70 DAS	
BC_1	[2(*G. arboreum*)×*G. hirsutum*]	19	Symptomless	$19(18^0+1^1)$	$19(19^0)$	1.3	0	0.05	0	0	0	R
	(*G. hirsutum*×*G. arboreum*)	15	35–40	$15(14^0+1^1)$	$15(15^0)$	1.6	0	0.06	0	0–1	0	R
	CIM-496 (Std.)	31	14	$31(1^2+5^3+25^4)$	$31(4^3+27^4)$	94.35	96.7E*	3.7	3.9	2–4E	3–4E	HS
BC_2	[2(*G. arboreum*)×*G. hirsutum*]	14	30–35	$14(13^0+1^1)$	$14(12^0+2^1)$	1.8	3.5	0.07	0.1	0–1	0–1	HT
	G. hirsutum×*G. arboreum*	161	25–30	$161(146^0+7^1+5^2+3^3)$	$161(138^0+9^1+8^2+5^3+1^4)$	4	6.8	0.17	0.2	0–3	0–4	HT
	CIM-496 (Std.)	168	13	$168(4^2+7^3+157^{4E})$	$168(4^2+11^3+153^{4E})$	97.7	97.1	3.9	3.9	2–4E	2–4E	HS
BC_3	[2(*G. arboreum*)×*G. hirsutum*]	12	25–30	$12(11^0+1^2)$	$12(9^0+2^1+1^2)$	4.2	8.3	0.17	0.3	0–2	0–2	HT
	G. hirsutum×*G. arboreum*	225	25–30	$225(200^0+5^1+8^2+6^3+6^4)$	$225(176^0+14^1+12^2+11^3+9^4)$	7	12	0.28	0.5	0–4	0–4	HT
	CIM-496 (Std.)	190	15	$190(4^2+6^3+180^{4E})$	$190(6^2+15^3+169^{4E})$	95	96.4	3.8	3.8	2–4E	2–4E	HS

[A]Infection type range is based on the 0–4 scale described in Table 1;
*Enation where observed; R Resistant; HT Highly tolerant; HS Highly susceptible;
[#]Disease reaction based on disease index 70 DAS.

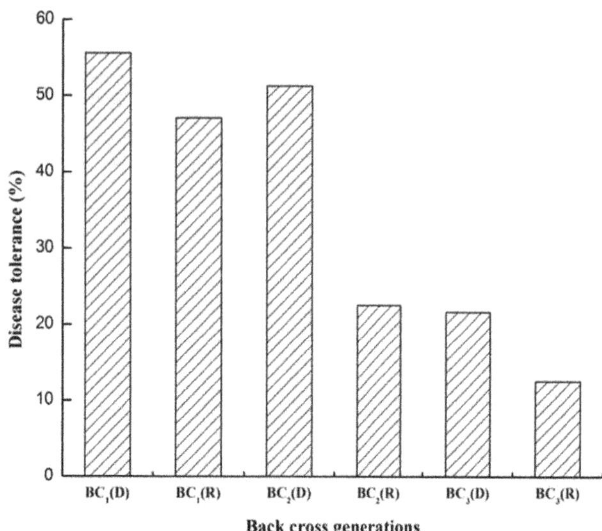

Figure 8. Percent disease tolerance across different generations (BC_1 to BC_3). D. Direct cross [2(*G. arboreum*)×*G. hirsutum*]; R. Reciprocal cross [*G. hirsutum*×*G. arboreum*].

CLCuD resistance genes from *G. arboreum* to *G. hirsutum* through conventional hybridization and backcrossing. As the backcross progressed, the disease incidence also increased, from BC_1 (1.3–1.6%) to BC_2 (1.8–4.0%) to BC_3 (4.2–7.0%). However, the disease incidence was much lower than that of the commercial cultivar CIM-496, which exhibited a very high incidence of CLCuD (97.7%). The disease incidence was lower in combination 2(*G. arboreum*)×*G. hirsutum* than in *G. hirsutum*×*G. arboreum*. As "A" genome is an invaluable genetic resource for improving modern tetraploid cotton (*G. hirsutum*). We observed very wide genetic variability among BC_1 to BC_3 progenies, which will certainly facilitate improvement of cotton resistances to diseases. And various scientists also utilized *G. arboreum* L. for introgression of desirable resistant genes into cultivated tetraploid cotton for drought [34–35], heat [36], root rot, cotton leaf curl virus [37,[44] and insect pests (bollworms and aphids) [12]. Therefore, the introgression lines of *G. arboreum* developed with or wothout resistance in this study can be employed to map the resistance gene(s)/loci, which will be very useful for future diverse (a)biotic-tolerant cotton breeding.

Acknowledgments

We thank Mr. Abdul Latif Sheikh for providing technical assistance.

Author Contributions

Conceived and designed the experiments: WN SA. Performed the experiments: ALT KM. Analyzed the data: BZ. Contributed reagents/materials/analysis tools: BZ SA. Contributed to the writing of the manuscript: WN BZ SA. Provided the breeding material: AM SA.

References

1. Tiendrébéogo F, Lefeuvre P, Hoareau M, Villemot J, Konaté G, et al. (2010) Molecular diversity of Cotton leaf curl Gezira virus isolates and their satellite DNAs associated with okra leaf curl disease in Burkina Faso. Virology Journal 7: 48.
2. Nazeer W, Ahmad S, Mahmood K, Tipu A, Mahmood A, et al. (2014) Introgression of genes for cotton leaf curl virus resistance and increased fiber strength from Gossypium stocksii into upland cotton (Gossypium hirsutum). Genetics and molecular research 13: 1133–1143.
3. Rajagopalan PA, Naik A, Katturi P, Kurulekar M, Kankanallu RS, et al. (2012) Dominance of resistance-breaking cotton leaf curl Burewala virus (CLCuBuV) in northwestern India. Archives of virology 157: 855–868.
4. Tahir MN, Amin I, Briddon RW, Mansoor S (2011) The merging of two dynasties–identification of an African cotton leaf curl disease-associated begomovirus with cotton in Pakistan. PloS One 6: e20366.
5. Ahuja SL, Monga D, Dhayal LS (2007) Genetics of resistance to cotton leaf curl disease in Gossypium hirsutum L. under field conditions. Journal of heredity 98: 79–83.
6. Briddon RW (2003) Cotton leaf curl disease, a multicomponent begomovirus complex. Molecular Plant Pathology 4: 427–434.
7. Mansoor S, Zafar Y, Briddon RW (2006) Geminivirus disease complexes: the threat is spreading. Trends in plant science 11: 209–212.
8. Azhar MT, Amin I, Anjum ZI, Arshad M, Briddon RW, et al. (2010) Both malvaceous and non-malvaceous betasatellites are associated with two wild cotton species grown under field conditions in Pakistan. Virus genes 41: 417–424.
9. Hussain T, Ali M (1975) A review of cotton diseases of Pakistan. Pakistan Cottons 19: 71–86.
10. Hussain T, Mahmood T (1988) A note on leaf curl disease of cotton. Pakistan Cotton 32: 248–251.
11. Thakur P (2002) Virus diseases of cotton. Diseases of Field Crops: 398.
12. Mansoor S, Amin I, Iram S, Hussain M, Zafar Y, et al. (2003) Breakdown of resistance in cotton to cotton leaf curl disease in Pakistan. Plant pathology 52: 784–784.
13. Zafar Y, Brown J (2011) Genome characterization of whitefly-transmitted geminivirus of cotton and development of virus-resistant plants through genetic engineering and conventional breeding. The ICAC Recorder 29: 7–12.
14. Briddon R, Markham P (2000) Cotton leaf curl virus disease. Virus research 71: 151–159.
15. Mann R (2011) Bemisia tabaci Interaction with Cotton Leaf Curl Virus. In: Thompson WMO (ed) The Whitefly, Bemisia tabaci (Homoptera: Aleyrodidae) Interaction with Geminivirus-Infected Host Plants. Springer Netherlands, 69–88.
16. Singh D, Gill J, Gumber R, Singh R, Singh S (2013) Yield and fibre quality associated with cotton leaf curl disease of Bt-cotton in Punjab. Journal of Environmental Biology 34: 113–116.
17. Rahman M, Hussain D, Malik T, Zafar Y (2005) Genetics of resistance to cotton leaf curl disease in Gossypium hirsutum. Plant pathology 54: 764–772.
18. Ahmad S, Hussain A, Hanif M, Mahmood K, Nazeer W, et al. (2012) CRSM-38, a new high yielding coupled with CLCuV tolerance cotton (Gossypium hirsutum L.) variety. African Journal of Biotechnology 11: 4368–4677.
19. Mahmood T, Arshad M, Gill MI, Mahmood HT, Tahir M, et al. (2003) Burewala strain of cotton leaf cur l virus: A threat to CLCuV cotton resistance varieties. Asian Journal Plant Sciences 2: 968–970.
20. Tahir M, Tariq M, Mahmood H, Hussain S (2004) Effect of sowing dates on incidence of cotton leaf curl virus on different cultivars of cotton. Plant Pathology Journal 3: 61–64.
21. Amrao L, Akhter S, Tahir MN, Amin I, Briddon RW, et al. (2010) Cotton leaf curl disease in Sindh province of Pakistan is associated with recombinant begomovirus components. Virus research 153: 161–165.
22. Akhtar K, Haidar S, Khan M, Ahmad M, Sarwar N, et al. (2010) Evaluation of Gossypium species for resistance to cotton leaf curl Burewala virus. Annals of Applied Biology 157: 135–147.
23. Mahmood T, Arshad M, Gill MI, Mahmood HT, Tahir M, et al. (2003) Burewala strain of cotton leaf curl virus: a threat to CLCuV cotton resistant varieties. Asian Journal of Plant Sciences 2: 968–970.
24. Cai J, Xie K, Lin L, Qin B, Chen B, et al. (2010) Cotton leaf curl Multan virus newly reported to be associated with cotton leaf curl disease in China. Plant pathology 59: 794–795.
25. Sattar MN, Kvarnheden A, Saeed M, Briddon RW (2013) Cotton leaf curl disease–an emerging threat to cotton production worldwide. Journal of General Virology 94: 695–710.
26. Azhar MT, Akhtar S, Mansoor S (2012) Letter to the Editor: Cotton leaf curl Multan betasatellite strains cloned from Gossypium barbadense further supports selection due to host resistance. Virus genes 45: 402–405.
27. Farooq A, Farooq J, Mahmood A, Shakeel A, Rehman A, et al. (2011) An overview of cotton leaf curl virus disease (CLCuD) a serious threat to cotton productivity. Australian Journal of Crop Science 5: 1823–1831.
28. Zaffalon V, Mukherjee SK, Reddy VS, Thompson JR, Tepfer M (2012) A survey of geminiviruses and associated satellite DNAs in the cotton-growing areas of northwestern India. Archives of virology 157: 483–495.
29. Amin K (1940) Interspecific hybridization between Asiatic and New World cottons. Indian Journal of Agricultural Science 10: 404–413.
30. Blank L, Lathers C (1963) Environmental and other factors influencing development of south western cotton rust. Phytopatholgy 53: 921–928.
31. Nelson RR (1973) Breeding plants for disease resistance concepts and applications. University Park, Penn.: The Pennsylvania State University Press.

32. Kalloo G (1992) Utilization of Wild Species. In: Kalloo G, Chowdhury JB, editors. Distant Hybridization of Crop Plants: Springer Berlin Heidelberg. 149–167.
33. Azhar MT, Aftab S, Zafar Y, Mansoor S (2010) Utilization of natural and genetically-engineered sources in Gossypium hirsutum for the development of tolerance against cotton leaf curl disease and fiber characteristics. International Journal of Agriculture and Biollogy 12: 744–748.
34. Maqbool A, Abbas W, Rao AQ, Irfan M, Zahur M, et al. (2010) Gossypium arboreum GHSP26 enhances drought tolerance in Gossypium hirsutum. Biotechnology Progress 26: 21–25.
35. Zhang L, Li FG, Liu CL, Zhang CJ, Zhang XY (2009) Construction and analysis of cotton (Gossypium arboreum L.) drought-related cDNA library. BMC research notes 2: 120.
36. Zahur M, Maqbool A, Irfan M, Barozai MYK, Qaiser U, et al. (2009) Functional analysis of cotton small heat shock protein promoter region in response to abiotic stresses in tobacco using Agrobacterium-mediated transient assay. Molecular Biology Reports 36: 1915–1921.
37. Azhar M, Anjum Z, Mansoor S (2013) Gossypium gossypioides: A source of resistance against cotton leaf curl disease among D genome diploid cotton species. JAPS, Journal of Animal and Plant Sciences 23: 1436–1440.
38. Cao Z, Wang P, Zhu X, Chen H, Zhang T (2013) SSR marker-assisted improvement of fiber qualities in Gossypium hirsutum using Gossypium barbadense introgression lines. Theoretical and Applied Genetics: 1–8.
39. Ahmad S, Khan N, Mahmood A, Mahmood K, Sheikh AL, et al. (2011) Exploring potential sources for leaf curl virus resistance in cotton (Gossypium hirsutum L.). 5th meeting of Asian Cotton Research and Development network Lahore Pakistan Lahore, Pakistan: International Cotton Advisory Committee (ICAC).
40. Guo W, Wang W, Zhou B, Zhang T (2006) Cross-species transferability of G. arboreum-derived EST-SSRs in the diploid species of Gossypium. Theoretical and Applied Genetics 112: 1573–1581.
41. Sacks EJ, Robinson AF (2009) Introgression of resistance to reniform nematode (Rotylenchulus reniformis) into upland cotton (Gossypium hirsutum) from Gossypium arboreum and a Gossypium hirsutum/Gossypium aridum bridging line. Field Crops Research 112: 1–6.
42. Chen Y, Wang Y, Wang K, Zhu X, Guo W, et al. (2014) Construction of a complete set of alien chromosome addition lines from Gossypium australe in Gossypium hirsutum: morphological, cytological, and genotypic characterization. Theoretical and Applied Genetics 127: 1105–1121.
43. Mergcai G, Baudoin JP, Vroh Bi I (1997) Exploitation of trispecific hybrids to introgress the glandless seed and glanded plant trait of Gossypium sturtianum Willis into Gossypium hirsutum L. Biotechnologie Agronomie Societe et Environment 1: 272–277.
44. Ahmad S, Mahmood K, Hanif M, Nazeer W, Malik W, et al. (2011) Introgression of cotton leaf curl virus-resistant genes from Asiatic cotton (Gossypium arboreum) into upland cotton (G. hirsutum). Genetics and Molecular Research 10: 2404–2414.
45. Akhtar KP, Khan AI, Hussain M, Khan MSI (2002) Comparison of resistance level to cotton leaf curl virus(CLCuV) among newly developed cotton mutants and commercial cultivars. Plant Pathollogy Journal 18: 179–186.
46. Ullah R, Akhtar KP, Moffett P, Mansoor S, Briddon RW, et al. (2014) An analysis of the resistance of Gossypium arboreum to cotton leaf curl disease by grafting. European Journal of Plant Pathology: 1–11.
47. Gill MS, Bajaj Y (1987) Hybridization between diploid (Gossypium arboreum) and tetraploid (Gossypium hirsutum) cotton through ovule culture. Euphytica 36: 625–630.
48. Sikka S, Joshi A (1960) Breeding. Cotton in India-a monograph Indian Central Cotton Committee, Bombay: 137–235.
49. Thengane S, Paranjpe S, Khuspe S, Mascarenhas A (1986) Hybridization of Gossypium species through in ovulo embryo culture. Plant cell, tissue and organ culture 6: 209–219.
50. Borole V, Dhumale D, Rajput J (2000) Embryo culture studies in interspecific crosses between arboreum and hirsutum cotton. Indian Journal of Genetics and Plant Breeding 60: 105–110.
51. Pundir N (1972) Experimental embryology of Gossypium arboreum L. and Gossypium hirsutum L. and their reciprocal crosses. Botanical gazette: 7–26.
52. Brubaker C, Brown A, Stewart JM, Kilby M, Grace J (1999) Production of fertile hybrid germplasm with diploid Australian Gossypium species for cotton improvement. Euphytica 108: 199–214.
53. Gerstel D, Sarvella PA (1956) Additional observations on chromosomal translocations in cotton hybrids. Evolution: 408–414.
54. Jafari Mofidabadi A, Soltanloo H, Ranjbran A (2011) Development of Genetic Broadening System in Cotton through Artificial Crosses between 2x and 4x Species. Cotton Genomics and Genetics 2.
55. Kovaleva L, Zakharova E, Minkina YV, Timofeeva G, Andreev I (2005) Germination and in vitro growth of petunia male gametophyte are affected by exogenous hormones and involve the changes in the endogenous hormone level. Russian Journal of Plant Physiology 52: 521–526.
56. Altman D (1988) Exogenous hormone applications at pollination for in vitro and in vivo production of cotton interspecific hybrids Plant Cell Reports 7: 257–261.
57. Sitch L, Snape J (1987) Factors affecting haploid production in wheat using the Hordeum bulbosum system. 1. Genotypic and environmental effects on pollen grain germination, pollen tube growth and the frequency of fertilization. Euphytica 36: 483–496.
58. Gordillo L, Jolley V, Horrocks R, Stevens M (2003) Interactions of BA, GA3, NAA, and surfactant on interspecific hybridization of Lycopersicon esculentum×Lycopersicon chilense. Euphytica 131: 15–23.
59. Liang Z, Sun C (1982) The significant effect of endosperm development on the interspecific hybridization of cotton. ActaGenetSinica (in Chinese) 9: 441–454.
60. Liang CL, Sun CW, Liu TL, Chiang JC (1978) Studies on interspecific hybridization in cotton. Scientia Sinica 21: 545–555.
61. Mofidabadi A (2009) Producing triploid hybrids plants through induce mutation to broaden genetic base in cotton. The ICAC Recorder 27: 10–11.
62. Jorgensen RB, Andersen B (1994) Spontaneous hybridization between oilseed rape (Brassica napus) and weedy B. campestris (Brassicaceae): a risk of growing genetically modified oilseed rape. American Journal of Botany: 1620–1626.
63. Roy N (1980) Species crossability and early generation plant fertility in interspecific crosses of Brassica. Sabrao Journal 12: 43–53.
64. Ali M, Lewis C (1962) Effects of Reciprocal Crossing on Cytological and Morphological Features of Interspecific Hybrids of Gossypium hirsutum L. and G. barbadense L. Crop Science 2: 20–22.
65. Stephens S (1954) Interspecific homologies between gene loci in Gossypium. I. Pollen color. Genetics 39: 701–711.
66. Silow RA (1941) The comparative genetics of Gossypium anomalum and the cultivated Asiatic cottons. Journal of Genetics 42: 259–358.
67. Harland SC (1929) The genetics of cotton. Part II. The inheritance of pollen colour in New World cottons. Journal of Genetics 20: 387–399.
68. Deshpande L, Kokate R, Kulkarni U, Nerkar Y (1991) Cytomorphological studies in induced tetraploid G. arboreum and its interspecific hybrid with tetraploid G. hirsutum L. The Indian Journal of Genetics and Plant Breeding 51: 194–202.
69. Matthews R (1991) Plant virology: Elsevier, San Diego:Academic Press, Inc.
70. Akhtar K, Khan A, Hussain M, Haq M, Khan M (2003) Upland cotton varietal response to cotton leaf curl virus (CLCuV). Tropical Agricultural Research & Extension 5: 29–34.
71. Akhtar KP, Khan MSI, Khan AI (2002) Improved bottle shoot grafting technique/method for the transmission of cotton leaf curl virus (CLCuV). FAO. Available: http://agris.fao.org/agris-search/search.do?recordID=PK2004000384#. 115–117.
72. Ahmad S, Khan N, Mahmood A, Nazeer W, Ahmad S, et al. (2011) Screening of cotton germplasm against cotton leaf curl virus. Pakistan Journal Botany 43: 725.

4

Overexpression of Rice NAC Gene *SNAC1* Improves Drought and Salt Tolerance by Enhancing Root Development and Reducing Transpiration Rate in Transgenic Cotton

Guanze Liu[1,2], Xuelin Li[3], Shuangxia Jin[1], Xuyan Liu[1], Longfu Zhu[1], Yichun Nie[1], Xianlong Zhang[1]*

1 National Key Laboratory of Crop Genetic Improvement, Huazhong Agricultural University, Wuhan, Hubei, P. R. China, **2** College of Tobacco Science, Yunnan Agricultural University, Kunming, Yunnan, P. R. China, **3** Agricultural College, Henan University of Science and Technology, Luoyang, Henan, P. R. China

Abstract

The *SNAC1* gene belongs to the stress-related NAC superfamily of transcription factors. It was identified from rice and overexpressed in cotton cultivar YZ1 by *Agrobacterium tumefaciens*-mediated transformation. *SNAC1*-overexpressing cotton plants showed more vigorous growth, especially in terms of root development, than the wild-type plants in the presence of 250 mM NaCl under hydroponic growth conditions. The content of proline was enhanced but the MDA content was decreased in the transgenic cotton seedlings under drought and salt treatments compared to the wild-type. Furthermore, *SNAC1*-overexpressing cotton plants also displayed significantly improved tolerance to both drought and salt stresses in the greenhouse. The performances of the *SNAC1*-overexpressing lines under drought and salt stress were significantly better than those of the wild-type in terms of the boll number. During the drought and salt treatments, the transpiration rate of transgenic plants significantly decreased in comparison to the wild-type, but the photosynthesis rate maintained the same at the flowering stage in the transgenic plants. These results suggested that overexpression of *SNAC1* improve more tolerance to drought and salt in cotton through enhanced root development and reduced transpiration rates.

Editor: Jinfa Zhang, New Mexico State University, United States of America

Funding: This work was supported by the National High-Tech Program of China (2013AA102601-4) and the project from Ministry of Agriculture of China (2013ZX08005-004). The funders had no role in study design, data cellection and analysis, decision to publish or preparation of the manuscript.

Competing Interests: The authors have declared that no competing interests exist.

* E-mail: xlzhang@mail.hzau.edu.cn

Introduction

Abiotic stresses such as drought, salinity and extreme temperatures have a crucial impact on agricultural productivity and yields. Climate models indicated that abiotic stresses would increase in the near future because of global climate change [1]. Therefore, drought and salinity would be the two major factors that adversely affect crop growth and productivity. Cotton (*Gossypium hirsutum L.*) is an important commercial crop worldwide, and used as a source of fiber and edible oil. Cotton showed higher drought and salt tolerance than other major crops such as rice, wheat and maize. Although that, abiotic stress still greatly affected cotton in growth and yield. Farmers plan to enlarge the planted area of cotton in western China, which is not suitable for food crops because of salinity and water shortage. Therefore, improved drought and salt tolerance of cotton through biotechnology has become an urgent task.

The NAC transcription factors have been characterized for their roles in plant growth, development, and stress tolerance. NAC was originally derived from the names of the first three proteins containing NAM (no apical meristem), ATAF1-2 and CUC2 (cup-shaped cotyledon), that contain a similar DNA-binding domain. NAC proteins appeared to be widespread in plants such as Arabidopsis, rice, wheat, soya bean and cotton [2,3,4,5], and comprised a large plant-specific family, which included 110 members in Arabidopsis and 140 members in rice, and only a few had been identified with diverse functions in plants [6]. Recently, Several studies reported that a subfamily of NAC transcription factors played a pivotal role in various abiotic stresses including salinity, drought and low temperature [4,7,8]. Previous works indicated that the overexpression of *ANAC019*, *ANAC055*, and *ANAC072* caused increased drought tolerance in transgenic plants through changing the transcription of a limited number of non-specific salt- and drought-responsive genes [9,10]. The overexpression of *SNAC1* in rice was another important example, which enhanced salt and drought tolerance and grain yield in the field test [11]. Recent results showed that overexpression of *OsNAC5* significantly enlarged roots and enhanced drought tolerance and grain yield under field conditions, as well [12]. At present, improved drought and salt tolerance could also be achieved by the overexpression of diverse NAC factors in species ranging from Arabidopsis, rice, chickpea, wheat, and tomato [4,13,14,15,16]. Constitutive overexpression of NAC gene in transgenic plants occasionally led to detrimental consequences such as dwarfing, late flowering and lower seed yield, which could be overcomed by preferentially employed stress-inducible or tissue-specific promoters such as *OsNAC6* or *RCc3* [12,17]. Thus, the research gave the evident functional of stress-responsive NAC

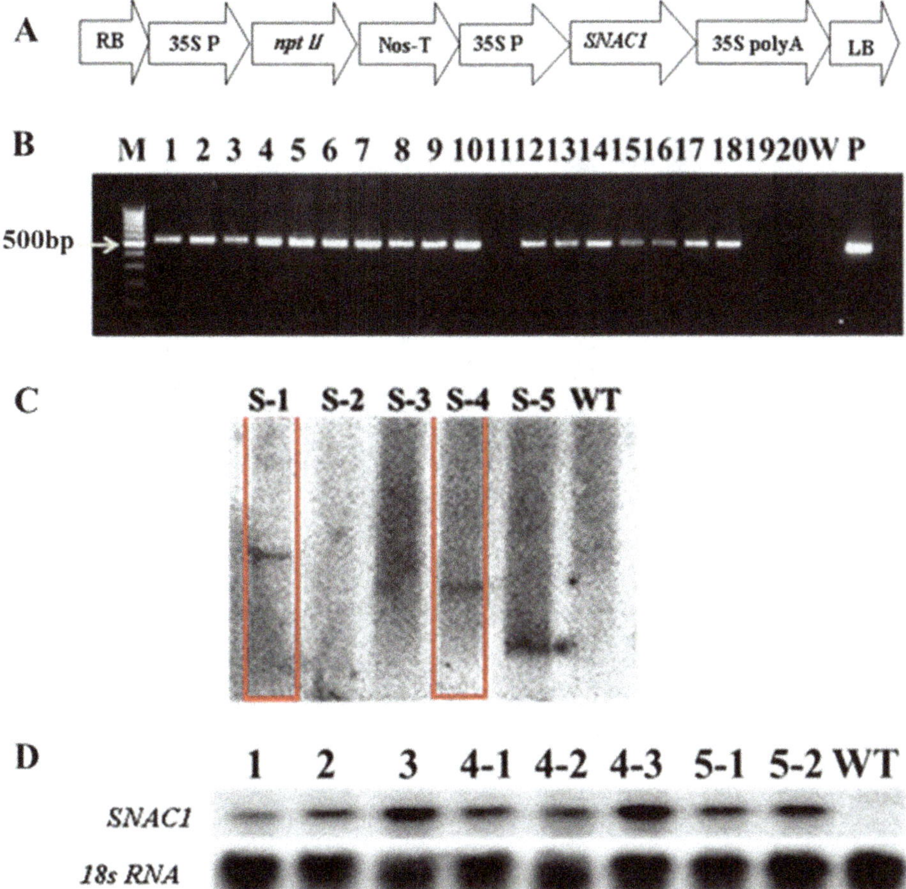

Figure 1. Overexpression vector and molecular identification of transgenic plants. (A): Map of SNAC1-overexpressing vector with the *npt II* gene as a screening marker. (B): Molecular analysis of *SNAC1* transgenic plants. M: Marker; 1–20: T_1 generation of transgenic plants (1–4 from S-1, 5–8 from S-2, 9–12 from S-3, 13–16 from S-4, 17–20 from S-5); W: Wild-type plant; P: Plasmid; (C): Southern blotting analysis of transgenic plants. S-1: *SNAC1* transgenic line-1; S-2: *SNAC1* transgenic line-2; S-3: *SNAC1* transgenic line-3; S-4: *SNAC1* transgenic line-4; S-5: *SNAC1* transgenic line-5; WT: wild-type plant. (**D**) Northern blot analysis of *SNAC1*-overexpressing transgenic plants; 1: Transgenic plant from line S-1; 2: Transgenic plant from line S-2; 3: Transgenic plant from line S-3; 4-1, 4-2, 4-3: Three individual transgenic plants from line S-4; 5-1, 5-2: Two individual transgenic plants from line S-5; WT: Wild-type plant. Red box: The selected lines.

genes, a meticulous characterization of their response to abiotic stress is crucial in conferring broad stress tolerance to plants.

A previous study suggested that *SNAC1*, which encoded an NAC transcription factor, was predominantly induced in guard cells by drought in rice. Further results showed that *SNAC1*-overexpression in rice significantly improved drought tolerance under field conditions and strong tolerance to salt stress. The increased drought tolerance might be partly explained by the reduced transpiration rate and an increased sensitivity to ABA. In addition, *SNAC1* overexpression did not result in unwanted dwarf phenotype in transgenic rice [11]. It was also important to be considered that the new transgenic lines with acquired stress-resistance phenotypes did not cause yield penalty and/or growth retardation. And breeders might be able to use *SNAC1* to increase stress tolerance by employing biotechnology in rice [18]. It was unclear whether *SNAC1* could be employed in other crops to improve tolerance to abiotic stress.

We reported that ectopic *SNAC1* expression in cotton led to improved drought and salt tolerance at the vegetative and reproductive stages in transgenic plants. Our results also showed that *SNAC1*-overexpression significantly enlarged root biomass and increased the boll number during drought and salt treatment in the greenhouse. This findings implied that *SNAC1* in improving agronomic traits and economic characteristics of cotton by ectopic expression would be an efficient way to accelerate the cotton breeding program.

Materials and Methods

Vector Construction and Cotton Transformation

The full-length of *SNAC1* from rice was kindly donated by Professor Lizhong Xiong (Huazhong Agricultural University). The coding sequence was ligated into a site within the binary vector pCAMBIA2300S that contains the neomycin phosphotransferase gene, or *NPT II*, which was used as the selective marker [19]. The overexpression vector (Figure 1A) was then introduced into *Agrobacterium tumefaciens* strain LBA4404, which was used to transform cotton according to an established method [20]. Segments of hypocotyls from *G. hirsutum cv* YZ1 were used as explants for transformation. The callus induction and plant regeneration protocols were previously described by Jin et al. [21].

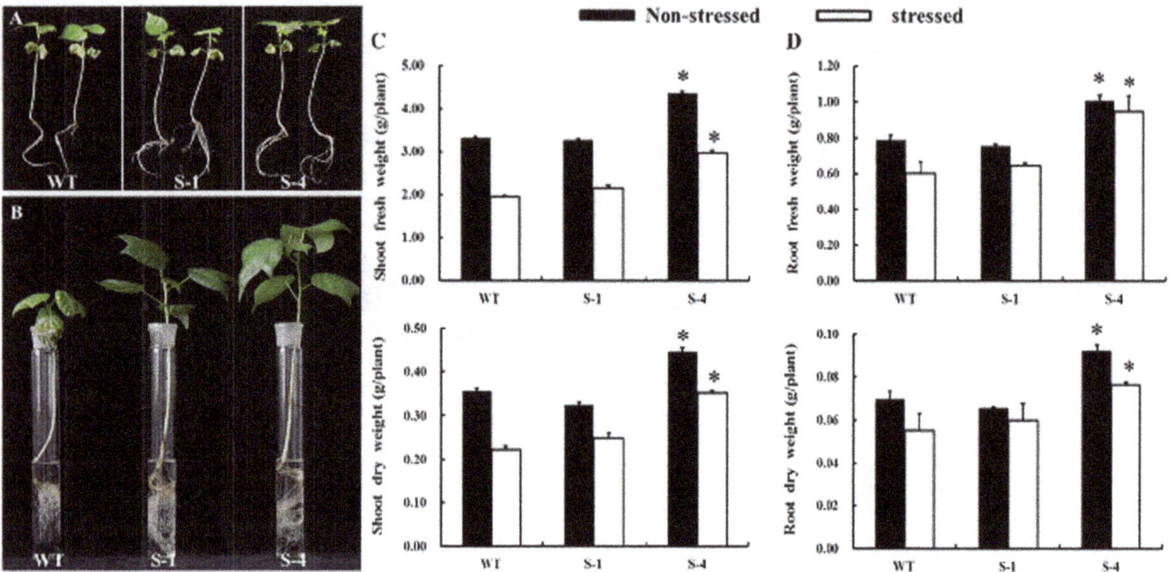

Figure 2. Salt treatment of *SNAC1*-overexpressing plants under hydroponic conditions. (A): Phenotype of wild-type and *SNAC1*-overexpressing transgenic plants before salt treatment; (B): Phenotype of wild-type and *SNAC1*-overexpressing transgenic plants after salt stress with 250 mmol/L NaCl for 1 week. (C, D): Biomass analysis of wild-type and *SNAC1*-overexpressing transgenic plants under normal conditions and with salt treatment; WT: wild-type plant; S-1: *SNAC1*-overexpressing transgenic plant from transgenic line S-1; S-4: *SNAC1*-overexpressing transgenic plant from transgenic line S-4. *statistically significant at 5%.

DNA Isolation and Southern Blot Analysis

Genomic DNA was isolated from the young leaves of transgenic and wild-type plants using a Plant Genomic DNA Kit (Tiangen Biotech, China). The PCR amplification was carried out with the specific primers of the *SNAC1* gene. Sequences of the forward primer and the reverse primer were 5′-AGCGAGAAGCAAG-CAAGA AGCG-3′and 5′-ACAGCACCCAATCATCCAACCT-3′, respectively. The predicted PCR product was 509 bp in length. For Southern hybridization, cotton genomic DNA was digested with *Hind III*, electrophoresed on 0.8% agarose gel and blotted onto a Hybond N^+ membrane for hybridization [22]. The fragment of *SNAC1* gene from PCR reaction mentioned above was used as a gene-specific probe and prepared using the Prime-a-Gene Labeling System (Promega, USA). After hybridization and stringent washing, the radioactive membranes were exposed to an imaging plate (Fuji Photo Film, Japan) for 5 h or overnight to record the image.

Total RNA Isolation for Northern Blot Analysis

Total RNA was extracted from the young leaves of transgenic and wild-type cotton plants [22]. For the northern blot analysis, 10 μg of total RNA was separated on a 1.2% formaldehyde-containing agarose gel. After electrophoresis, the gel was rinsed with DEPC-H_2O, equilibrated with 20× SSC and then transferred to a Hybond N^+ nylon membrane by capillary blotting. The membranes were air dried and baked for 2 h at 80°C in an oven. Hybridization was performed at 65°C with a *SNAC1* probe. The same blot was then stripped and reprobed with an 18s RNA probe. The washing and imaging steps were the same as for the Southern blot procedure.

Hydroponic Cotton Growth under High-salt Conditions

Cotton seeds from wild-type and transgenic lines were sterilized and germinated on sterile Stewart's germination media [23]. Transgenic seeds were grown on a medium containing 50 mg L^{-1} kanamycin to screen the transgenic plants. Positive transgenic plants with complete rooting systems, including lateral roots, were used for this experiments. At the two-leaf stage, wild-type and transgenic seedlings were transferred to Hoagland solution. After 21 days at normal conditions, six seedlings of each replicate from both wild-type and transgenic lines were exposed to different salinity conditions in which NaCl was added to the Hoagland solution at 50, 100, 150, and 200 mM increments every 72 h until a final concentration of 250 mM was reached and maintained for one week. All wild-type and transgenic lines were then removed from the tubs and dried with paper towels to measure the fresh shoot and fresh root weights. The total plant dry weights were calculated for a biomass determination by drying the samples for 72 h in an oven at 80°C.

Proline and Malondialdehyde (MDA) Content Measurements

Cotton seeds from wild-type and transgenic lines were sterilized and germinated on sterile Stewart's germination media to screen the transgenic plants as described. Each plant was transferred to a small pot of uniform soil mix and then placed in a larger tub container. Every tub container included fifteen plants (each tub had five wild-type, five transgenic line S-1 and five S-4 plants). The eighteen-day-old plants were grown in a greenhouse and then subjected to drought and salt stress. For the control conditions, every tub container received 4000 ml of water. For drought stress, the plants were subjected to water stress by withholding half the water for two weeks. For salt stress, plants were watered with 250 mM NaCl solution for two weeks. The plants' proline and MDA contents were recorded. The levels of proline content in leaf tissue were determined according to the method outlined in reference [24]. Samples weighing approximately 100 mg were collected from the first fully expanded leaf of stressed (drought-stress and salt-stress) or control plants, and they were then placed in 10 ml microcentrifuge tubes containing 3 ml of 3% sulfosa-

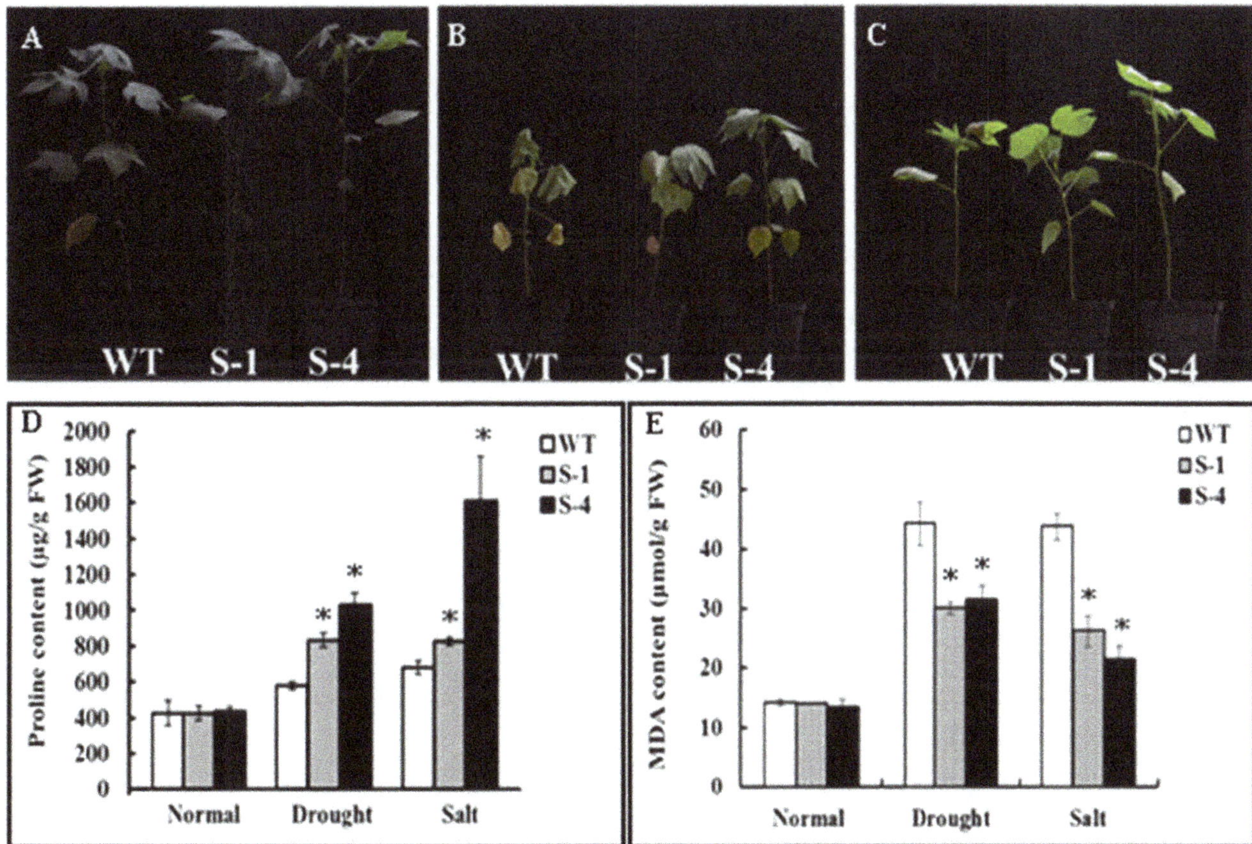

Figure 3. Detection of stress tolerance in *SNAC1*-overexpression transgenic plants at the seedling stage. (A): Phenotype of wild-type and *SNAC1*-overexpressing transgenic plants under normal conditions; (B): Phenotype of wild-type and *SNAC1*-overexpressing transgenic plants after two weeks of drought stress; (C): Phenotype of wild-type and *SNAC1*-overexpressing transgenic plants after two weeks of salt stress with a 250 mmol/L NaCl solution; (D): Proline content analysis under normal conditions and stress treatments; (E): MDA content analysis under normal conditions and stress treatments; WT: wild-type plant; S-1: *SNAC1*-overexpression transgenic plant from transgenic line S-1; S-4: *SNAC1*-overexpression transgenic plant from transgenic line S-4. *statistically significant at 5%.

licylic acid for grinding. Ground samples were centrifuged, and 1 ml of the supernatant was mixed with 1 ml of acid ninhydrin solution (1.25 g ninhydrin in 30 ml of glacial acetic acid and 20 ml of 6 M phosphoric acid) and 1 ml of glacial acetic acid in a fresh glass tube. The tubes were capped and incubated at 100°C in a water bath for 30 min. The reaction was terminated by chilling the tubes on ice. Then, 2 ml of toluene was added to each tube, and the mixture was vortexed for 10 s. One milliliter of the upper toluene phase containing the chromophore was aspirated and read at 520 nm in a quartz cuvette. The proline concentrations were estimated based on a standard curve for proline.

In addition, the same cotton seedling was used to analyze leaf lipid peroxidation by estimating the formation of malondialdehyde (MDA), an end product of lipid peroxidation. A 100 mg leaf piece was collected from the first fully expanded leaf (excluding the midrib) and homogenized in 5 ml of 10% TCA (trichloroacetic acid). The extract was centrifuged at 10,000 rpm for 10 min. The reaction was initiated by adding 2 ml of supernatant to 2 ml of 0.6% TBA (thiobarbituric acid dissolved in 10% TCA), and the mixture was then incubated at room temperature for 2 h. The reaction mixture was boiled at 100°C for 1 h. After cooling to room temperature, the OD was measured at 450 nm, 532 nm and 600 nm. The MDA content was estimated using the formula reported in citation [25].

Plant Performance under Salt and Drought Treatment in the Greenhouse

Cotton seeds from the wild-type and two transgenic lines were planted in soil mix. These plants were allowed to grow under normal conditions until the budding stage, and uniform plants were then divided into three groups: a well-watered control group, a drought stress group and a salt stress group. Each group consisted of eighteen plants. Stress treatment was applied at the budding and flowering stages. During the budding stage, the well-watered plants were irrigated with 2000 ml of water every ten days. For drought stress, plants were irrigated with 1000 ml of water per pot every ten days for 20 days. For the salt stress group, plants were irrigated with 2000 ml of 100 mM NaCl solution every ten days for 20 days. After 20 days of treatment, the photosynthesis, stomatal conductance and transpiration rate were measured using a portable photosynthesis system (Li-6400XT, LI-COR Inc, Lincoln, NE, USA). The photosynthesis, stomatal conductance and transpiration rate were assessed at a CO_2 concentration of 400 μmol mol^{-1}, a relative humidity of 50%, a chamber temperature of 28°C, an air flow of 500 μmol s^{-1} and a photon flux density of 1500 μmol m^{-2}s^{-1}. The instrument was stabilized according to the manufacturer's guidelines [26]. Steady-state levels of reference CO_2 and H_2O were observed before taking the measurements. When all the plants were nearly

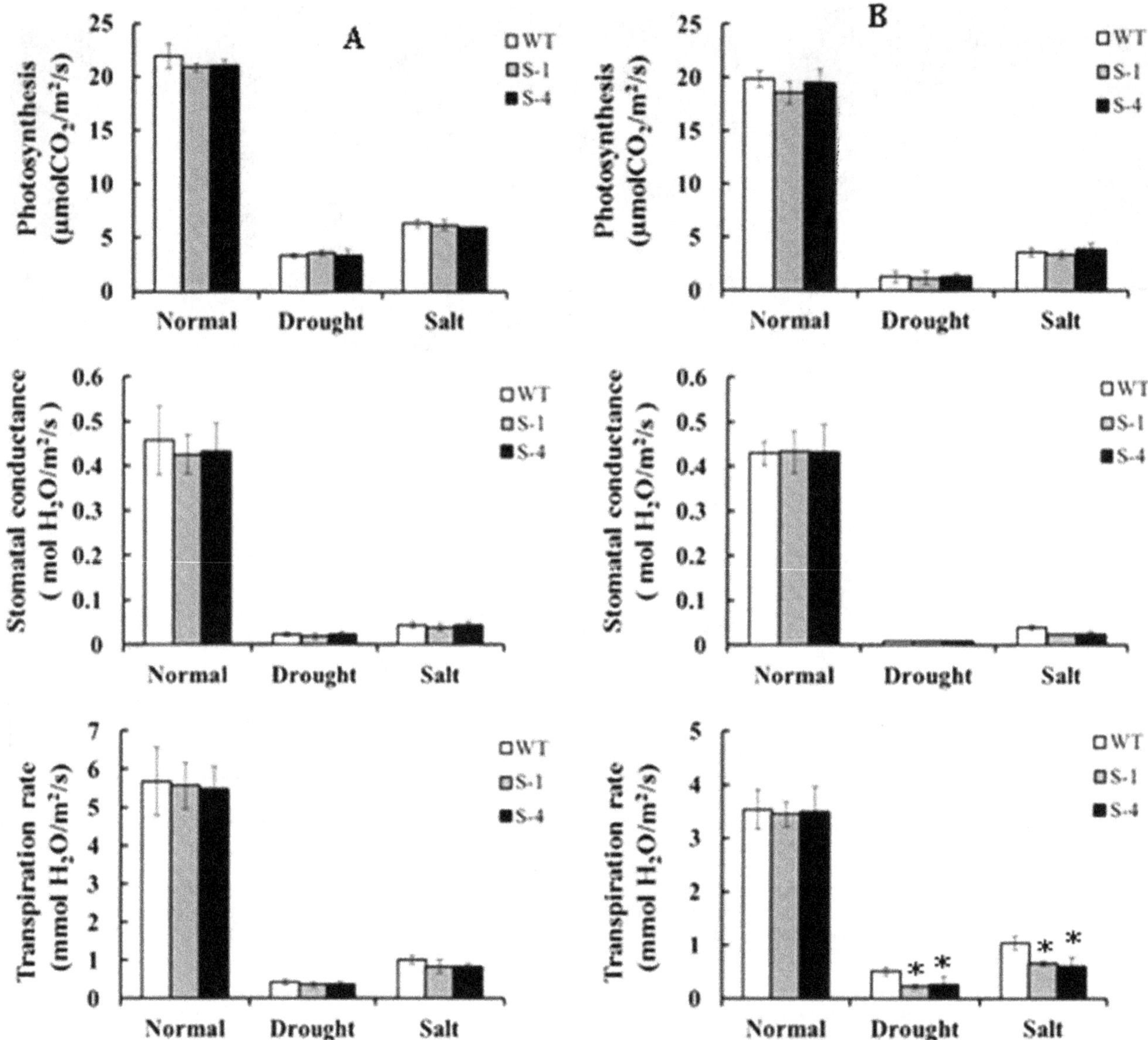

Figure 4. Photosynthetic performance of wild-type and *SNAC1*-overexpressing transgenic plants at different stages under stress treatment. (A): Photosynthetic performance of wild-type and *SNAC1*-overexpressing transgenic plant after 20 days of drought stress or salt stress under 100 mmol/L NaCl solution during the budding stage; (B): Photosynthetic performance of wild-type and *SNAC1*-overexpressing transgenic plants after 20 days of drought stress or salt stress in 250 mmol/L NaCl solution during the flowering stage; WT: wild-type plant; S-1: *SNAC1*-overexpressing transgenic plant from transgenic line S-1; S-4: *SNAC1*-overexpressing transgenic plant from transgenic line S-4. *statistically significant at 5%.

flowering, the second treatment was applied. During the flowering stage, the well-watered plants were irrigated with 2000 ml of water every five days. For drought treatment, irrigation was reduced to 1000 ml of water per pot every ten days. For salt treatment, the NaCl concentration was increased gradually to 250 mM and irrigated with 2000 ml of NaCl solution every five days. After 20 days of treatment, the photosynthesis, stomatal conductance and transpiration rate were measured as described earlier. When the flowering stage stress was finished, the plants were re-watered with 3000 ml of water per pot, and the biomass analysis was carried out after one week. All stress experiments were conducted in a greenhouse at 35±3°C/28±3°C (day/night) with a relative humidity of 50–70% and a photoperiod of 14/10 h (light/dark).

Rate of Water Loss from Excised Leaves

The water loss from detached leaves of wild-type and transgenic was measured by monitoring the fresh weight loss at the indicated time points.

Results

Transformation and Molecular Analyses of *SNAC1*-overexpressing Cotton Plants

The pCAMBIA2300S-*SNAC1* construct was introduced into cotton via *Agrobacterium*-mediated transformation, and 5 independent transgenic lines were produced. The T1 population (progeny of T0 plants) was then employed for molecular analysis. The presence and integrity of the transgene were confirmed using PCR

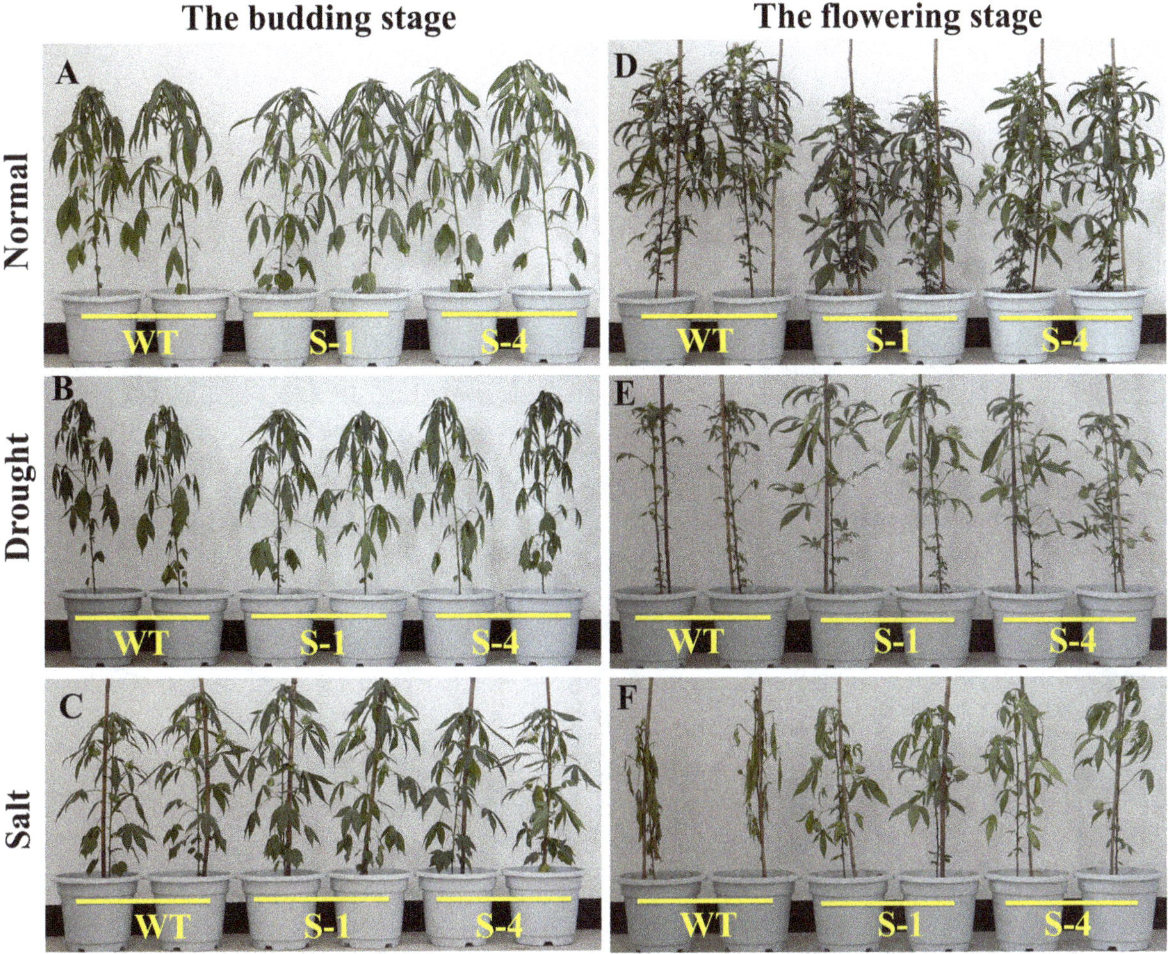

Figure 5. Phenotype of *SNAC1* transgenic and wild-type plants at budding and flowering stages during stress treatments. (A), (B), (C): Phenotype of wild-type and *SNAC1*-overexpressing transgenic plants under stress at the budding stage; (A): normal conditions after 20 d; (B): drought stress after 20 d; (C): 100 mmol/L NaCl stress after 20 d; (D),(E),(F): Phenotype of wild-type and *SNAC1*-overexpressing transgenic plants under stress treatments at the flowering stage; (D): normal conditions after 20 d at the flowering stage; (E): Drought stress after 20 d at the flowering stage; (F): Salt stress in 250 mmol/L NaCl solution for 20 d at the flowering stage. WT: Wild-type; S-1: *SNAC1* plants from transgenic line S-1; S-4: *SNAC1* plants from transgenic line S-4.

analysis on the genomic DNA with specific primers for *SNAC1* (Figure 1B). The results of southern blot showed three transgenic lines that appeared to contain a single-copy insertion, and they were identified as S-1, S-4 and S-5 (Figure 1C). Transgenic lines S-2 and S-3 had been proved as positive transgenic plants through PCR analysis, but did not show hybridization signals in southern blotting analysis. We deduced that the reason might due to the poor quality of DNA from transgenic plant, which influenced the efficiency of membrane blotting (Figure 1B, C). To check the expression level of *SNAC1*, northern blotting was employed with the T2 transgenic cotton lines. The results showed that mRNA transcription level of *SNAC1* was diverse among the five transgenic lines (Figure 1D). Two independent lines, S-1 and S-4, with high expression level of *SNAC1*, were selected and used for drought and salt-tolerance analysis.

SNAC1-overexpression Increases Tolerance to Salinity under Hydroponic Conditions in Cotton

To identify whether *SNAC1*-overexpression enhances salt tolerance in transgenic cotton, we performed hydroponic experiments in a greenhouse. Transgenic lines S-1, S-4 and wild-type plants were grown hydroponically in a saline medium and under normal conditions. No obvious phenotype differences were found between the wild-type and transgenic plants before salt treatment (Figure 2A). Uniform seedlings of transgenic lines S-1 and S-4 and wild-type plants were cultured in hydroponic solution with NaCl concentrations from 50 to 250 mM, with a 50 mM addition every 72 hrs. After 7 days under the treatment with 250 mM NaCl, the wild-type plants exhibited growth retardation and leaf wilting, but the transgenic line grew better than the wild-type plants (Figure 2B). Moreover, transgenic lines S-1 and S-4 had considerably larger shoots and roots than those of the wild-type plants (Figure 2B), particularly transgenic line S-4, which produced significantly more shoot and root biomass than the wild-type plants (Figure 2C, D). The S-1 and S-4 lines produced 12% and 59% more dry shoot mass and 9% and 38% more dry root mass, respectively, compared to the wild-type plants under the salt treatment. All the plants were also measured under controlled conditions. Interestingly, the transgenic plants of line S-4 had increased biomass production under normal conditions (Figure 2C, D).

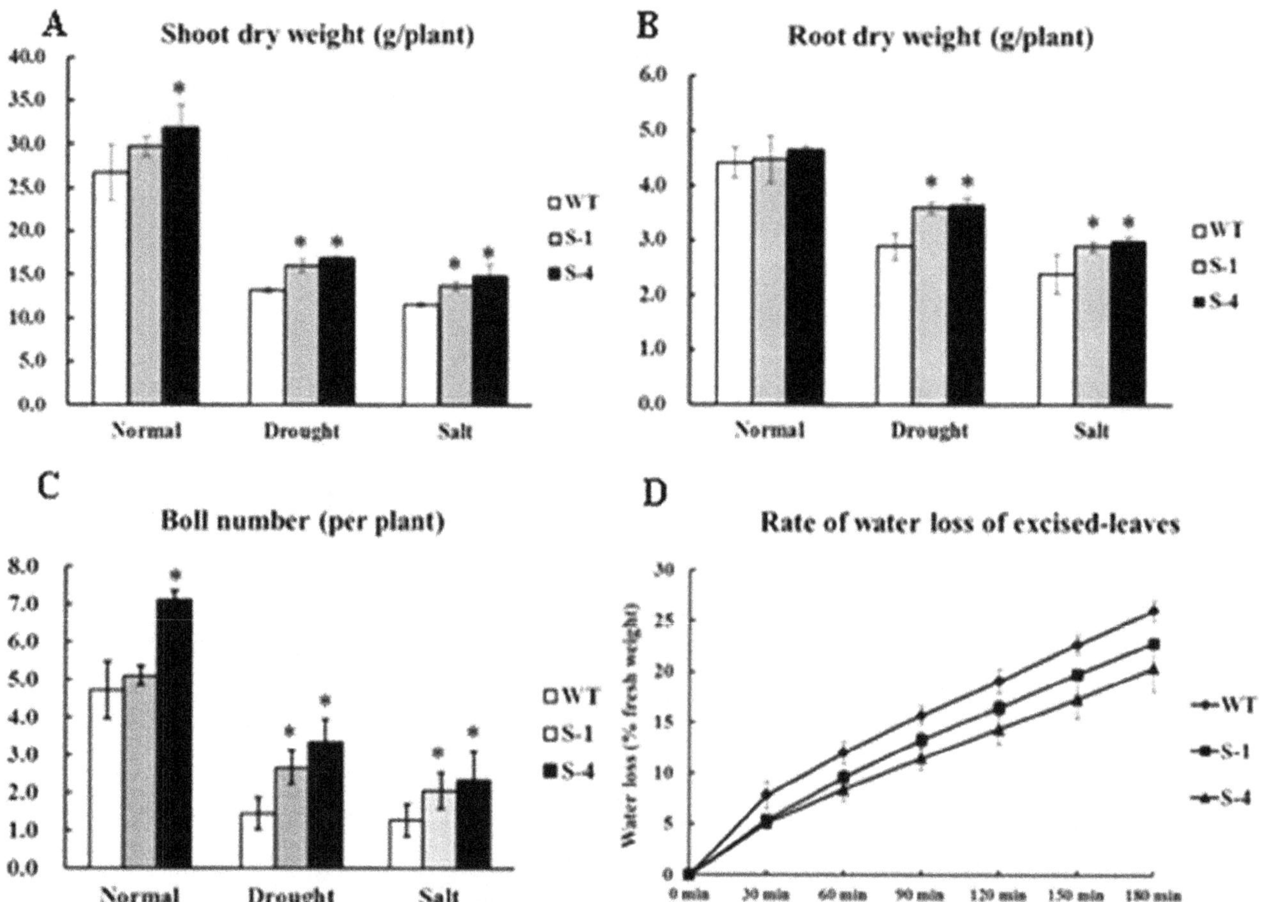

Figure 6. Biomass analysis of wild-type and *SNAC1*-overexpressing transgenic plants under normal condition and stress treatments. (A): Shoot dry weight analysis of wild-type and *SNAC1* plants after different stress treatments; (B): Root dry weight analysis of wild-type and *SNAC1* plants after different stress treatments; (C): Final boll number of plant under different stress treatment in the greenhouse; (D): Rate of water loss from excised leaves. WT: wild-type plants; S-1: *SNAC1* plants from transgenic line S-1; S-4: *SNAC1*-overexpression plant from transgenic line S-4. *statistically significant at 5%.

Overexpressing *SNAC1* Elevated Osmoprotectant Levels in the Transgenic Seedlings

SNAC1-overexpressing and wild-type plants were grown under drought and salt treatment conditions to examine stress tolerance in soil. At two-leaf stage, each plant was transplanted into a large plastic container containing fifteen pots with five pots for each line and one plant in every pot. All the plants of transgenic lines S-1, S-4 and wild-type plants were divided into three groups including normal, drought and salt treatments. All of the wild-type and transgenic plants grew well under normal conditions. While, the leaves of the wild-type plants withered more significantly than the transgenic lines after two weeks treatment of drought stress. The leaves of S-1 experienced medium wilting, and the leaves of S-4 showed slight wilting (Figure 3A, B). Under salt treatment, transgenic lines S-1 and S-4 shrank slightly, but the wild-type plants clearly withered and had shrunken leaves (Figure 3C). Meanwhile, we also estimated the levels of osmoprotectants such as proline and MDA in the leaves of wild-type and transgenic plants in response to different treatments. No significantly different in proline content was observed between the wild-type and transgenic plants under normal conditions. Under the drought and salt treatments, the content of proline was elevated in all of the plants, but it was remarkedly higher in two transgenic lines. Under drought treatment, the proline content increased up to 1.4-fold and 1.8-fold in the S-1 and S-4 transgenic lines, respectively, which was significantly higher than that of the wild-type plants. Similar results were also found in the two transgenic lines with 1.2-fold and 2.4-fold in proline content, which was significantly higher than that in the wild-type plants, after 250 mM NaCl treatment (Figure 3D).

A remarkable high level of MDA was detected in the wild-type and transgenic plants after drought and salt treatment when compared with the normal condition plants. A maximum of 3.1 times in the MDA level was observed in wild-type plants, but an increase of 2.1-fold for S-1 and 2.3-fold for S-4 in the MDA content was measured under drought stress. The MDA levels were also significantly increased under salt treatment in all seedlings, while the amounts of MDA in lines S-1 and S-4 were remarkablely less than that in the wild-type (Figure 3E). There were no significant differences in the MDA levels between wild-type and transgenic plants under normal conditions (Figure 3E). These results indicated that *SNAC1*-overexpression in cotton could alleviate cell membrane injury under both drought and salt stresses.

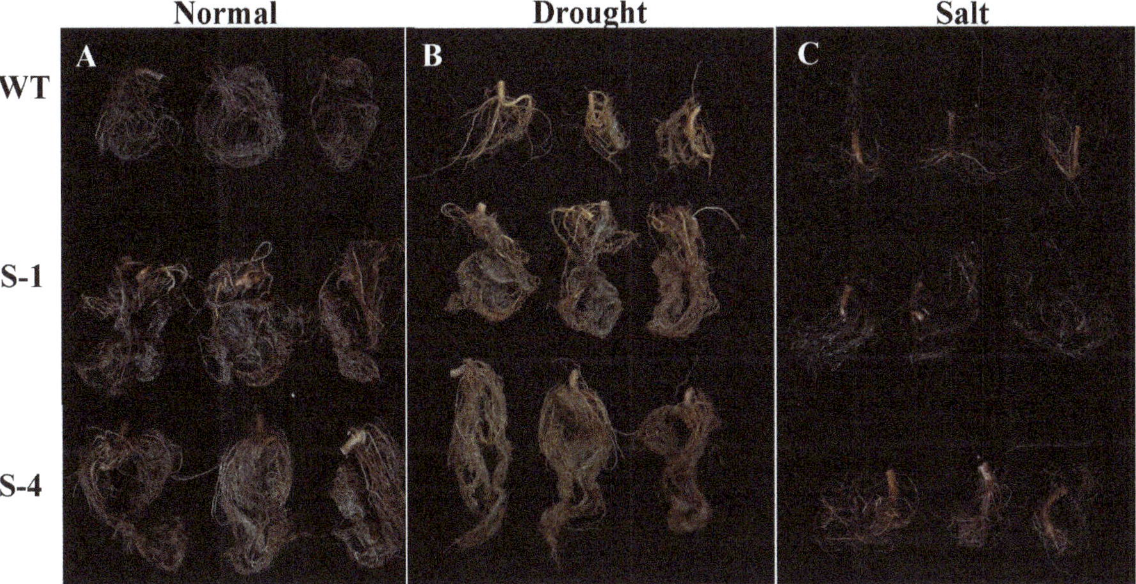

Figure 7. The root growth phenotype of wild-type and *SNAC1*-overexpressing transgenic plants after normal condition and stress treatments. (A): Normal conditions of wide-type and *SNAC1* plants; (B): Drought stress of wide-type and *SNAC1* plants; (C): Salt stress of wide-type and *SNAC1* plant. WT: Wild-type; S-1: *SNAC1* plants from transgenic line S-1; S-4: *SNAC1* plants from transgenic line S-4.

SNAC1-overexpression Increases Cotton Tolerance to Drought and Salt Stress in the Greenhouse

To evaluate the drought and salt tolerance of transgenic cotton at the budding and flowering stages, transgenic and wild-type plants were subjected to drought and salt stress in greenhouse. The results showed that no significant difference was found in the photosynthesis, stomatal conductance, and transpiration rate between the transgenic and wild-type plants during the budding stage under both treatments (Figure 4A). Although photosynthesis, stomatal conductance and transpiration rate decreased rapidly in all plants under the drought and salt stress compared to normal conditions, the transgenic and wild-type plants did not show significant phenotype differences at the budding stage under different conditions (Figure 5A, B, C). When the plants were subjected to stress for twenty days at the flowering stage, the photosynthesis and stomatal conductance values decreased greatly under drought or salt stress treatments in all plants. While, no difference was detected between the wild-type and transgenic plants. However, the transpiration rate in transgenic plants significantly decreased compared to the wild-type plants under drought and salt treatments (Figure 4B), which might be inferred that overexpression of *SNAC1* in cotton could improve efficiency in water usage. In conclusion, the transgenic plants had a higher tolerance than the wild-type at the reproductive stage under both stresses (Figure 5D, E, F).

Furthermore, the biomass of the plants was calculated for all treatments. The transgenic cotton produced significantly more shoots and roots under the stress treatments (Figure 6A, B). When subjected to drought, lines S-1 and S-4 produced 22% and 23% more dry shoot mass and 24% and 26% more dry root mass than the wild-type plants, respectively. Under the salt treatment, transgenic lines S-1 and S-4 had 18% and 28% more dry shoot mass and 21% and 26% more dry root mass, respectively. In addition, the total boll number was also calculated from these plants and the results showed that 85% and 131% more bolls under drought stress and 61% and 83% more bolls under salt stress were produced in lines S-1 and S-4 than the wild-type plants, respectively (Figure 6C). The rate of water loss (RWL) from excised leaves was also monitored at the indicated time points. The RWL of wild-type plants increased to 26%, but only 20% and 22% were found in S-1 and S-4 (Figure 6D). The result suggested that the water-holding ability of transgenic plant was higher than that of wild-type plants. And the transgenic plants showed more roots than wide-type plants under normal conditions and stress treatments (Figure 7). Generally, overexpression of *SNAC1* in cotton significantly improved abiotic stress tolerance with increased biomass and boll numbers under reduced irrigation and salinity conditions.

Discussion

One major challenge in modern agricultural production is the growing demand for food paralleled by dramatic losses of arable land as a consequence of increasingly severe soil destruction by abiotic environmental conditions [27]. Transcription factors played important roles in the regulation of gene expression in response to abiotic stresses and could be employed as candidate genes for the genetic engineering to improve stress tolerance in crops [18]. By increasing the understanding of the NAC transcription factor class in controlling abiotic stress responses, practical approaches had been developed for engineering stress tolerance in crops [27]. *SNAC1* was predominantly induced in guard cells by drought, and overexpression of *SNAC1* in rice showed significantly improved drought resistance and salt tolerance with higher seed setting than the control under severe drought stress conditions at the reproductive stage [11].

As expected, *SNAC1*-overexpressing cotton plants performed better than wild-type plants under drought and salinity conditions at the vegetative and reproductive stages. More importantly, transgenic *SNAC1* cotton showed no obvious differences from the wild-type plants in all investigated traits under normal condition, which in accordance with the results in transgenic rice. In our study, transgenic cotton grew much better than wide-type plants in the presence of 250 mM NaCl under hydroponic growth conditions. The increased salinity tolerance was measured by quantifying the biomass of cotton plants. The dry shoot and dry root masses of all *SNAC1*-overexpressing seedlings were significantly higher than those of wild-type plants grown under 250 mM NaCl conditions. In particular, the dry root masses of transgenic S-4 seedlings were higher than those of wild-type seedlings. The soil-grown *SNAC1*-overexpressing cotton also displayed significantly improved salt and drought tolerance in the greenhouse. The dry biomasses of both the shoots and roots of the *SNAC1*-overexpressing plants were significantly higher than those of the wide-type plants. In rice, *SNAC1*-overexpressing plants were no different from wild-type plants in terms of root depth and volume. Therefore, it was possible that this gene had a very limited contribution to root development. Recently, *OsNAC9* (which is identical to the *SNAC1* gene) was overexpressed under the control of root-specific promoter *RCc3* enhanced drought resistance in rice, and increased root diameters were found in the transgenic plants [28]. In a previous study, 18 NAC domain factors were identified from expression profiling of stress-treated rice. *OsNAC10* and *SNAC1/OsNAC9*, which were members of subgroup I, were believed to have similar stress response function because of their sequence similarity. *OsNAC10*-overexpressing in rice caused enlarged roots, enhanced drought tolerance, and significantly increased grain yield under field drought conditions [29]. These results suggested the *SNAC1* may confer drought resistance through the altered root architecture. A longer root system should have facilitated water absorption from deeper soils, and thus strengthened drought tolerance and increased biomass under water-deficit conditions [30]. In our study, the root biomass of transgenic lines was higher than that of the wild-typ whether in normal condition or stress. The overexpression of *SNAC1* in cotton significantly increased root development, which suggested that the development of larger roots should be favorable for drought resistance breeding.

In a recent report, *OsSRO1c* was characterized as a direct target of SNAC1 [31]. Moreover, *SNAC1*-overexpressing and *OsSRO1c*-overexpressing plants both increased stomatal closure and reduced water loss under drought stress [11,31]. In our study, photosynthesis was not significantly affected in the transgenic plants in relative to the wild-type under different conditions. While, the transpiration rate was lower in the transgenic plants than the wild-type under drought and salt treatments. Further investigation of the RWL from excised leaves in transgenic and wild-type plants showed that transgenic leaves had less water loss. These results implied that the enhanced tolerance of *SNAC1*-overexpressing plants was related to water usage efficiency.

Osmotic adjustment is a fundamental cell tolerance response to osmotic stress and could be realized by the accumulation of osmoprotectants. The osmotic potential is a direct reflection of the osmotic adjustment capability at the physiological level and has been used as an effective index for assessing crop genotypes for osmotic stress tolerance [32]. Numerous studies had shown that free proline was the most widely distributed multifunctional osmolyte in many organisms, and it played important roles in enhancing osmotic stress tolerance [33]. In our study, the proline content increased more significantly in transgenic plant than in the wild-type plant under drought and salt stress. In addition, transgenic cotton maintained less MDA in content than wild-type plants during stress treatment, which was in accordance with the improved tolerance to stress in transgenic cotton.

In summary, *SNAC1*-overexpressing cotton plants significantly enhanced drought resistance and salinity tolerance at the vegetative and reproductive stages. The resuls suggested that this gene may show great promise for the genetic improvement of stress tolerance in cotton.

Acknowledgments

The authors thank Dr. Lizhong Xiong for his generosity in providing the *SNAC1* cDNA vector.

Author Contributions

Conceived and designed the experiments: GZL XLZ. Performed the experiments: GZL XLL. Analyzed the data: GZL XLL LFZ. Contributed reagents/materials/analysis tools: XYL YCN. Wrote the paper: XLZ GZL SXJ.

References

1. Ahuja I, de Vos RCH, Bones AM, Hall RD (2010) Plant molecular stress responses face climate change. Trends in Plant Sci 15: 664–674.
2. Meng CM, Cai CP, Zhang TZ, Guo WZ (2009) Characterization of six novel NAC genes and their responses to abiotic stresses in *Gossypium hirsutum L.* Plant Sci 176: 352–359.
3. Puranik S, Sahu PP, Srivastava PS, Prasad M (2012) NAC proteins: regulation and role in stress tolerance. Trends in Plant Sci 17: 369–381.
4. Mao X, Zhang H, Qian X, Li A, Zhao G, et al. (2012) TaNAC2, a NAC-type wheat transcription factor conferring enhanced multiple abiotic stress tolerances in *Arabidopsis*. J Exp Bot 63: 2933–2946.
5. Wu A, Allu AD, Garapati P, Siddiqui H, Dortay H, et al. (2012) JUNGBRUNNEN1, a reactive oxygen species-responsive NAC transcription factor, regulates longevity in *Arabidopsis*. Plant Cell 24: 482–506.
6. Fang YJ, You J, Xie KB, Xie WB, Xiong LZ (2008) Systematic sequence analysis and identification of tissue-specific or stress-responsive genes of NAC transcription factor family in rice. Mol Genet Genomics 280: 547–563.
7. Hao YJ, Wei W, Song QX, Chen HW, Zhang YQ, et al. (2011) Soybean NAC transcription factors promote abiotic stress tolerance and lateral root formation in transgenic plants. Plant J 68: 302–313.
8. Hu HH, You J, Fang YJ, Zhu XY, Qi ZY, et al. (2008) Characterization of transcription factor gene *SNAC2* conferring cold and salt tolerance in rice. Plant Mol Biol 67: 169–181.
9. Tran LSP, Nakashima K, Sakuma Y, Simpson SD, Fujita Y, et al. (2004) Isolation and functional analysis of *Arabidopsis* stress-inducible NAC transcription factors that bind to a drought-responsive cis-element in the early responsive to dehydration stress 1 promoter. Plant Cell 16: 2481–2498.
10. Bu QY, Jiang HL, Li CB, Zhai QZ, Zhang JY, et al. (2008) Role of the *Arabidopsis thaliana* NAC transcription factors ANAC019 and ANAC055 in regulating jasmonic acid-signaled defense responses. Cell Res 18: 756–767.
11. Hu HH, Dai MQ, Yao JL, Xiao BZ, Li XH, et al. (2006) Overexpressing a NAM, ATAF, and CUC (NAC) transcription factor enhances drought resistance and salt tolerance in rice. Pro Natl Acad Sci USA103: 12987–12992.
12. Jeong JS, Kim YS, Redillas MCFR, Jang G, Jung H, et al. (2013) OsNAC5 overexpression enlarges root diameter in rice plants leading to enhanced drought tolerance and increased grain yield in the field. Plant Biotechnol J 11: 101–114.
13. Peng H, Cheng HY, Yu XW, Shi QH, Zhang H, et al. (2009) Characterization of a chickpea (*Cicer arietinum L.*) NAC family gene, CarNAC5, which is both developmentally- and stress-regulated. Plant Physiol Bioch 47: 1037–1045.
14. Yokotani N, Ichikawa T, Kondou Y, Matsui M, Hirochika H, et al. (2009) Tolerance to various environmental stresses conferred by the salt-responsive rice gene ONAC063 in transgenic *Arabidopsis*. Planta 229: 1065–1075.
15. Han Q, Zhang J, Li H, Luo Z, Ziaf K, et al. (2012) Identification and expression pattern of one stress-responsive NAC gene from *Solanum lycopersicum*. Mol Biol Rep 39: 1713–1720.
16. Xue GP, Way HM, Richardson T, Drenth J, Joyce PA, et al. (2011) Overexpression of TaNAC69 leads to enhanced transcript levels of stress up-regulated genes and dehydration tolerance in bread wheat. Mol Plant 4: 697–712.
17. Nakashima K, Tran LSP, Van Nguyen D, Fujita M, Maruyama K, et al. (2007) Functional analysis of a NAC-type transcription factor OsNAC6 involved in abiotic and biotic stress-responsive gene expression in rice. Plant J 51: 617–630.
18. Yang SJ, Vanderbeld B, Wan JX, Huang YF (2010) Narrowing down the targets: towards successful genetic engineering of drought-tolerant crops. Mol Plant 3: 469–490.
19. Munis MFH, Tu LL, Deng FL, Tan JF, Xu L, et al. (2010) A thaumatin-like protein gene involved in cotton fiber secondary cell wall development enhances resistance against Verticillium dahliae and other stresses in transgenic tobacco. Biochem Bioph Res Co 393: 38–44.

20. Jin SX, Zhang XL, Liang SG, Nie YC, Guo XP, et al. (2005) Factors affecting transformation efficiency of embryogenic callus of Upland cotton (*Gossypium hirsutum*) with Agrobacterium tumefaciens. Plant Cell Tiss Org 81: 229–237.
21. Jin SX, Zhang XL, Nie YC, Guo XP, Liang SG, et al. (2006) Identification of a novel elite genotype for in vitro culture and genetic transformation of cotton. Biol Plantarum 50: 519–524.
22. Liu GZ, Jin SX, Liu XY, Tan JF, Yang XY, et al. (2012) Overexpression of *Arabidopsis* cyclin D2;1 in cotton results in leaf curling and other plant architectural modifications. Plant Cell Tiss Org 110: 261–273.
23. Stewart J, Hsu C (1977) In-ovulo embryo culture and seedling development of cotton (*Gossypium hirsutum L.*). Planta 137: 113–117.
24. Divya K, Jami SK, Kirti PB (2010) Constitutive expression of mustard annexin, AnnBj1 enhances abiotic stress tolerance and fiber quality in cotton under stress. Plant Mol Biol 73: 293–308.
25. Lv SL, Yang A, Zhang K, Wang L, Zhang JR(2007) Increase of glycinebetaine synthesis improves drought tolerance in cotton. Mol Breeding 20: 233–248.
26. Yan J, He C, Wang J, Mao Z, Holaday SA, et al. (2004) Overexpression of the *Arabidopsis* 14-3-3 protein GF14{lambda} in cotton Leads to a "Stay-Green" phenotype and improves stress tolerance under moderate drought conditions. Plant Cell Physiol 45: 1007–1014.
27. Lawlor DW (2013) Genetic engineering to improve plant performance under drought: physiological evaluation of achievements, limitations, and possibilities. J Exp Bot 64: 83–108.
28. Redillas MCFR, Jeong JS, Kim YS, Jung H, Bang SW, et al. (2012) The overexpression of OsNAC9 alters the root architecture of rice plants enhancing drought resistance and grain yield under field conditions. Plant Biotechnol J 10: 792–805.
29. Jeong JS, Kim YS, Baek KH, Jung H, Ha SH, et al. (2010) Root-specific expression of OsNAC10 improves drought tolerance and grain yield in rice under field drought conditions. Plant Physiol 153: 185–197.
30. Werner T, Nehnevajova E, Koellmer I, Novak O, Strnad M, et al. (2010) Root-specific reduction of cytokinin causes enhanced root growth, drought tolerance, and leaf mineral enrichment in *Arabidopsis* and Tobacco. Plant Cell 22: 3905–3920.
31. You J, Zong W, Li XK, Ning J, Hu HH, et al. (2013) The SNAC1-targeted gene *OsSRO1c* modulates stomatal closure and oxidative stress tolerance by regulating hydrogen peroxide in rice. J Exp Bot 64: 569–583.
32. Zhu JK (2002) Salt and drought stress signal transduction in plants. Annu Rev Plant Biol 53: 247–273.
33. Krasensky J, Jonak C (2012) Drought, salt, and temperature stress-induced metabolic rearrangements and regulatory networks. J Exp Bot 63: 1593–1608.

5

GhWRKY40, a Multiple Stress-Responsive Cotton WRKY Gene, Plays an Important Role in the Wounding Response and Enhances Susceptibility to *Ralstonia solanacearum* Infection in Transgenic *Nicotiana benthamiana*

Xiuling Wang, Yan Yan, Yuzhen Li, Xiaoqian Chu, Changai Wu, Xingqi Guo*

State Key Laboratory of Crop Biology, College of Life Sciences, Shandong Agricultural University, Taian, Shandong, PR China

Abstract

WRKY transcription factors form one of the largest transcription factor families and function as important components in the complex signaling processes that occur during plant stress responses. However, relative to the research progress in model plants, far less information is available on the function of WRKY proteins in cotton. In the present study, we identified the *GhWRKY40* gene in cotton (*Gossypium hirsutum*) and determined that the GhWRKY40 protein is targeted to the nucleus and is a stress-inducible transcription factor. The *GhWRKY40* transcript level was increased upon wounding and infection with the bacterial pathogen *Ralstonia solanacearum*. The overexpression of *GhWRKY40* down-regulated most of the defense-related genes, enhanced the wounding tolerance and increased the susceptibility to *R. solanacearum*. Consistent with a role in multiple stress responses, we found that the *GhWRKY40* transcript level was increased by the stress hormones salicylic acid (SA), methyl jasmonate (MeJA) and ethylene (ET). Moreover, GhWRKY40 interacted with the MAPK kinase GhMPK20, as shown using yeast two-hybrid and bimolecular fluorescence complementation systems. Collectively, these results suggest that *GhWRKY40* is regulated by SA, MeJA and ET signaling and coordinates responses to wounding and *R. solanacearum* attack. These findings highlight the importance of WRKYs in regulating wounding- and pathogen-induced responses.

Editor: Sara Amancio, ISA, Portugal

Funding: This work was financially supported by the Genetically Modified Organisms Breeding Major Projects of China (2009ZX08009-092B) and the National Science Foundation of China (grant no. 31171837; 30970225). The funders had no role in study design, data collection and analysis, decision to publish, or preparation of the manuscript.

Competing Interests: The authors have declared that no competing interests exist.

* E-mail: xqguo@sdau.edu.cn

Introduction

Stress is perceived and transduced through a series of signaling molecules that ultimately affect the regulation of stress-inducible genes to initiate the synthesis of different types of proteins, including transcription factors, enzymes, molecular chaperones, ion channels, and transporters, or to alter their activities [1]. Among these proteins, transcription factors (TFs) are crucial in eliciting stress responses by modulating the expression of specific target genes in a temporal and spatial manner; they are also necessary for normal development and proper responses to physiological or environmental stimuli [2–5]. WRKY proteins are a class of zinc finger-containing TFs that are encoded by large gene families in all higher plants and are reported to play a pivotal role in many physiological processes. WRKY TFs share a highly conserved sequence of approximately 60 amino acids called the WRKY domain, which contains the conserved amino acid sequence motif WRKYGQK at the N-terminus and a novel zinc finger-like motif at the C-terminus. Based on their domain structures, WRKY TFs are classified into three major groups (I, II, and III) [6]. Additionally, WRKY TFs act as transcriptional regulators by binding to the W-box, which is present in the promoter regions of various stress-related genes, thus regulating the expression of many genes, resulting in stress tolerance [7].

WRKY TFs have mostly been studied with respect to their participation in the regulation of defense against biotic stresses or against tolerance of abiotic stresses [8]. Some WRKY proteins are reported to be involved in the coordination of multiple biological processes. For example, *AtWRKY33* regulates disease resistance, NaCl tolerance and thermotolerance [9–11], while Ca*WRKY40* modulates tolerance to heat stress and resistance to *Ralstonia solanacearum* infection [12]. This suggests that some WRKY proteins serve as nodes in a crosstalk between different physiological processes. However, the functions of the majority of WRKY family members and their possible roles in signaling crosstalk are limited.

Previous studies have shown that several wound-responsive WRKY genes are also regulated by pathogen infection [13–16]. These reports have provided useful research methods and broadened our knowledge on the function of plant WRKYs. However, so far, few reports have addressed the mechanistic details of the crosstalk between wounding and pathogen infection.

The prominent role of WRKYs in stress signaling indicates a promising target for applied studies in crop species. Moreover, dissecting the crosstalk between different pathways is critical to understanding the plant response to environmental cues. Wounding presents a threat to plant survival because it not only physically destroys plant tissues but also provides a pathway for pathogen invasion. Therefore, it is necessary to map the interaction between the wounding response and pathogen infection and to identify novel genes involved in these processes. In particular, the molecular mechanisms involved in these processes should be examined in different genetic backgrounds.

WRKY proteins have been linked to the MAP kinase (MAPK) cascade in *Arabidopisis*; *AtWRKY22* and *AtWRKY29* are thought to function downstream of *MPK3/MPK6* [17]. The MAPK cascade is the basic module for transmitting signals from upstream ligand receptors to downstream substrates in response to various biotic and abiotic stress signals, hormones, and growth and developmental processes [18]. In tobacco, WRKY8 is a physiological substrate of SIPK, NTF4, and WIPK [19]. In rice, OsWRKY30 interacts with and is phosphorylated by OsMPK3 [20]. Thus, only a limited number of upstream WRKY components have been identified, particularly in crops, and whether MAPKs interact with WRKYs in cotton should be investigated.

Cotton is one of the most economically important crops worldwide and is an excellent source of fiber and oil. However, its growth and yield are severely inhibited under various biotic and abiotic stress conditions. The applied study of cotton WRKYs will provide new insight that may aid in creating cotton plants that are better able to adapt to environmental challenges. However, the majority of WRKY TFs in cotton have not been characterized. In the present study, a group II a WRKY gene from cotton (*Gossypium hirsutum*), *GhWKRY40*, was isolated and characterized. *GhWRKY40* expression was induced by various abiotic and biotic stresses. We obtained information on the ability of *GhWRKY40* overexpression to alter the responses to wounding and infection with the bacterial pathogen *R. solanacearum* in *Nicotiana benthamiana*. Moreover, we demonstrated that GhWRKY40 interacted with GhMPK20 but not GhMK6a, two MAPKs that were previously identified by our group, using a yeast two-hybrid system and bimolecular fluorescence complementation (BiFC). This study suggests that the transcriptional responses of *GhWRKY40*-overexpressing plants to wounding may be related to defense signaling pathways.

Materials and Methods

Cloning of the full-length *GhWRKY40* cDNA

Total RNA was extracted from the leaves of seven-day old cotton seedlings using a modified cetyltrimethylammonium bromide (CTAB) protocol [21]. Reverse transcription-PCR (RT-PCR) and RACE-PCR were used to amplify the full-length *GhWRKY40* cDNA. A pair of degenerated primers (MP1/MP2) was designed to isolate WRKY family members from the cotton cotyledons. According to the obtained fragment, specific primers (5P1/5P2, 3P1/3P2, and QC1/QC2) were used for 5′ rapid amplification of cDNA ends (RACE), 3′ RACE and the identification of the full-length cDNA sequence. The general PCR procedures and primers are shown in Table S1 and Table S2, respectively. The PCR product was purified, cloned into the pEasy-T1 vector, and transformed into competent *Escherichia coli* cells for sequencing. The amino acid sequence alignment of GhWRKY40 and its homologues was conducted using BLAST (http://www.ncbi.nlm.gov/blast) and DNAman software 5.2.2. The phylogenetic tree was performed in MEGA version 4.1 (http://megasoftware.net) using the neighbour-joining method.

Amplification of the *GhWRKY40* genomic sequence and promoter

For the amplication of *GhWRKY40* genomic sequence, one pair of primers (QG1 and QG2), which was designed and synthesized based on the full-length *GhWRKY40* cDNA, was used. Genomic DNA was isolated from seedling leaves using the CTAB method. Inverse-PCR (I-PCR) was performed to obtain the promoter sequence. Three restriction endonucleases (NdeI, SspI and VspI) were used to digest the cotton seedling genomic DNA, and T4 DNA ligase was used to self-ligate the DNA fragments into circles, which were used as templates to amplify the promoter region. Three promoter fragments were amplified using six pairs of primers (Nde1/2 and Nde3/4, Ssp1/2 and Ssp3/4, Vsp1/2 and Vsp3/4). The deduced portion of the promoter was subsequently verified using the special primers WP1 and WP2. The sequences of the primers are provided in Table S2. The programme PlantCARE (http://bioinformatics.psb.ugent.be/webtools/plantcare/html) and PLACE (http://www.dna.affrc.go.jp/PLACE/) was used to analyze the *GhWRKY40* promoter sequence.

Subcellular localization of GhWRKY40

The coding sequence of *GhWRKY40* without the termination codon was amplified by PCR using the primers Wgf1 (5′-GGATCCATGGATACTTCTTCATGGGTGG-3′, *Bam*HI site underlined) and Wgf2 (5′-CTCGAGCTTATAGTTGACAAAATCATAGAAAC-3′, *Xho*I site underlined). Then, the coding sequence was ligated into the binary vector pBI121-GFP, which contains the *Cauliflower mosaic virus* (CaMV) 35S promoter, to yield the expression vector p35S::GhWRKY40-GFP. The resulting expression plasmid, pBI121-GhWRKY40-GFP, or the pBI121-GFP control plasmid was transformed into onion (Allium cepa) epidermis cells via biolistic bombardment transformation using the Biolistic PDS-1000/He system (Bio-Rad, USA) with gold particles (1.0 μl) and a helium pressure of 1,100 psi. And then plated on MS agar medium in the dark condition at 28°C for 24 h, the nuclei were stained with 100 μg/ml of 4′,6-diamidino-2-phenylindole (DAPI) in phosphate-buffered saline for 4 min, and the fluorescence signal of the GhWRKY40-GFP fusion protein was imaged using a fluorescence microscope using excitation wavelength of 488 nm and 350 nm, respectively. The vector p35S::GFP was used as a control.

Transactivation assay

The transactivation activity of the GhWRKY40 protein was investigated in the Y2HGold yeast strain, which contains the *HIS3*, *ADE2* and *MEL1* reporter genes and the GAL4 promoter. The coding region of *GhWRKY40* was amplified using the primers Wbd1 and Wbd2, which possess *EcoR*I and *BamH*I sites (underlined), 5′-GAATTCATGGATACTTCTTCATGGGTGG-3′ and 5′- GGATCCTTCAACTGGACTTTGCTGAAAC-3′ to build the pGBKT7-GhWRKY40 vector (Clontech, TaKaRa) containing the GAL4 DNA-binding domain. The plasmid pGBKT7-GhWRKY40 and pGBKT7 (negative control) was transformed into Y2HGold yeast cells. The transformed yeast cells were streaked on SD/-Trp and SD/-Trp-Ade-His medium plates to observe yeast growth at 30°C for 3–4 days. An assay of α-galactosidase activity was performed using X-α-gal.

Cotton growth conditions and *GhWRKY40* expression assay

Cotton (*Gossypium hirsutum* L. cv. lumian 22) seeds were grown under greenhouse conditions at 25±1°C with a 16 h light/8 h dark cycle (relative humidity of 60–75%), and seven-day-old seedlings were used for the following treatments. For the pathogen treatment, cotton seedlings were inoculated with *R. solanacearum* suspensions using the root dip method. For the H_2O_2 treatment, cotton seedlings were sprayed with 10 mM H_2O_2. The wounding, MeJA, SA and ET treatments were performed as described previously [22]. The treated cotyledons were collected, frozen directly in liquid nitrogen and stored at −80°C for RNA extraction and further analysis. Each treatment was repeated at least three times.

Total RNA was extracted from the treatment samples using a modified cetyltrimethylammonium bromide (CTAB) protocol [21] and then treated with RNase-free DNaseI to remove any potential genomic DNA contamination. First-strand cDNA was synthesized using the EasyScript First-Strand cDNA Synthesis SuperMix. The *GhWRKY40* (KC414679) gene primer pairs WRT1/WRT2 and *Ghubiquitin* (EU304080) primer pairs Ub1/Ub2 were used for quantitative real-time PCR (qPCR) with the SYBR PrimeScript RT-PCR Kit in a CFX96TM Real-time System. The PCR programme was as follows: predenaturation at 95°C for 30 s; 40 cycles of 95°C for 5 s, 55°C for 15 s and 72°C for 15 s; and a melt cycle from 65°C to 95°C. The data was analyzed using the CFX Manager software, version 1.1, and significant differences were determined using the Statistical Analysis System (SAS) software (version 9.1). All reactions were performed with three technical replicates.

Wounding analysis of transgenic plants

The *GhWRKY40* coding region was amplified with primers WOE1 (5′-GGATCCATGGATACTTCTTCATGGGTGG-3′, *BamHI* site underlined) and WOE2 (5′-GAGCTCTTCAACTGGACTTTGCTGAAAC-3′, *SacI* site underlined). Then this fragment was subcloned into pBI121 under the control of the CaMV35S promoter. The recombinant plasmid was introduced into the *Agrobacterium tumefaciens* (strain LBA4404) for *Nicotiana benthamiana* (*N. benthamiana*) transformation using the leaf disc method as described previously [23]. The transgenic seedlings were screened on Murashige and Skoog (MS) agar medium containing 100 mg/L of kanamycin and further confirmed by PCR. Eight independent transgenic T_1 native tobacco lines were obtained. Three independent *GhWRKY40-OE* lines (OE1, OE2 and OE3) and wild-type plants were used for the following experiments. The transgenic T_2 lines were used in the experiments.

Transgenic *N. benthamiana* seeds were surface sterilized and planted on MS medium for germination under greenhouse conditions. Four-leaf stage seedlings were transplanted into soil and maintained under greenhouse conditions. For the wounding treatment, the third leaf from the top of 8-week old seedlings was cut with scissors. After wounding, the leaves were harvested at the indicated timepoints for histochemical staining and the preparation of RNA. The experiment was repeated at least three times.

After wounding treatment, the OE and WT leaves were incubated with 3,3′-Diaminobenzidine (DAB, 1 mg/mL, pH 3.8) or Nitro Blue tetrazolium (NBT, 0.1 mg/mL) for 36 h at 25°C in the dark. Then, the leaves were boiled in ethanol (95%) for 5 min. After cooling, the leaves were soaked and preserved in fresh ethanol at room temperature and photographed. To examine oxidative tolerance, leaf discs (1.3 cm in diameter) were detached from healthy, fully expanded leaves of OE and WT plants, floated in methyl viologen (MV) (0, 200, 400 or 600 μM) solutions for 64 h, and immersed in 95% ethanol for 40 h to extract chlorophyll for spectrophotometric measurement.

Disease resistance analysis of transgenic plants

For the disease resistance analysis, *R. solanacearum* strains were cultured at 200 rpm and 37°C in Luria-Bertani (LB) broth. The bacterial cell density was diluted to OD_{600} = 0.6–0.8. A total of 20 μL of the resulting *R. solanacearum* suspension was injected into the third leaf from the top of each plant of 8-week old seedlings using a syringe with a needle. The leaves were harvested at the indicated timepoints for the preparation of RNA. The experiment was repeated at least three times.

Expression analysis of defense-related genes in transgenic and WT lines

Total RNA of all samples was extracted with TRIzol reagent. qPCR was performed using cDNA as the template, which was synthesized from total RNA extracted from the transgenic or WT *N. benthamiana* lines after wounding and disease treatments. *N. benthamiana β-actin* (*Nbβ-actin*) genes were used as the standard controls. The GenBank accession numbers of the defense-related genes examined in the qPCR analysis are as follows: JQ256516.1 (*β-actin*), ACY30445.1 (*JAZ1*), BAG68657.1 (*JAZ3*), X84040.1 (*LOX1*), AF392978 (*ACS6*), U15933.1 (*APX*), AB093097 (*SOD*), D10524 (*GST*), X12485.1 (*PR1a*) M60460.1 (*PR2*), EH365959.1 (*PR4*), and Y07563 (*HIN1*). The primers of the defense-related genes examined in the qPCR are listed in Table S3.

Yeast two-hybrid and BiFC assay

For the yeast two-hybrid assay, the *GhWRKY40* cDNA fragment was cloned into the pGADT7 vector in-frame with the GAL4 activation domain. The *GhMPK6a* and *GhMPK20* cDNA fragments were cloned into the pGBKT7 vector in-frame and proximal to the binding domain. These vectors were co-transformed into the Y2HGold yeast strain using the Matchmaker Gold Yeast Two-Hybrid System. Positive clones were plated onto selective SD medium (DDO: SD/-Leu/-Trp, QDO: SD/-Ade/-His/-Leu/-Trp and QDO/X/A: QDO with X-α-gal and aureobasidin A). For the BiFC assay, the *GhWRKY40* and *GhMPK20* cDNA fragments were cloned into pUC-SPYCE-35S and pUC-SPYNE-35S, respectively. These two BiFC constructs were co-transformed into onion epidermal cells using the particle bombardment method, and the fluorescent signal of the resultant proteins was detected using a confocal microscope.

Statistical analysis

Data were shown as the mean ± standard deviation (SD) with n = 3. The results were analyzed using multiple comparisons by analysis of variance (ANOVA), and means were separated by the Duncan's Multiple Range test. ANOVA was performed using Statistical Analysis System (SAS) version 9.1 software.

Results

Characterization of *GhWRKY40*

The full-length cDNA of *GhWRKY*40 was determined to be 1463 bp. It contained an open reading frame (ORF) of 945 bp, encoding 314 amino acids, an untranslated region of 203 bp at the 5′ end and 315 bp of the noncoding region at the 3′ end. The relative molecular mass and theoretical *pI* of the predicted protein were 34.4 kDa and 8.45, respectively. According to the nomenclature for plant WRKYs as well as the alignment this WRKY sequence with related sequences, it was found to share 48.0% and 47.48% identity with AtWRKY40 (NP_178199.1) and BnWRKY40 (ACQ76806.1), respectively. Therefore, we termed

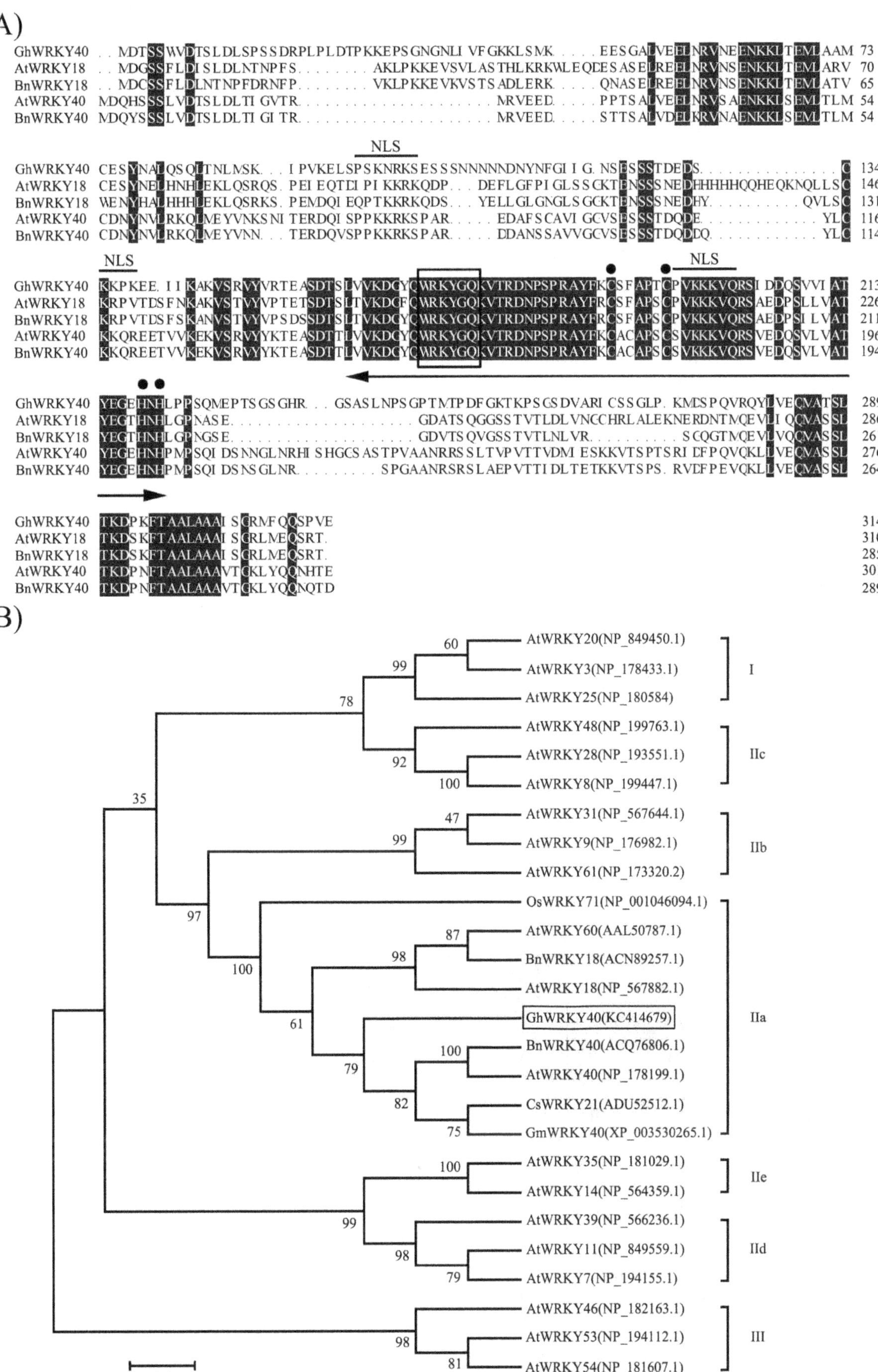
(A)
GhWRKY40
AtWRKY18
BnWRKY18
AtWRKY40
BnWRKY40
NLS
NLS
NLS
(B)
AtWRKY20(NP_849450.1)
AtWRKY3(NP_178433.1)
AtWRKY25(NP_180584)
I
AtWRKY48(NP_199763.1)
AtWRKY28(NP_193551.1)
AtWRKY8(NP_199447.1)
IIc
AtWRKY31(NP_567644.1)
AtWRKY9(NP_176982.1)
AtWRKY61(NP_173320.2)
IIb
OsWRKY71(NP_001046094.1)
AtWRKY60(AAL50787.1)
BnWRKY18(ACN89257.1)
AtWRKY18(NP_567882.1)
GhWRKY40(KC414679)
BnWRKY40(ACQ76806.1)
AtWRKY40(NP_178199.1)
CsWRKY21(ADU52512.1)
GmWRKY40(XP_003530265.1)
IIa
AtWRKY35(NP_181029.1)
AtWRKY14(NP_564359.1)
IIe
AtWRKY39(NP_566236.1)
AtWRKY11(NP_849559.1)
AtWRKY7(NP_194155.1)
IId
AtWRKY46(NP_182163.1)
AtWRKY53(NP_194112.1)
AtWRKY54(NP_181607.1)
III
0.1

Figure 1. Characterization and sequence analysis of GhWRKY40. (A) Alignment of the amino acid sequences of GhWRKY40 and the representative related proteins AtWRKY18 (NP_567882), BnWRKY18 (ACN89257), AtWRKY40 (NP_178199) and BnWRKY40 (ACQ76806). Amino acids with 100% identity are shaded in black. The approximately 60-amino acid WRKY domain and the C and H residues in the zinc-finger motif (C-X_{4-5}-C-X_{22-23}-H-X_1-H) are marked by a two-headed arrow and dot, respectively. The highly conserved amino acid sequence WRKYGQK in the WRKY domain is boxed. The putative nuclear localization signals are marked by lines. (B) Phylogenetic relationship between GhWRKY40 and other plant WRKY proteins. A neighbor-joining phylogenetic tree was created using MEGA 4.1 software. GhWRKY40 is boxed. Each gene name is followed by its protein ID. The abbreviations of the gene names are indicated as follows: Gh, *Gossypium hirsutum*; At, *Arabidopsis thaliana*; Os, *Oryza sativa*; Bn, *Brassica napus*; Cs, *Cucumis sativus* and Gm, *Glycine max*.

the cDNA clone as *GhWRKY40* (GenBank accession number: KC414679).

According to our multiple alignment and phylogenetic analyses of the WRKY proteins, GhWRKY40 was placed into group IIa of the WRKY superfamily (Fig. 1). The GhWRKY40 protein possesses fully canonical motif structures, including a typical DNA binding domain, the WRKY domain and a putative zinc finger structure (C-X_{4-5}-C-X_{22-23}-H-X_1-H). The phylogenetic relationships among various WRKY IIa subgroup members from different organisms were further analyzed by comparing the protein sequences of their conserved WRKY domains.

To analyze the structure of GhWRKY40, we isolated a 1936 bp genomic fragment (GenBank accession number: KC414680) from cotton genomic DNA. Sequence comparison revealed that *GhWRKY40* has four introns. This intron number differs from that of *AtWRKY40* but is similar to that of *AtWRKY18*, *AtWRKY60* and is characteristic of group IIa of the WRKY superfamily.

GhWRKY40 is localized to the nucleus

The PSORT program predicted that GhWRKY40 is localized to the nucleus. Sequence analysis using WoLF PSORT (http://wolfpsort.org/) indicated that the predicted GhWRKY40 protein contains three putative nuclear localization signals (^{98}PSKNRKS^{104}, ^{135}KKPK^{138}, ^{195}PVKKKVQ^{201}; Fig. 1A). To confirm its nuclear localization, we generated a construct for the expression of GhWRKY40 fused to green fluorescent protein (GFP) under the control of the constitutive CaMV35S promoter (Fig. 2A). Typical results indicated the exclusive localization of GhWRKY40-GFP in the nucleus, whereas GFP alone occurred throughout in the cell (Fig. 2B). This result suggests that GhWRKY40 has a nuclear localization.

GhWRKY40 functions as a potential transcriptional activator

To determine whether the GhWRKY40 protein has transcriptional activity in eukaryotic cells, a plasmid containing the GAL4 DNA binding domain and the whole ORF of *GhWRKY40* was constructed (pGBKT7-GhWRKY40). The plasmid pGBKT7-GhWRKY40 or pGBKT7 (negative control) was transformed into Y2HGold yeast cells. All transformants containing pGBKT7-GhWRKY40 and pGBKT7 grew well on selective medium without tryptophan (SD/-Trp). As shown in Fig. 2C, yeast transformed with pGBKT7-GhWRKY40 grew on selective medium without tryptophan, histidine and adenine (SD/-Trp-His-Ade). In addition, α-galactosidase activity was detected in these cultures, indicating that the expression of the reporter genes

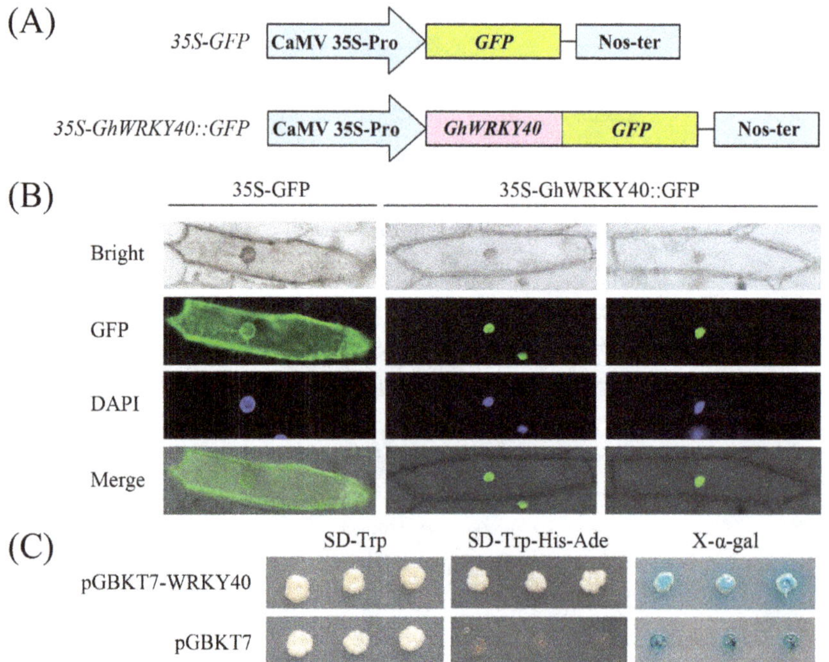

Figure 2. Subcellular localization of the GhWRKY40 protein and transcriptional activation of the *GhWRKY40* gene. (A) Schematic diagram of the 35S-GhWRKY40::GFP fusion protein construct and the 35S-GFP construct. (B) Transient expression of the 35S-GFP and 35S-GhWRKY40::GFP constructs in onion epidermal cells. Green fluorescence corresponding to the expressed proteins was observed with a fluorescence microscope 24 h after particle bombardment. The nuclei of the onion cells were visualized by DAPI staining. (C) Transactivation of the GhWRKY40 gene in yeast. The vector pGBKT7 was used as a control. The transformed yeast culture was streaked onto SD/-Trp or SD/-Trp-His-Ade medium, and the α-galactosidase activity was determined. Three independent experiments were performed.

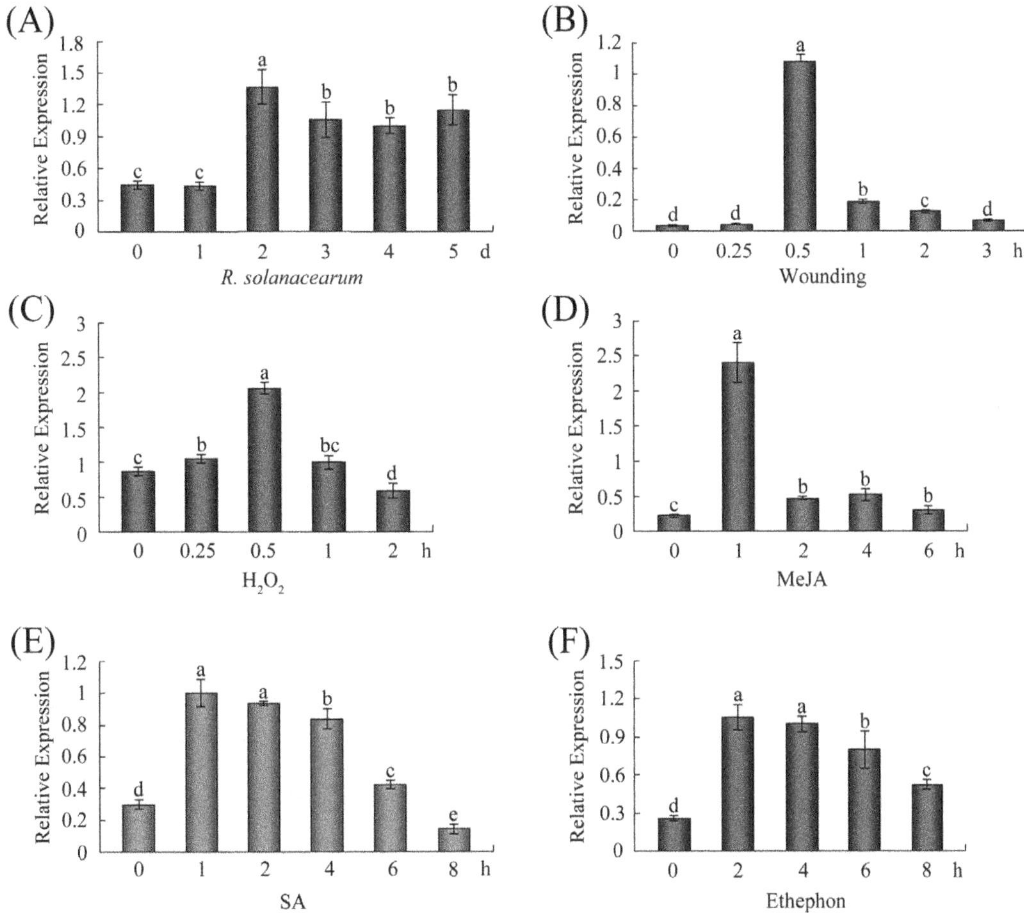

Figure 3. Expression of the *GhWRKY40* gene in response to stress. Seven-day-old cotton seedlings in hydroponic culture were treated with *R. solanacearum* (A), wounding (B), 100 μM H_2O_2 (C), 100 μM MeJA (D), 2 mM SA (E) or 5 mM ET released from ethephon (F). Total RNA was isolated at the indicated times after the treatment and subjected to qPCR analysis. The *ubiquitin* gene was employed as an internal control. This experiment was repeated at least twice. The values indicated by the different letters are significantly different at $P<0.01$, as determined using Duncan's multiple range tests.

(*HIS3*, *ADE2* and *MEL1*) was activated, whereas neither nor *HIS3* were activated in yeast transformants containing the negative control plasmid (pGBKT7). These results indicated that GhWRKY40 is a transcriptional activator.

Expression profile of *GhWRKY40* under stress conditions

To test if GhWRKY40 is involved in the plant response to abiotic and biotic stresses, the transcript levels of *GhWRKY40* were measured by qPCR after the treatment of cotton seedlings with *R. solanacearum*, wounding or H_2O_2. The *GhWRKY40* transcript level was up-regulated in response to infection with the bacterial pathogen *R. solanacearum* (Fig. 3A). The *GhWRKY40* transcript level was also increased at 0.5–2 h after wounding, with peak expression 0.5 h (Fig. 3B). During H_2O_2 treatment, the *GhWRKY40* transcript level was noticeably elevated at 0.5 h (Fig. 3C). These results suggest that the expression of the *GhWRKY40* gene is induced by environmental stimuli and may play an important role in the stress response.

Phytohormones, such as SA, JA and ET, serve as significant signalling molecules in the regulation of plant defense responses against biotic and abiotic stresses and play vital roles in mediating the expression of downstream defense genes [24]. To determine the possible involvement of *GhWRKY40* in the signaling cascades, the *GhWRKY40* transcript levels were determined by qPCR in seven-day old cotton seedlings that were exogenously treated with MeJA, SA or ET. In response to 100 μM MeJA, the transcript level of *GhWRKY40* was enhanced from 1–6 h, with maximum expression at approximately 1 h (Fig. 3D). *GhWRKY40* mRNA was also induced and reached a peak at 1–2 h with 2 mM SA, (Fig. 3E). Application of 5 mM ET increased the *GhWRKY40* transcript level at 2–8 h, reaching maximal levels from 2–4 h (Fig. 3F). The strong induction of GhWRKY40 expression by these signaling molecules suggests that this gene is involved in signaling pathways in stress resistance.

GhWRKY40 promoter analysis

To determine whether the *GhWRKY40* gene is induced by stress, we isolated a 782 bp fragment from the upstream region of the *GhWRKY40* gene by I-PCR. Analysis of this region using the PlantCARE and PLACE databases revealed many putative *cis*-elements, suggesting that *GhWRKY40* plays a role in the plant response to environmental stress. The elements in this region include pathogen/elicitor-related elements, such as ARE, RAV1AAT and WBOXATRNPR1 and WBOX71OS, abiotic stress responsive elements, such as MBS, CCAAT-box, WBOXNTERF3, OSE2ROOTNODULE and CURECORECR, and

Table 1. Putative *cis*-elements in the *GhWRKY40* promoter.

cis-element	Position	Sequence (5′-3′)
Abiotic stress response elements		
MBS	−116(+)	TAACTG
CURECORECR	−490(+)	GTAC
CCAAT-box	−615(−)	CAACGG
WBOXNTERF3	−304 (−),−372(+)	TGACY
Pathogen/elicitor response elements		
ARE	−411(−)	TGGTTT
RAV1AAT	−307(+)	CAACA
WBOXATNPR1	−371(+),−305(−)	TTGAC
WRKY71OS	−55(+),−141(+),−372(+),−305(−)	TGAC
Tissue-specific and development-related elements		
circadian	−683(−)	CAANNNNATC
Skn-1_motif	−53(−),−139(−)	GTCAT
OSE2ROOTNODULE	−216(+),−323(−),−530(−),−652(−),−342(−),−352(−)	CTCTT
POLLEN1LELAT52	−1004(+),−828(+),−604(+),−569(+),−271(+),−253(+),−28(+),−9(+),−861(−)	AGAAA
WBOXHVISO1	−372(+),−304(−)	TGACT
Light regulation elements		
AE-box	−654(−)	AGAAACAA
GA-motif	−347(+)	AAGGAAGA
I-box	−578(−)	GATAAGAATA
Sp1	−327(+)	CC(G/A)CCC

tissue-specific and development-related elements, such as the circadian, Skn-1 motif, POLLEN1LELAT52 and WBOXH-VISO1. All of the identified *cis*-elements are listed in Table 1.

Overexpression of *GhWRKY40* in transgenic plants affects the expression of defense-related genes in response to wounding

The *GhWRKY40* transcript patterns suggest a role for this gene in defense against biotic and various abiotic stresses. To assess the

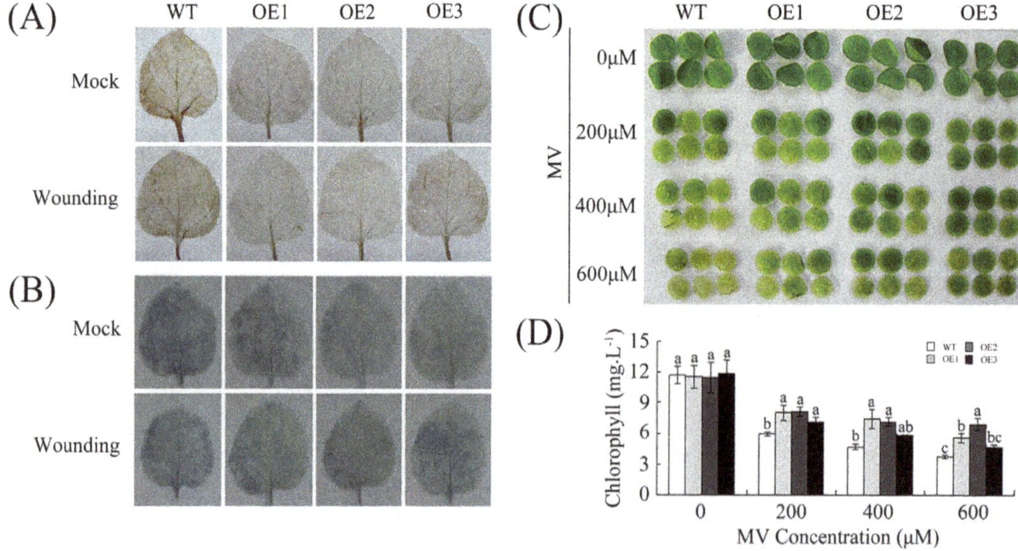

Figure 4. Analysis of ROS accumulation after wounding in WT and OE plants. (A–B) Wound-induced H_2O_2 and O_2^- accumulation, as detected via DAB staining and NBT staining, respectively. (C) The phenotype of leaf disks from WT and OE plants that were incubated in different concentrations of MV (0, 200, 400 or 600 mM). (D) Relative chlorophyll content in the leaf disks from (C). Disks floated in water were used as a control. The presented data are the means ± standard error of three independent experiments. The different letters above the columns indicate significant differences ($P<0.01$) according to Duncan's multiple range test, which was performed using SAS version 9.1 software.

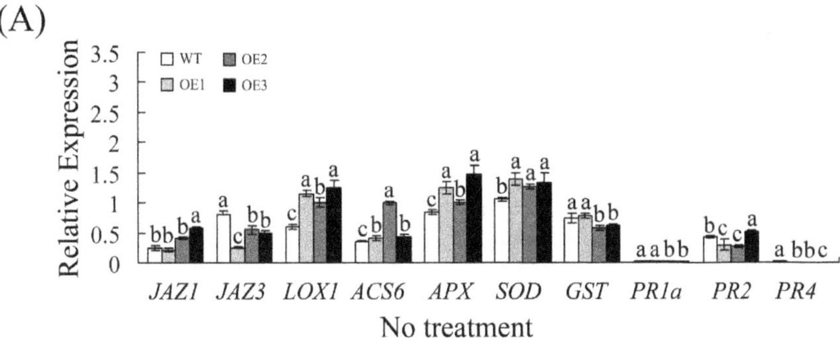

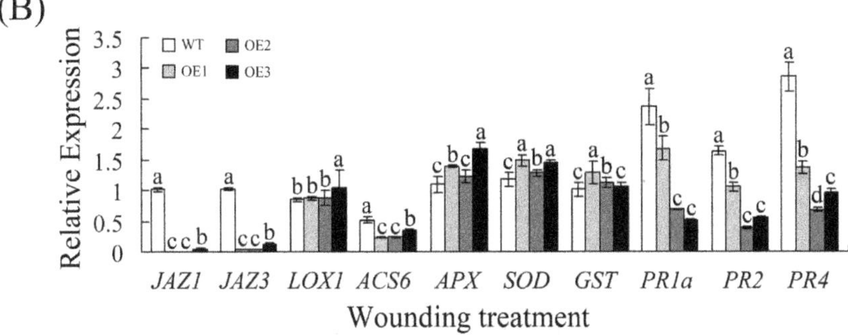

Figure 5. qPCR analysis of stress-related gene expression in WT and OE plants under normal conditions (A) and after wounding (B). The data are presented as the mean ± standard error of three independent experiments. The values indicated by the different letters are significantly different at P<0.01, as determined using Duncan's multiple range tests.

significance of this gene in the response to these stresses, we generated transgenic native tobacco T_2 lines that overexpress *GhWRKY40* driven by the CaMV35S promoter. Except for the later germination of the transgenic plants relative to the WT plants (Fig. S1), we observed no differences between the plants. Wounding is a common plant injury and presents a potential threat to plant survival because it not only damages tissues but also provides means for pathogen invasion. Three independent *GhWRKY40-OE* lines and wild-type plants were used to understand the wounding response.

We first examined the wounding response by the DAB staining of H_2O_2 accumulation and the NBT staining of O_2^- accumulation in wounded leaves. Compared with WT plants, the *GhWRKY40-OE* lines exhibited clearly decreased DAB staining intensities in their wounded leaves, reflecting low levels of H_2O_2 accumulation (Fig. 4A). These lines also exhibited decreased NBT intensities staining in the treated leaves compared with the WT plants (Fig. 4B). Meanwhile, the *GhWRKY40-OE* lines exhibited slightly lower H_2O_2 and O_2^- accumulation compared with the WT plants in the absence of wounding treatment. Additionally, leaf discs from the overexpression lines were used to illustrate whether *GhWRKY40* plays a role in oxidative resistance. Leaf discs were soaked in solutions containing various concentrations (0, 200, 400 or 600 μM) of methyl viologen (MV). The results presented in Fig. 4C and 4D show that the transgenic plants exhibited intense oxidative resistance. Leaf discs from both the OE and WT plants showed signs of chlorosis, but neither of these plants showed abnormalities in water. However, MV treatment led to more severe damage in the leaf discs from the WT plants. This result was further confirmed by measuring the leaf disc chlorophyll content before and after MV treatment. Taken together, *GhWRKY40* overexpression appears to enhance the defense response to wounding, resulting in decreased H_2O_2 and O_2^- levels.

To further confirm the role of *GhWRKY40* in the wounding response and to elucidate its possible mechanism of action, the transcriptional responses of known defense genes to *GhWRKY40* overexpression were investigated by qPCR (Fig. 5). We examined the transcript levels of the JA-responsive genes *JAZ1*, *JAZ3*, and *LOX1*, the ET production-associated gene *ACS6*, the reactive oxygen species (ROS) detoxification-associated genes *APX*, *GST*, and *SOD*, and the SA-responsive genes *PR1a*, *PR2*, and *PR4*. Previous studies have shown that each of the tested genes is up-regulated in response to wounding [16,25]. We found that the transcript levels of the two *JAZ* genes were clearly decreased in the OE plants after wounding compared with the WT plants. In contrast, we did not find a difference in the *LOX1* transcript level between the WT and OE plants in response to wounding. The transcript expression of the ET production-associated genes, which have been identified as early wound-response genes [26–28], was inhibited in the OE plants in response to wounding. Interestingly, *GhWRKY40* does not appear to significantly enhance the expression of *APX*, *GST*, or *SOD*. In the WT plants, the transcript levels of *PR1a*, *PR2*, and *PR4* were up-regulated in response to wounding. However, this induction was attenuated by the overexpression of *GhWRKY40*. Based on the above analysis, it was proposed that wounding may elicit the activation of pathways that interact with the defense response and possibly other signaling pathways. Most importantly, *GhWRKY40* may play a crucial role in the wounding response.

Overexpression of *GhWRKY40* increases the susceptibility to *R. Solanacearum* in transgenic plants

The up-regulation of the *GhWRKY40* transcript in response to *R. solanacearum* suggests a role for this gene in the defense response.

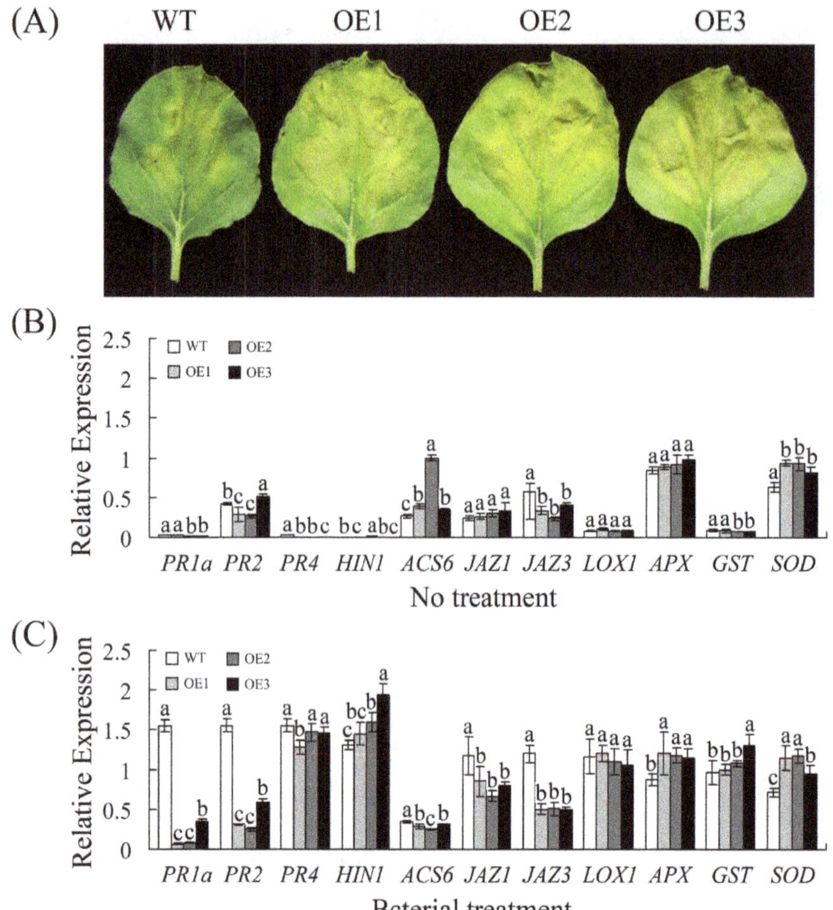

Figure 6. *GhWRKY40* overexpression enhances susceptibility to *R. Solanacearum* in transgenic plants. (A) Phenotype of WT and OE lines after 5 days of incubation with *R. solanacearum*. (B–C) Relative transcript levels of defense-related genes in non-infected and infected WT and OE plants were analyzed by qPCR. The data are presented as the mean ± standard error of three independent experiments. The values indicated by the different letters are significantly different at P<0.01, as determined using Duncan's multiple range tests.

To analyze the role of *GhWRKY40* in plant basal defense, the bacterial pathogen *R. solanacearum* was used to infect the OE and WT plants. Five days after infection, all three of the tested transgenic lines exhibited enhanced wilting symptoms and chlorosis (Fig. 6A), indicating that *GhWRKY40* overexpression enhances the susceptibility of native tobacco plants to *R. solanacearum*.

Consistent with the enhanced susceptibility of the *GhWRKY40-OE* plants, they also exhibited reduced transcript levels of the SA production-associated genes *PR1a* and *PR2* relative to the WT plants after *R. solanacearum* infection. Similarly, the transcript level of ET-responsive gene *ACS6* was decreased. After *R. solanacearum* infection, the JA-responsive genes *JAZ1* and *JAZ3* were lower in the OE plants than in the WT plants. The ROS detoxification-associated genes *APX*, *GST* and *SOD* exhibited increased transcript levels in the OE plants after *R. solanacearum* infection, and their transcripts accumulated to higher levels in the OE plants relative to the WT plants. Expression of the HR-associated gene *HIN1* was obviously induced, as its transcript accumulated to higher levels in the OE plants than in the WT plants. However, the transcript levels of *PR4* and *LOX1* did not show any significant difference in the OE plants relative to the WT plants (Fig. 6 B–C).

GhWRKY40 interacts with GhMPK20 but not with GhMPK6a

Yeast two-hybrid and BiFC systems were used to determine the interactions between GhWRKY40 and GhMPK6a (HM055511)/GhMPK20 (HQ828072). In the yeast two-hybrid system, positive clones expressing GhMPK20 and GhWRKY40 were able to grow on QDO and QDO/X/A plates, indicating that GhWRKY40 interacted with GhMPK20 due to the activation of the reporter genes *AbA* and *MEL1*. However, yeast cells cotransformed with GhMPK6a and GhWRKY40 were unable to grow on QDO and QDO/X/A plates (Fig. 7A). These interactions were further confirmed with the BiFC system, in which two plasmids were constructed for the expression of GhWRKY40-yellow fluorescent protein (YFP)^{C} and GhMPK20-YFP^{N}, reseparately, and then cotransformed into onion epidermal cells by particle bombardment. The YFP fluorescence signal in the onion epidermal cells transfected with GhWRKY40-YFP^{C} and GhMPK20-YFP^{N} was exclusively nuclear (Fig. 7B). Notably, GhMPK20 was detected in both the cytoplasm and nucleus (unpublished data). These data demonstrate that GhWKY40 interacts with GhMPK20 in the nucleus.

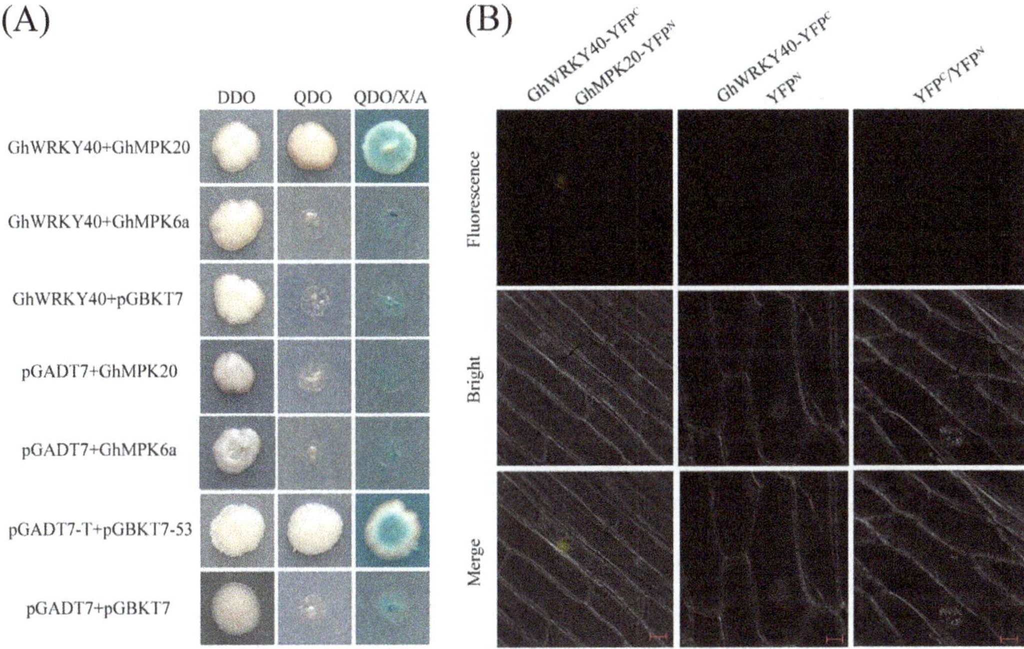

Figure 7. Interaction between GhWRKY40 and GhMPK20/GhMPK6a in yeast and onion epidermal cells. (A) Transformants grown on DDO, QDO or QDO/X/A. (B) *In vitro* BiFC analysis of the GhWRKY40 -interacting protein GhMPK20 in co-transformed into onion epidermal cells. The yellow fluorescence indicates the interaction between GhWRKY40 and GhMPK20. The fluorescent signals were observed using a confocal microscope. Scale bar = 20 μm.

Discussion

To gain an increased understanding of WRKY transcription factors in cotton, we cloned a *WRKY* gene from *G. hirsutum*. Our sequence and phylogenetic tree analyses indicated that the *GhWRKY40* gene belongs to subgroup IIa (Fig. 1B). Our subcellular localization experiment with GhWRKY40-GFP indicated that GhWRKY40 is located in the nucleus (Fig. 2B), which is consistent with previous studies on WRKY transcription factors from other species [29]. Moreover, consistent with the putative role of WRKY proteins as transcription factors, three nuclear targeting sequences were identified (Fig. 1A). Transcriptional activation analysis in yeast showed that the full-length GhWRKY40 protein is transcriptionally active (Fig. 2C). These results suggest that *GhWRKY40* is a member of the WRKY family in cotton and may serve as a transcriptional activator.

Because plants are sessile, they are constantly affected by their environmental conditions in the form of different abiotic and biotic stresses. Wounding is a common injury in plants that occurs as a result of abiotic stress factors, such as wind, rain and hail, and biotic factors, especially insect feeding. A previous microarray study that focused on the transcriptional profiling of genes after wounding indicated that wounding induces the expression of WRKY family proteins in *Arabidopsis* [16]. In this study, we found that the *GhWRKY40* transcript level is induced by wounding treatment (Fig. 3B) and that *GhWRKY40* overexpression affects the expression of defense-related genes in response to wounding (Fig. 5B). The transcriptional induction of *JAZ1* and *JAZ3* in wounded *GhWRKY40-OE* plants is lower than in wild-type plants, suggesting that *GhWRKY40* negatively controls the expression of certain *JAZ* genes. Jasmonate ZIM-domain (JAZ) genes, key repressors of JA signaling, are primary response genes in the JA signaling pathway. Most members of the JAZ gene family are highly expressed in response to mechanical wounding [30]. The promoters of *JAZ* genes contain several W boxes, which can be bound by WRKY genes. In addition, the overexpression of *GhWRKY40* was also found to significantly inhibit the expression of *PR* genes. *PR*s, defense-related genes often associated with SA-mediated defense responses. Moreover, the expression of *APX*, *SOD* and *GST* is slightly enhanced in response to wounding. *PR*s and *GST* are induced by wounding and have been identified as late response genes [16]. In addition, the ROS levels were lower in OE lines. Thus, the wounding tolerance of OE plants might be correlated with oxidative tolerance. Both wounding and pathogenic attack induce the expression of WRKYs, and the responses to wounding and pathogenic infection in plants share a number of signal transduction pathway components [31]. *PR*s and *JAZ*s have been reported to be such components. The expression of *PR*s and *JAZ*s were found to be significantly inhibited by the overexpression of *GhWRKY40* after *R. solanacearum* infection (Fig. 6C), and lead to the susceptibility of transgenic plants to *R. solanacearum*. Similarly, *AtWRKY40* affects JA signaling by directly controlling the expression of a subset of *JAZ*s upon plant-pathogen interaction, transcriptional reprogramming regulated by *WRKY40* facilitates powdery mildew infection of *Arabidopsis* [32]. Taken together, *GhWRKY40* may be a key component in response to wounding and *R. solanacearum* attack, and *JAZ*s and *PR*s may play roles in *GhWRKY40*-mediated crosstalk between wounding and pathogen defense responses.

As discussed earlier, the signaling molecules SA, JA and ET play important roles in the regulation of the complex defense mechanisms [33]. Previous studies have revealed that responses against biotrophic pathogens are generally regulated by SA, while responses to necrotrophs are mediated by JA and ET [34–35]. SA, JA and ET have been shown to activate different sets of plant *PR* genes and to act either synergistically or antagonistically during defense signaling [36–38]. In numerous plants, the transcription of WRKY genes is strongly and rapidly upregulated in response to pathogen infection or defense-related plant hormones, such as SA and JA. In cotton, *GhWRKY40* was upregulated by the *R. solanacearum*, MeJA and SA (Fig. 3). Moreover, the overexpression of *GhWRKY40* decreased the resistance of transgenic plants to *R. solanacearum* infection, as well as *PR* and *JAZ* gene transcripts (Fig. 6). SA accumulates in infected leaves after infection with biotrophic pathogens and mediates the induced expression of defense genes, resulting in an enhanced state of defense known as systemic acquired resistance (SAR) [39]. *PRs* are often used as molecular markers for SAR. In *Arabidopsis*, the enhanced susceptibility of transgenic plants overexpressing *WRKY8* to *Pseudomonas syringae* was associated with reduced expression of *PR1* [40]. Moreover, *Arabidopsis WRKY8* is a wounding-induced WRKY gene. In addition to their involvement in disease resistance signaling, SA, JA and ET have been reported to be involved in the wounding response [16,30 and 41]. We showed that *GhWRKY40* is transcriptionally inducible by wounding (Fig. 3) and that the overexpression of *GhWRKY40* represses the expression of the SA-dependent genes *PR1a*, *PR2* and *PR4* and the JA-responsive genes *JAZ1* and *JAZ3* upon wounding (Fig. 5). Therefore, we speculate that SA/JA induce *GhWRKY40* expression which leads to the reduction in expression of downstream defense genes.

WRKY TFs exhibit autoregulation and crossregulation activities and also interact with different proteins, such as MAP kinases, to carry out diverse plant functions [8,42]. The last decade of research has shown that MPKs regulate WRKY TFs in response to multiple stresses. In previous studies, we functionally identified two MAPK genes from cotton, *GhMPK6a* [43] and *GhMPK20* (unpublished). The results obtained in the present study indicate that GhWRKY40 interacts with GhMPK20 but not with GhMPK6a (Fig. 7). In *Arabidopsis*, the group II WRKY proteins of WRKY6 and WRKY22 were found to interact with MPK10 and MPK3/MPK6, respectively [44,42]. However, the identification of the upstream components that regulate WRKY TFs is difficult due to cellular interactions, redundancy, and functional pleiotropy. Our results provide valuable information that aids our understanding of the relationship between MAPKs and the WRKY family proteins in cotton and enhances our understanding of the molecular mechanism of signal transduction in cotton plants under stress.

In conclusion, our results suggest that *GhWRKY40* responds to a variety of stresses and that the overexpression of *GhWRKY40* in *N. benthamiana* affects defense-related gene express, enhances the resistance to wounding and the susceptibility to bacterial pathogen. Furthermore, we show that GhWRKY40 interacts with GhMPK20 both *in vivo* and *in vitro*. The elucidation of the regulatory mechanism of *GhWRKY40* overexpression may reveal a converging node in the regulatory pathways involved the plant responses to wounding and pathogenic infection. Understanding the biological function of *GhWRKY40* in cotton enriches our knowledge concerning WRKY function in crops. As we learn more about WRKY regulation, potential applications in genetic improvement should become possible in crops.

Supporting Information

Figure S1 Comparison of the seed germination and post germination of WT and OE plants. (A) Seeds germination phenotype of WT and OE lines on MS medium. (B) The germination rate (greening cotyledon ratio) of the seeds under normal condition. Germination was scored daily. (C) The mass of thousand grains of WT and OE plants. (D) The fresh weight (weight of twenty seedlings) of the seedlings was recorded 10 d after sowing. The data shown indicate the means ± standard errors of three independent experiments. Different letters above the columns indicate significant differences ($P<0.01$) according to Duncan's multiple range test using SAS version 9.1 software.

Figure S2 The full-length cDNA sequence and primers on the sequence of *GhWRKY40*. The primers mentioned in the text were underlined. The initiation codon and termination codon was bold.

Table S1 Polymerase chain reaction amplification conditions.

Table S2 Primers used for gene cloning.

Table S3 The primers used for qPCR.

Author Contributions

Conceived and designed the experiments: XW XG. Performed the experiments: XW YY XC. Analyzed the data: XW. Contributed reagents/materials/analysis tools: YL CW XG. Wrote the paper: XW.

References

1. Mukhopadhyay A, Vij S, Tyagi AK (2004) Overexpression of zinc-finger protein gene from rice confers tolerance to cold, dehydration, and salt stress in transgentic tobacco. Proc Nat Acad Sci U S A 101:6309–6314.
2. Singh K, Foley RC, Onate-Sanchez L (2002) Transcription factors in plant defense and stress responses. Curr Opin Plant Biol 5:430–436.
3. Vinocur B, Altman A (2005) Recent advances in engineering plant tolerance to abiotic stress: achievements and limitations. Curr Opin Plant Biol 16:123–132.
4. Yamaguchi-Shinozaki K, Shinozaki K (2006) Transcriptional regulatory networks in cellular responses and tolerance to dehydration and cold stresses. Annu Rev Plant Biol 57:781–803.
5. Ren X, Chen Z, Liu Y, Zhang H, Zhang M, et al. (2010) ABO3, a WRKY transcription factor, mediates plant responses to abscisic acid and drought tolerance in *Arabidopsis*. Plant J 3:417–429.
6. Eulgem T, Rushton PJ, Robatzek S, Somssich IE (2000) The WRKY superfamily of plant transcription factors. Trends Plant Sci 5:199–206.
7. Agarwal PK, Agarwal P, Reddy MK, Sopory SK (2006) Roles of DREB transcription factors in abiotic and biotic stress tolerance in plants. Plant Cell Rep 25:1263–1274.
8. Rushton PJ, Somssich IE, Ringler P, Shen QJ (2010) WRKY transcription factors. Trends in Plant Science 15:247–258.
9. Birkenbihl RP, Diezel C, Somssich IE (2012) Arabidopsis *WRKY33* is a key transcriptional regulator of hormonal and metabolic responses toward *Botrytis cinerea* infection. Plant Physiol 159(1):266–285.
10. Jiang Y, Deyholos MK (2009) Functional characterization of Arabidopsis NaCl-inducible WRKY25 and WRKY33 transcription factors in abiotic stresses. Plant Mol Biol 69(1–2):91–105.
11. Li S, Fu Q, Chen L, Huang W, Yu D (2011) Arabidopsis thaliana WRKY25, WRKY26, and WRKY33 coordinate induction of plant thermotolerance. Planta 233(6):1237–1252.
12. Dang FF, Wang YN, Yu L, Eulgem T, Lai Y, et al. (2012) *CaWRKY40*, a WRKY protein of pepper, plays an important role in the regulation of tolerance to heat stress and resistance to *Ralstonia solanacearum* infection. Plant Cell Environ 36(4):757–774.
13. Chen W, Provart N, Glazebrook J, Katagiri F, Chang HS, et al. (2002) Expression profile matrix of Arabidopsis transcription factor genes suggests their putative functions in response to environmental stresses. Plant Cell 14:559–574.
14. Maleck K, Levine A, Eulgem T, Morgan A, Schmid J, et al. (2000) The transcriptome of *Arabidopsis thaliana* during systemic acquired resistance. Nat Genet 26:403–409.
15. Schenk PM, Kazan K, Wilson I, Anderson JP, Richmond T, et al. (2000) Coordinated plant defense responses in *Arabidopsis* revealed by microarray analysis. Proc Natl Acad Sci U S A 97:11655–11660.
16. Cheong YH, Chang HS, Gupta R, Wang X, Zhu T, et al. (2002) Transcriptional profiling reveals novel interactions between wounding, pathogen, abiotic stress, and hormonal responses in *Arabidopsis*. Plant Physiol 129(2):661–677.

17. Asai T, Tena G, Plotnikova J, Willmann MR, Chiu WL, et al. (2002) MAP kinase signalling cascade in *Arobidopisis* innate immunity. Nature 415:977–983.
18. Nakagami H, Soukupova H, Schikora A, Zarsky V, Hirt H (2006) A mitogen-activated protein kinase kinase kinase mediates reactive oxygen species homeostasis in *Arabidopsis*. J Biol Chem 281:38697–38704.
19. Ishihama N, Yamada R, Yoshioka M, Katou S, Yoshioka H (2011) Phosphorylation of the *Nicotiana benthamiana* WRKY8 transcription factor by MAPK functions in the defense response. Plant Cell 23:1153–1170.
20. Shen H, Liu C, Zhang Y, Meng X, Zhou X, et al. (2012) OsWRKY30 is activated by MAP kinases to confer drought tolerance in rice. Plant Mol Biol 80(3):241–253.
21. Wang X, Xiao H, Chen G, Zhao X, Huang C, et al. (2011) Isolation of high-quality RNA from *Reaumuria soongorica*, a desert plant rich in secondary metabolites. Mol Biotechnol 48:165–172.
22. Yu F, Huaxia Y, Lu W, Wu C, Cao X, et al. (2012) *GhWRKY15*, a member of the WRKY transcription factor family identified from cotton (*Gossypium hirsutum* L.), is involved in disease resistance and plant development. BMC Plant Biol 12:144.
23. Horsch RB, Rogers SG, Fraley RT (1985) Cold Spring Harbor Symposia on Quantitative Biology. Transgenic plants 50:433–437.
24. Fujita M, Fujita Y, Noutoshi Y, Takahashi F, Narusaka Y, et al. (2006) Crosstalk between abiotic and biotic stress responses: a current view from the points of convergence in the stress signaling networks. Curr Opin Plant Biol 9(4):436–442.
25. León J, Rojo E, Sanchez-Serrano JJ (2001) Wounding signaling in plants. J Exp Bot 52(354):1–9.
26. O'Donnell PJ, Calvert C, Atzorn R, Wasternack C, Leyser HMO, et al. (1996) Ethylene as a signal mediating the wound response of tomato plants. Science 274:1914–1917.
27. Ecker JR (1995) The ethylene signal transduction pathway in plants. Science 268(5211): 667–675.
28. Reymond P, Farmer EE (1998) Jasmonate and salicylate as global signals for defense gene expression. Curr Opin Plant Biol 1: 404–411.
29. Zhang CQ, Xu Y, Lu Y, Yu HX, Gu MH, et al. (2011) The WRKY transcription factor *OsWRKY78* regulates stem elongation and seed development in rice. Planta 234(3):541–554.
30. Chung HS, Koo AJ, Gao X, Jayanty S, Thines B, et al. (2008) Regulation and function of Arabidopsis JASMONATE ZIM-Domain genes in response to wounding and herbivory. Plant Physiol 146(3):952–964.
31. Du L, Chen Z (2000) Identification of genes encoding receptor-like protein kinases as possible targets of pathogen- and salicylic acid-induced WRKY DNA binding proteins in *Arabidopsis*. Plant J 24: 837–847.
32. Pandey SP, Roccaro M, Schön M, Logemann E, Somssich IE (2010) Transcriptional reprogramming regulated by *WRKY18* and *WRKY40* facilitates powdery mildew infection of *Arabidopsis*. Plant J 64(6):912–923.
33. Spoel SH, Johnson JS, Dong X (2007) Regulation of tradeoffs between plant defenses against pathogens with different lifestyles. Proc Natl Acad Sci U S A 104:18842–18847.
34. Vlot AC, Dempsey DMA, Klessig DF (2009) Salicylic acid, a multifaceted hormone to combat disease. Annual Review of Phytopathology 47:177–206.
35. Farmer EE, Alméras E, Krishnamurthy V (2003) Jasmonates and related oxylipins in plant responses to pathogenesis and herbivory. Curr Opin Plant Biol 6:372–378.
36. Leon-Reyes A, Du Y, Koornneef A, Proietti S, Körbes AP, et al. (2010) Ethylene signaling renders the jasmonate response of *Arabidopsis* insensitive to future suppression by salicylic acid. Mol Plant-Microbe Interact 23:187–197.
37. Mur LA, Kenton P, Atzorn R, Miersch O, Wasternack C (2006) The outcomes of concentration-specific interactions between salicylate and jasmonate signaling include synergy, antagonism, and oxidative stress leading to cell death. Plant Physiol 140:249–262.
38. Koornneef A, Pieterse CMJ (2008) Cross talk in defense signaling. Plant Physiol 146: 839–844.
39. Glazebrook J (2005) Contrasting mechanisms of defense against biotrophic and necrotrophic pathogens. Annu Rev Phytopathol 43:205–227.
40. Chen L, Zhang L, Yu D (2010) Wounding-induced *WRKY8* is involved in basal defense in *Arabidopsis*. Mol Plant Microbe Interact 23(5):558–565.
41. Reymond P, Farmer EE (1998) Jasmonate and salicylate as global signals for defense gene expression. Curr Opin. Plant Biol 1:404–411.
42. Popescu SC, Popescu GV, Bachan S, Zhang Z, Gerstein M, et al. (2009) MAPK target networks in *Arabidopsis thaliana* revealed using functional protein microarrays. Genes Dev 1:23(1):80–92.
43. Li Y, Zhang L, Wang X, Zhang W, Hao L, et al. (2013) Cotton GhMPK6a negatively regulates osmotic tolerance and bacterial infection in transgenic *Nicotiana benthamiana*, and plays a pivotal role in development. FEBS J 280(20):5128–5144.
44. Robatzek S, Somssich IE (2002) Targets of *AtWRKY6* regulation during plant senescence and pathogen defense. Genes Dev 16:1139–1149.

6

Genetic and DNA Methylation Changes in Cotton (*Gossypium*) Genotypes and Tissues

Kenji Osabe, Jenny D. Clement, Frank Bedon, Filomena A. Pettolino, Lisa Ziolkowski, Danny J. Llewellyn, E. Jean Finnegan, Iain W. Wilson*

CSIRO, Plant Industry, ACT, Australia

Abstract

In plants, epigenetic regulation is important in normal development and in modulating some agronomic traits. The potential contribution of DNA methylation mediated gene regulation to phenotypic diversity and development in cotton was investigated between cotton genotypes and various tissues. DNA methylation diversity, genetic diversity, and changes in methylation context were investigated using methylation-sensitive amplified polymorphism (MSAP) assays including a methylation insensitive enzyme (*Bsi*SI), and the total DNA methylation level was measured by high-performance liquid chromatography (HPLC). DNA methylation diversity was greater than the genetic diversity in the selected cotton genotypes and significantly different levels of DNA methylation were identified between tissues, including fibre. The higher DNA methylation diversity (CHG methylation being more diverse than CG methylation) in cotton genotypes suggest epigenetic regulation may be important for cotton, and the change in DNA methylation between fibre and other tissues hints that some genes may be epigenetically regulated for fibre development. The novel approach using *Bsi*SI allowed direct comparison between genetic and epigenetic diversity, and also measured CC methylation level that cannot be detected by conventional MSAP.

Editor: Tianzhen Zhang, Nanjing Agricultural University, China

Funding: This study was supported by CSIRO, Plant Industry, Australia. The funder has permitted to publish the manuscript.

Competing Interests: The authors have declared that no competing interests exist.

* E-mail: Iain.Wilson@csiro.au

Introduction

Cytosine methylation is a flexible epigenetic regulatory mechanism that controls gene expression by inhibiting proteins binding to DNA and by changing the structure of the associated chromatin. In plants, DNA methylation can occur on cytosines in any context (CG, CHG and asymmetric CHH, where H is A, C or T) with CG being the most commonly methylated dinucleotide [1,2]. CG and non-CG methylation can silence transposons and pseudogenes, and regulate plant development and tissue specific gene expression [3,4]. CG, CHG and CHH methylation are established through *de novo* methylation dependent on small RNAs, but maintained through different processes [5]. In *Arabidopsis*, CG methylation is maintained by METHYLTRANSFERASE1(-MET1), CHG methylation is maintained by DOMAINS REARRANGED METHYLTRANSFERASE1/2 (DRM1/2) and CHROMOMETHYLASE3 (CMT3), while DECREASE in DNA METHYLATION1 (DDM1) is required for both CG and non-CG methylation [6–10].

Changes in DNA methylation levels between developmental stages or tissues can indicate the involvement of epigenetic regulation. The total DNA methylation level measured by Methylation-sensitive amplified polymorphism (MSAP) in different tissues or developmental stages in maize [11], rice [12], sorghum [13], or Arabidopsis [14] is around 16–40%. Generally, the methylation level increases as the tissue matures [14,15], and endosperm tissue is often hypomethylated [16–19]. In cotton, changes in the levels of DNA methylation between cotyledon, seedling leaf, mature plant leaf, and roots were observed [20–22], but the relative level of DNA methylation in cotton fibre compared to other tissues is not known. Changes in DNA methylation level between developmental stages or tissues suggest that epigenetic regulation may be important in creating phenotypic diversity.

The requirement for DNA methylation in plant development has been demonstrated by the pleiotropic phenotypes observed when the epigenome was disrupted by down-regulating genes such as *DDM1* and *MET1* that are required for DNA methylation [10,23,24], or by chemical treatment [25,26]. Loss of DNA methylation can influence plant traits such as yield, fruit ripening, seed size, flowering time, plant size, plant stature, sex determination, and pathogen resistance [10,27–32]. In cotton, there have been reports of DNA methylation changes related to response to light quality, heterosis, salt-tolerance, alkali stress, and annual habit [20–22,33–35].

Epigenetic changes can occur more frequently than spontaneous genetic mutations [36,37], allowing phenotypic plasticity and divergence. Higher DNA methylation diversity compared to the genetic diversity has been reported previously in *Viola cazorlensis* [38] and *Brassica oleracea* [39], which suggests the potential involvement of epigenetic regulation of phenotypic traits. Cotton has limited genetic diversity [40–43] due to a relatively recent polyploidization event [44] and subsequent domestication, but the extent of DNA methylation polymorphism is greater compared to the genetic polymorphisms in *G. hirsutum* accessions collected from different geographical regions around the world [45]. DNA methylation may be contributing to increased phenotypic diversity in cotton, including in fibre traits that have been selected during domestication and breeding.

Cotton fibres, which are widely used for textile production, are elongated single cell seed trichomes growing from the epidermis of the outer integument of ovules. Cotton ovules have two layers of integuments, outer and inner integuments, which develop into the seed coat after fertilisation. About 30% of the epidermal cells in the outer integument form fibre initials [46], which expand, elongate, and thicken over about 50 days post anthesis (dpa) to form mature fibres. *G. hirsutum* L. and *G. barbadense* are the most common cultivated cotton species. *G. hirsutum* dominates global cotton production being grown for its high fibre yield, whereas *G. barbadense* is grown for its high fibre quality. Both species are allotetraploids (AADD-genome) derived from diploid progenitors similar to present day G. *arboreum* (A-genome) and *G. raimondii* (D-genome) species, and have superior fibre yield and quality relative to their ancestors [44,47]. Complex genomic and epigenomic change affecting gene expression is thought to accompany polyploid formation [48–51], and this is likely to have contributed to the improved fibre traits of the polyploids. Many fibre-related genes change expression during fibre development [52], but it is not known whether these genes are epigenetically regulated.

To understand the potential contribution of DNA methylation regulation to the phenotypic diversity and plant development in cotton, the change in DNA methylation between cotton genotypes and various tissues was investigated. The DNA methylation diversity and genetic diversity was compared using methylation-sensitive amplified polymorphism (MSAP) assays and the total methylation level was measured by high-performance liquid chromatography (HPLC). MSAP analyses demonstrated higher DNA methylation diversity than genetic diversity in the selected cotton genotypes, and significantly different levels of DNA methylation were observed between tissues of the *G. hirsutum* cultivar Coker 315-11.

Materials and Methods

Plant Materials

Gossypium hirsutum L. genotypes Namcala, Delta Pine 16 (DP16), Sicot 75, Sicot 71, Coker 315-11 and three advanced breeding lines CSX6280, CSX5150, and CSX4184, and *G. barbadense* genotypes Sipima 280 and CPX12 were used for genotype comparisons. The genotypes were selected to represent a range of short/long fibre length and weak/strong fibre strength (Fig. S1). *G. barbadense* genotypes were used as a genetic outlier for our analysis. Pedigree analysis (data not shown) shows Namcala as the most distant genotype to other *G. hirsutum* genotypes. DP16 and Coker 315 are closely related, and Sicot 71, Sicot 75, CSX6280, CSX5150, and CSX4184 form a separate group.

Field experiments were grown at the Australian Cotton Research Institute (ACRI) near Narrabri, NSW, Australia (30°S; 150°E). The soil type was heavy grey clay, Vertosol classified as Ug5.2 [53]. Genotypes were planted in three rows by 12 meter plots with three replications. Field experiments were sown in early October 2010 in rows 100 cm apart. Crops were managed with full irrigation, spraying for insect pests as required and weeds controlled by pre-planting application of herbicides such as trifluralin and Fluorometuron followed by inter-row cultivation prior to flowering.

Cotton leaf samples were taken in February 2011 with the crop being near the cutout stage of development. Five fully expanded leaves from twenty individual plants were sampled from the inner row of each plot. The samples were placed immediately in liquid nitrogen, and transferred to −80°C for storage. Care was taken to select leaves at a similar developmental stage to minimize potential epigenetic variability. After harvesting, cotton was ginned on a 20 saw laboratory gin, and fibre quality was analysed with a High Volume Instrument (HVI; Uster technologies Inc., Charlotte, NC) for fibre length and strength.

A separate glasshouse experiment was performed from July to September 2011 for tissue analyses in Coker 315 grown at Canberra, ACT, Australia. Three-week-old plants were used to collect cotyledon, stems and total root tissues, while 6-month-old mature plants were used to collect fully expanded (mature) leaves, primary roots, 0 dpa and 3 dpa ovules, and 35 dpa fibre (manually separated from seeds). The outer integument (OI) and inner integument (II) were dissected from 0 dpa ovules harvested between 13:00–15:00 [54]. Two flowers were combined as one replicate for OI and II, and for each tissue, three or four biological replicates were collected from individual plants.

DNA Extraction

All DNA extraction was performed using modified a DNeasy mini plant DNA extraction kit (Qiagen, Melbourne, Australia). DNA extraction for genotype comparisons was performed by adding polyvinylpyrrolidone (20 $mgml^{-1}$) to Buffer AP1, and followed the manufacturer's instructions. DNA extraction from tissues was performed by adding polyvinylpyrrolidone (20 $mgml^{-1}$) to buffer AP1, and including a 10 minute incubation with 10 mM dithiothreitol, and 0.5 $mgml^{-1}$ Proteinase K (final concentration) at 65°C, after the RNase A incubation step, and followed the manufacturer's instructions. DNA quality was determined by the 260/280 ratio from the Nanodrop spectrophotometer (Thermo Scientific, Melbourne, Australia) and visual integrity of the DNA bands by gel electrophoresis in 1.2% TAE agarose.

HPLC

DNA digest for quantifying methylcytosine was performed as previously described [55], and the digested DNA was separated using a reverse-phase HPLC [56] with modifications. Modifications were made on the HPLC run as follows: Hold at methanol/50 mM KH2PO4 [2.5/97.5](v/v) for 5 min, linear gradient to methanol/KH2PO4 [25/75] over 8 min, and linear gradient back to methanol/KH2PO4 [2.5/97.5] over 2 min. HPLC System Gold (Beckman Coulter, Sydney, Australia) fitted with ZORBAX Eclipse XBD-C18 column, 4.6×150 mm, 5-micron (Agilent, Sydney, Australia) was used for DNA separation, and absorbance at 280 nm (and reference absorbance at 320 nm) was measured by a diode array detector.

Standards containing cytidine, uridine, guanosine, adenosine, 2′-deoxy cytidine, 5-methyl-2′-deoxycytidine, thymidine, deoxy-guanosine, deoxy-adenosine at concentration of 1 µg/ml each, were used to determine retention times of each nucleoside. The average area under the peak were standardised to six 2-fold dilution series of the standards, and the area of deoxy cytidine (dC) and methyl dC was used to calculate the percentage of methyl dC (%mdC) to the amount of total cytosine (methyl dC+dC). Oligonucleotides (5′-TCGAATTCGGCCATGGCC-GAATTCGA-3′) containing 0%, 28.5%, and 57.1% methylcytosine (substituted at two or four C's with methyldeoxycytosine) were synthesized (Sigma, Sydney, Australia) and used as controls for DNA digest and methylcytosine quantification. Peaks were analysed using System Gold (Beckman Coulter, Sydney, Australia) and Microsoft Excel, and error rates were calculated for DNA extraction, digest, and HPLC runs (Table S1).

MSAP

MSAP was performed for each genotype in three biological replicates [57], with modifications. Modification made were: 500 ng of template DNA, double-digest using combinations of *Eco*RI and *Bsi*SI (Jenabioscience, Jena, Germany), *Hpa*II (New England Biolabs, Arundel, Australia) or *Msp*I (New England Biolabs, Gold Coast, Australia), PCR using FastStart Taq polymerase (Roche diagnostics, Sydney, Australia), fluorescently labelled reactions (FAM, VIC, NED, PET) were mixed, and peaks were separated using an ABI3130×l capillary sequencer (Applied Biosystems, Melbourne, Australia). Traces were analysed using GeneMarker software (SoftGenetics, Pennsylvania, USA). Adaptors and oligonucleotides used are listed in Table S2. The pre-amplification and selective amplification cycling conditions were performed as manufacturer's instruction with 40 cycles for the selective amplification.

Scoring of each CCGG site was automated to assess the presence ("1") or absence ("0") of peaks using GeneMarker. A panel for each oligonucleotide pair was constructed based on all methylation insensitive (*Eco*RI/*Bsi*SI) data of *G. hirsutum* and *G. barbadense* genotypes. The panel was manually refined by selecting the peaks that were strong and consistent in at least two of the three replicates. The panel constructed using methylation insensitive data was applied to all genotype/tissue samples to produce binary data for genetic analysis. To minimize biological and technical scoring errors that occur between each replicate, a consensus score was constructed for each site, producing a single binary data point for each genotype/tissue. All three enzyme combinations recognize the same CCGG site, hence the panels constructed from the methylation insensitive data was applied to the methylation sensitive EcoRI/*Hpa*II and *Eco*RI/*Msp*I data to automate binary data for DNA methylation analysis. The inclusion of *Bsi*SI is a novel modification of the conventional MSAP method that allowed analysis of additional sites that could not otherwise be assessed, and to directly compare the genetic diversity to the DNA methylation diversity at the same CCGG site.

MSAP data analysis. A total of 28 oligonucleotide pairs were used to generate 389 bands that could be scored reliably across the tissues, and 44 primer pairs were used to generate 1120 bands that could be scored reliably across the ten genotypes, with 1084 bands (subtracting the *G. barbadense* specific bands) from just the *G. hirsutum* genotypes. The polymorphism ratio within each *Eco*RI/*Bsi*SI data set was determined by calculating the total number of polymorphic sites within the genotypes divided by the total number of sites analysed. The percentage polymorphism of the methylation sensitive enzyme was calculated by total number of DNA methylation polymorphic sites identified, divided by the total number of sites analysed. The calculated DNA methylation polymorphic sites do not exclude sites that are polymorphic both genetically and by DNA methylation.

DNA methylation level was quantified for each genotype and tissue using the MSAP binary data. Presence of peaks in the *Eco*RI/*Msp*I and absence in *Eco*RI/*Hpa*II was considered CG methylated site, presence of peaks in *Eco*RI/*Hpa*II and absence in *Eco*RI/*Msp*I was considered CHG methylation site, and absence of peaks in both were considered CC methylation (Table S3). However, when CHG is methylated on both strands *Hpa*II and *Msp*I cannot cleave the site, and is represented in the CC methylation (i.e. CC is inclusive of double-strand CHG methylation). The CG, CHG, and CC methylation level was calculated by the number of absent peaks in *Eco*RI/*Hpa*II or *Eco*RI/*Msp*I divided by the total number of peaks analysed in the *Eco*RI/*Bsi*SI data. Genetically polymorphic sites were excluded in the analysis.

The similarity coefficient (Simple matching), cluster analysis, Mantel's test, and the principal component analysis (PCA) were performed using NTSYS v2.2 [58]. Simple matching method considers the double-absence of peaks as additional information in a pair-wise comparison for closely related species [59]. This is also appropriate for assessing the *G. hirsutum* genotypes as these genotypes are expected to have low heterozygosity, and the presence/absence of the bands are likely due to homology rather than homoplasy (different DNA fragments from different ancestral origin comigrating). Use of Jaccard and Dice coefficients [60,61] for closely related species is commonly used when it is not known whether the double-absence of peaks in pair-wise comparison are due to DNA sequence polymorphism or homoplasy [62]. However, in this study, the methylation insensitive (*Eco*RI/*Bsi*SI) data had bands present across most genotypes and showed that the absence of peaks were not due to sequence polymorphism, and the potential contribution of homoplasy is expected to be very small. Absence of bands in pair-wise comparison of genotypes in both *Eco*RI/*Hpa*II and *Eco*RI/*Msp*I data indicates that this region is likely to share the same methylation state. Unweighted pair group method with arithmetic mean (UPGMA) dendrogram was constructed using the simple matching similarity coefficient and DendroUPGMA [63]. Principal component analysis (PCA) was performed using a correlation matrix of the polymorphic fragments of the genotypes as previously described [45], and used to visualize the relatedness of the genotypes using the "PCA batch module" of NTSYS v2.21 to represent the relatedness between each sample by its spacial distance.

Statistical Analysis

Significant differences between each sample's methylation level determined by HPLC or MSAP were analysed by One-way ANOVA and Tukey's test using BrightStat [64]. The dendrograms generated from MSAP were supported by Mantel's test with 1000 permutations (Table S4) and Bootstrap test with 2000 permutations (95% accuracy) [65] using NTSYS v2.21 and Winboot [66], respectively. The average error rate per locus was calculated for each tissue/genotype and each MSAP enzyme combination [67] and are included in Table S4.

Comparison of Fibre Phenotypic Diversity to MSAP Diversity

The Euclidean distance of fibre length and strengths was calculated using NTSYS v2.21 as a measure of distance between each genotype. The computed Euclidean distance matrix and the dissimilarity matrix computed from the simple matching correlation coefficient were assessed for correlation using Mantel's test with 1000 permutations (Table S5).

Results

DNA Methylation Analysis of Tissues

HPLC and MSAP assays were used to monitor DNA methylation level and context in different tissues of cotton that were grown in the glasshouse. DNA methylation level of tissues harvested from 3-week-old plantlet (cotyledon, stems, roots) and 6-month-old mature plant (mature leaves, stem internode, mature roots, 0 dpa ovules, 3 dpa ovules, 35 dpa fibres, outer integument and inner integument of 0 dpa ovules) were assessed by HPLC (Figure 1). The 0 dpa ovules, 3 dpa ovules, and 35 dpa fibres represent fibre initiation, early elongation, and late elongation stage [47,68], respectively. Both inner integument and outer integument are derived from the ovule primordium, and the inner integument develops more slowly and independently of the outer integument [69,70]. The outer integument gives rise to fibre initials (visible at 0 dpa) that develop into fibres and the remaining epidermal cells form the epidermis of the seed coat. The separation of fibre initials from surrounding epidermal cells is difficult, and recently, a method has been developed to isolate the outer integument of ovules that enriches for cells (~30% of epidermal cells) that produce fibre [54]. The comparison of the fibre forming cells on the outer integument to the non-fibre forming inner integument may be useful to understand the changes during fibre development.

Plantlet roots had the lowest percentage of total methylation (17%mdC) and cotyledon, stem internodes, mature leaves, 35 dpa fibre, and inner and outer integument had the highest methylation (~23–25%mdC). DNA methylation levels in plantlet cotyledon, stems and roots were significantly different from each other, and the level of methylation in cotyledons was comparable to that of mature leaves and stem internodes of mature plants. Plantlet roots had lower methylation (17%mdC) than mature roots (20%mdC). Both 0 dpa ovules and 3 dpa ovules possessed significantly lower methylation (20%mdC) than the outer and inner integument harvested from ovules at 0 dpa (23%mdC). The 35 dpa fibre (~23%mdC) was comparable to 0 dpa outer and inner integument.

A subset of tissues used in HPLC analysis was selected for MSAP analysis to investigate the methylation context in fibre-developing tissues relative to other tissues (Figure 2). MSAP assays examine the context of DNA methylation using the methylation-sensitive isoschizomers *Hpa*II and *Msp*I, allowing discrimination between CG and CHG methylation. The addition of *BsiSI* (a methylation insensitive isoschizomer of *Hpa*II and *Msp*I) permitted the identification CC methylation of the CCGG sites that were not cleaved by either *Hpa*II or *Msp*I. Methylation was divided into three categories, scored as CG, CHG and CC methylation according to MSAP data (Table S3). Significant differences were found only in the CC methylation context (Figure 2). Lowest CC methylation was found in 3 dpa ovules and the highest were 35 dpa fibre and 0 dpa inner integument.

To visualise the relationship of CG and CHG methylation between different tissues, the MSAP data were used to construct a dendrogram (Fig. S2). The overall relationship between the tissues for CG and CHG was similar, indicating a potential relationship in the CG and CHG methylation pattern in different tissues.

The number of methylation polymorphic sites was determined for outer integument, inner integument, and 35 dpa fibre to identify any DNA methylation changes that occur between these tissues during fibre development. Using *Bsi*SI, 389 fragments were detected; of these 28 CG (7.2%) and 23 CHG (5.9%) methylation polymorphisms were found between outer and inner integuments. There were 55 CG (14.1%) and 56 CHG (14.4%) methylation polymorphisms between outer integument and 35 dpa fibre fragments, and fibre had the most number of unique polymorphisms amongst the three tissues (Fig S3).

DNA Methylation Comparison between Genotypes

HPLC. The methylation level of DNA for two biological replicates was measured by HPLC for each cotton genotype

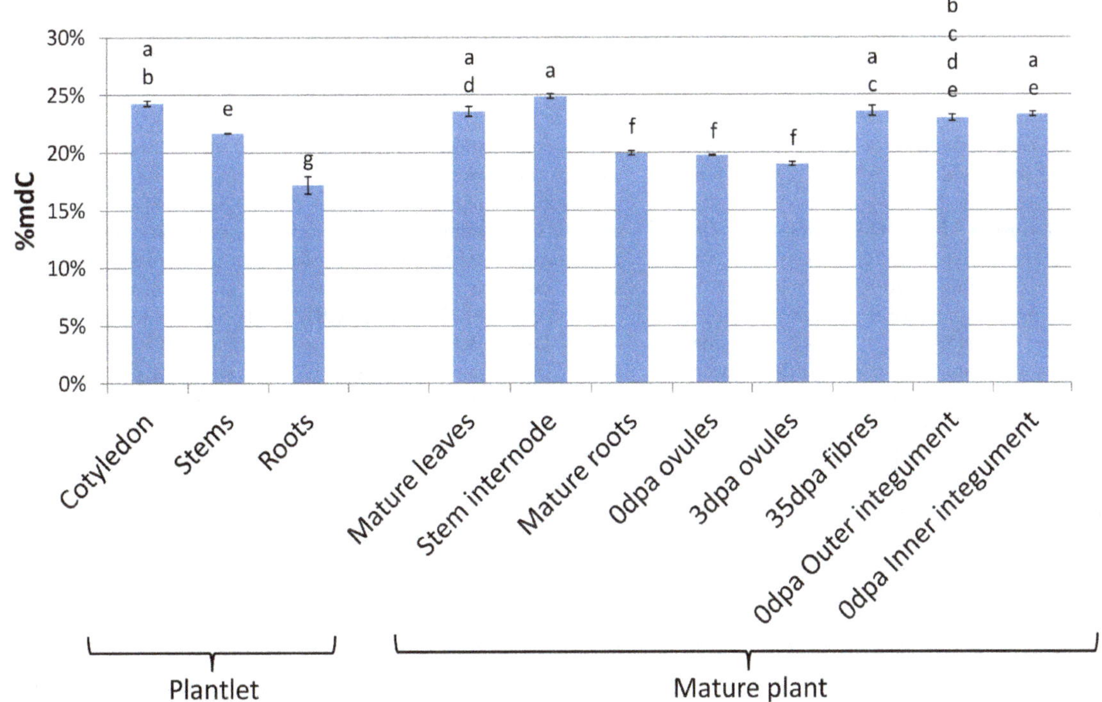

Figure 1. Percentage of total methylation (%mdC) level of Coker 315-11 tissues determined by HPLC. Different letters above the bars denotes samples that have significantly different levels of methylation (p-value of <0.05). The error bars represent the standard error of the mean. Different tissues from 3-week-old plantlet and 6-months-old mature plant were used.

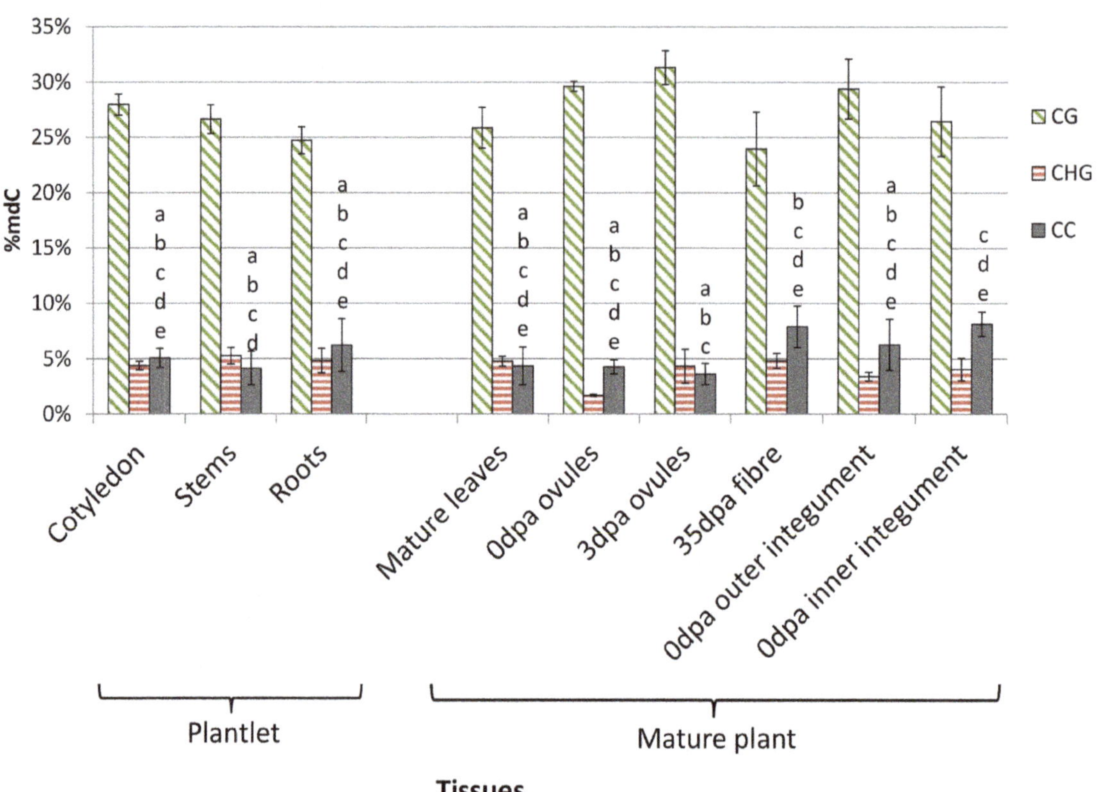

Figure 2. Methylation level of Coker 315-11 tissues determined by MSAP. Methylation was categorized into CG, CHG and CC methylation and represented as a bar graph. Significant differences are denoted by different letters for CC methylation. The error bars represent the standard error of mean from at least three biological replicates. No significant differences for CG and CHG methylation were found.

(Figure 3). There were no significant differences of %mdC between genotypes within *G. hirsutum*, and the average methylation level between species was very similar (24.8% and 24.2%) for *G. hirsutum* and *G. barbadense*, respectively.

Level of DNA methylation by MSAP (CG, CHG, CC). The ten genotypes, including both *G. hirsutum* and *G. barbadense*, were selected to represent diversity of fibre length and strength (Fig. S1). The changes in DNA methylation context between the ten genotypes were measured by MSAP to compare the genetic diversity and fibre quality diversity. The use of *Bsi*SI allowed a direct comparison between genetic and DNA methylation diversity of the selected cotton genotypes.

The extent of methylation at CCGG sites was determined using MSAP data (Figure 4). The average CG, CHG, and CC methylation level of the eight *G. hirsutum* genotypes were 37.8%, 5.2%, and 6.7% (49.7% total methylation), respectively. The average CG, CHG, and CC methylation level for the two *G. barbadense* genotypes was comparable to that of *G. hirsutum*. Within *G. hirsutum*, the total methylation of CCGG sites assessed was 7.5–8 percentage points different between Sicot 75 and CSX5150/CSX6280 ($p<0.05$). The amount of CHG methylation differed between Sicot 75 and CSX4184/CSX5150/CSX6280/Namcala ($p<0.05$), but no significant differences were observed across *G. hirsutum* and *G. barbadense*.

Diversity analysis by Simple Matching (SM) coefficient. The proportion of polymorphic sites (number of polymorphic sites within the total number of assessed bands) for all ten genotypes and within *G. hirsutum* genotypes were calculated for each of the enzyme combinations (Table S6) Within *G. hirsutum* genotypes, the number of CHG polymorphisms was about 1.5-fold and CG polymorphism was 3-fold more than the genetic polymorphism. The similarity coefficient determined using the simple matching method for the ten genotypes averaged 0.878 (range = 0.729–0.976) for *Eco*RI/*Bsi*SI, 0.877 (range = 0.797–0.937) for *Eco*RI/*Hpa*II, and 0.837 (range = 0.709–0.945) for *Eco*RI/*Msp*I. Considering the *G. hirsutum* genotypes only, the similarity coefficient averaged 0.932 (range = 0.85–0.975) for *Eco*RI/*Bsi*SI, 0.9 (range = 0.834–0.935) for *Eco*RI/*Hpa*II, and 0.88 (range = from 0.768–0.945) for *Eco*RI/*Msp*I. As expected, *G. barbadense* genotypes were genetically more distant than any of the *G. hirsutum* genotypes. Irrespective of whether they were compared across all genotypes or within *G. hirsutum*, the genetic diversity was very low and the DNA methylation diversity was greater than the genetic diversity. Comparing CG and CHG methylation, the CHG methylation was more diverse for both species.

Simple matching similarity coefficients were used to construct dendrograms to represent the relationship between genotypes and their DNA methylation state (Figure 5). Genotypes methylation analysis formed four clades, three clades differentiated by the genetic or DNA methylation state and one differentiating *G. barbadense* and Coker 315-11. Coker 315-11 is the most distant

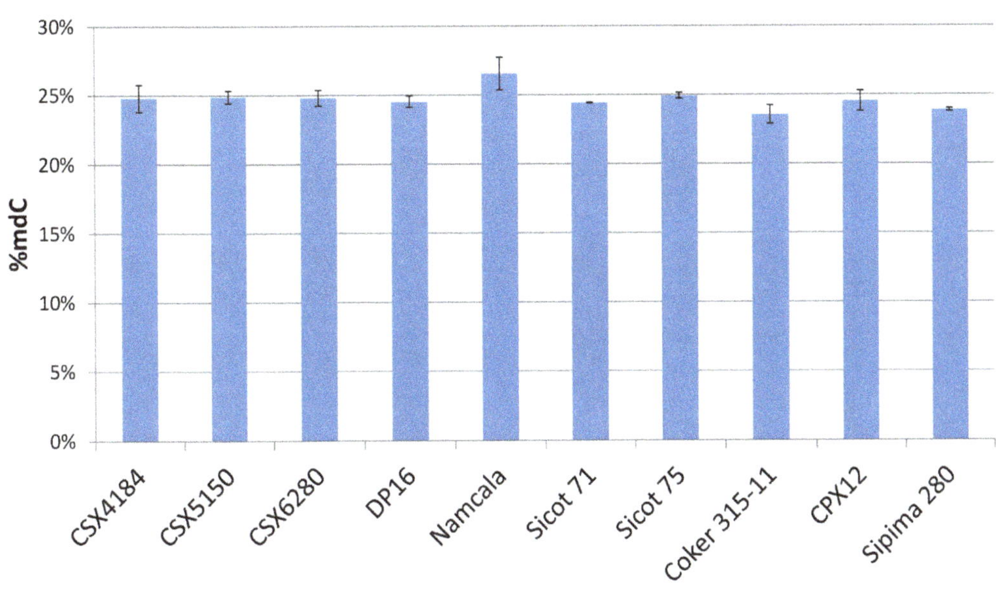

Figure 3. Methylation level of leaves for each genotype quantified by HPLC. The average global methylation level measured by HPLC for *G. hirsutum* and *G. barbadense* were 24.8% and 24.2%, respectively. The error bars represent the standard error of mean calculated from two biological replicates. No significant differences were identified between genotypes.

genotype both genetically and in DNA methylation to the other *G. hirsutum* genotypes. The dendrogram relationship pattern of the *G. hirsutum* genotypes (except Coker 315-11) differs within each genetic or DNA methylation clade, indicating that the genetic relation between each genotype is distinct from the DNA methylation relation.

The genetic and DNA methylation dendrograms were used to compare with the fibre length or strength-based dendrograms to assess whether DNA methylation was contributing more than the genetic component to fibre quality. There were no statistically significant relations between fibre lengths to the genetic or DNA methylation diversity. Weak but statistically significant positive correlation at $p<0.05$ level between fibre strength to the genetic and DNA methylation diversity was found, but the DNA methylation was no more correlated to fibre strength than the genetic component.

Visualizing diversity by Principal Component Analysis (PCA). Principal component analysis (PCA) was performed to visualise the relative distance of each genotype, the genetic and DNA methylation state, using the similarity coefficient (Figure 6). The first and second dimension contributes 27.3% and 14.3% (cumulates to 41.6%), respectively, of the relationship in the multivariate space. CHG methylation was closer to the genetic similarity of *G. hirsutum* genotypes, and CG methylation was distant to both genetic and CHG methylation. Coker 315-11 and the two *G. barbadense* genotypes were distant from the other genotypes in both their genetic and methylation relationships.

Bootstrap analysis indicated that the divergence of Coker 315-11 and the two *G. barbadense* genotypes (CPX12, and Sipima 280) from the others are statistically significant ($p<0.05$) in their genetic relationship. Within the *G. hirsutum* only, the divergence of Coker 315-11 and other *G. hirsutum* genotypes was statistically significant for genetic, CG, and CHG methylation relationships (p-value of <0.05).

Discussion

Epigenetic regulation is known to be involved in some traits in cotton [20–22,33,34], and has the potential to create phenotypic diversity that can improve agronomical performance. Spontaneous DNA methylation changes resulted in epigenomic divergence over as little as 30 generations in *Arabidopsis thaliana* [36,37], suggesting that the 1–2 million years since allopolyploidization of cotton [44] would be sufficient to allow significant genetic, epigenetic, and phenotypic divergence between cotton species. We also found changes in absolute DNA methylation levels between various tissues, including fibre, but no significant difference was found between DNA methylation in leaf tissues across the ten genotypes representing a range of fibre length and strength. By contrast, MSAP analyses showed that DNA methylation diversity of the ten genotypes was higher than the genetic diversity. The higher level of DNA methylation diversity may change gene expression, adding to the higher phenotypic diversity that cannot be explained by the limited genetic diversity in cotton [41–43,71,72]. Our results demonstrate that epigenetic regulation is likely to be involved in cotton development and phenotypic diversity, and therefore has potential value for improving agronomic performance.

A unique feature of the MSAP method presented for the first time in this study is the inclusion of the enzyme, *BsiSI*, which is an

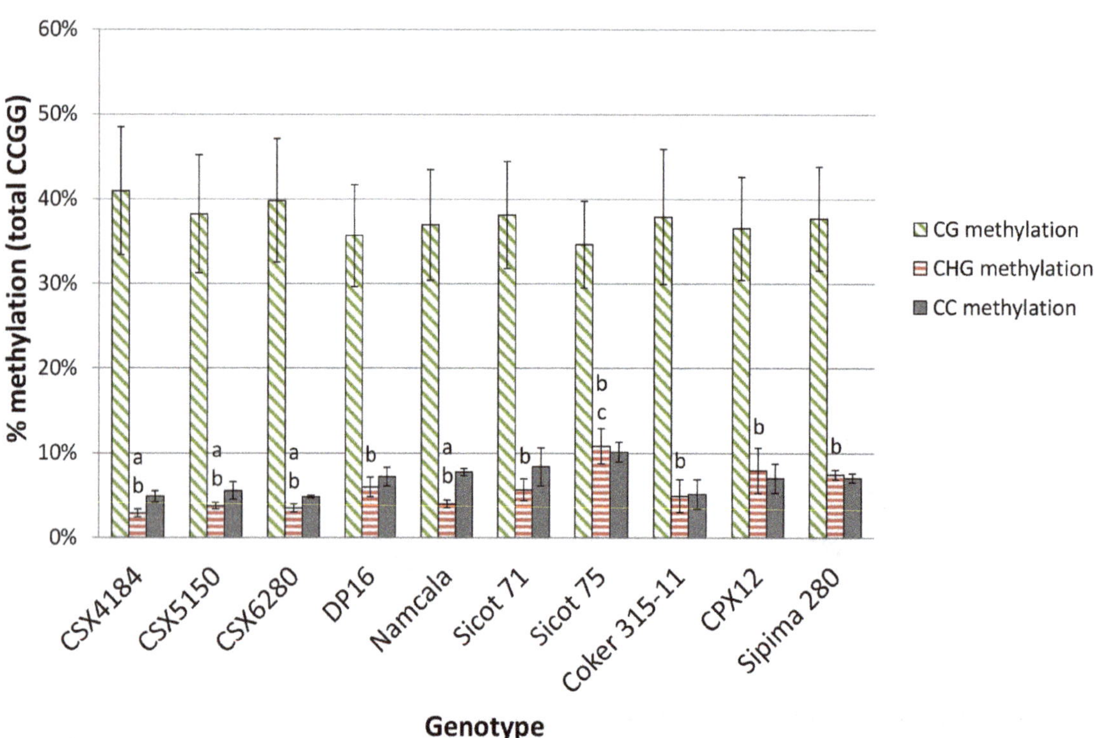

Figure 4. Methylation level of cotton genotypes determined by MSAP. The average genome-wide methylation level measured by MSAP for *G. hirsutum* and *G. barbadense* was 49.7% and 51.9%, respectively. No significant differences in CG or CC methylation were found between genotypes. Significant differences found in CHG methylation is denoted by different letters.

isoschizomer of *Hpa*II and *Msp*I that recognises CCGG sequence, but is not methylation sensitive. With the aid of *Bsi*SI, the fully methylated sites containing both CG and CHG methylation have been classified as CC (also includes double-strand CHG methylation, because *Hpa*II cannot cleave CCGG methylated on the outer C on both strands), adding further sites that could not be assessed by the conventional MSAP method. More importantly, the use of *Bsi*SI allowed us to directly compare the genetic diversity to the DNA methylation diversity at the same CCGG site. The genetic diversity of *G. hirsutum* genotypes was very low, as expected in this species that has gone through a number of genetic bottlenecks during polyploidization, domestication and modern breeding [41–43], but the CG and CHG methylation diversity was always higher than the genetic diversity. Notably, the CHG methylation polymorphism was less frequent (thus less differentiated from the genetic polymorphisms) but occurred more randomly across the genotypes for each site, leading to high diversity and provides some evidence for possible small RNA mediated regulation of phenotypes. Consistent with this, CHG methylation can be guided and maintained by small RNAs, CMT3 and DRMs [73], and small RNAs have been shown to be involved in fibre development and altered by viral infection during fibre development in cotton [74–78].

Although no differences were observed for total DNA methylation level measured by HPLC, the pattern of DNA methylation, as determined by MSAP was different between some genotypes. HPLC was technically more accurate than MSAP in determining the total methylation level, probably due to technical errors of the MSAP method [67,79]. However, while HPLC quantifies the methylated cytosine content of DNA, it cannot distinguish between the different methylation contexts. MSAP can measure CG and CHG methylation context changes, but generally underestimates the methylation level as it: does not detect hypermethylated sites, and the dominant nature of the AFLP detects mixed methylated/non-methylated loci as non-methylated site. Despite this, the total methylation level estimated by MSAP (50%) was approximately double the HPLC measurement. Similar results were seen in *Brassica oleracea* where more than 3-times global methylation was detected by MSAP compared to HPLC [39]. The difference between HPLC and MSAP may partly be caused by the bias of assessing more methylated region of the genome (CCGG sites) in the MSAP method, where CG methylation is the most common context of DNA methylation [1,2], whereas the HPLC method assesses the total methylated cytosines of the genome (including the non-methylated organellar genomes).

Higher DNA methylation diversity than the genetic diversity has also been demonstrated in another study where cotton plants from different geographic regions were sampled to monitor CG methylation patterns using conventional MSAP [45]. In this study of 20 *G. hirsutum* accessions grown in different geographical regions, the level of CG methylation polymorphism was 67%. This was somewhat higher than we observed in the eight *G. hirsutum* genotypes (59.2% CG polymorphism), although the reasons for

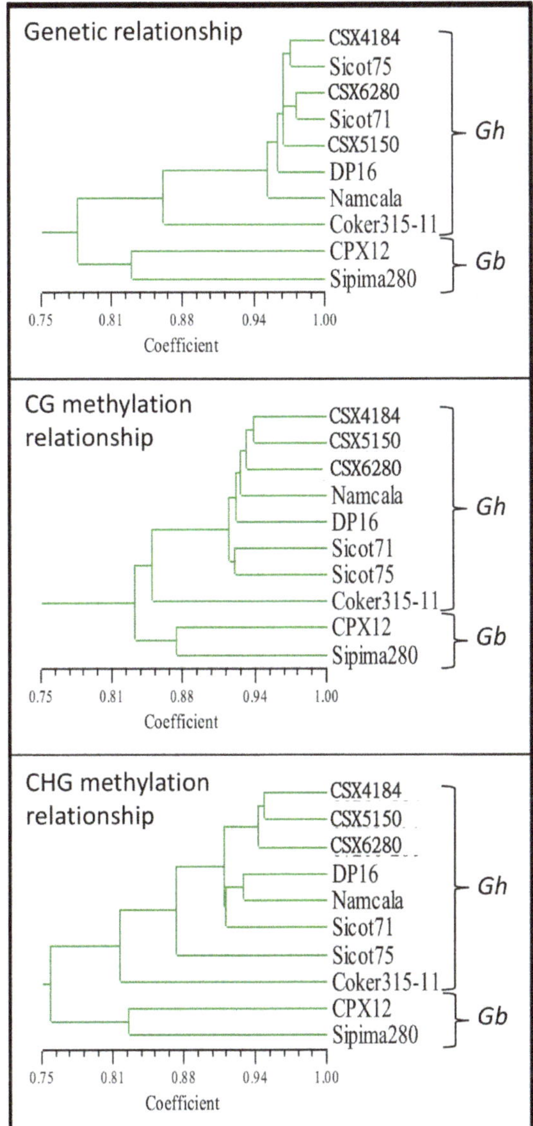

Figure 5. Dendrogram of the ten genotypes constructed using the similarity coefficient. *G. hirsutum* species are indicated by "*Gh*" and *G. barbadense* species are indicated by "*Gb*". The separation of species and Coker 315-11 from other *G. hirsutum* genotypes is consistent in all dendrograms, but the relationships between G. hirsutum genotypes are different. Separation of genetic, CHG and CG methylation clusters in *G. hirsutum* genotypes (except Coker 315-11) show clear genetic/DNA methylation divergence in cultivated cotton.

this are unclear. Our study demonstrated that the DNA methylation diversity remains high even in cotton genotypes that were grown in the same environment over many generations. Similar results have been reported in other cultivated plants that suggest the involvement of epigenetic variation compensating for the lack of genetic variation [80,81].

G. hirsutum (CSX4184, CSX5150, CSX6280, DP16, Namcala, Sicot 71, Sicot 75, and Coker 315) and *G. barbadense* (CPX12, and Sipima 280) total DNA methylation level determined by MSAP were 43% and 44.8% (CG and CHG, without CC methylation), respectively, and falls within the DNA methylation level range (16–60%) for other plant species [39,48,82,83]. A large range in total methylation level in leaf DNA has been reported from other studies of *G. hirsutum* (19–37%), and the total methylation level measured in our study was about 6 percentage points higher than the upper level of these studies [20,21,33,45]. The higher methylation level may result from the difference of technical errors, genotypes and/or environmental conditions used.

The level and context of DNA methylation of selected cotton genotypes and various tissues of Coker 315–11 were assessed using HPLC and MSAP. Higher methylation in mature tissues compared to developing tissues has been reported in other plant species [84–87], and this was also observed in cotton stems and roots, perhaps reflecting the accumulation of methylation over time. Care is needed in the interpretation of methylation changes between tissues as the number of plastids can vary depending on the tissue [88], and plastid DNA is generally not (or very lowly) methylated [89,90], leading to underestimation of the total methylation level for plastid rich tissues (e.g. leaf). Isolation of nuclei is more accurate for quantifying DNA methylation level, but is technically difficult especially in fibre where limited amount of tissue was available (e.g. only small amount of outer and inner integument material can be obtained from each flower).

The significant increase of total DNA methylation in 35 dpa fibre compared to 0 dpa, 3 dpa ovules, stems and roots indicates a possible involvement of epigenetic regulation during fibre development. The relatively low DNA methylation level in 0 dpa ovules compared to the dissected outer and inner integuments suggests that nucellar tissues are hypomethylated, causing an overall decrease in ovule DNA methylation level. By 3 dpa, the nucellus appears smaller and the endosperm is undergoing rapid development [91]. The endosperm of Arabidopsis and rice is known to be hypomethylated [16,19], which may also decrease the total DNA methylation level of 3 dpa ovules relative to other tissues.

There were no statistical differences between the total DNA methylation level between the outer integument and 35 dpa fibres, but MSAP analyses showed that there was considerable methylation polymorphism (97 loci). This contrasts strongly with the similarity of methylation level and fewer polymorphism seen between inner and outer integument (31). The polymorphism between 35 dpa fibre and outer integument may be associated with genes involved in fibre development, and the polymorphism between outer integument to inner integument may represent candidate loci involved in fibre initiation that are epigenetically regulated. However, as the outer integument consists of both fibre initials (about 30%) and epidermal cells, changes in methylation may not necessarily (only) be associated with fibre development. Nevertheless, the change in DNA methylation between fibre and other tissues hint that some genes may be epigenetically regulated for fibre development, supported by other studies that show potential involvement of small RNA directed DNA methylation [35,76,78]. Sequencing of differentially amplified fragments may provide further insight into the role of DNA methylation and gene expression during these fibre development stages.

DNA methylation changes during fibre development have shown the potential involvement of epigenetic regulation that may influence fibre quality. However, the DNA methylation pattern (of leaves) did not show any more correlation to fibre length and fibre strength, compared to the genetic pattern. The number of cotton genotypes that was assessed was too small to identify an association between DNA methylation and fibre quality. It is important to note that DNA methylation is only one level of multi-layered epigenetic regulation (such as histone modifications), and the DNA methylation diversity measured in this study may be underestimating the epigenetic diversity. Further work investigating DNA

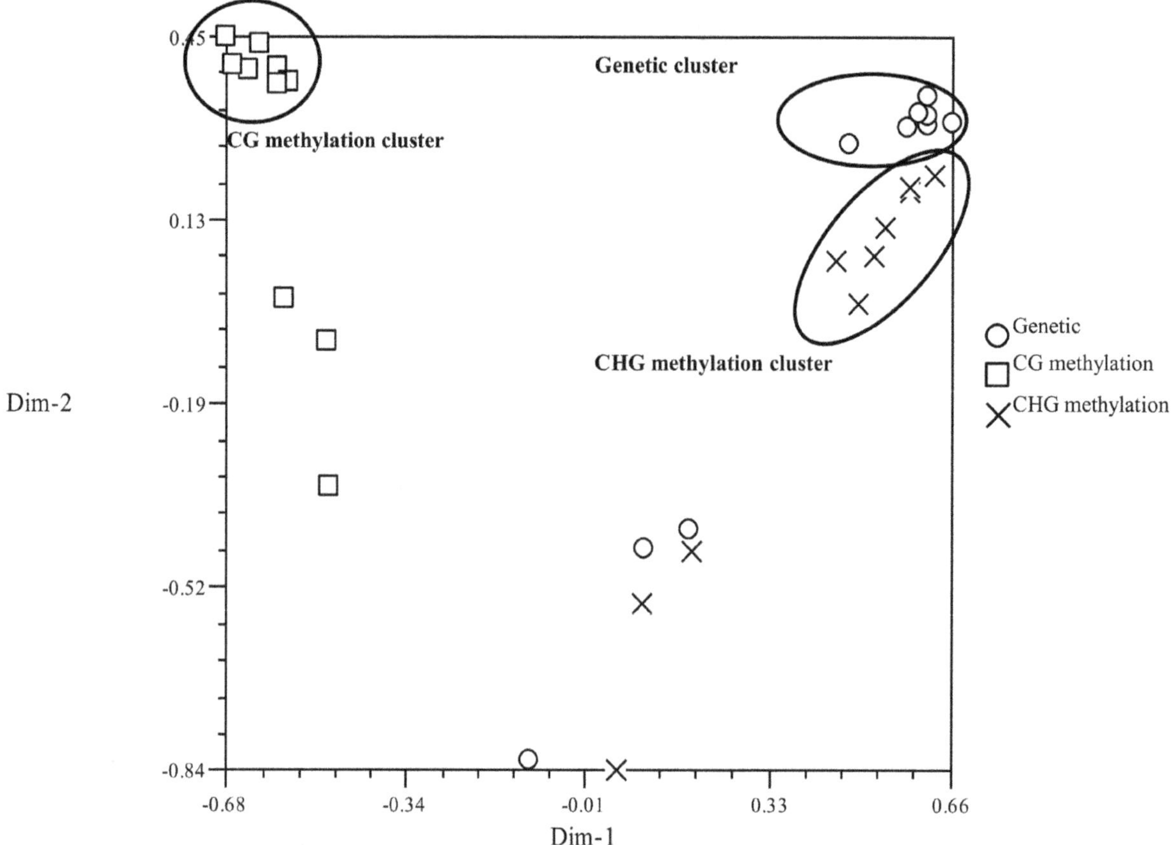

Figure 6. PCA of the ten genotypes for each enzyme combinations plotted in two dimensions. The spatial distance on the graph represents the genetic/DNA methylation relationship between each genotype. Genetic and DNA methylation relationship between the genotypes is more distant between the genetic and DNA methylation state forming genetic, CG methylation, CHG methylation clusters. The CHG methylation and genetic component are closely related, whereas the CG methylation is more distant. The *G. barbadense* genotypes (CPX12 and Sipima 280) and Coker 315-11 are outliers that are distant from the three clusters, genetically and epigenetically.

methylation and other epigenetic changes during different fibre developmental stages or fibre cells across multiple cultivars may provide an understanding of the epigenetic regulation of fibre traits. The high DNA methylation polymorphism may provide sufficient diversity for epigenome based breeding, even in crops with limited genetic diversity, with further investigation to link the polymorphisms to traits of interest.

Supporting Information

Figure S1 Statistically significant differences are denoted by different letters. Fibre length and strength of the ten genotypes measured using HVI. Both graphs are arranged in ascending order. The *G. hirsutum* genotypes represent a range of fibre lengths and strengths, and *G. barbadense* represents longer and stronger fibre compared to *G. hirsutum.*

Figure S2 Dendrogram representing the relationship between tissues and CG/CHG methylation. "EH" represents *Eco*RI/*Hpa*II, and "EM" represents *Eco*RI/*Msp*I. The tissues are each represented by; 0 dpa = 0 dpa ovules, 3 dpa = 3 dpa ovules, Cot = Cotyledons, St = plantlet stems, RT = plantlet roots, ML = mature (fully expanded) leaf from mature plant, OI = 0 dpa ovule outer integument, II = 0 dpa ovule inner integument, and 35F = 35 dpa fibres. Mantel's test supports the reliability of the dendrogram (r = 0.985). The error rate for each tissue comparison was determined by the number of absent peaks in either of the tissue in the *Eco*RI/*Bsi*SI data (i.e. all tissues should be genetically identical). Comparison of Outer integument-35 dpa fibre had 1.8% error rate, and Outer integument-Inner integument had 2.57% error rate.

Figure S3 Venn diagram representing the number of CG and CHG polymorphic fragments that are unique to each tissue and the number of polymorphic fragments that are shared between tissues. The 35 dpa fibre was unique with the highest numbers of specific polymorphism between the three tissues, for both CG and CHG methylation context.

Table S1 Error rate of HPLC method. The standard error of mean of different steps of the HPLC method was evaluated from six DNA extraction replicates, two sets of four DNA digest replicates, five HPLC run replicates (using the same DNA sample), and four day-to-day independent runs of standards. The %mdC and standard error of mean was calculated for each trial. The highest variation was observed in the day-to-day run with a standard error of mean at +/−0.54%.

Table S2 List of oligonucleotides used for MSAP. Selective oligonucleotides with fluorescent labels indicated beside primer name (FAM, VIC, NED or PET).

Table S3 Classification of methylation type for MSAP. Example of four types of methylation classification and the possible polymorphisms is represented by comparing two genotypes. "1" represents presence of bands and "0" represents absence of bands. Example of determining the methylation state is shown from "Cultivar A" and the polymorphism determined from comparing "Cultivar A" and "Cultivar B".

Table S4 Statistical validation of dendrogram and calculated error rate for MSAP. The r-value was determined for each constructed dendrogram (r-value >0.9 indicates good reliability of data). The average error rate per locus was calculated from the three biological replicates for *Eco*RI/*Bsi*SI, *Eco*RI/*Hpa*II, and *Eco*RI/*Msp*I from all genotypes.

Table S5 Matrix correlation (Mantel's test) between fibre quality and genetic/methylation relationship of cotton genotypes. When comparing the correlation coefficient between matrices with n = 10, coefficient above 0.282 is statistically significant at the 5% level and 0.445 at the 1% level (Lapointe & Legendre, 1992).

Table S6 Percentage of polymorphisms identified in each enzyme combination. The percentage represents the number of polymorphic sites within the total number of sites analysed. The percentage polymorphism in the *Eco*RI/*Hpa*II and *Eco*RI/*Msp*I does not include the polymorphic sites identified in *Eco*RI/*Bsi*SI.

Acknowledgments

We would like to acknowledge Walter Tate for his guidance during the optimization of the MSAP protocol. We would like to thank Dr. Shiming Liu, Dr. Greg Constable, and Dr. Elizabeth Dennis for their advices and valuable discussions throughout this study. Kenji Osabe was supported by CSIRO Office of Chief Executive (OCE) postdoctoral fellowship scheme.

Author Contributions

Conceived and designed the experiments: KO EJF DJL IWW. Performed the experiments: KO JDC LZ. Analyzed the data: KO. Contributed reagents/materials/analysis tools: KO JDC LZ FB FAP. Wrote the paper: KO EJF DJL IWW.

References

1. Cokus SJ, Feng S, Zhang X, Chen Z, Merriman B, et al. (2008) Shotgun bisulphite sequencing of the Arabidopsis genome reveals DNA methylation patterning. Nature 452: 215–219.
2. Lister R, O'Malley RC, Tonti-Filippini J, Gregory BD, Berry CC, et al. (2008) Highly integrated single-base resolution maps of the epigenome in Arabidopsis. Cell 133: 523–536.
3. Schöb H, Grossniklaus U (2006) The First High-Resolution DNA "Methylome". Cell 126: 1025–1028.
4. Zhang M, Kimatu JN, Xu K, Liu B (2010) DNA cytosine methylation in plant development. Journal of Genetics and Genomics 37: 1–12.
5. Hauser M-T, Aufsatz W, Jonak C, Luschnig C (2011) Transgenerational epigenetic inheritance in plants. Biochimica et Biophysica Acta (BBA) - Gene Regulatory Mechanisms 1809: 459–468.
6. Henderson IR, Jacobsen SE (2007) Epigenetic inheritance in plants. Nature 447: 418–424.
7. Chan SWL, Henderson IR, Jacobsen SE (2005) Gardening the genome: DNA methylation in Arabidopsis thaliana. Nature reviews Genetics 6: 351–360.
8. Vongs A, Kakutani T, Martienssen RA, Richards EJ (1993) Arabidopsis thaliana DNA methylation mutants. Science (New York, NY) 260: 1926–1928.
9. Kakutani T, Kato M, Kinoshita T, Miura A (2004) Control of development and transposon movement by DNA methylation in Arabidopsis thaliana. Cold Spring Harbor Symposia on Quantitative Biology 69: 139–143.
10. Finnegan EJ, Peacock WJ, Dennis ES (1996) Reduced DNA methylation in Arabidopsis thaliana results in abnormal plant development. 93: 8449–8454.
11. Lu Y, Rong T, Cao M (2008) Analysis of DNA methylation in different maize tissues. Journal of Genetics 35: 41–48.
12. Xiong LZ, Xu CG, Maroof MAS, Zhang QF (1999) Patterns of cytosine methylation in an elite rice hybrid and its parental lines, detected by a methylation-sensitive amplification polymorphism technique. Molecular and General Genetics 261: 439–446.
13. Zhang M, Xu C, von Wettstein D, Liu B (2011) Tissue-Specific Differences in Cytosine Methylation and their Association with Differential Gene Expression in Sorghum bicolor. Plant physiology.
14. Ruiz-Garcia L, Cervera MT, Martinez-Zapater JM (2005) DNA methylation increases throughout Arabidopsis development. Planta 222: 301–306.
15. Messeguer R, Ganal MW, Steffens JC, Tanksley SD (1991) Characterization of the level, target sites and inheritance of cytosine methylation in tomato nuclear-DNA. Plant Molecular Biology 16: 753–770.
16. Zemach A, Kim MY, Silva P, Rodrigues JA, Dotson B, et al. (2010) Local DNA hypomethylation activates genes in rice endosperm. Proceedings of the National Academy of Sciences of the United States of America 107: 18729–18734.
17. Zhang MS, Yan HY, Zhao N, Lin XY, Pang JS, et al. (2007) Endosperm-specific hypomethylation, and meiotic inheritance and variation of DNA methylation level and pattern in sorghum (Sorghum bicolor L.) inter-strain hybrids. Theoretical and Applied Genetics 115: 195–207.
18. Lauria M, Rupe M, Guo M, Kranz E, Pirona R, et al. (2004) Extensive Maternal DNA Hypomethylation in the Endosperm of Zea mays. Society 16: 510–522.
19. Hsieh T-f (2011) Genome-Wide Demethylation of Arabidopsis endosperm. Science 1451.
20. Zhao Y, Yu S, Xing C, Fan S, Song M (2008) Analysis of DNA methylation in cotton hybrids and their parents. Molecular Biology 42: 169–178.
21. Cao DH, Gao X, Liu J, Kimatu JN, Geng SJ, et al. (2011) Methylation sensitive amplified polymorphism (MSAP) reveals that alkali stress triggers more DNA hypomethylation levels in cotton (Gossypium hirsutum L.) roots than salt stress. African Journal of Biotechnology 10: 18971–18980.
22. Zhao Y-l, Yu S-x, Ye W-w, Wang H-m, Wang J-j, et al. (2010) Study on DNA Cytosine Methylation of Cotton (Gossypium hirsutum L.) Genome and Its Implication for Salt Tolerance. Agricultural Sciences in China 9: 783–791.
23. Jacobsen SE, Sakai H, Finnegan EJ, Cao X, Meyerowitz EM (2000) Ectopic hypermethylation of flower-specific genes in Arabidopsis. Current Biology 10: 179–186.
24. Kakutani T, Jeddeloh JA, Richards EJ (1995) Characterization of an Arabidopsis thaliana DNA hypomethylation mutant. Nucleic acids research 23: 130–137.
25. Bossdorf O, Arcuri D, Richards CL, Pigliucci M (2010) Experimental alteration of DNA methylation affects the phenotypic plasticity of ecologically relevant traits in Arabidopsis thaliana. Evolutionary Ecology 24: 541–553.
26. Amoah S, Kurup S, Lopez CMR, Welham SJ, Powers SJ, et al. (2012) A Hypomethylated population of Brassica rapa for forward and reverse Epi-genetics. Bmc Plant Biology 12.
27. Manning K, Tör M, Poole M, Hong Y, Thompson AJ, et al. (2006) A naturally occurring epigenetic mutation in a gene encoding an SBP-box transcription factor inhibits tomato fruit ripening. Nature genetics 38: 948–952.
28. Martin A, Troadec C, Boualem A, Rajab M, Fernandez R, et al. (2009) A transposon-induced epigenetic change leads to sex determination in melon. Nature 461: 1135–1138.
29. Miura K, Agetsuma M, Kitano H, Yoshimura A, Matsuoka M, et al. (2009) A metastable DWARF1 epigenetic mutant affecting plant stature in rice. Proceedings of the National Academy of Sciences 106: 11218–11223.
30. Hauben M, Haesendonckx B, Standaert E, Van Der Kelen K, Azmi A, et al. (2009) Energy use efficiency is characterized by an epigenetic component that can be directed through artificial selection to increase yield. Proceedings of the National Academy of Sciences of the United States of America 106: 20109–20114.
31. Zhong S, Fei Z, Chen Y-R, Zheng Y, Huang M, et al. (2013) Single-base resolution methylomes of tomato fruit development reveal epigenome modifications associated with ripening. Nat Biotech 31: 154–159.
32. Finnegan EJ, Genger RK, Kovac K, Peacock WJ, Dennis ES (1998) DNA methylation and the promotion of flowering by vernalization. Proceedings of the National Academy of Sciences of the United States of America 95: 5824–5829.
33. Li TC, Fan HH, Li ZP, Wei J, Cai YP, et al. (2011) Effect of different light quality on DNA methylation variation for brown cotton (Gossypium hirstum). African Journal of Biotechnology 10: 6220–6226.
34. Li X-L, Lin Z-X, Nie Y-C, Guo X-P, Zhang X-L (2009) Methylation-Sensitive Amplification Polymorphism of Epigenetic Changes in Cotton Under Salt Stress. Acta Agronomica Sinica 35: 588–596.
35. Jin X, Pang Y, Jia F, Xiao G, Li Q, et al. (2013) A Potential Role for CHH DNA Methylation in Cotton Fiber Growth Patterns. PLoS ONE 8: e60547.

36. Becker C, Hagmann J, Muller J, Koenig D, Stegle O, et al. (2011) Spontaneous epigenetic variation in the Arabidopsis thaliana methylome. Nature 480: 245–249.
37. Schmitz RJ, Schultz MD, Lewsey MG, O'Malley RC, Urich MA, et al. (2011) Transgenerational Epigenetic Instability Is a Source of Novel Methylation Variants. Science 334: 369–373.
38. Herrera CM, Bazaga P (2010) Epigenetic differentiation and relationship to adaptive genetic divergence in discrete populations of the violet Viola cazorlensis. New Phytologist 187: 867–876.
39. Salmon A, Clotault J, Jenczewski E, Chable V, Manzanares-Dauleux MJ (2008) Brassica oleracea displays a high level of DNA methylation polymorphism. Plant Science 174: 61–70.
40. Campbell BT, Williams VE, Park W (2009) Using molecular markers and field performance data to characterize the Pee Dee cotton germplasm resources. Euphytica 169: 285–301.
41. Abdalla AM, Reddy OUK, El-Zik KM, Pepper AE (2001) Genetic diversity and relationships of diploid and tetraploid cottons revealed using AFLP. TAG Theoretical and Applied Genetics 102: 222–229.
42. Wendel JF, Brubaker CL, Percival AE (1992) Genetic diversity in gossypium-hirsutum and the origin of upland cotton. American Journal of Botany 79: 1291–1310.
43. Iqbal MJ, Aziz N, Saeed NA, Zafar Y, Malik KA (1997) Genetic diversity evaluation of some elite cotton varieties by RAPD analysis. Theoretical and Applied Genetics 94: 139–144.
44. Wendel JF, Cronn RC (2003) Polyploidy and the evolutionary history of cotton. Advances in Agronomy 78: 139–186.
45. Keyte AL, Percifield R, Liu B, Wendel JF (2006) Intraspecific DNA methylation polymorphism in cotton (Gossypium hirsutum L.). The Journal of heredity 97: 444–450.
46. Lovell D, Wu Y, White R, Machado A, Llewellyn DJ, et al. (2007) Phenotyping cotton ovule fibre initiation with spatial statistics. Australian Journal of Botany 55: 608–608.
47. Lee JJ, Woodward AW, Chen ZJ (2007) Gene expression changes and early events in cotton fibre development. Annals of botany 100: 1391–1401.
48. Shaked H, Kashkush K, Ozkan H, Feldman M, Levy AA (2001) Sequence elimination and cytosine methylation are rapid and reproducible responses of the genome to wide hybridization and allopolyploidy in wheat. Plant Cell 13: 1749–1759.
49. Kashkush K, Feldman M, Levy AA (2002) Gene loss, silencing and activation in a newly synthesized wheat allotetraploid. Genetics 160: 1651–1659.
50. Wang J, Tian L, Madlung A, Lee H-S, Chen M, et al. (2004) Stochastic and epigenetic changes of gene expression in Arabidopsis polyploids. Genetics 167: 1961–1973.
51. Xu YH, Zhong L, Wu XM, Fang XP, Wang JB (2009) Rapid alterations of gene expression and cytosine methylation in newly synthesized Brassica napus allopolyploids. Planta 229: 471–483.
52. Mansoor S, Paterson AH (2012) Genomes for jeans: cotton genomics for engineering superior fiber. Trends in Biotechnology 30: 521–527.
53. Isbell RF, editor (1996) The Australian soil classification: CSIRO publishing.
54. Bedon F, Ziolkowski L, Osabe K, Venables I, Machado A, et al. (2013) Separation of integument and nucellar tissues from cotton ovules (*Gossypium hirsutum* L.) for both high- and low-throughput molecular applications. BioTechniques 54: 44–46.
55. Quinlivan EP, Gregory JF III (2008) DNA digestion to deoxyribonucleoside: A simplified one-step procedure. Analytical Biochemistry 373: 383–385.
56. Johnston JW, Harding K, Bremner DH, Souch G, Green J, et al. (2005) HPLC analysis of plant DNA methylation: a study of critical methodological factors. Plant Physiology and Biochemistry 43: 844–853.
57. Vos P, Hogers R, Bleeker M, Reijans M, Vandelee T, et al. (1995) AFLP - a new technique for DNA-fingerprinting. Nucleic Acids Research 23: 4407–4414.
58. Rohlf FJ (2008) NTSYSpc: Numerical Taxonomy System, ver. 2.21. Setauket, NY: Exeter Publishing, Ltd.
59. Halldén C, Nilsson NO, Rading IM, Säll T (1994) Evaluation of RFLP and RAPD markers in a comparison of Brassica napus breeding lines. Theoretical and Applied Genetics 88: 123–128.
60. Dice LR (1945) Measures of the amount of ecologic association between species. Ecology 26: 297–302.
61. Jaccard P (1901) Etude comparative de la distribution florale dans une portion des Alpes et du Jura. Bulletin de la Société vaudoise des Sciences Naturelles 37: 547–579.
62. Laurentin H (2009) Data analysis for molecular characterization of plant genetic resources. Genetic Resources and Crop Evolution 56: 277–292.
63. Garcia-Vallve S, Palau J, Romeu A (1999) Horizontal gene transfer in glycosyl hydrolases inferred from codon usage in Escherichia coli and Bacillus subtilis. Molecular Biology and Evolution 16: 1125–1134.
64. Stricker D (2008) BrightStat.com: Free statistics online. Comput Methods Prog Biomed 92: 135–143.
65. Hedges SB (1992) The number of replications needed for accurate estimation of the bootstrap-p value in phylogenetic studies. Molecular Biology and Evolution 9: 366–369.
66. Nelson R (1995) WinBoot: A program for performing bootstrap analysis of binary data to determine the confidence limits of UPGMA-based dendrograms. IRRI Discussion Paper Series No 14.
67. Pompanon F, Bonin A, Bellemain E, Taberlet P (2005) Genotyping errors: causes, consequences and solutions. Nat Rev Genet 6: 847–846.
68. Walford S-A, Wu Y, Llewellyn DJ, Dennis ES (2011) GhMYB25-like: a key factor in early cotton fibre development. The Plant journal : for cell and molecular biology 65: 785–797.
69. Gore UR (1932) Development of the female gametophyte and embryo in cotton. American Journal of Botany 19: 795–807.
70. Lintilha PM, Jensen WA (1974) Differentiation, organogenesis, and tectonics of cell-wall orientation.1. Preliminary observations on development of ovule in cotton. American Journal of Botany 61: 129–134.
71. Brubaker CL, Bourland FM, Wendel JF (1999) The origin and domestication of cotton. In: Smith W, editor. Cotton: origin, history, technology, and production. New York: John Wiley & Sons. 3–31.
72. Applequist WL, Cronn R, Wendel JF (2001) Comparative development of fiber in wild and cultivated cotton. Evolution & Development 3: 3–17.
73. Cao X, Aufsatz W, Zilberman D, Mette MF, Huang MS, et al. (2003) Role of the DRM and CMT3 Methyltransferases in RNA-Directed DNA Methylation. Current 13: 2212–2217.
74. Romanel E, Silva TF, Correa RL, Farinelli L, Hawkins JS, et al. (2012) Global alteration of microRNAs and transposon-derived small RNAs in cotton (Gossypium hirsutum) during Cotton leafroll dwarf polerovirus (CLRDV) infection. Plant Molecular Biology 80: 443–460.
75. Kwak P, Wang Q, Chen X, Qiu C, Yang Z (2009) Enrichment of a set of microRNAs during the cotton fiber development. BMC Genomics 10: 457.
76. Wang Z-M, Xue W, Dong C-J, Jin L-G, Bian S-M, et al. (2011) A Comparative miRNAome Analysis Reveals Seven Fiber Initiation-Related and 36 Novel miRNAs in Developing Cotton Ovules. Molecular plant.
77. Pang MX, Xing CZ, Adams N, Rodriguez-Uribe L, Hughs SE, et al. (2011) Comparative expression of miRNA genes and miRNA-based AFLP marker analysis in cultivated tetraploid cottons. Journal of Plant Physiology 168: 824–830.
78. Li Q, Jin X, Zhu YX (2012) Identification and Analyses of miRNA Genes in Allotetraploid Gossypium hirsutum Fiber Cells Based on the Sequenced Diploid G-raimondii Genome. Journal of Genetics and Genomics 39: 351–360.
79. Meudt HM, Clarke AC (2007) Almost forgotten or latest practice? AFLP applications, analyses and advances. Trends in Plant Science 12: 106–117.
80. Fang J, Song C, Zheng Y, Qiao Y, Zhang Z, et al. (2008) Variation in cytosine methylation in Clementine mandarin cultivars. Journal of Horticultural Science & Biotechnology 83: 833–839.
81. Fang JG, Song CN, Qian JL, Zhang XY, Shangguan LF, et al. (2010) Variation of cytosine methylation in 57 sweet orange cultivars. Acta Physiologiae Plantarum 32: 1023–1030.
82. Ashikawa I (2001) Surveying CpG methylation at 5′-CCGG in the genomes of rice cultivars. Plant Molecular Biology 45: 31–39.
83. Takata M, Kishima Y, Sano Y (2005) DNA methylation polymorphisms in rice and wild rice strains: Detection of epigenetic markers. Breeding Science 55: 57–63.
84. Sakowicz T, Olszewska MJ, Luchniak P, Kazmierczak J (1998) Tissue-specific DNA methylation in Haemanthus katharinae Bak. (amaryllidaceae). Acta Societatis Botanicorum Poloniae 67: 175–180.
85. Kazmierczak J (1998) Effect of DNA methylation on potential transcriptional activity in different tissues and organs of Vicia faba ssp. minor. Folia Histochemica Et Cytobiologica 36: 45–49.
86. Palmgren G, Mattsson O, Okkels FT (1991) Specific levels of dna methylation in various tissues, cell-lines, and cell-types of daucus-carota. Plant Physiology 95: 174–178.
87. Jia F, Fu Y, Liu W, Du Z, Zhao Y (2011) Quantitative determination of DNA Methylation in tobacco leaves by HPLC. Journal of Agricultural Research 6: 1545–1548.
88. Scott NS, Tymms MJ, Possingham JV (1984) Plastid-DNA Levels In The Different Tissues Of Potato. Planta 161: 12–19.
89. Marano MR, Carrillo N (1991) Chromoplast formation during tomato fruit ripening - no evidence for plastid dna methylation. Plant Molecular Biology 16: 11–19.
90. Fojtova M, Kovarik A, Matyasek R (2001) Cytosine methylation of plastid genome in higher plants. Fact or artefact? Plant Science 160: 585–593.
91. Schulz P, Jensen WA (1977) Cotton Embryogenesis: The Early Development of the Free Nuclear Endosperm. American Journal of Botany 64: 384–394.

7

Demographics and Genetic Variability of the New World Bollworm (*Helicoverpa zea*) and the Old World Bollworm (*Helicoverpa armigera*) in Brazil

Natália A. Leite[1], Alessandro Alves-Pereira[2], Alberto S. Corrêa[1], Maria I. Zucchi[3], Celso Omoto[1]*

1 Departamento de Entomologia e Acarologia, Escola Superior de Agricultura "Luiz de Queiroz", Universidade de São Paulo, Piracicaba, São Paulo, Brazil, **2** Departamento de Genética, Escola Superior de Agricultura "Luiz de Queiroz", Universidade de São Paulo, Piracicaba, São Paulo, Brazil, **3** Agência Paulista de Tecnologia dos Agronegócios, Piracicaba, São Paulo, Brazil

Abstract

Helicoverpa armigera is one of the primary agricultural pests in the Old World, whereas *H. zea* is predominant in the New World. However, *H. armigera* was first documented in Brazil in 2013. Therefore, the geographical distribution, range of hosts, invasion source, and dispersal routes for *H. armigera* are poorly understood or unknown in Brazil. In this study, we used a phylogeographic analysis of natural *H. armigera* and *H. zea* populations to (1) assess the occurrence of both species on different hosts; (2) infer the demographic parameters and genetic structure; (3) determine the potential invasion and dispersal routes for *H. armigera* within the Brazilian territory; and (4) infer the geographical origin of *H. armigera*. We analyzed partial sequence data from the cytochrome c oxidase subunit I (COI) gene. We determined that *H. armigera* individuals were most prevalent on dicotyledonous hosts and that *H. zea* were most prevalent on maize crops, based on the samples collected between May 2012 and April 2013. The populations of both species showed signs of demographic expansion, and no genetic structure. The high genetic diversity and wide distribution of *H. armigera* in mid-2012 are consistent with an invasion period prior to the first reports of this species in the literature and/or multiple invasion events within the Brazilian territory. It was not possible to infer the invasion and dispersal routes of *H. armigera* with this dataset. However, joint analyses using sequences from the Old World indicated the presence of Chinese, Indian, and European lineages within the Brazilian populations of *H. armigera*. These results suggest that sustainable management plans for the control of *H. armigera* will be challenging considering the high genetic diversity, polyphagous feeding habits, and great potential mobility of this pest on numerous hosts, which favor the adaptation of this insect to diverse environments and control strategies.

Editor: João Pinto, Instituto de Higiene e Medicina Tropical, Portugal

Funding: This work was partially supported by Conselho Nacional de Desenvolvimento Científico e Tecnológico (CNPq) (Grant 308150/2009-0) and Comitê Brasileiro de Ação a Resistência a Inseticidas (IRAC-BR). The funders had no role in study design, data collection and analysis, decision to publish, or preparation of the manuscript.

Competing Interests: The authors have declared that no competing interests exist.

* Email: celso.omoto@usp.br

Introduction

The Heliothinae (Lepidoptera: Noctuidae) subfamily has 381 described species, many of which are important agricultural pests from the *Helicoverpa* Hardwick and *Heliothis* Ochsenheimer genera [1]. The *Helicoverpa* genus contains two of the primary Heliothinae pest species: *Helicoverpa armigera* (Hübner) (Old World bollworm) and *Helicoverpa zea* (Boddie) (New World bollworm). Although the exact evolutionary relationship between *H. armigera* and *H. zea* remains uncertain, these insects are considered to be 'twin' or 'sibling' species, and they are able to copulate and produce fertile offspring under laboratory conditions [2–5]. Some hypotheses propose that *H. zea* evolved from a small portion of the larger *H. armigera* population (i.e., a "founder effect") that reached the American continent approximately 1.5 million years ago, which is consistent with previous phylogeographic analyses of *H. armigera* and *H. zea* individuals [6,7].

H. armigera is considered to be one of the most important agricultural pests in the world. This insect is widely distributed throughout Asia, Africa, Europe, and Australia, and it has been shown to attack more than 100 host species from 45 different plant families [8–10]. In contrast, *H. zea* is restricted to the American continent and is of lesser economic importance; it is a secondary pest of cotton, tomato, and, most significantly, maize crops [11]. However, the scenario in Brazil changed in 2013 when *H. armigera* individuals, which are considered to be A1 quarantine pests, were officially reported within the Brazilian territory [12–14]. This situation increased in severity due to the great dispersal ability of this insect as well as the steady reports from several regions of the world that described new *H. armigera* lineages showing tolerance/resistance to insecticides and genetically modified plants [15,16]. It is estimated that *H. armigera* will cause a loss of more than US$2 billion to the 2013/14 Brazilian agriculture crop because of direct productivity losses and resources spent on phytosanitary products for soybean, cotton, and maize, which are the main crops of Brazilian agribusinesses. Therefore, *H. armigera* is now one of the most important pest species with respect to agriculture in Brazil [17].

High population densities of *Helicoverpa* spp. and the resulting economic damages to cultivated plants have been reported in different regions of Brazil, in particular in the Western state of Bahia [18]. Therefore, these reports suggest the existence of an invasion period prior to the first official report of *H. armigera* in Brazil. This atypical and confusing scenario was likely caused by the significant morphological similarities between *H. zea* and *H. armigera* [9,19] and by major changes in pest management programs over recent years. In addition, these population changes may have been related to the release and increased cultivation of crops that express *Bacillus thuringiensis* (Bt) genes in Brazil.

Aside from the identification of *H. armigera* individuals within the Brazilian territory, many basic pieces of information concerning this species, including its geographical distribution, the types of hosts it attacks, its invasion source, and its dispersal routes, remain poorly understood or completely unknown. Therefore, we attempted to address some of these outstanding questions using a phylogeographic approach by analyzing genetic sequence data from a portion of the cytochrome c oxidase subunit I (COI) gene of *Helicoverpa* spp. specimens isolated from different hosts and regions of Brazil. This study was performed with the following goals in mind: (1) to confirm and evaluate the occurrence of *H. armigera* and *H. zea* individuals from different hosts and regions of Brazil; (2) to assess the demographic parameters and genetic structure of *H. armigera* and *H. zea* populations within the Brazilian territory, with a focus on the region, season, and host; (3) to assess the potential invasion (single or multiple) and dispersal routes for *H. armigera* within the Brazilian territory; and (4) to determine the geographical origin of the *H. armigera* populations present in Brazil. This information will be essential for understanding the genetic diversity and population dynamics of these pests as well as for guiding both immediate control strategies (legal and/or phytosanitary) and subsequent long-term integrated management programs for the *Helicoverpa* spp. complex in Brazil.

Results

Identification of *Helicoverpa* spp., hosts, and geographic locations

One hundred thirty-nine individuals from the 274 *Helicoverpa* spp. specimens initially sampled were identified as *H. armigera* (98–100% homology) and 134 individuals were identified as *H. zea* (98–100% homology) (GenBank Accession numbers KM274936–KM275209 are listed in Table 1). *H. armigera* was primarily found on soybean, bean, and cotton crops, and these insects were widely distributed throughout the Midwest and Northeast of Brazil during both crop periods (winter and summer) (Figure 1). *H. armigera* was also found on sorghum, millet, and maize crops. However, for maize, *H. armigera* individuals were only found at one site during the summer growing season in Northeastern Brazil (state of Bahia). *H. armigera* was not found on maize crops in the Midwest, Southeast, or South of Brazil. *H. zea* was primarily found on maize crops and was present in all sampled regions during both the winter and summer growing seasons. Of the winter crops, millet and cotton were exceptional in that they could simultaneously support *H. zea* and *H. armigera* (Figure 1). We found no correlations between specific *H. armigera* mitochondrial lineages (haplotypes) and specific hosts (Figure 1).

Dataset assembly, haplotypes, and demographic analysis

Following alignment and editing, we were unable to identify indels or stop codons in the sequences from either species. However, using the most common haplotype for each species as a reference, eight non-synonymous substitutions were observed in 17 *H. armigera* individuals, and four non-synonymous substitutions were observed in eight *H. zea* individuals. However, considering the relatively high mutation rate reported for the COI gene in the *Helicoverpa* genus [20], as well the absence of indels and stop codons, it is unlikely that these sequences represent numts (nuclear mitochondrial DNA).

Twenty-six polymorphic sites were found among the 139 *H. armigera* individuals sampled, which yielded 31 haplotypes with a haplotype diversity (Hd) of 0.821 and a nucleotide diversity (Pi) of 0.0028. Sequence analysis of the 134 sampled *H. zea* individuals identified 19 polymorphic sites, which yielded 20 haplotypes with an Hd of 0.420 and a Pi of 0.0011 (Table 2). No significant differences in Hd or Pi were found for either species when the individuals were separated by growing season according to the sampled crops (Table 2). The results from Tajima's D test were only not significant for *H. armigera* individuals ($p = 0.07$) sampled on summer crops; however, Fu's Fs test was significant ($p<0.01$). The Tajima's D and Fu's Fs test results for both *H. armigera* and *H. zea* were negative and significant when the individuals were tested as a single group and when the individuals were split into groups based on the crop on which they were sampled (summer or winter; temporally). These results indicate an excess of low frequency polymorphisms and are consistent with either population expansion or purifying selection (Table 2). In addition, the model of sudden expansion [21] did not reject the hypothesis of expansion demographics for *H. armigera* (SSD = 0.0012, $p = 0.48$; Raggedness = 0.0433, $p = 0.61$) or *H. zea* (SSD = 0.0002, $p = 0.90$; Raggedness = 0.1492, $p = 0.72$).

Statistical analysis of population structure

The results of the analysis of molecular variance (AMOVA) with two hierarchical levels showed that the greatest amount of total variation was accounted for by differences among individuals within populations: 92.89% for *H. armigera* ($\Phi_{ST} = 0.071$) and 94.22% for *H. zea* ($\Phi_{ST} = 0.058$) (Table S1). For the AMOVA with three hierarchical levels for *H. armigera*, the largest percentage of variation occurred within populations, separating individuals into groups by time (winter and summer crops; 93.17%, $\Phi_{CT} = 0.006$; $\Phi_{SC} = 0.074$; $\Phi_{ST} = 0.068$), host group (mono- and dicotyledonous; 99.24%, $\Phi_{CT} = -0.01$; $\Phi_{SC} = 0.018$; $\Phi_{ST} = 0.007$), and each host type (crop; 93.19%, $\Phi_{CT} = -0.042$; $\Phi_{SC} = 0.105$; $\Phi_{ST} = 0.068$) (Table S1). The group separation for *H. armigera* was not significant for any of the three tested groups ($p>0.10$). The AMOVA with three hierarchical levels divided the *H. zea* individuals into groups by time (winter and summer crops), which showed a larger variation within populations (93.76%, $\Phi_{CT} = 0.010$; $\Phi_{SC} = 0.052$; $\Phi_{ST} = 0.062$); the group division was not significant ($p>0.10$) (Table S1).

Network analysis and Bayesian phylogeny

Analysis of the genetic connections between the *Helicoverpa* spp. represented in the haplotype network revealed a close genetic relation between *H. armigera* and *H. zea*, which were separated by only 13 mutational steps (Figure 2). By separately analyzing the connections between the genetic haplotypes of each species, we inferred the existence of two predominant maternal lineages for *H. armigera*: H1 (31.65%) and H3 (23.02%), which were located at the center of the haplotype network. The other haplotypes of *H. armigera,* with the exception of haplotype H2 (15.83%), all had

Table 1. Sampling sites for *Helicoverpa armigera* and *Helicoverpa zea* in Brazil, including the sites where these insects were sampled for this study, abbreviations, sample sizes for the mitochondrial genes (COI), crops sampled, geographic coordinates, dates sampled, and GenBank Accession.

Sites (City, State)	Abbreviation (Site, Crop)	Crop	Sample size		Lat. (S)	Lon. (W)	Date	GenBank Accession
			H. armigera	*H. zea*				
Winter cropping								
Barreiras, Bahia	BA1Co	Cotton	3	-	12°08′54′′	44°59′33″	05.22.12	KM274936–KM274938
Luís E. Magalhães, Bahia	BA2Co	Cotton	11	1	12°05′58″	45°47′54″	05.24.12	KM274939–KM274950
Balsas, Maranhão	MA1Co	Cotton	10	-	07°31′59″	46°02′06″	06.23.12	KM274987–KM274996
Luís E. Magalhães, Bahia	BA3Be	Bean	23	-	12°05′58″	45°47′54″	06.12.12	KM274979–KM274986, KM275038–KM275052
Luís E. Magalhães, Bahia	BA4Mi	Millet	6	3	12°05′58″	45°47′54″	05.10.12	KM274951–KM274959
Luís E. Magalhães, Bahia	BA5Sr	Sorghum	16	-	12°05′58″	45°47′54″	05.10.12	KM274960–KM274975
Capitólio, Minas Gerais	MG1Ma	Maize	-	14	20°36′17″	46°04′19″	06.08.12	KM274997–KM275010
Luís E. Magalhães, Bahia	BA6Ma	Maize	-	13	12°05′58″	45°47′54″	06.12.12	KM274976–KM274978, KM275053–KM275062
Itapira, São Paulo	SP1Ma	Maize	-	7	22°26′11″	46°49′20″	06.12.12	KM275011–KM275017
Assis, São Paulo	SP2Ma	Maize	-	7	22°39′40″	50°23′58″	06.15.12	KM275018–KM275024
São Gabriel do Oeste, Mato Grosso do Sul	MS1Ma	Maize	-	13	19°23′37″	54°33′49″	06.27.12	KM275025–KM275038
Rondonópolis, Mato Grosso	MT1Ma	Maize	-	7	16°28′17″	54°38′14″	08.01.12	KM275063–KM275069
Summer cropping								
Riachão das Neves, Bahia	BA7Sy	Soybean	8	-	12°08′54″	44°59′33″	10.21.12	KM275070–KM275077
Luís E. Magalhães, Bahia	BA8Sy	Soybean	5	-	12°05′58″	45°47′54″	10.31.12	KM275078–KM275082
Rondonópolis, Mato Grosso	MT2Sy	Soybean	13	-	16°28′17″	54°38′14″	11.08.12	KM275083–KM275092, KM275156–KM275158
Chapadão do Sul, Mato Grosso do Sul	MS2Sy	Soybean	6	-	18°46′44″	52°36′59″	11.29.12	KM275097–KM275102
Balsas, Maranhão	MA2Sy	Soybean	10	-	07°31′59″	46°02′06″	01.06.13	KM275103–KM275112
São Desidério, Bahia	BA9Sy	Soybean	10	-	12°21′08″	44°59′03″	01.15.13	KM275127–KM275136
Limoeiro do Norte, Ceará	CE1Co	Cotton	-	4	05°08′56′′	38°05′52′′	10.08.12	KM275093–KM275096
São Desidério, Bahia	BA10Co	Cotton	14	-	12°21′08″	44°59′03″	01.15.13	KM275147–KM275155, KM275202–KM275206
Cândido Mota, São Paulo	SP3Ma	Maize	-	7	22°44′46″	50°23′15″	01.14.13	KM275113–KM275119
Jardinópolis, São Paulo	SP4Ma	Maize	-	7	21°03′47″	47°45′05″	03.04.13	KM275120–KM275126
Barreiras, Bahia	BA11Ma	Maize	4	-	11°33′33″	46°19′47″	02.21.13	KM275137–KM275140
Luís E. Magalhães, Bahia	BA12Ma	Maize	-	9	12°05′58″	45°47′54″	03.28.13	KM275141–KM275146, KM275207–KM275209
Rolândia, Paraná	PR1Ma	Maize	-	12	23°19′13′′	51°29′01′′	01.24.13	KM275159–KM275170
Passo Fundo, Rio Grande do Sul	RS1Ma	Maize	-	10	28°16′08′′	52°37′15′′	01.30.13	KM275171–KM275180
Montividiu, Goiás	GO1Ma	Maize	-	10	17°19′19′′	51°14′51′′	02.05.13	KM275181–KM275190
Capitólio, Minas Gerais	MG2Ma	Maize	-	11	20°36′17″	46°04′19″	03.10.13	KM275191–KM275201
Total			**139**	**135**				

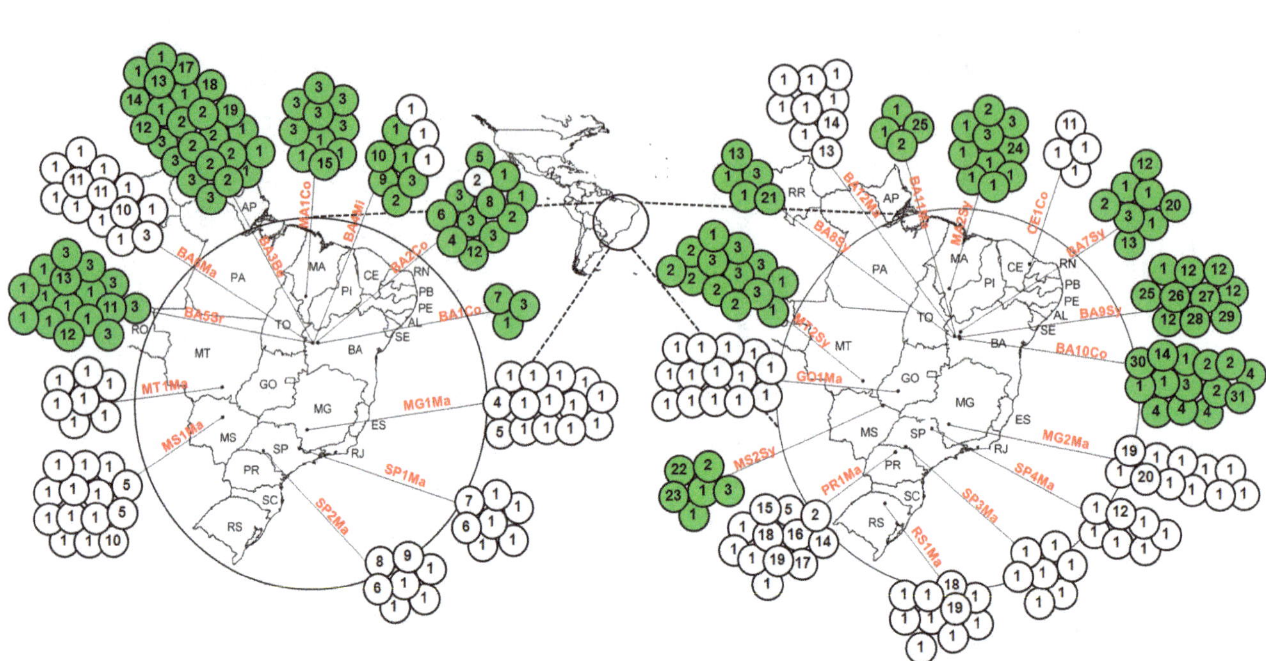

Figure 1. Geographic distributions of COI haplotypes of *H. armigera* and *H. zea*. One hundred and thirty nine and 135 COI haplotypes were analyzed for these species, respectively. The samples were separated into two temporal groups (winter crops and summer crops). Each circle represents the haplotypes identified in a given population; a number within a circle denotes the COI haplotypes identified in that population. Colored circles refer to *H. armigera* specimens, and white circles refer to *H. zea* specimens. The abbreviations refer to the sampled locations and crops (Table 1).

frequencies below 5%. Haplotypes H19, H18, H16, H12, H21, and H25 formed an outer cluster within the haplotype network of *H. armigera* (Figure 2). The haplotype network for *H. zea* revealed a genetic haplotype relationship with a single central high-frequency lineage (H1 = 76.30%) surrounded by low-frequency haplotypes (<5%) (Figure 2).

The optimal nucleotide substitution model identified by the MODELTEST 2.3 software program was the GTR+I+G model (Generalized time reversible + Proportion of invariable sites + Gamma distribution model). The estimated model parameters were based on empirical base frequencies (A = 0.3092, C = 0.1463, G = 0.1312, and T = 0.4133), with the proportion of invariable sites (I) set to 0.7393 and the gamma distribution shape parameter set to 0.5778. The consensus tree generated by the Bayesian analysis divided the *Helicoverpa* spp. specimens sampled in Brazil into two monophyletic clades (*H. armigera* and *H. zea*) with an associated probability of 99% (Figure 3; Figure S1). The probabilities separating the *H. zea* individuals into groups within this species were not significant. A single *H. armigera* individual (MS2Sy6) was separated from the other individuals with an associated probability of 98%. Finally, *Helicoverpa gelotopoeon* showed a closer phylogenetic relationship to *H. armigera* and *H. zea* compared with *H. assulta* (Figure 3; Figure S1).

Network analysis: Brazilian vs. Old World *Helicoverpa armigera*

The haplotype network constructed using the edited sequences collected in Brazil, along with numerous Old World sequences, identified 38 distinct haplotypes (Figure 4). H1 (28%) and H2 (24%), which are widely distributed throughout Brazil, Europe, and China, were the most frequent haplotypes and occupied the central region of the haplotype network. All other haplotypes, with the exception of H3 and H10, showed frequencies below 5%. Finally, the majority of haplotypes with low frequencies represented by singletons were located at the network extremities (Figure 4).

Discussion

Our results indicate a widespread distribution for *H. armigera* throughout the Midwest and Northeast of Brazil on a variety of crops, particularly dicotyledons, beans, soybeans, and cotton as well as, to a lesser extent, millet, sorghum, and maize. This pest was not found on maize crops in the Midwest, Southeast, or South of Brazil, despite the fact that these crops were initially identified as sources of *H. armigera* in this system. *H. armigera* individuals associated with maize crops were only found at a single sampling site in the Northeast (state of Bahia) during February 2013. In contrast, *H. zea* individuals were essentially found only on maize crops, with the exception of a few individuals collected from millet and cotton crops, where *H. zea* individuals were found alongside *H. armigera* individuals. Before the documentation of *H. armigera* in Brazil in 2013, we had hypothesized that major source of *Helicoverpa* spp. attacking different host plant was maize crops. However, our findings showed that targeting the control of *H. armigera* on maize crops may not be effective because *H. zea* was the predominant species in this host plant. The possibility of the formation of hybrid individuals between these two species, which has been reported under laboratory conditions [3,4], needs to be investigated under field conditions to improve our pest management programs.

Table 2. Number of individuals, haplotype designation, and genetic diversity for the sampled populations grouped according to geographical origin.

Group	N. Individuals (samples)	N. haplotypes	Distribution of Haplotypes (n)	Haplotype Diversity (Hd)	Nucleotide diversity (Pi)	Tajima's D test (*p* value)	Fu's Fs test (*p* value)
H. armigera							
Pooled	**139 (14)**	**31**	-	**0.821**	**0.0028**	**−1.729 (<0.01)**	**−26.361 (<0.01)**
Winter cropping	69 (6)	19	H1(22); H2(9); H3(21); H4(1); H5(1); H6(1); H7(1); H8(1); H9(1); H10(1); H11(1); H12(2); H13(2); H14(1); H15(1); H16(1); H17(1); H18(1); H19(1).	0.805	0.0028	−1.608 (=0.03)	−11.891 (<0.01)
Summer cropping	70 (8)	19	H1(22); H2(13); H3(11); H4(4); H12(4); H13(2); H14(1); H20(1); H21(1); H22(1); H23(1); H24(1); H25(2); H26(1); H27(1); H28(1); H29(1); H30(1); H31(1).	0.835	0.0028	−1.353 (=0.07)	−11.254 (<0.01)
H. zea							
Pooled	**135 (16)**	**20**	-	**0.420**	**0.0011**	**−2.190 (<0.01)**	**−22.912 (<0.01)**
Winter cropping	65 (8)	11	H1(50); H2(1); H3(1); H4(1); H5(3); H6(2); H7(1); H8(1); H9(1); H10(2); H11(2).	0.408	0.0009	−2.156 (<0.01)	−9.735 (<0.01)
Summer cropping	70 (8)	13	H1(53); H2(1); H5(1); H11(1); H12(1); H13(1); H14(2); H15(1); H16(1); H17(1); H18(2); H19(3); H20(2).	0.427	0.0012	−1.967 (<0.01)	−10.411 (<0.01)

Demographic analyses using neutrality tests and a Mismatch Distribution Analysis indicated an expansion of the *H. armigera* and *H. zea* populations within the Brazilian territory. Population expansions were also consistent with the Haplotype network structure, which was characteristic of species undergoing processes of demographic expansion [22]. Brazilian *H. armigera* individuals showed two primary maternal lineages, whereas *H. zea* showed a single primary lineage, all of which were surrounded by numerous lower-frequency haplotypes. Therefore, these central high-frequency haplotypes represent the ancestral haplotypes, with the low-frequency haplotypes more recently derived [23]. Furthermore, signs of the *H. armigera* population expansion are likely because of the recent introduction of this pest into Brazil. Following the founder event, during which a portion of the overall genetic diversity of the species was introduced to Brazil, the *H. armigera* population further propagated. According to Nibouche et al. [24], *H. armigera* can migrate as far as 2,000 km, which likely facilitated the colonization of a variety of crops. The migration and colonization of crop areas by a small group of individuals can cause bottleneck effects, which, combined with plague population-suppression strategies (e.g., insecticide use that kills all but a small portion of the population), can lead to the types of demographic expansions observed for *H. zea* and *H. armigera* in Brazil [25–27]. In addition, the expansion of maize, soybean, and cotton crops into the North and Northeast of Brazil over the previous decade may also be responsible, in part, for the demographic expansion of these species, specifically *H. zea*. Additionally, assuming that not all COI variation is neutral, *Helicoverpa* spp. populations could be suffering selection, especially considering that populations have colonized new environments recently. However, further studies using a larger number of molecular markers from nuclear and mitochondrial genome regions would answer these questions. The *H. armigera* and *H. zea* population genetics were not structured according to space, time (winter and summer crops), or host (crops). Unstructured genetic networks have been reported for other populations of these two pest species in other parts of the world, which were based on several molecular markers, including mtDNA, allozymes, and microsatellites [7,24,25,28,29,30]. Both species showed wide spatial haplotype distributions, and no genetic relationships were identified using a haplotype network analysis or an AMOVA. This scenario may be because these populations have a polyphagous feeding habit and migratory characteristics.

The unstructured population of *H. armigera* and the wide distribution of the two ancestral maternal lineages within the Brazilian territory did not allow us to infer any hypothetical invasion or dispersal routes for this species within the region. However, we noted that the haplotype and nucleotide diversities found for *H. armigera* in Brazil are similar to or greater than those reported for natural *H. armigera* populations in the Old World [7,20]. For example, one outer branch of the *H. armigera* haplotype network, formed by haplotypes H19, H18, H16, H12, H21, and H25, is noteworthy for having the greatest genetic distance from the central haplotypes (H1 and H3), and these haplotypes have yet to be identified in Old World populations [7,20]. In addition, joint analysis of the haplotypes from Brazil and the Old World yielded an overall structure that was similar to the haplotype network obtained only from the Brazilian individuals. In particular, the two most frequent haplotypes were identified throughout Brazil, Europe, China, and India, whereas the majority of the singletons were from Brazil and China. The cited literature, along with our results that showed a wide geographic distribution for *H. armigera* during the first half of 2012, support the hypothesis of an invasion period prior to the first reports of this

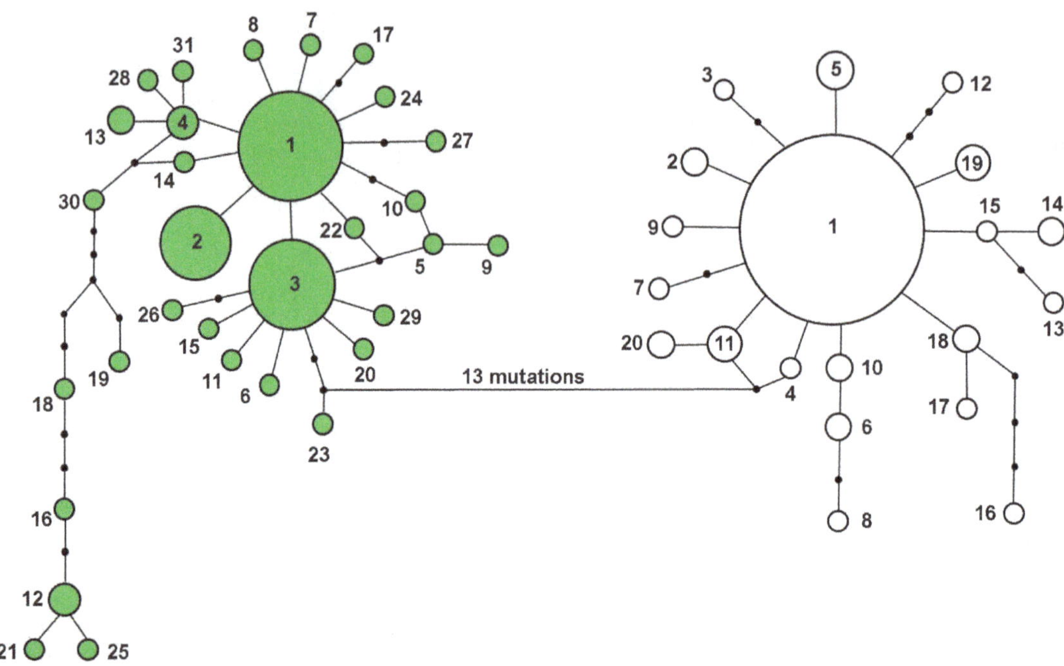

Figure 2. Haplotype network based COI sequences from *H. armigera* and *H. zea* samples collected in Brazil. Partial mtDNA COI (658 bp) sequences from *H. armigera* (colored circles) and *H. zea* (white circles) were analyzed from samples collected in Brazil. Each haplotype is represented by a circle and is identified by a number from 1–31. The *H. armigera* and *H. zea* COI haplotypes are shown as described in Table 2. The numbers of nucleotide substitutions between the haplotypes are indicated by black circles. The total number of nucleotide substitutions separating the *H. armigera* specimens from the *H. zea* specimens is shown.

species in Brazil. Alternatively, these findings are also consistent with a more recent invasion that involved a large gene pool, multiple invasion events, or some combination of these events.

The low genetic divergence observed between *H. armigera* and *H. zea* in the haplotype network analysis and the Bayesian phylogeny confirms the close genetic relatedness of these two species. Therefore, the reported co-occurrence of these species in time and space, as well as on the same hosts (as described here), could allow for the formation of hybrid individuals, which has been reported under laboratory conditions [3,4]. Although the existence of hybrids in the wild remains unconfirmed, this scenario is of significant concern. In particular, recombination or introgression phenomena between *H. armigera*, which is reportedly resistant to control methods, and *H. zea*, which has adapted to the environmental conditions of the American continent, may enable gene transfer and fixation in some individuals. Therefore, hybridization may enable the selection of breeds with enhanced hybrid vigor and the ability to rapidly adapt to current management and suppression methods.

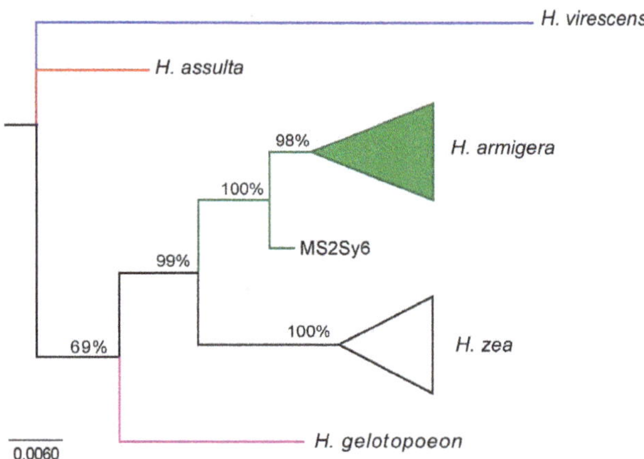

Figure 3. Bayesian phylogenetic tree of *H. armigera* and *H. zea* individuals sampled in Brazil. This phylogenetic tree is based on partial COI haplotype sequences and includes *H. assulta* and *H. gelotopoeon* sequences. Numbers near the interior branches indicate posterior probability (×100) values. The outgroup used was *Heliothis virescens*. *H. armigera* COI haplotypes and Genbank Accession numbers can be found in Table S2.

The population studies described in this study indicate a recent demographic expansion and a high mitochondrial genetic diversity for *H. armigera* and *H. zea* in Brazil. Therefore, the sustainable management of *H. armigera* will likely become a significant challenge for Brazilian entomology in the coming years, especially considering the polyphagous feeding habit, the great dispersal ability, and the numerous reports of resistance to insecticides and Bt crops for this insect [8,24,31–35]. This scenario requires immediate attention, as there is an imminent risk of *H. armigera* expanding throughout the American territory and perhaps reaching agricultural areas in Central and North America. However, it was not possible to trace the invasion and dispersal routes of *H. armigera* in the Brazilian territory. Nevertheless, the hypotheses of an invasion period prior to the first reports in the literature and/or an invasion that involved a diverse gene pool are both consistent with the observed high incidence and rapid adaptation of *H. armigera* in the Brazilian territory. Our confirmation that the predominant maternal lineages in the Brazilian territory are the same compared with those in Europe and Asia may represent a starting point to guide *H. armigera* management programs. Indeed, control strategies have a greater

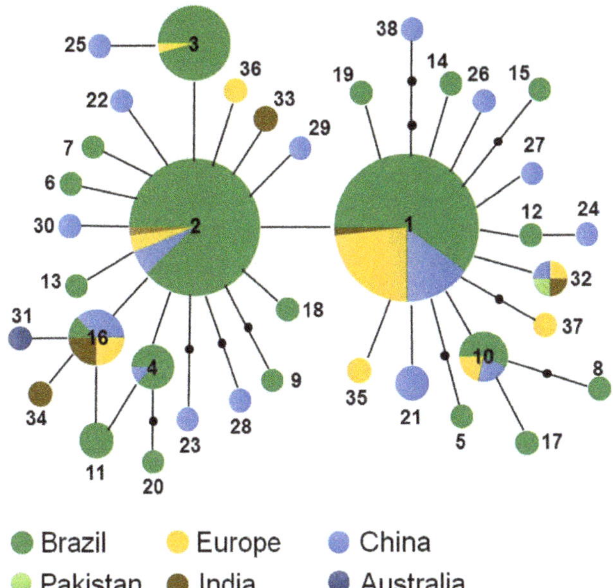

Figure 4. Haplotype network based COI sequences from *H. armigera* samples from Brazil and Old World specimens. Partial mtDNA COI (590 bp) sequences from this species were analyzed. Thirty-eight haplotypes were identified from 212 individuals sampled from China (n = 35), Thailand (n = 1), Australia (n = 1), Pakistan (n = 2), Europe (n = 28), India (n = 6), and Brazil (n = 139). *H. armigera* COI haplotypes are shown as described in Table S2. Each circle represents a haplotype and its number. The colors represent the frequency of each haplotype in the country/continent, with dark green (Brazil), light green (Pakistan), yellow (Europe), brown (India), light blue (China), and dark blue (Australia).

chance of success when reliable information is gathered in the regions where the pests, their hosts, and their natural enemies have co-evolved over a significant period of time.

Materials and Methods

Sampling procedures

Permit access to collect material used in our research at various crop sites was granted by respective growers. GPS coordinates of each location are listed in Table 1.

Brazilian agriculture has shown successive and overlapping crops in space and time, and these crops can be largely separated into two harvest groups that are primarily characterized by their rainfall needs. In particular, winter crops are grown between May and September and require low rainfall, whereas summer crops are grown between October and April and require high rainfall. Our initial sampling design was directed at understanding the *H. zea* population dynamics and primarily involved maize fields. However, attacks on soybean, cotton, bean, sorghum, and millet crops were also reported between May 2012 and April 2013 (Brazilian agricultural year). Therefore, we directed our sampling efforts towards a variety of crops and regions throughout Brazil. We also focused on the Western region of Bahia State, Brazil, which was the site of numerous *Helicoverpa* spp. attacks, to determine whether maize crops were the main source of *H. zea* in the Brazilian agricultural system. A total of 274 *Helicoverpa* caterpillars were collected at 19 sampling sites from six different crops (Table 1). In the absence of morphological characters or nuclear markers to reliably distinguish between *H. zea* and *H. armigera*, species identification was carried out using the sequence fragment of COI mitochondrial gene by comparing with *H. zea* and *H. armigera* species barcodes [7,18,19,39] and determining homology with BlastN tool.

DNA extraction, PCR amplification, and gene sequencing

Genomic DNA was isolated from the thorax of each adult using an Invisorb Spin Tissue Kit (STRATEC Molecular, Berlin, Germany), according to the manufacturer's protocol. A fragment of the COI mitochondrial gene was amplified by polymerase chain reaction (PCR) with the primers LCO(F) (5′ - GGT CAA CAA ATC ATA AAG ATA TTG G - 3′) and HCO(R) (5′ - TAA ACT TCA GGG TGA CCA AAA AAT CA - 3′) [36]. Amplification reactions were performed using 10 ng genomic DNA, 50 mM $MgCl_2$, 0.003 $mg.mL^{-1}$ BSA, 6.25 mM dNTPs, 10 pmol each primer, 1 U Taq DNA Polymerase (Life Technologies, Carlsbad, CA, USA), and 10% 10× Taq Buffer in a final volume of 25 µL. The PCR program consisted of an initial denaturation step at 94°C for 3 min, followed by 35 cycles of denaturation at 94°C for 30 s, annealing at 45°C for 30 s, and polymerization at 72°C for 1.5 min, with a final extension step at 72°C for 10 min. Following amplification, the aliquots were visually inspected using agarose gel (1.5% w/v) electrophoresis. The amplicons were purified by ethanol precipitation, and a second round of amplification was performed using the Big Dye Terminator v3.1 Cycle Sequencing system (Applied Biosystems, Foster City, CA, USA), which was followed by further purification. DNA sequencing was performed using the ABI3500xl automated genetic analyzer (Applied Biosystems, Foster City, CA, USA) at the State University of Campinas (Universidade Estadual de Campinas, Campinas, São Paulo, Brazil).

Dataset assembly, haplotypes, and demographic analysis

All sequences were manually edited using the Chromas Lite version 2.01 [37] software program and were aligned using the ClustalW tool from the BioEdit version 7.0 [38] software program. After editing and aligning the COI sequences, we determined the 658 bp consensus sequence, which was then posteriorly compared with the *H. zea* and *H. armigera* species barcodes [41] to determine homology using the BlastN tool, which is available online at NCBI [40].

The MEGA version 4 [41] software program was used to inspect the COI sequences from each species individually for the presence of numts [42]. In particular, we searched for the following numt signatures: (i) insertions/deletions (*indels*); (ii) stop codons leading to premature protein termination; and (iii) increased rates of non-synonymous mutations. The presence of signatures (i) and (ii) was considered sufficient to regard a sequence as a COI numt. In the presence of signatures (i) or (ii), signature (iii) was used to confirm the sequence as a numt. The presence of signature (iii) alone was not considered sufficient to define a sequence as a numt.

Haplotype and nucleotide diversity parameters for each species were estimated using the DnaSP version 5 [43] software program. Neutrality tests using Tajima's D [44] and Fu's Fs [45] were performed using the Arlequin version 3.1 [46] software program, and significance was determined using 1,000 random samples in coalescent simulations. Based on the recommendations in the Arlequin manual, we activated the "Infer from distance matrix" option under "Haplotype definition", and the Fu's Fs statistical values were considered to be significant at a level of 5% only when the *P*-value was below 0.02. The diversity estimates and neutrality tests were performed using all sampled individuals from each species, which were divided into winter-crop and summer-crop groups. A Mismatch Distribution Analysis using a spatial expansion model [21] was also performed using the Arlequin version 3.1 software program, and significance was determined using 1,000 bootstrap replicates. We used the goodness-of-fit of the observed mismatch distribution to the expected distribution from the spatial expansion model and the sum of square deviations (SSD) as a test statistic (*P*-value support).

Population structure analysis

Using Arlequin 3.1, we also performed an AMOVA at the two- and three-hierarchy levels [47]. For the three-hierarchy AMOVA, we first separated the samples depending on whether they were collected on winter or summer crops and then further divided them by host plant (monocotyledonae or dicotyledonae).

Network analysis and Bayesian phylogenies

Genetic differences and connections among *Helicoverpa* spp. haplotypes were determined by constructing a maximum parsimony network [48] using the TCS 1.21 software program [49]. To resolve ambiguities present in the haplotype network, we used the criteria of coalescence theory and population geography proposed by Crandall and Templeton [23].

We used the distance matrix option in the PAUP *4.0 software program to calculate the inter- and intra-species genetic distances, which were inferred using the nucleotide substitution model and the Akaike Information Criteria [50] selected by MODELTEST 2 [51]. The MrBayes v3.2 software program [52] was used to estimate Bayesian phylogenies. In particular, the Bayesian analysis was performed with 10 million generations using one cold and three heated chains. *Helicoverpa assulta* (Guenée) (GenBank Accession number: EU768937), *H. gelotopoeon* Dyar (EU768938), and *H. virescens* (IN799050) sequences were included as outgroups for the Bayesian analysis. We obtained a 50%-majority-rule consensus tree with posterior probabilities that were equal to the bipartition frequencies.

Network analysis: Brazil vs. Old World

Seventy-three sequences from a variety of Old World sites that were present in GenBank were included with the 139 *H. armigera* sequences we collected in Brazil. In particular, 73 sequences were obtained from specimens collected in China (N = 35) [GenBank Accession numbers GQ892840 - GQ892855, GQ995232 - GQ995244 [20], HQ132369 (Yang, 2010), JX392415, and JX392497 (not published)], Thailand (1) [(EU768935)], Australia (1) [(EU768936) [5]], Pakistan (2) [(JN988529 and JN988530) (not published)], Europe (28) [(FN907979, FN907980, FN907988, FN907989, FN907996 - FN907999, FN908000 - FN908003, FN908005, FN908006, FN908011, FN908013 - FN908018, FN908023, FN908026, GU654969, GU686757, GU686955, and JF415782) (not published)] and India (6) [(HM854928-HM854932 and JX32104) (not published)] (Table S2). This new data set was edited and aligned as follows. The sequences were different lengths; thus, the editing and alignment processes generated a total of 212 sequences 590 bp in length, excluding indels. The sequences from individuals collected in Brazil, which were previously analyzed using a fragment length of 658 bp, as entered into GenBank (see Table 1), were edited by removing the first 36 bp and the last 32 bp. Using the TCS 1.21 software program [49], we subjected this data set to haplotype network analysis using a maximum parsimony network [48] to investigate the genetic connections between haplotypes from Brazil and the Old World as well as to infer the origins of maternal lineages within *H. armigera* populations in Brazil.

Supporting Information

Figure S1 Bayesian phylogenetic tree of *H. armigera* and *H. zea* individuals sampled in Brazil. This phylogenetic tree is based on partial COI haplotype sequences and includes *H. assulta* and *H. gelotopoeon* sequences. Numbers near the interior branches indicate the posterior probability (×1,000) values. The outgroup used was *Heliothis virescens*. *H. armigera* COI haplotypes and Genbank Accession numbers can be found in Table S2.

Table S1 Hierarchical analysis of molecular variance (AMOVA), for population genetics structure of *Helicoverpa armigera* and *H. zea* with a mithocondrial (COI) region marker.

Table S2 Global *Helicoverpa armigera* including the Brazilian *H. armigera* haplotypes, and relevant GenBank Accession numbers. Numbers of individuals sequenced from each locality are indicated in parentheses.

Acknowledgments

We thank Celito Breda, Diego Miranda, Fábio Wazne, Germison Tomquelski, José Wilson de Souza, Marcos Michelotto, Milton Ide, Paulo Saran, Pedro Brugnera, Pedro Matana Junior, Rodrigo Franciscatti, Rodrigo Sorgatto, Rubem Staudt, Sérgio de Azevedo, and SGS Gravena (SISBIO License #18018-1) for helping to collect insect samples in different Brazilian regions. We also thank Jaqueline Campos for technical assistance and Prof. José Baldin Pinheiro for providing laboratory space and equipment.

Author Contributions

Conceived and designed the experiments: NAL AAP MIZ CO. Performed the experiments: NAL AAP. Analyzed the data: NAA ASC AAP MIZ CO. Contributed reagents/materials/analysis tools: CO MIZ ASC NAL AAP. Wrote the paper: NAL ASC AAP MIZ CO.

References

1. Pogue MG (2013) Revised status of *Chloridea* Duncan and (Westwood), 1841, for the *Heliothis virescens* species group (Lepidoptera: Noctuidae: Heliothinae) based on morphology and three genes. Syst Entomol 38: 523–542.
2. Mitter C, Poole RW, Matthews M (1993) Biosystematics of the Heliothinae (Lepidoptera: Noctuidae). Annu Rev Entomol 38: 207–225.
3. Laster ML, Hardee DD (1995) Intermating compatibility between north american *Helicoverpa zea* and *Heliothis armigera* (Lepidoptera: Noctuidae) from Russia. J Econ Entomol 88: 77–80.
4. Laster ML, Sheng CF (1995) Search for hybrid sterility for *Helicoverpa zea* in crosses between the north american *H. zea* and *H. armigera* (Lepidoptera: Noctuidae) from China. J Econ Entomol 88: 1288–1291.
5. Cho S, Mitchell A, Mitter C, Regier J, Matthews M, et al. (2008) Molecular phylogenetics of heliothine moths (Lepidoptera: Noctuidae: Heliothinae), with comments on the evolution of host range and pest status. Syst Entomol 33: 581–594.
6. Mallet J, Korman A, Heckel D, King P (1993) Biochemical genetics of *Heliothis* and *Helicoverpa* (Lepidoptera: Noctuidae) and evidence for a founder event in *Helicoverpa zea*. Ann Entomol Soc Am 86: 189–197.
7. Behere GT, Tay WT, Russell DA, Heckel DG, Appleton BR, et al. (2007) Mitochondrial DNA analysis of field populations of *Helicoverpa armigera* (Lepidoptera: Noctuidae) and of its relationship to *H. zea*. BMC Evol Biol 7: 117.
8. Fitt GP (1989) The ecology of *Heliothis* species in relation to agroecosystems. Annu Rev Entomol 34: 17–52.

9. Pogue M (2004) A new synonym of *Helicoverpa zea* (Boddie) and differentiation of adult males of *H. zea* and *H. armigera* (Hübner) (Lepidoptera: Noctuidae: Heliothinae). Ann Entomol Soc Am 97: 1222–1226.
10. Wu KM, Lu YH, Feng HQ, Jiang YY, Zhao JZ (2008) Suppression of cotton bollworm in multiple crops in China in areas with Bt toxin-containing cotton. Science 321: 1676–1678.
11. Degrande PE, Omoto C (2013) Estancar prejuízos. Cultivar Grandes Culturas Abril: 32–35.
12. Czepack C, Albernaz KC, Vivan LM, Guimarães HO, Carvalhais T (2013) Primeiro registro de ocorrência de *Helicoverpa armigera* (Hübner) (Lepidoptera: Noctuidae) no Brasil. Pesq Agropec Trop 43: 110–113.
13. Specht A, Sosa-Gómez DR, Paula-Moraes SV de, Yano SAC (2013) Identificação morfológica e molecular de *Helicoverpa armigera* (Lepidoptera: Noctuidae) e ampliação de seu registro de ocorrência no Brasil. Pesq Agropec Bras 48: 689–692.
14. Agropec Consultoria (2013) Pragas quarentenárias: consulta a dados sobre pragas quarentenárias presentes e ausentes no Brasil. Available: http://spp.defesaagropecuaria.com/. Accessed 2014 Jan 29.
15. Yang Y, Li Y, Wu Y (2013) Current status of insecticide resistance in *Helicoverpa armigera* after 15 years of Bt cotton planting in China. J Econ Entomol 106: 375–381.
16. Martin T, Ochou GO, Djihinto A, Traore D, Togola M, et al. (2005) Controlling an insecticide-resistant bollworm in West Africa. Agric Ecosyst Environ 107: 409–411.
17. MAPA (2014) Combate à praga *Helicoverpa armigera*. Brasilia: MAPA.
18. Tay WT, Soria MF, Walsh T, Thomazoni D, Silvie P, et al. (2013) A brave new world for an old world pest: *Helicoverpa armigera* (Lepidoptera: Noctuidae) in Brazil. Plos One 8: e80134.
19. Behere GT, Tay WT, Russell DA, Batterham P (2008) Molecular markers to discriminate among four pest species of *Helicoverpa* (Lepidoptera: Noctuidae). Bull Entomol Res 98: 599–603.
20. Li QQ, Li DY, Ye H, Liu XF, Shi W, et al. (2011) Using COI gene sequence to barcode two morphologically alike species: the cotton bollworm and the oriental tobacco budworm (Lepidoptera: Noctuidae). Mol Biol Rep 38: 5107–5113.
21. Rogers AR, Harpending H (1992) Population growth makes waves in the distribution of pairwise genetic differences. Mol Biol Evol 9: 552–569.
22. Excoffier L, Hofer T, Foll M (2009) Detecting loci under selection in a hierarchically structured population. Heredity 103: 285–298.
23. Crandall KA, Templeton AR (1993) Empirical tests of some predictions from coalescent theory with applications to intraspecific phylogeny reconstruction. Genetics 134: 959–969.
24. Nibouche S, Bues R, Toubon JF, Poitout S (1998) Allozyme polymorphism in the cotton bollworm *Helicoverpa armigera* (Lepidoptera: Noctuidae): comparison of African and European populations. Heredity 80: 438–445.
25. Endersby NM, Hoffmann AA, McKechnie SW, Weeks AR (2007) Is there genetic structure in populations of *Helicoverpa armigera* from Australia? Entomol Exp Appl 122: 253–263.
26. Albernaz KC, Silva-Brandao KL, Fresia P, Consoli FL, Omoto C (2012) Genetic variability and demographic history of *Heliothis virescens* (Lepidoptera: Noctuidae) populations from Brazil inferred by mtDNA sequences. Bull Entomol Res 102: 333–343.
27. Domingues FA, Silva-Brandão KL, Abreu AG, Perera OP, Blanco CA, et al. (2012) Genetic structure and gene flow among Brazilian populations of *Heliothis virescens* (Lepidoptera: Noctuidae). J Econ Entomol 105: 2136–2146.
28. Zhou XF, Faktor O, Applebaum SW, Coll M (2000) Population structure of the pestiferous moth *Helicoverpa armigera* in the Eastern Mediterranean using RAPD analysis. Heredity 85: 251–256.
29. Han Q, Caprio MA (2002) Temporal and spatial patterns of allelic frequencies in cotton bollworm (Lepidoptera: noctuidae). Environ Entomol 31: 462–468.
30. Asokan R, Nagesha S, Manamohan M, Krishnakumar N, Mahadevaswamy H, et al. (2012) Molecular diversity of *Helicoverpa armigera* Hübner (Noctuidae: Lepidoptera) in India. Orient Insects 46: 130–143.
31. Gunning RV, Dang HT, Kemp FC, Nicholson IC, Moores GD (2005) New resistance mechanism in *Helicoverpa armigera* threatens transgenic crops expressing *Bacillus thuringiensis* Cry1Ac toxin. Appl Environ Microbiol 71: 2558–2563.
32. Zhang X, Liang Z, Siddiqui ZA, Gong Y, Yu Z, et al. (2009) Efficient screening and breeding of *Bacillus thuringiensis* subsp. kurstaki for high toxicity against *Spodoptera exigua* and *Heliothis armigera*. J Ind Microbiol Biotechnol 36: 815–820.
33. Martin T, Ochou GO, Hala-N'Klo F, Vassal J-M, Vaissayre M (2000) Pyrethroid resistance in the cotton bollworm, *Helicoverpa armigera* (Hübner), in West Africa. Pest Manag Sci 56: 549–554.
34. Achaleke J, Brevault T (2010) Inheritance and stability of pyrethroid resistance in the cotton bollworm *Helicoverpa armigera* (Lepidoptera: Noctuidae) in Central Africa. Pest Manag Sci 66: 137–141.
35. Nair R, Kalia V, Aggarwal KK, Gujar GT (2013) Variation in the cadherin gene sequence of Cry1Ac susceptible and resistant *Helicoverpa armigera* (Lepidoptera: Noctuidae) and the identification of mutant alleles in resistant strains. Curr Sci 104: 215.
36. Folmer O, Black M, Hoeh W, Lutz R, Vrijenhoek R (1994) DNA primers for amplification of mitochondrial cytochrome c oxidase subunit I from diverse metazoan invertebrates. Mol Mar Biol Biotechnol 3: 294–299.
37. Technelysium Pty Ltd (1998–2005) Chromas lite version 2.01. Available: http://www.technelysium.com.au/chromas_lite.html.
38. Hall TA (1999) Bioedit: a user-friendly biological sequence alignment editor and analysis program for Windows 95/98/NT. Nucleic Acids Symposium Series.
39. Encyclopedia of Life. Available: http://www.eol.org. Accessed 2013 July 4.
40. Matten T (2002) The BLAST Sequence Analysis Tool. In: McEntyre J, Ostell J, editors. The NCBI Handbook [Internet]. Bethesda (MD): National Center for Biotechnology Information (US).
41. Tamura K, Dudley J, Nei M, Kumar S (2007) MEGA4: Molecular Evolutionary Genetics Analysis (MEGA) software version 4.0. Mol Biol Evol 24: 1596–1599.
42. Lopez JV, Yuhki N, Masuda R, Modi W, O'Brien SJ (1994) Numt, a recent transfer and tandem amplification of mitochondrial DNA to the nuclear genome of the domestic cat. J Mol Evol 39: 174–190.
43. Librado P, Rozas J (2009) DnaSP v5: A software for comprehensive analysis of DNA polymorphism data. Bioinformatics 25: 1451–1452.
44. Tajima F (1989) Statistical method for testing the neutral mutation hypothesis by DNA polymorphism. Genetics 123: 585–595.
45. Fu YX (1997) Statistical tests of neutrality of mutations against population growth, hitchhiking and background selection. Genetics 147: 915–925.
46. Excoffier L, Laval G, Schneider S (2005) Arlequin v. 3.0: an integrated software package for population genetics data analysis. Evol Bioinform Online 1: 47–50.
47. Excoffier L, Smouse PE, Quattro JM (1992) Analysis of molecular variance inferred from metric distances among DNA haplotypes - application to humam mitochondrial - DNA restriction data. Genetics 131: 479–491.
48. Templeton AR, Crandall KA, Sing CF (1992) A cladistic analysis of phenotypic associations with haplotypes inferred from restriction endonuclease mapping and DNA sequence data. III. Cladogram estimation. Genetics 132: 619–633.
49. Clement M, Posada D, Crandall KA (2000) TCS: a computer program to estimate gene genealogies. Mol Ecol 9: 1657–1659.
50. Akaike H (1974) A new look at the statistical model identification. IEEE T Automat Contr 19: 716–723.
51. Nylander JAA (2004) MrModeltest v2. Program distributed by the author. Uppsala University: Evolutionary Biology Centre.
52. Ronquist F, Teslenko M, van der Mark P, Ayres DL, Darling A, et al. (2012) MrBayes 3.2: efficient Bayesian phylogenetic inference and model choice across a large model space. Syst Biol 61: 539–542.

8

AtWuschel Promotes Formation of the Embryogenic Callus in *Gossypium hirsutum*

Wu Zheng[1], Xueyan Zhang[1], Zuoren Yang[1], Jiahe Wu[1,2], Fenglian Li[1], Lanling Duan[1], Chuanliang Liu[1], Lili Lu[1], Chaojun Zhang[1*], Fuguang Li[1*]

1 State Key Laboratory of Cotton Biology, Institute of Cotton Research, Chinese Academy of Agricultural Sciences, Anyang, Henan, China, **2** Institute of Microbiology, Chinese Academy of Sciences, Beijing, China

Abstract

Upland cotton (*Gossypium hirsutum*) is one of the most recalcitrant species for *in vitro* plant regeneration through somatic embryogenesis. Callus from only a few cultivars can produce embryogenic callus (EC), but the mechanism is not well elucidated. Here we screened a cultivar, CRI24, with high efficiency of EC produce. The expression of genes relevant to EC production was analyzed between the materials easy to or difficult to produce EC. Quantitative PCR showed that CRI24, which had a 100% EC differentiation rate, had the highest expression of the genes *GhLEC1*, *GhLEC2*, and *GhFUS3*. Three other cultivars, CRI12, CRI41, and Lu28 that formed few ECs expressed these genes only at low levels. Each of the genes involved in auxin transport (*GhPIN7*) and signaling (*GhSHY2*) was most highly expressed in CRI24, with low levels in the other three cultivars. WUSCHEL (WUS) is a homeodomain transcription factor that promotes the vegetative-to-embryogenic transition. We thus obtained the calli that ectopically expressed *Arabidopsis thaliana Wus* (*AtWus*) in *G. hirsutum* cultivar CRI12, with a consequent increase of 47.75% in EC differentiation rate compared with 0.61% for the control. Ectopic expression of *AtWus* in CRI12 resulted in upregulation of *GhPIN7*, *GhSHY2*, *GhLEC1*, *GhLEC2*, and *GhFUS3*. *AtWus* may therefore increase the differentiation potential of cotton callus by triggering the auxin transport and signaling pathways.

Editor: David D. Fang, USDA-ARS-SRRC, United States of America

Funding: This study was supported by National Science Fund for Distinguished Young Scholars (Grant no. 31125020). The funders had no role in study design, data collection and analysis, decision to publish, or preparation of the manuscript.

Competing Interests: The authors have declared that no competing interests exist.

* E-mail: zcj1999@yeah.net (CZ); aylifug@163.com (FL)

These authors contributed equally to this work.

Introduction

Somatic embryogenesis (SE) is a principal model for studying the growth and development of zygotic embryos in higher plants. This process includes callus induction, embryogenic callus (EC) formation, embryo development, and plant regeneration. During the past three decades, much effort has attempted to determine important genes controlling SE [1,2].

The gene *WUSCHEL* (*Wus*) is essential for stem cell formation and maintenance in shoot and root apical meristems [3,4]. *Wus* mediates stem cell homeostasis by regulating cell division and differentiation [5–7]. In *wus* mutants, apical meristems are unable to preserve the pool of undifferentiated cells [3]. The maintenance function of WUS can be repressed by inducing *AGAMOUS* (AG) expression and floral meristem differentiation [8]. *Wus* was first reported as the key gene promoting SE in *pga6* mutants in *Arabidopsis*, with overexpression of *PGA6/Wus* causing the vegetative-to-embryonic transition [9]. *Wus* is crucial for EC renewal during SE in *Arabidopsis* [10]. Overexpression of *Wus* in *Coffea canephora* can also promote SE [11].

Auxin is necessary for SE [12,13], but the auxin transport and signaling pathways during SE are not well understood. *PIN-FORMED* (*PIN*) genes encoding efflux carrier proteins are involved in auxin transport [14,15]. Genetic analysis indicates that *PIN1* is a major regulatory factor for auxin gradients in EC and embryo [16]. Auxin regulates auxin-responsive genes via the Aux/IAA (SHY)-ARF module. At sufficiently low auxin concentration, auxin response factors (*ARFs*) are repressed by Aux/IAA. At sufficiently high auxin concentration, Aux/IAA is degraded by SCF^{TIR1}, and *ARFs* are activated [17–19]. *ARFs* positively or negatively control downstream genes, resulting in responses to auxin signaling. Transcript profiling reveals that the auxin signaling pathway may play a vital role during SE in cotton [13].

LEC1, *LEC2*, and *FUS3* are key genes that control SE progression [20,21]. The capacity for SE is completely repressed in double (*lec1 lec2*, *lec1 fus3*, *lec2 fus3*) or triple (*lec1 lec2 fus3*) mutants of *Arabidopsis thaliana* [21]. *LEC2* expression changes rapidly during auxin responses [22], suggesting that *LEC/FUS* may be downstream genes in the auxin signaling pathway [21,23].

The majority of cotton cultivars are incapable of undergoing SE [24] because of their difficulty in inducing callus differentiation to form EC. Thus, most cultivars are not used for molecular breeding using transgenic technologies with *Agrobacterium*-mediated transformation via SE. Therefore, it is essential to study the mechanism of SE in cotton so as to improve regeneration of various cotton cultivars.

Here, we report the ectopic expression of *A. thaliana Wus* (*AtWus*) in *G. hirsutum* cv. CRI12, a cultivar that shows poor SE ability under established tissue culture methods. *AtWus* promoted differentiation of transgenic callus. Furthermore, ectopic expres-

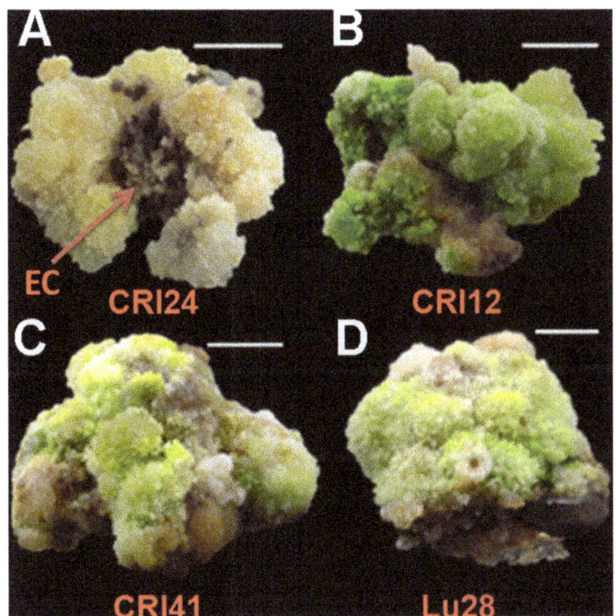

Figure 1. Cotton cultivars have different differentiation rates in EC. After 90 days of culture in NEIM, all calli of CRI24 produced EC, whereas calli of CRI12, CRI41, and Lu28 were dark green and tight and unable to differentiate into EC. Bar, 1 cm.

sion of *AtWus* could upregulate *GhLEC1* (*G. hirsutum LEC1*), *GhLEC2*, and *GhFUS3* expression during SE and alter auxin transport and signaling mechanisms. *AtWus* therefore promotes the efficiency of EC differentiation in cotton callus.

Materials and Methods

Plant Materials and Tissue Culture Conditions

We selected four cotton cultivars, CRI24, CRI12, CRI41 and Lu28, as experiment materials. CRI24 has a 100% EC differentiation rate and is the main transgenic material used for *Agrobacterium*-mediated method in our lab. CRI12 used to be an important basic breeding material because of traits of its relatively high yield and disease resistance, yet it has a low rate of differentiation during SE. CRI41 and Lu28, the main cultivars planted in China, can not undergo SE because of failure in EC induction.

Seeds of the four cotton cultivars were sterilized with 0.1% (w/v) mercuric chloride for 3 min. The seeds were then washed five times with sterilized distilled water and then germinated on modified Murashige and Skoog (MS) medium (25 g l^{-1} sucrose, 50 ml l^{-1} MSI, 5.6 g l^{-1} agar) for hypocotyl induction. Sterilized seeds were cultured at 28°C with a 14 h/10 h light/dark photoperiod. The hypocotyls from 7-day-old sterile seedlings were cut into 2 cm segments. For transgenic experiments using an *Agrobacterium*-mediated method [25], the hypocotyl cuts of CRI12 were transferred to 250 ml flasks and placed on callus-induction medium (CIM; MS medium plus B5 vitamins, supplemented with 0.05 mg l^{-1} 3-indole acetic acid (IAA), 0.05 mg l^{-1} kinetin, 0.05 mg l^{-1} 2,4-dichlorophexoxyacetic acid, 25 g l^{-1} glucose, 2 g l^{-1} gelrite gellan gum, 50 mg l^{-1} kanamycin, 100 mg l^{-1} cefotaxime, pH 5.8). The medium was changed once per month. After 2 months of culture, all calli were transferred onto EC induction medium (EIM; MSB supplemented with 25 g l^{-1} glucose, 2 g l^{-1} gelrite, 0.5 g l^{-1} $MgCl_2$, 0.16 mg l^{-1} kinetin, 0.08 mg l^{-1} IAA, 50 mg l^{-1} kanamycin, 100 mg l^{-1} cefotaxime, pH 6.5). The medium was changed monthly. After 4 months of culture, ECs were transferred to somatic embryo induction medium (SIM; MSB supplemented with 25 g l^{-1} glucose, 2 g l^{-1} gelrite, 0.5 g l^{-1} $MgCl_2$, 0.08 g l^{-1} kinetin, 0.12 mg l^{-1} 6-benzylaminopurine, 50 mg l^{-1} kanamycin, 100 mg l^{-1} cefotaxime, pH 6.8), and the medium was refreshed monthly. For non-transgenic experiments, the hypocotyl cut explants of CRI24, CRI12, CRI41, and Lu28 were transferred onto NCIM (CIM lacking kanamycin and cefotaxime), and the medium was changed once per month. After 30 days of culture, all calli were transferred onto NEIM (EIM lacking kanamycin and cefotaxime), and the medium was changed monthly. After 90 days of culture, all calli of CRI24 differentiated into EC, most calli of CRI12 and all calli of CRI41 and Lu28 did not differentiate into ECs. To confirm the lack of capacity for EC differentiation among the latter three cultivars, the calli which can not differentiate in ECs were cultured in NEIM for another 120 days, with the medium being refreshed monthly. Indeed, those calli did not differentiate into ECs, and thus no further experiments were conducted with these non-transgenic cultivars.

Gene Cloning and Vector Construction

The nucleotide sequences of *GhLEC1*, *GhLEC2*, and *GhFUS3* were obtained from the D subgenome database of *Gossypium. raimondii* by comparing with amino acid sequences of *AtLEC1*, *AtLEC2*, and *AtFUS3* using the tblastn tool. The three genes were then amplified from a full-length cDNA library of CRI24 with specific primers (Table S1). For ectopic expression of *AtWus*, the full-length coding regions (CDS) of the gene was cloned from wild-type *Arabidopsis* (Columbia ecotype) (Table S1). The full-length CDS of *AtWus* was amplified via PCR with specific primers (Table S1) and ligated into vector pMD18-T. After verifying the sequence, each of the *AtWus* fragment and Vector pBI121 was digested with *Bam*H I and *Sac* I. and the *AtWus* fragment was

Table 1. EC induction in four cotton cultivars.

	CRI12 (WT)			CRI24 (WT)			CRI41 (WT)			Lu28 (WT)		
Replication	**C[a]**	**EC[b]**	**EC rate (%)**	**C[a]**	**EC[b]**	**EC rate (%)**	**C[a]**	**EC[b]**	**EC rate (%)**	**C[a]**	**EC[b]**	**EC rate (%)**
1	175	1	0.57	371	371	100	177	0	0.00	209	0	0.00
2	221	3	1.36	204	204	100	303	0	0.00	242	0	0.00
3	286	0	0.00	261	261	100	319	0	0.00	138	0	0.00
Average	682	4	0.59	836	836	100	799	0	0.00	589	0	0.00

[a]Number of explants forming a callus.
[b]Number of EC forming from a callus. WT, wild type.

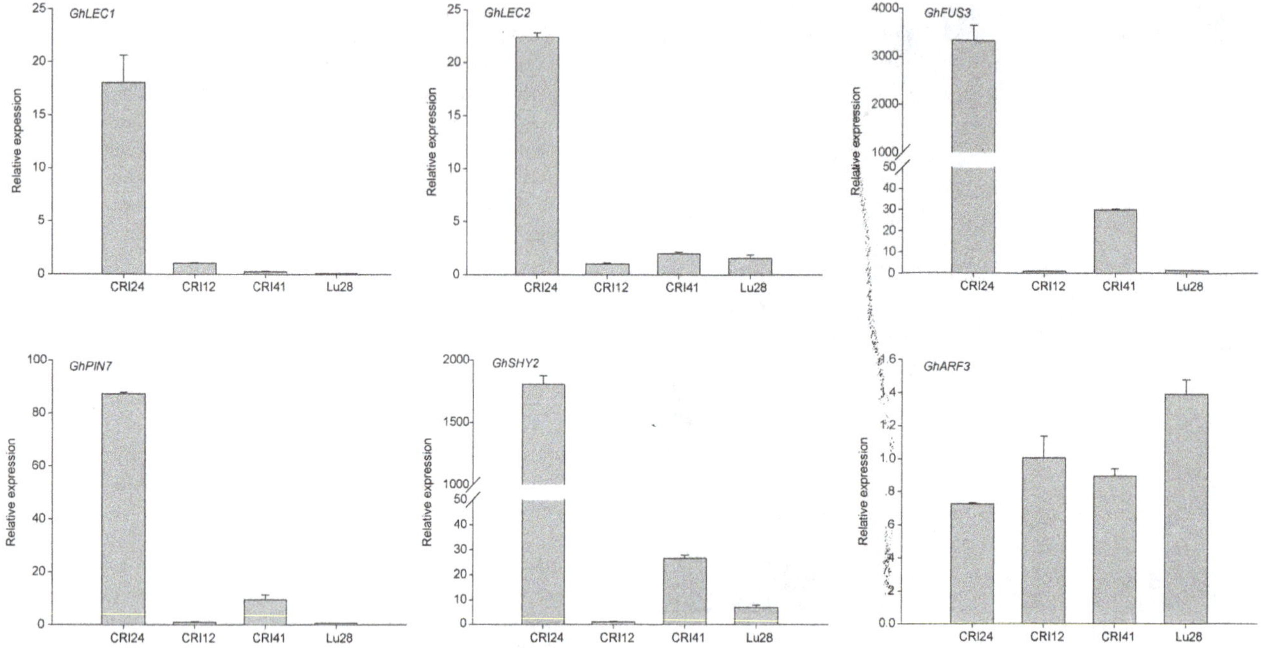

Figure 2. Analysis of gene expression in the non-transgenic calli of the four cultivars.

inserted into pBI121. The nucleotide sequences of *GhPIN7*, *GhSHY2*, and *GhARF3* were obtained from the D subgenome database of *G. raimondii* by comparing with amino acid sequences of *AtPIN1*, *AtSHY2*, and *AtARF1* using the tblastn tool.

RNA Extraction

All calli of CRI24, CRI12, CRI41 and Lu28 cultured for 90 days in NEIM and of 35S:WUS and CK lines cultured for 4 months in EIM were stored at −80°C. We extracted RNA of the above samples using a modified CTAB method [26]. RNA samples with A260/A280 ratios between 1.8 and 2.0 and A260/A230 ratios >1.5 were considered acceptable.

Quantitative Real Time PCR (QPCR)

Approximately 1 μg total RNA samples were reverse transcribed using the PrimeScript RT reagent kit with gDNA Eraser (Takara). The cDNA templates were diluted three times prior to amplification. The QPCR experiment was conducted according to the guidelines of SYBR Premix Ex Taq^{TM} kit (Takara). QPCR was performed in 96-well plates with a total volume of 20 μL containing 10 μL 2× SYBR Premix Ex Taq^{TM}, 6.8 μL PCR-grade water, 2 μL cDNA template, 0.4 μL 50× ROX reference dye I, and 0.4 μL each of forward and reverse primers (10 μM). All QPCRs were run with three technical replicates on an ABI 7900 Real-Time PCR system (Applied Biosystems). The thermal cycling conditions were as follows: an initial denaturation step of 30 s at 95°C, followed by 40 cycles of 95°C annealing for 5 s and 60°C extension for 30 s. The primers used for QPCR are shown in Table S2.

Scanning Electron Microscopy

Scanning electron microscopy was performed on somatic embryos obtained after 1.5 months culture on SIM. Samples were prefixed at room temperature for 12 h in 2.5% (v/v) glutaraldehyde (phosphate buffer, pH 7.2). After dehydration using a graded ethanol series, samples were dried with a CO_2

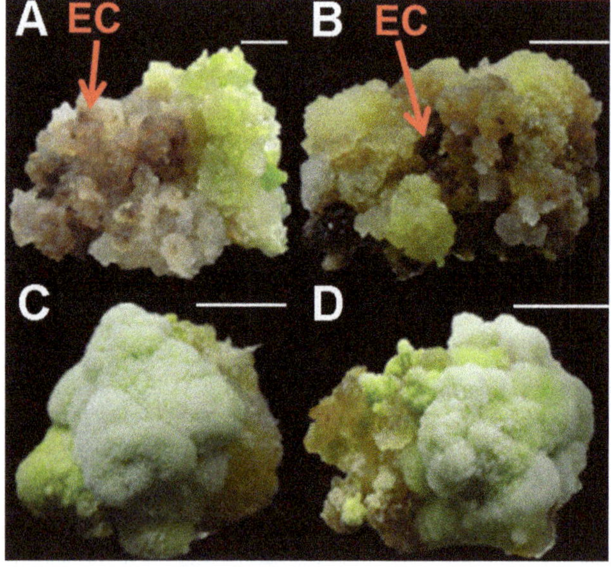

Figure 3. The callus of 35S:WUS and CK cultured for 1.5 months in EIM. A, B: Calli of 35S:Wus lines at the beginning of EC formation, **C, D:** Calli of CK lines were unable to differentiate. Bar, 1 cm.

critical-point drying system (HITACHI HCP-2). Subsequently, samples were sputtered with gold dust and observed under a HITACHI S-530 scanning electron microscopy.

Statistics

The rate of EC = number of EC/number of calli. After 45 days of culture in SIM, we determined the weight of the abnormal embryos in 35S:WUS lines and normal cotyledonary embryos in CK lines with three technical replicates. The average weight of each individual somatic embryo = weight of 10 somatic embryos/

Table 2. EC induction in CRI12.

	35S:WUS			Empty vector (CK)		
Replication	C[a]	EC[b]	EC rate (%)	C[a]	EC[b]	EC rate (%)
1	173	88	50.66	151	2	1.32
2	201	94	46.77	114	0	0.00
3	93	41	44.09	223	1	0.45
Average	467	223	47.75**	488	3	0.61

[a]Number of explants forming callus.
[b]Number of EC forming from a callus. **$p<0.01$.

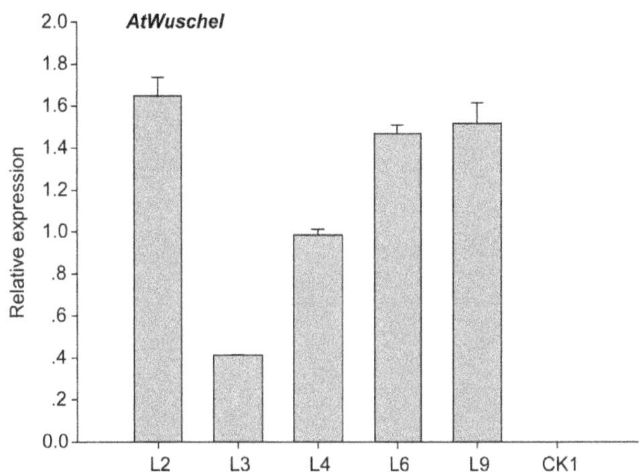

Figure 4. *AtWus* expression in calli of CRI12.

10. We conducted t-tests to determine significant differences ($p<0.05$ or $p<0.01$, depending on the experiment).

Results

Expression of *GhLEC* and *GhFUS3* in Cultivars with Diverse Differentiation Rates

We used four cotton cultivars (*G. hirsutum*) for tissue culture. The EC differentiation rate of CRI24 was 100% but only 0.59% for CRI12, whereas the calli of Lu28 and CRI41 could not form EC (Table 1). After culturing hypocotyl segments for 30 days on NCIM (callus induction medium lacking kanamycin and cefotaxime), the induced calli were transferred onto NEIM (EC induction medium lackling kanamycin and cefotaxime). After 90 days in NEIM, all the CRI24 calli were non-compacted and had differentiated into EC (Figure 1**A**). However, most calli of CRI12 and all calli of CRI41 and Lu28 were compacted, dark green, and did not differentiate (Figure 1**B**–**D**).

LEC1, *LEC2*, and *FUS3* are essential for SE [21], but there are few studies on the roles of *LEC* and *FUS3* in SE in plants other than *Arabidopsis*. Hence, we first isolated the homologs *GhLEC1*, *GhLEC2*, *GhFUS3* in CRI24 [27]. At the amino acid level, these homologs share high sequence similarity with those of in *Arabidopsis* (Figure S1). The results implied that *GhLEC1*, *GhLEC2*, and *GhFUS3* may have functions similar to those of the *Arabidopsis* homologs that control the capacity for SE [21].

To investigate the expression of *GhLEC1*, *GhLEC2*, and *GhFUS3* in calli of the four cultivars, we carried out QPCR on calli cultured for 90 days on NEIM. The expression levels of *GhLEC1*, *GhLEC2*, and *GhFUS3* were significantly higher in the calli of CRI24 than in CRI12, Lu28, and CRI41, with barely detectable expression of *GhFUS3* in CRI12 and Lu28 (Figure 2).

Expression of Genes Involved in Auxin Transport or Signaling Pathways

Auxin plays an important role in SE of cotton [13,28], but genes involved in auxin transport (*PINs*) and auxin signaling are not well studied in cotton SE. QPCR analysis of expression levels of these genes in SE used the same samples as for expression patterns of *LECs* and *FUS3*. *GhPIN7* and *GhSHY2* (*AUX/IAA2*) levels in calli of CRI24 were much higher than in the other three cultivars (Figure 2). In contrast, *GhARF3* expression was lower in CRI24 than in other cultivars (Figure 2), suggesting that *GhARF3* expression may be inhibited by the *GhSHY2* gene product. These results indicated that the auxin transport and signaling pathways were more active in calli with high EC differentiation rates than those with low EC differentiation rates.

Ectopic Expression of *AtWus* Improves EC Induction

CRI12 is one of the most recalcitrant cotton cultivars for plant regeneration via SE. To study the function of *AtWus* in SE, we overexpressed *AtWus* in CRI12 (35S:WUS). An empty vector was used as the control (CK) for parallel transformation. Hypocotyl segments transformed with 35S:WUS or the control CK were cultured on CIM containing kanamycin and cefotaxime to induce resistant calli. One month later, these segments were transferred to fresh CIM. There were no apparent differences between 35S:WUS and CK within the first 2 months. Then, calli that formed on segments were transferred to EIM containing kanamycin and cefotaxime for inducing EC. Calli were transferred to fresh EIM monthly. After culture for 1.5 months on EIM, most 35S:WUS calli were non-compacted and light green, and some began to differentiate into EC (Figure 3**A, B**). However, the CK calli were compact and dark green and did not undergo EC differentiation (Figure 3**C, D**). After culturing for 4 months on EIM, the differentiation rate of 35S:WUS transformants was 47.75% compared with 0.61% for CK (Table 2).

QPCR was used to determine *AtWus* expression level in transformed calli. The calli of transgenic cultures (L2, L3, L4, L6, and L9) cultured for 4 months in EIM were selected for analysis, whereas the calli of CK that did not differentiate into EC served as the control (CK1). The data revealed *AtWus* was overexpressed in 35S:WUS cultures (Figure 4).

Ectopic Expression of *AtWus* Regulates Auxin Transport and Signal Transduction

AtWus induction by IAA can regulate *AtPIN* expression during SE in *Arabidopsis* [10]. Hence, the role of *AtWus* in auxin transport and signal transduction in cotton was studied. Transgenic calli of L2, L3, L4, L6, and L9 cultured for 4 months in EIM were analyzed by QPCR, with CK1 calli serving as a control. *GhPIN7* expression was higher in 35S:WUS transformed callus lines than in CK (Figure 5).

Transcript profiling during SE in cotton has been used to establish the association between the auxin signaling pathway and callus differentiation [13]. The above-mentioned data also demonstrated an interaction between the auxin signaling pathway and EC induction whereby *GhSHY2* transcripts were increased in 35S:WUS lines compared with the CK1 line. However, *GhARF3* transcripts were reduced in 35S:WUS, possibly owing to suppression by *GhSHY2* (Figure 5).

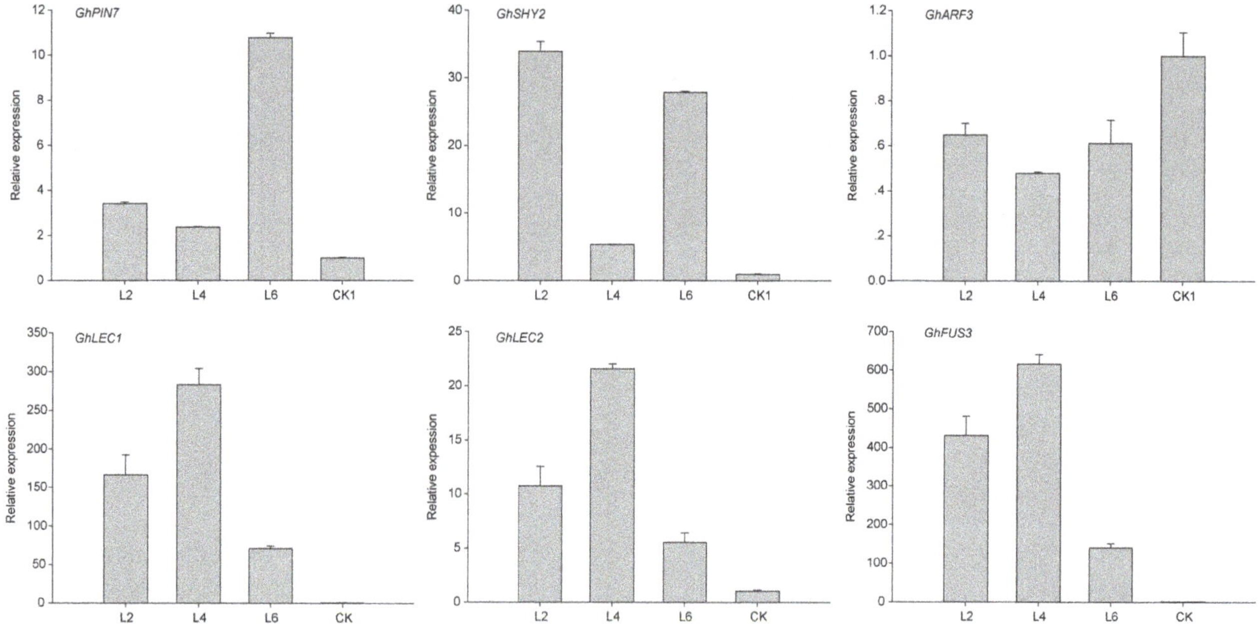

Figure 5. Analysis of gene expression in the transgenic calli carrying *AtWus*.

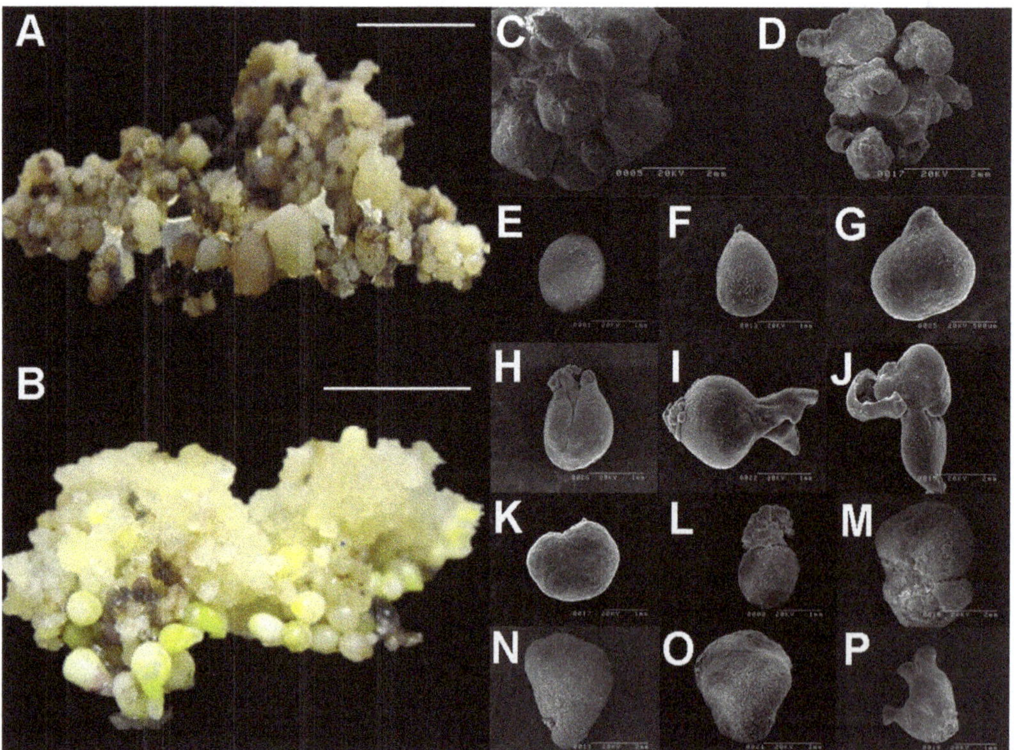

Figure 6. AtWus overexpression results in abnormal development of somatic embryos. A: Many abnormal somatic embryos were produced in 35S:WUS lines, and the somatic embryos were inflated and lacked cotyledons. **B:** Formation of normal somatic embryos in CK lines at different stages. Scanning electron microscopy: Holistic perspective of somatic embryos in 35S:WUS lines (**C**) and CK lines (**D**). **E–J:** Normal somatic embryos at different stages. **E, F:** globular embryo. **G:** heart-shape embryo. **H–J:** cotyledonary embryo. **K–P:** Abnormal somatic embryos having various appearance. **O:** leaf-like embryo. **P:** multiple-cotyledon embryo. Bar in **A** or **B,** 1 cm.

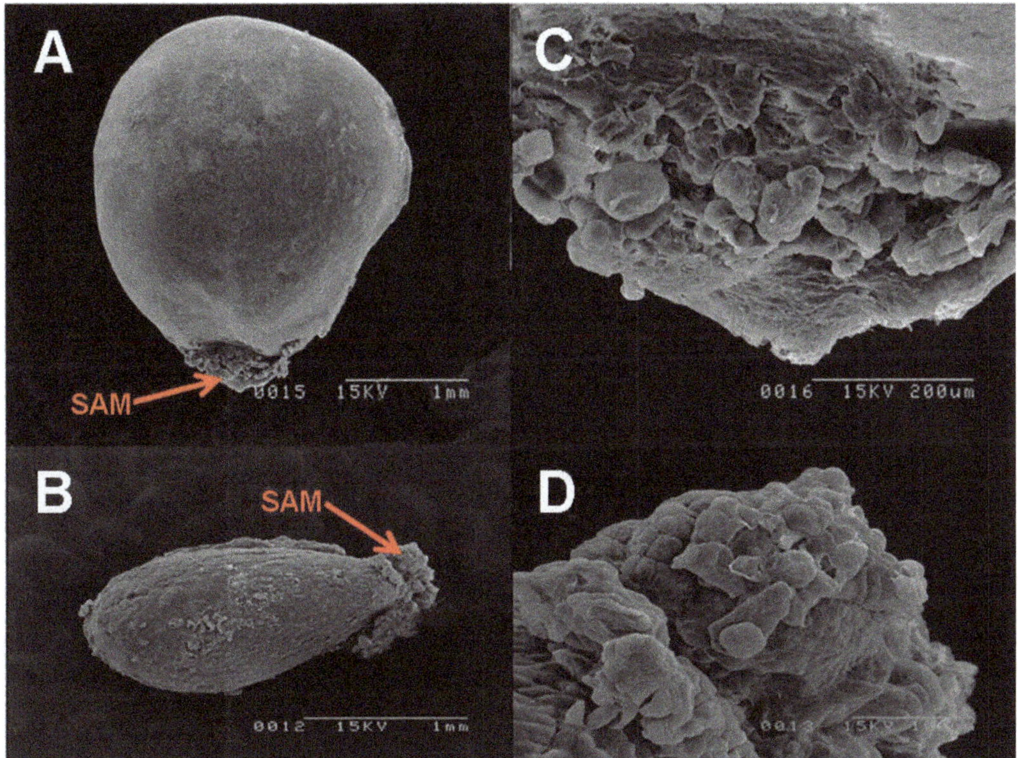

Figure 7. SAM structure of somatic embryos in transgenic lines observed with scanning electron microscopy. A: Abnormal somatic embryo of 35S:WUS lines. **B:** Normal somatic embryo of CK lines. **C:** Enlarged image of SAM in **A**. **D:** Enlarged image of SAM in **B**.

AtWus Activates the Expression of *LEC* and *FUS3* during SE in Cotton

GhLEC1, *GhLEC2*, and *GhFUS3* were expressed in calli of CRI24 but only barely detected in CRI12. To characterize the regulatory relation between *AtWus* and *LEC/FUS3*, we selected L2, L4, and L6 callus lines cultured for 4 months in EIM for further study. QPCR analysis revealed that *GhLEC1*, *GhLEC2*, and *GhFUS3* transcript levels were low in calli of CK1. In calli of L2, L4 and L6, however, higher expression levels of the three genes were detected owing to upregulation by *AtWus* (Figure 5). Hence, *AtWus* promoted the expression of *GhLEC1*, *GhLEC2*, and *GhFUS3* during SE in cotton, similar to *Arabidopsis* [10].

AtWus Overexpression Results in Abnormal Development of Somatic Embryos

For somatic embryo induction, 4-month-old EC of 35S:WUS and CK lines were transferred to SIM containing kanamycin and cefotaxime for somatic embryo induction. ECs were transferred onto a fresh SIM once per month and cultured for 50 days on SIM. EC formed from calli of CK lines grew into normal-looking globular, heart-shaped, and cotyledonary embryos (Figure 6 **B**). However, ECs that developed from 35S:WUS callus lines grew into various abnormal translucent embryos (Figure 6**A**) that could not form the normal-looking cotyledonary embryos except for some leaf-like or multi-cotyledon embryos, although they produced more somatic embryos than CK (Figure 6**C, D**). Scanning electron microscopy was used to investigate the structural abnormalities of these embryos. Somatic embryos from 35S:WUS lines were much larger and heavier than those from CK lines (Figure 6**C**, Table 3). In CK lines, globular, heart-shaped and cotyledonary embryos were clearly observed (Figure 6**E–J**). However, somatic embryos of 35S:WUS lines exhibited several abnormal morphologies, such as leaf-like or multi-cotyledon embryos (Figure 6**K–P**). Cotyledons are derived from the shoot apical meristem (SAM) region. SAM cells were arranged in an organized manner in CK somatic embryos (Figure 7**B, D**), but the SAM cells of 35S:WUS lines were disorganized and rather unstructured (Figure 7**A, C**).

Table 3. Somatic embryo weight in CRI12.

Replication	35S:WUS (g)	CK (g)
1	0.0138	0.00764
2	0.0124	0.00445
3	0.0181	0.00530
Average	0.0148	0.00580

*p<0.05.

Discussion

Wus is a key gene that controls renewal of stem cells in the apical meristem [29]. However, because *AtWus* promotes the vegetative-to-embryonic transition during SE in *Arabidopsis*, numerous studies have elucidated its functions during SE in several plant species [30]. During preparation of the current report, *AtWus* was shown to promote SE differentiation in *G. hirsutum* Coker cotton [31], a cultivar that easily undergoes SE. However, it is unknown if these results apply to other cotton genotypes. Furthermore, little work has been done on the regulatory role of *Wus* in the auxin signaling pathway during SE.

Here, we examined the role and molecular mechanism of *Wus*-promoted SE in *G. hirsutum*. Cultivar CRI12, a genotype that is difficult to regenerate. We constitutively overexpressed *AtWus* in CRI12 using an *Agrobacterium*-mediated transformation and found that *AtWus* can promote the formation of non-compact light green calli that induce EC easily. The EC differentiation rate was higher in 35S:WUS of CRI12 (47.75%) in comparison with <1% in CK. Hence, *AtWus* is a possible candidate that could promote the nonembryonic-to-embryonic transition during SE in cotton.

Transcript analysis has revealed that the auxin transport and signaling pathways may play a substantive role in EC induction [13]. Our QPCR results showed that *GhPIN*s and *GhSHY*/*GhARF* may regulate the efficiency of callus differentiation into EC. *PINs* play a fundamental role in embryonic auxin distribution in plant embryos [16], and *AtPIN1* is associated with the establishment of auxin gradients during SE in *Arabidopsis*. Previous studies revealed diverse changes in the endogenous auxin levels when EC was induced [13]. The antisense cDNA of *AtWus* suppresses *PIN*s during SE in *Arabidopsis* [10]. When auxin levels are low, *SHY* (*AUX/IAA*) expression is induced and certain *ARF* activity is repressed, but at high auxin levels SHY is degraded by $SCF^{TIR1/AFBs}$ and the *ARF* inhibitory action is abolished [17,19]. Our results revealed enhanced *GhSHY2* expression and repressed *GhARF3* expression in CRI12 calli overexpressing *AtWus*. However, *AtWus* is unable to modify the amount of auxin in cotton calli [31]. Hence, *AtWus* may play an important role in upregulating *GhPIN*s to redistribute auxin gradients, which may alter expression patterns of *SHY-ARF* at low levels of auxin.

LEC1, *LEC2*, and *FUS3* are crucial for SE. The capacity of SE is almost completely repressed in double and triple mutants of the three genes, indicating that *LEC* genes may function downstream of endogenous auxin-induced SE in *Arabidopsis* [21,23]. In our present study, *GhLEC1*, *GhLEC2*, and *GhFUS3* transcript levels were extremely low in callus that was unable to differentiate into EC, but levels were high in callus producing EC. Hence, *AtWus* positively regulated *LECs* and *FUS3*. In *Arabidopsis*, WUSCHEL and PGA37/MYB118 promote SE and activate the expression of *LEC1*, *LEC2*, and *FUS3* [32]. In our study, *GhLEC1*, *GhLEC2*, and *GhFUS3* were also upregulated in callus of 35S:WUS lines. These results suggest that *AtWus* may alter *PIN* expression, which leads to the establishment of new auxin gradients in the callus. Subsequently, a new auxin response was formed and stimulated *GhLEC1*, *GhLEC2*, and *GhFUS3* in the callus of CRI12. *AtWus* may provoke the ability of differentiation in the callus by reactivating *GhLECs* and *GhFUS3* expression through auxin transport and signaling mechanisms.

Although *AtWus* improved EC induction, the observed abnormal somatic embryos were an unexpected consequence of *AtWus* overexpression and prevented seedling generation. *AtWus* expression is limited to the SAM because auxin accumulates in the cotyledon primordial cells during somatic embryo development in *Arabidopsis* [10]. Therefore, ectopic expression of *Wus* could cause loss of expression specificity in somatic embryos, leading to asymmetric growth. In *Arabidopsis*, *LEC1* and *FUS3* may control multiple aspects of seed development [33]. Constitutive expression of *LEC1* leads to occasional formation of somatic embryo like structures [34]. Therefore, constitutive expression of *AtWus* in CRI12 may have led to constitutive expression of *GhLEC1*, *GhLEC2* and *GhFUS3* in embryos, and then the expression of *GhLECs* and *GhFUS3* may have resulted in the observed abnormal embryos and failure of seedling regeneration. Using inducible promoters such as the estradiol-inducible promoter [35] rather than the 35S promoter during SE in cotton may avoid abnormal embryo formation. Estradiol could be added into the CIM and EIM for EC induction with subsequent transfer of ECs onto SIM without estradiol for normal embryo development. This promoter has been successfully applied for SE in several species [30].

Cotton is an important source of textile fiber and edible oil, but cotton yield is adversely affected by abiotic or biotic stresses [36]. Therefore, efforts to improve cotton resistance against such stresses by genetic modification may play a vital role in efforts to increase production. *Agrobacterium*-mediated transformation via SE has been the most popular transgenic technology in cotton. Most genotypes cannot undergo EC induction or have low rates of differentiation, although many of those recalcitrant genotypes have certain positive agronomic characters [37]. Thus, the difficulty of EC induction always restricts the application of transgenic breeding and *in vitro* regeneration in additional cultivars. For example, CRI12 used to be an important elite cultivar widely cultured in China for its disease resistance, high yeild and superior fiber quality. However, SE production in this cultivar is not easy, making it difficult to improve traits using transgenic technology. In our study, the introduction of *AtWus* into the recalcitrant cotton cultivars enhanced their somatic embryogenesis. With this foundation established, we may now construct a vector to overexpress *AtWus* with an estradiol-inducible promoter and transfer it into CRI12 or other cultivars. This will require a simple addition of estradiol to CIM and EIM to ensure EC induction with high frequency and production of transgenic seedlings. This will then enable us to improve the rate of cotton transformation for many foreign genes or cotton genes. Such transgenic plants can be used directly as germplasm for cotton breeding. This protocol will achieve our goals of creating more germplasm resources that facilitate SE and expand the scope of transgenic breeding in more cultivars.

Supporting Information

Figure S1 Characterization of *GhLEC1*, *GhLEC2* and *GhFUS3*.

Table S1 Specific primers for cloning *GhLEC1*, *GhLEC2*, *GhFUS3* and *AtWuschel*.

Table S2 Primers used for quantitative real time PCR.

Acknowledgments

We are grateful to Lihua Ma and Tianping Suo (Cotton Research Institute, Chinese Academy of Agricultural Sciences) for their technical assistance in scanning electronic microscopy. We thank Dr. Jianru Zuo (State Key Laboratory of Plant Genomics, Institute of Genetics and Developmental Biology, Chinese Academy of Sciences) for helpful suggestions and discussions.

Author Contributions

Conceived and designed the experiments: Fuguang Li CZ WZ. Performed the experiments: WZ. Analyzed the data: WZ Fuguang Li CZ JW XZ ZY. Contributed reagents/materials/analysis tools: CZ LD Fuguang Li CL Fenglian Li ZY. Wrote the paper: WZ. Check writing: LL JW.

References

1. Schmidt ED, Guzzo F, Toonen MA, de Vries SC (1997) A leucine-rich repeat containing receptor-like kinase marks somatic plant cells competent to form embryos. Development 124: 2049–2062.
2. Thomas TL (1993) Gene expression during plant embryogenesis and germination: an overview. Plant Cell 5: 1401.
3. Laux T, Mayer KF, Berger J, Jurgens G (1996) The WUSCHEL gene is required for shoot and floral meristem integrity in *Arabidopsis*. Development 122: 87–96.
4. Kamiya N, Nagasaki H, Morikami A, Sato Y, Matsuoka M (2003) Isolation and characterization of a rice WUSCHEL-type homeobox gene that is specifically expressed in the central cells of a quiescent center in the root apical meristem. Plant J 35: 429–441.
5. Yadav RK, Reddy GV (2011) WUSCHEL-mediated cellular feedback network imparts robustness to stem cell homeostasis. Plant Signal Behav 6: 544–546.
6. Yadav RK, Perales M, Gruel J, Girke T, Jonsson H, et al. (2011) WUSCHEL protein movement mediates stem cell homeostasis in the *Arabidopsis* shoot apex. Genes Dev 25: 2025–2030.
7. Yadav RK, Tavakkoli M, Reddy GV (2010) WUSCHEL mediates stem cell homeostasis by regulating stem cell number and patterns of cell division and differentiation of stem cell progenitors. Development 137: 3581–3589.
8. Lenhard M, Bohnert A, Jurgens G, Laux T (2001) Termination of stem cell maintenance in *Arabidopsis* floral meristems by interactions between WUSCHEL and AGAMOUS. Cell 105: 805–814.
9. Zuo J, Niu QW, Frugis G, Chua NH (2002) The WUSCHEL gene promotes vegetative-to-embryonic transition in *Arabidopsis*. Plant J 30: 349–359.
10. Su YH, Zhao XY, Liu YB, Zhang CL, O'Neill SD, et al. (2009) Auxin-induced WUS expression is essential for embryonic stem cell renewal during somatic embryogenesis in *Arabidopsis*. The Plant Journal 59: 448–460.
11. Arroyo-Herrera A, Gonzalez AK, Moo RC, Quiroz-Figueroa FR, Loyola-Vargas VM, et al. (2008) Expression of WUSCHEL in *Coffea canephora* causes ectopic morphogenesis and increases somatic embryogenesis. Plant Cell Tissue and Organ Culture 94: 171–180.
12. Ikeda-Iwai M, Satoh S, Kamada H (2002) Establishment of a reproducible tissue culture system for the induction of *Arabidopsis* somatic embryos. J Exp Bot 53: 1575–1580.
13. Yang X, Zhang X, Yuan D, Jin F, Zhang Y, et al. (2012) Transcript profiling reveals complex auxin signalling pathway and transcription regulation involved in dedifferentiation and redifferentiation during somatic embryogenesis in cotton. BMC Plant Biol 12: 110.
14. Wisniewska J, Xu J, Seifertova D, Brewer PB, Ruzicka K, et al. (2006) Polar PIN localization directs auxin flow in plants. Science 312: 883.
15. Zazimalova E, Krecek P, Skupa P, Hoyerova K, Petrasek J (2007) Polar transport of the plant hormone auxin - the role of PIN-FORMED (PIN) proteins. Cell Mol Life Sci 64: 1621–1637.
16. Weijers D, Sauer M, Meurette O, Friml J, Ljung K, et al. (2005) Maintenance of embryonic auxin distribution for apical-basal patterning by PIN-FORMED-dependent auxin transport in *Arabidopsis*. Plant Cell 17: 2517–2526.
17. Goh T, Kasahara H, Mimura T, Kamiya Y, Fukaki H (2012) Multiple AUX/IAA-ARF modules regulate lateral root formation: the role of *Arabidopsis* SHY2/IAA3-mediated auxin signalling. Philos Trans R Soc Lond B Biol Sci 367: 1461–1468.
18. Sauer M, Balla J, Luschnig C, Wisniewska J, Reinohl V, et al. (2006) Canalization of auxin flow by Aux/IAA-ARF-dependent feedback regulation of PIN polarity. Genes Dev 20: 2902–2911.
19. Weijers D, Benkova E, Jager KE, Schlereth A, Hamann T, et al. (2005) Developmental specificity of auxin response by pairs of ARF and Aux/IAA transcriptional regulators. EMBO J 24: 1874–1885.
20. Fambrini M, Durante C, Cionini G, Geri C, Giorgetti L, et al. (2006) Characterization of LEAFY COTYLEDON1-LIKE gene in *Helianthus annuus* and its relationship with zygotic and somatic embryogenesis. Development genes and evolution 216: 253–264.
21. Gaj MD, Zhang S, Harada JJ, Lemaux PG (2005) Leafy cotyledon genes are essential for induction of somatic embryogenesis of *Arabidopsis*. Planta 222: 977–988.
22. Stone SL, Braybrook SA, Paula SL, Kwong LW, Meuser J, et al. (2008) *Arabidopsis* LEAFY COTYLEDON2 induces maturation traits and auxin activity: implications for somatic embryogenesis. Proceedings of the National Academy of Sciences 105: 3151–3156.
23. Stone SL, Kwong LW, Yee KM, Pelletier J, Lepiniec L, et al. (2001) LEAFY COTYLEDON2 encodes a B3 domain transcription factor that induces embryo development. Proceedings of the National Academy of Sciences 98: 11806–11811.
24. Hu L, Yang X, Yuan D, Zeng F, Zhang X (2011) GhHmgB3 deficiency deregulates proliferation and differentiation of cells during somatic embryogenesis in cotton. Plant Biotechnol J 9: 1038–1048.
25. Jin S, Zhang X, Nie Y, Guo X, Huang C (2005) Factors affecting transformation efficiency of embryogenic callus of upland cotton (*Gossypium hirsutum*) with *Agrobacterium tumefaciens*. Plant cell, tissue and organ culture 81: 229–237.
26. Wan C-Y, Wilkins TA (1994) A Modified Hot Borate Method Significantly Enhances the Yield of High-Quality RNA from Cotton (*Gossypium hirsutum L.*). Analytical biochemistry 223: 7–12.
27. Wang K, Wang Z, Li F, Ye W, Wang J, et al. (2012) The draft genome of a diploid cotton *Gossypium raimondii*. Nat Genet 44: 1098–1103.
28. Xu Z, Zhang C, Zhang X, Liu C, Wu Z, et al. (2013) Transcriptome Profiling Reveals Auxin and Cytokinin Regulating Somatic Embryogenesis in Different Sister Lines of Cotton Cultivar CCRI24. J Integr Plant Biol.
29. Mayer KFX, Schoof H, Haecker A, Lenhard M, Jurgens G, et al. (1998) Role of WUSCHEL in regulating stem cell fate in the *Arabidopsis* shoot meristem. Cell 95: 805–815.
30. Solís-Ramos LY, González-Estrada T, Nahuath-Dzib S, Zapata-Rodriguez LC, Castaño E (2009) Overexpression of WUSCHEL in *C. chinense* causes ectopic morphogenesis. Plant Cell, Tissue and Organ Culture (PCTOC) 96: 279–287.
31. Bouchabke-Coussa O, Obellianne M, Linderme D, Montes E, Maia-Grondard A, et al. (2013) Wuschel overexpression promotes somatic embryogenesis and induces organogenesis in cotton (*Gossypium hirsutum L.*) tissues cultured in vitro. Plant Cell Rep 32: 675–686.
32. Wang X, Niu QW, Teng C, Li C, Mu J, et al. (2009) Overexpression of PGA37/MYB118 and MYB115 promotes vegetative-to-embryonic transition in *Arabidopsis*. Cell Res 19: 224–235.
33. Parcy F, Valon C, Kohara A, Miséra S, Giraudat J (1997) The ABSCISIC ACID-INSENSITIVE3, FUSCA3, and LEAFY COTYLEDON1 loci act in concert to control multiple aspects of *Arabidopsis* seed development. The Plant Cell Online 9: 1265–1277.
34. Lotan T, Ohto M-a, Yee KM, West MA, Lo R, et al. (1998) *Arabidopsis* LEAFY COTYLEDON1 Is Sufficient to Induce Embryo Development in Vegetative Cells. Cell 93: 1195–1205.
35. Zuo J, Niu QW, Chua NH (2000) Technical advance: An estrogen receptor-based transactivator XVE mediates highly inducible gene expression in transgenic plants. Plant J 24: 265–273.
36. Leelavathi S, Sunnichan V, Kumria R, Vijaykanth G, Bhatnagar R, et al. (2004) A simple and rapid *Agrobacterium*-mediated transformation protocol for cotton (*Gossypium hirsutum L.*): embryogenic calli as a source to generate large numbers of transgenic plants. Plant Cell Rep 22: 465–470.
37. Wu J, Zhang X, Nie Y, Jin S, Liang S (2004) Factors affecting somatic embryogenesis and plant regeneration from a range of recalcitrant genotypes of Chinese cottons (*Gossypium hirsutum L.*). In Vitro Cellular & Developmental Biology-Plant 40: 371–375.

9

Light Spatial Distribution in the Canopy and Crop Development in Cotton

Xiaoyu Zhi, Yingchun Han, Shuchun Mao, Guoping Wang, Lu Feng, Beifang Yang, Zhengyi Fan, Wenli Du, Jianhua Lu, Yabing Li*

Institute of Cotton Research of the Chinese Academy of Agricultural Sciences/State Key Laboratory of Cotton Biology, Anyang, 455000, Henan, China

Abstract

The partitioning of light is very difficult to assess, especially in discontinuous or irregular canopies. The aim of the present study was to analyze the spatial distribution of photosynthetically active radiation (PAR) in a heterogeneous cotton canopy based on a geo-statistical sampling method. Field experiments were conducted in 2011 and 2012 in Anyang, Henan, China. Field plots were arranged in a randomized block design with the main plot factor representing the plant density. There were 3 replications and 6 densities used in every replicate. The six plant density treatments were 15,000, 33,000, 51,000, 69,000, 87,000 and 105,000 plants ha^{-1}. The following results were observed: 1) transmission within the canopy decreased with increasing density and significantly decreased from the top to the bottom of the canopy, but the greatest decreases were observed in the middle layers of the canopy on the vertical axis and closing to the rows along the horizontal axis; 2) the transmitted PAR (TPAR) of 6 different cotton populations decreased slowly and then increased slightly as the leaves matured, the TPAR values were approximately 52.6–84.9% (2011) and 42.7–78.8% (2012) during the early cotton developmental stage, and were 33.9–60.0% (2011) and 34.5–61.8% (2012) during the flowering stage; 3) the Leaf area index (LAI) was highly significant exponentially correlated ($R^2=0.90$ in 2011, $R^2=0.91$ in 2012) with the intercepted PAR (IPAR) within the canopy; 4) and a highly significant linear correlation ($R^2=0.92$ in 2011, $R^2=0.96$ in 2012) was observed between the accumulated IPAR and the biomass. Our findings will aid researchers to improve radiation-use efficiency by optimizing the ideotype for cotton canopy architecture based on light spatial distribution characteristics.

Editor: Manuel Reigosa, University of Vigo, Spain

Funding: The study was supported by the National Natural Science Foundation of China(31371561). The funders had no role in study design, data collection and analysis, decision to publish, or preparation of the manuscript.

Competing Interests: The authors have declared that no competing interests exist.

* Email: criliyabing@163.com

Introduction

Crop yields depend on a canopy's capacity to intercept and efficiently use solar radiation. Photosynthetically active radiation (PAR) represents the solar radiation that can be absorbed by green plants [1] and used for photosynthesis to produce biomass [2–5]. Canopy architectural information is essential to a mechanistic description of radiation interception [6].

In 1953, Beer's law [7] was used to measure the leaf area and light intensity within each layer based on height in order to describe the spatial distribution of light. Since then, numerous investigations of radiation interception have been conducted using various approaches [8–11] and models such as CERES [12–13], GROPGRO [14], AFRCWHEAT [15] and CropSyst [16] based on their description of light extinction in plant canopies. Rosenthal [17] argued that the crop extinction coefficient had a negative linear relationship with the leaf area index (LAI). Campbell [18] expressed the extinction coefficient of the population using a function based on the angle of the sun and the leaf angle distribution. However, the crop population extinction coefficient is a variable that is sensitive to environmental factors and is difficult to measure. Furthermore, previous researchers [19–20] argued that Beer's law failed to fully consider the canopy spatial heterogeneity, which has often led to discrepancies between models and experimental data. Therefore, the application of these methods to estimate light distribution remains limited, and a more complete model is required.

Muchow et al. [21] suggested using four tube solarimeters in each plot to obtain estimates of radiation interception for sugarcane (*Saccharum officinarum*). Alados [22] studied measurements of solar global irradiance using a Kipp & Zonen model CM-11 solarimeter (Delft, Netherlands), and Singer et al. [23] measured the cumulatively intercepted PAR by deploying eight line quantum sensors in each of their experimental fields; however, the partitioning of light in different density systems is very difficult to assess, especially in discontinuous or irregular canopies. In theory, an accurate assessment could be achieved by placing a large number of sensors across the canopy to cope with spatial variability, but this solution is impractical due to the increased labor and capital costs involved [24–25]. Furthermore, Campillo et al. [26] used digital images and line quantum sensors to characterize light interception, which neglected the canopy architecture's effect on the spatial distribution of light. Previous research has focused on improving the efficiency of light utilization and exploring the spatial distribution of light [25,27–29]. Light interception has commonly been measured with expensive equipment or estimated with elaborate models; therefore, simpler and more economical methods, particularly techniques that consider spatial heterogeneity, are highly desirable. In addition, the different light intensities caused by different cotton plant densities have not yet been determined [30–31]. However, some studies have indicated that leaf area components had the greatest effect on light intensities [32], and Yang [33] concluded that different cotton canopy structures caused different light intensities. To determine the optimal plant density for biomass production, it is crucial to determine the spatial distribution of light in more detail.

Spatial heterogeneity effects should be considered in the study of light distribution characteristics in a canopy. The distribution of PAR in plant canopies is influenced not only by the radiation intensity but also by the plant density [34–38], the leaf angle [39–41], the nutritional status [3,42], and the LAI [3,35,43–46]. Spatial statistics are a versatile tool for environmental disciplines such as agriculture, geology, soil science, hydrology, ecology, oceanography, forestry, meteorology and climatology [35,47–49].

The objectives of this study were as follows: 1) to quantify the spatial distribution of light in heterogeneous cotton canopies; 2) and to explore the PAR variation and distribution characteristics under different plant densities using a geo-statistical sampling method to provide the theoretical and technical basis for optimizing the canopy architecture to intercept more radiation and improve the cotton lint yield.

Materials and Methods

1 Experimental design

Field assays were conducted in 2011 and 2012 at the Cotton Research Institute of the Chinese Academy of Agricultural Sciences in Anyang, Henan, China (36° 06 ′N, 114° 21′ E). During the cotton developmental stage, the average temperature was 21.2°C in 2011 and 23.2°C in 2012, and the total rainfall was 448.1 mm in 2011 and 421.3 mm in 2012. The same field was used in each year and was characterized by a medium loam soil with a total N 0.66 g kg^{-1}, P 0.01 g kg^{-1} and K 0.11 g kg^{-1}. The land was plowed and irrigated in early spring before planting. A randomized block design was used with 6 treatments and 3 replicates. The area of every plot was 64.0 m^2 with 8.0 m width, 8.0 m length and 0.8 m row spacing. The 6 plant density treatments were 15,000, 33,000, 51,000, 69,000, 87,000 and 105,000 plants ha^{-1}. The plants were sown by machine on April 19, 2011 and April 18, 2012. The sampling areas were free of weeds, and all of the plots received fertilizer at 225.0 kg ha^{-1} N, 150.0 kg ha^{-1} P_2O_5 and 225.0 kg ha^{-1} K_2O. Irrigation was applied at a volume of approximately 40 m^3 in total by flooding the furrows during the flowering stage. Weeds were manually controlled; pesticides were used to control insects and diseases.

2 Collection of PAR data

The incident PAR (PARi), PAR reflection (PARr) and agronomic characters (green leaf area, dry mass, boll weight and lint yield) were measured every ten days. The fluxes of PARi and PARr at each layer were measured with a spatial grid method [50], and the incident PAR at 20 cm (PARI) above the canopy was measured synchronously. The PARi and PARr between 2 rows of each plot were measured under clear or partly cloudy conditions at 10:00 am using a 100 cm line light quantum sensor (LI-191SA, LI-COR, Lincoln, NE, USA) and datalogger (LI-1400, LI-COR, Lincoln, NE, USA). The canopy was divided into 6 or 7 thin vertical layers according to plant height and 5 horizontal layers [51]. Then the PARi and PARr of 30 or 35 positions within the cotton canopy were measured. The volume within an area of 100 cm of row length and between 2 rows 80 cm apart up to a height of 100 cm was sampled every 10 days. The sensor was placed parallel to the row orientation and measured the light above a row at 100 cm, then moved 20 cm towards the adjacent row and measured the light again; this was performed again at 40 and 60 cm before being placed above the adjacent row (80 cm from the initial row). The instrument was then lowered to 80 cm and the process was repeated at 60, 40, 20 and 0 cm above the ground to provide a comprehensive set of spatial data of light intensities within the canopy. This process was repeated every 10 days, but only the data from 2 dates in each year is reported here. The model calculated, by interpolation, contours lines of equal light intensity. The canopy TPAR, RPAR and IPAR were calculated using the following equations [52]:

$$TPAR=\frac{PARi}{PARI} \tag{1}$$

$$RPAR=\frac{PARr}{PARI} \tag{2}$$

$$IPAR=\frac{(PARI-PARi-PARr)}{PARI} \tag{3}$$

where PARI is the incident PAR at 20 cm above the canopy ($\mu mol{\cdot}m^{-2}{\cdot}s^{-1}$), and PARi and PARr are the incident PAR and PAR reflection at each layer of the canopy, respectively.

3 Estimating cotton canopy PAR

The PARi, PARI and PARr in other positions in the canopy were calculated. Value estimates were calculated by spatial interpolation as follows:

$$Z(X_o)=\sum_{i=1}^{n}\lambda_i Z(X_i) \tag{4}$$

where $Z(x_O)$ =measured PAR values, λ_i = the coefficient of the sample, and the unbiased condition $\sum\lambda_i=1$ was employed. Based on the minimum variance, the Kriging equation [53–54] is stated as follows:

$$\sum_{i=1}^{n}\lambda_i r(x_i,x_j)+\varphi=r(x_i,x_o) \tag{5}$$

Where φ = Lagrangian, $r\ (x_i,\ x_j)$ = the measured value of the variation function, $r\ (x_i,\ x_O)$ = the measured and calculated PAR, and x_O is the estimated value of the calculated point as computed by the unbiased estimate.

The TPAR within the canopy was computed by the Simpson 3/8 rules [55]. Surfer software V12 (Golden Software Inc., USA) was used with the application of the following equation (6):

$$\begin{aligned} A_i&=\frac{3\Delta x}{8}(G_{i,1}+3G_{i,2}+3G_{i,3}+2G_{i,4}+,\ldots,+2G_{i,ncol-1}+Gi,ncol);\\ Volume&\approx\frac{3\Delta x}{8}(A_1+3A_2+3A_3+2A_4+,\ldots,+2A_{ncol-1}+A_{ncol}) \end{aligned} \tag{6}$$

Table 1. Simulation equations of transmitted PAR, reflected PAR and intercepted PAR: $y = Ax^2+Bx+C$.

Treatments (plants ha^{-1})		2011, n = 11					2012, n = 9				
		A	B	C	R^2	P>F	A	B	C	R^2	P>F
PAR transmittance	15000	1.608	−0.019	$0.793*10^{-4}$	0.979	<0.001	1.819	−0.026	$0.122*10^{-3}$	0.950	<0.001
	33000	1.577	−0.019	$0.775*10^{-4}$	0.978	<0.001	1.508	−0.021	$0.973*10^{-4}$	0.929	<0.001
	51000	1.593	−0.020	$0.831*10^{-4}$	0.972	<0.001	1.373	−0.018	$0.855*10^{-4}$	0.893	0.001
	69000	1.515	−0.019	$0.803*10^{-4}$	0.934	<0.001	1.333	−0.018	$0.867*10^{-4}$	0.975	<0.001
	87000	1.487	−0.020	$0.844*10^{-4}$	0.951	<0.001	1.223	−0.016	$0.779*10^{-3}$	0.972	<0.001
	105000	1.192	−0.020	$0.840*10^{-4}$	0.984	<0.001	1.255	−0.017	$0.816*10^{-4}$	0.936	<0.001
PAR reflectivity	15000	0.226	−0.004	$0.140*10^{-4}$	0.961	<0.001	0.189	−0.003	$0.137*10^{-4}$	0.910	<0.001
	33000	0.190	−0.003	$0.121*10^{-4}$	0.907	<0.001	0.166	−0.003	$0.132*10^{-4}$	0.845	0.004
	51000	0.196	−0.003	$0.138*10^{-4}$	0.947	<0.001	0.123	−0.002	$0.097*10^{-4}$	0.917	<0.001
	69000	0.183	−0.003	$0.130*10^{-4}$	0.936	<0.001	0.124	−0.002	$0.108*10^{-4}$	0.925	<0.001
	87000	0.176	−0.003	$0.126*10^{-4}$	0.930	<0.001	0.130	−0.002	$0.106*10^{-4}$	0.808	0.007
	105000	0.163	−0.003	$0.114*10^{-4}$	0.944	<0.001	0.102	−0.002	$0.084*10^{-4}$	0.920	<0.001
PAR interception	15000	−0.835	0.023	−13.330	0.984	<0.001	−1.008	0.029	−4.360	0.969	<0.001
	33000	−0.766	0.022	−12.980	0.982	<0.001	−0.674	0.023	−4.110	0.959	<0.001
	51000	−0.789	0.023	−13.670	0.975	<0.001	−0.496	0.020	−13.520	0.926	<0.001
	69000	−0.697	0.022	−9.3E−05	0.941	<0.001	−0.458	0.020	−13.750	0.980	<0.001
	87000	−0.521	0.020	−12.490	0.951	<0.001	−0.353	0.018	−12.860	0.976	<0.001
	105000	−0.656	0.022	−13.550	0.986	<0.001	−0.357	0.019	−13.000	0.955	<0.001

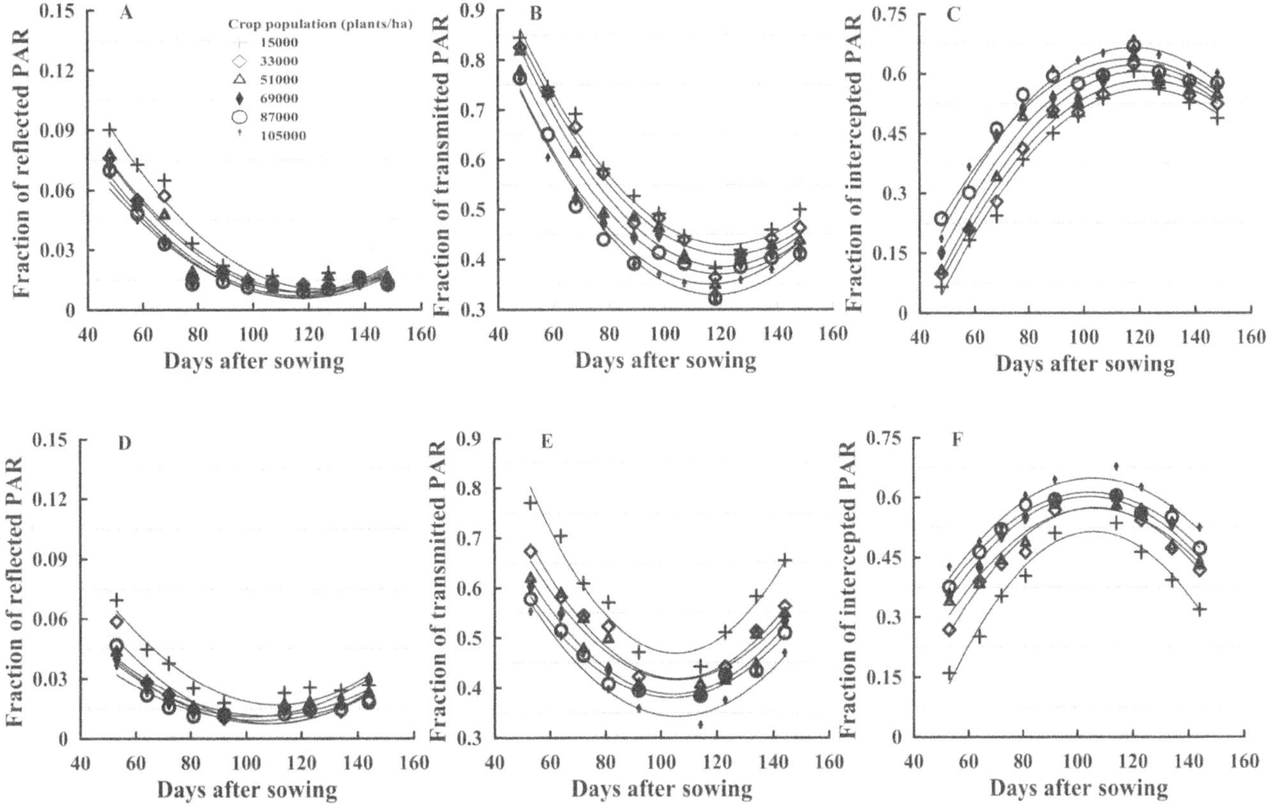

Figure 1. The variation of reflected, transmitted and intercepted PAR of all the plant densities over the growing period of cotton in 2011 (A, B, C) and 2012 (D, E, F).

where the coefficient vector is *[5,3,3,2,..., 3,3,2,1]*, Δx is the vertical distance of the grid, Δy is the horizontal distance; *G (I, j)* is the grid node number *(I, j)*, and volume is the total light volume of a certain cross-sectional area.

4 Agronomic traits of cotton

Two randomly selected plants from each plot were harvested every 10 days in 2011 and 2012. During sampling, at least two edge rows were excluded to avoid the boundary effects. These destructive samples were subdivided into leaves, stems, flowers and bolls depending on their developmental stage. The leaf area was determined using a scanner (Phantom 9800xl, MiCROTEK, Shanghai, China) [56] and was measured using Image-Pro Plus (Media Cybernetics, Inc.). After the leaf areas were measured, the leaves, stems and bolls were dried at 80°C to a constant weight, and the dry mass was determined.

Results

1 Cotton PAR of the entire canopy throughout the cotton growth period

To study the PAR spatial distribution, the values of the TPAR, RPAR and IPAR were calculated using the 3/8 Simpson and Quadratic relationships of the days after sowing for the PAR (Table 1, Fig. 1) with highly significant and determination coefficients all above 0.90. Furthermore, the values of "A" were positive for the TPAR and RPAR simulation equations and negative for the IPAR simulation equations (Table 1).

The TPAR of 6 treatments in 2 years presented quadratic tendencies, with highly significant correlation coefficients between 0.93–0.98 (Table 1). Over the cotton developmental stage, the TPAR values were approximately 52.6–84.9% (2011) and 47.7–78.8% (2012) before 64 days after sowing and were 33.9–60.0% (2011) and 34.5–61.8% (2012) during 65–120 days after sowing (Fig. 1). At the same development stage, the TPAR in 2012 was higher than in 2011, which demonstrated that cotton developed better in 2011 than in 2012. According to Table 1, the minimum TPARs of the different plant densities were 43.0, 41.0, 39.0, 37.0, 35.0 and 33.0% at 122, 123, 121, 119, 116 and 118 days after sowing in 2011, respectively. The minimum TPARs were 47.0, 42.0, 42.0, 39.0, 38.0 and 34.0% at 105, 106, 106, 104, 104 and 106 days after sowing in 2012, respectively.

The RPAR spatial distribution is shown in Fig. 1. The RPAR group declined rapidly and then increased slowly with time. The RPAR was approximately 9.0% in the early days after sowing and 1.0–2.0% in the late days after sowing in 2011. The RPAR of different plant densities decreased with the increase of plant density, particularly in the early developmental stage. At 48 days after sowing in 2011, the RPARs of the 6 different plant densities were 9.0, 7.6, 7.8, 7.3, 7.0 and 6.6% at 15,000, 33,000, 51,000, 69,000, 87,000 and 105,000 plants ha^{-1}, respectively. The minimum estimations of RPAR (Table 1) were 1.0% at 124 days after sowing at 15,000 plants ha^{-1} and 0.8% at 121 days after sowing at 33,000 plants ha^{-1} in 2011. In 2011, the RPARs of the other 4 plant densities were 0.9, 0.7, 0.6 and 0.6% at 117, 116, 116, and 117 days after sowing, respectively. In both years, the RPARs of different plant densities were 2.0–6.5% at 64 days after sowing. However, cotton senescence occurred later in 2011 than in 2012; therefore, the RPAR was higher in 2012 than in 2011 during the late developmental stage. In 2012, the minimum RPARs of different

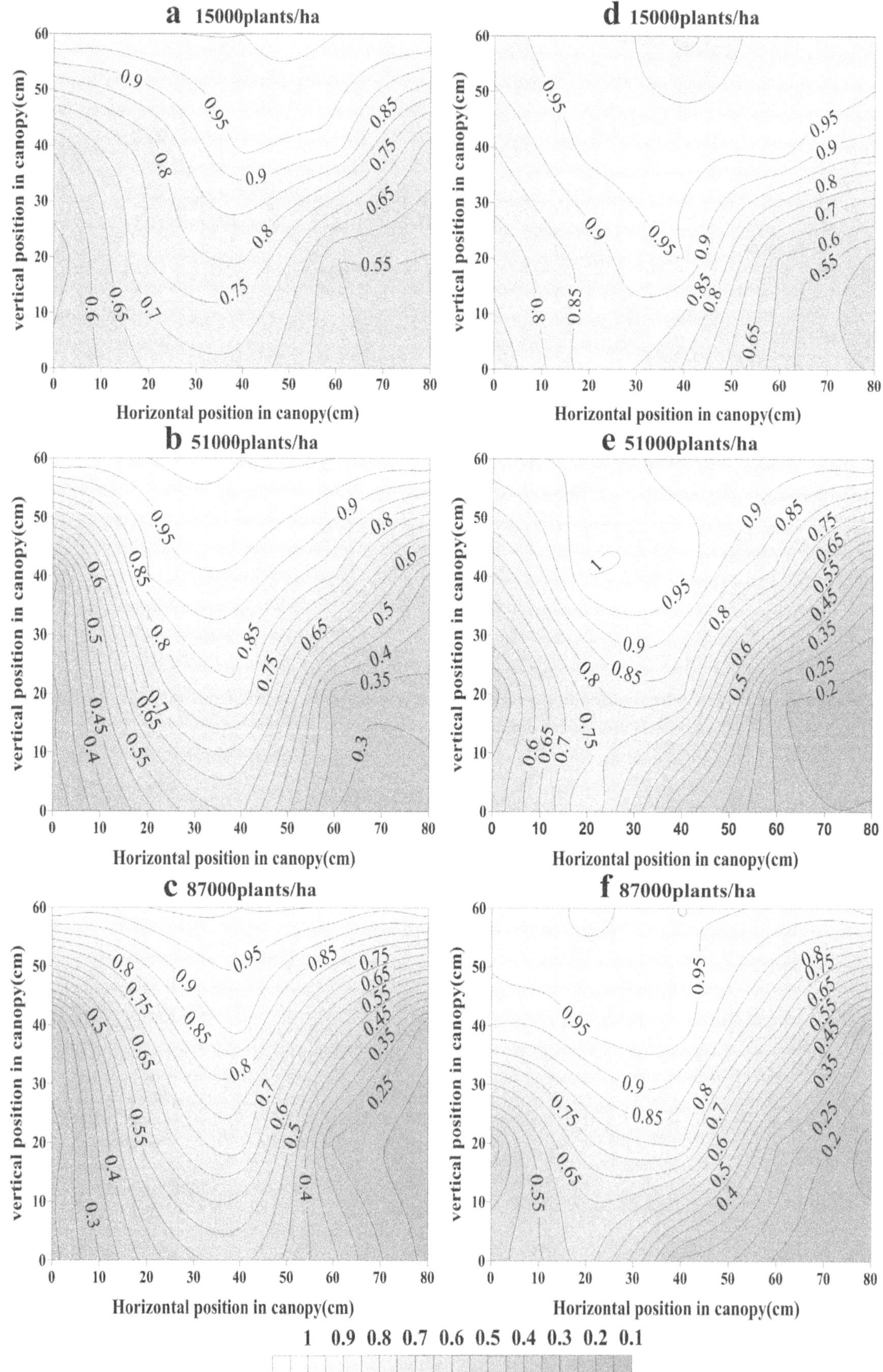

Figure 2. Vertical and horizontal distribution of TPAR at the early cotton developmental stage in 2011 (a, b, c) and 2012 (d, e, f).

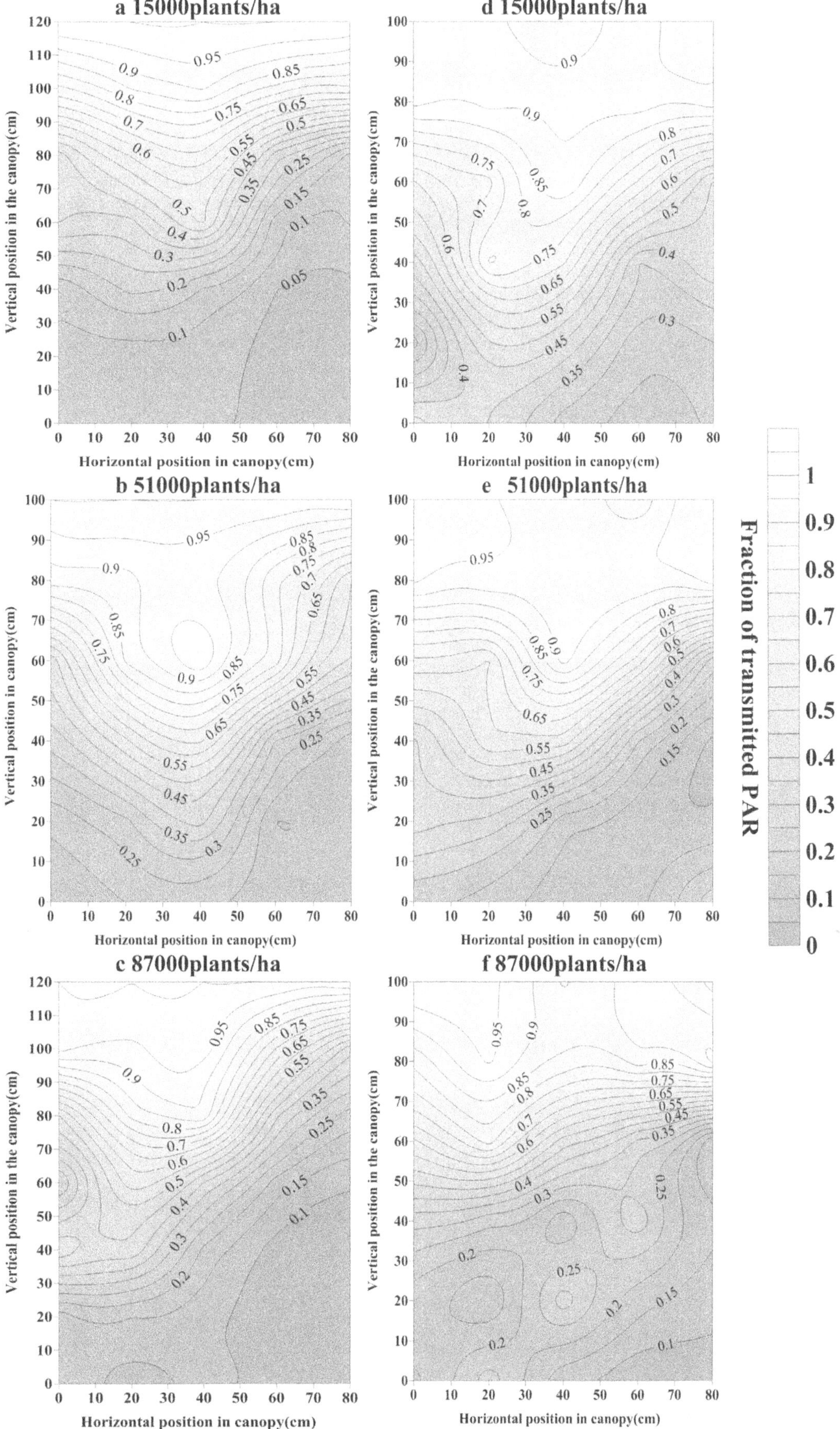
a 15000plants/ha
d 15000plants/ha
b 51000plants/ha
e 51000plants/ha
c 87000plants/ha
f 87000plants/ha
Vertical position in the canopy(cm)
Vertical position in canopy(cm)
Horizontal position in canopy(cm)
Fraction of transmitted PAR
1
0.9
0.8
0.7
0.6
0.5
0.4
0.3
0.2
0.1
0

Figure 3. Vertical and horizontal distribution of TPAR at the flowering stage in 2011 (a, b, c) and 2012 (d, e, f).

plant densities were 1.7, 0.7, 1.1, 1.1, 0.8 and 0.9%, at 111, 109, 107, 102, 107 and 105 days after sowing, respectively.

The IPAR spatial distribution was shown in Fig. 1. The IPAR of different plant densities was 10.0–70.0% over the cotton developmental stage. The maximum estimations of IPAR (Table 1) were 56.0, 58.0, 61.0, 62.0, 64.0 and 67.0% at 122, 123, 120, 119, 117 and 118 days after sowing in 2011, respectively. The maximum IPARs were 51.0, 57.0, 57.0, 60.0, 61.0 and 65.0% at 106, 106,106,104,104 and 106 days after sowing in 2012, respectively. The maximum IPAR in the different plant densities increased with the increase of plant density in both years (Fig. 1), but the difference in quadratic tendencies between the 2 years was caused by excessive cotton vegetative growth in 2012.

2 Spatial distribution of the transmitted PAR within the cotton canopy

This study analyzed the TPAR spatial distribution in two developmental stages (Figs. 2 and 3). A tendency toward less TPAR was observed in higher plant densities. For example, at the early cotton developmental stage in 2011 (Fig. 2), from 0–50 cm vertical position, the TPAR was approximately 60.0–80.0% (15,000 plants ha^{-1}), 35.0–65.0% (51,000 plants ha^{-1}) and 30.0–60.0% (87,000 plants ha^{-1}) near the cotton rows (10 cm horizontal position); and was 67.0–95.0% (15,000 plants ha^{-1}), 61.0–95.0% (51,000 plants ha^{-1}) and 53.0–93.0% (87,000 plants ha^{-1}) at the mid-point between rows (40 cm horizontal position). At the flowering stage in 2011 (Fig. 3), from 0–80 cm vertical position, the TPAR was 7.0–45.0% (15,000 plants ha^{-1}), 17.0–85.0% (51,000 plants ha^{-1}) and 10.0–65.0% (87,000 plants ha^{-1}) near the rows (10 cm horizontal position); and 6.0–62.0% (15,000 plants ha^{-1}), 22.0–100.0% (51,000 plants ha^{-1}) and 12.0–80.0% (87,000 plants ha^{-1}) at the mid-point between 2 rows (40 cm horizontal position).

Cotton branch development and leaf area expansion played prominent roles in explaining the TPAR spatial distribution during different years and in different spatial positions. For example, in the early cotton developmental stage (Fig. 2), at a density of 51,000 plants ha^{-1}, from 0–80 cm horizontal position, the TPAR ranged from 25.9 to 60.9% at the 0 cm vertical position, but the TPAR was a constant 100.0% at the 60 cm vertical position; however, from 0–60 cm vertical position, the TPAR decreased from 100.0 to 25.6% at the 0 cm horizontal position and to 60.9% at the 40 cm horizontal position in 2011. And at 51,000 plants ha^{-1} density, the TPAR ranged from 20.8 to 76.0% at the 0 cm vertical position and from 91.3 to 95.8% at the 60 cm vertical position; however, the TPAR decreased from 93.8 to 42.4% at the 0 cm horizontal position and from 97.7 to 41.9% at the 40 cm horizontal position in 2012.

3 Relationship between the intercepted PAR and leaf area index

The LAI exhibited a highly significant logarithmic correlation with the IPAR of different plant densities in both years (2011, $n=66$, $R^2=0.90$, $P>|t|: <0.001$; 2012, $n=54$, $R^2=0.91$, $P>|t|: <0.001$) (Fig. 4).

$$IPAR_{2011}=\frac{(\ln LAI-2.21)}{5.30}, n=66, R^2=0.90;$$
$$IPAR_{2012}=\frac{(\ln LAI-2.30)}{5.30}, n=54, R^2=0.91. \quad (7)$$

4 Relationship between the intercepted PAR and biomass

Across all plant densities, dry mass accumulation was linearly related to the cumulative IPAR in both years, employing all of the data from different populations and stages into two study years (2011, $n=66$, $P>F: <0.001$; 2012, $n=54$, $P>F: <0.001$) (Fig. 5).

$$Y_{2011}=213.18\times X-65.15, n=66, R^2=0.92;$$
$$Y_{2012}=201.22\times X-814.71, n=54, R^2=0.96. \quad (8)$$

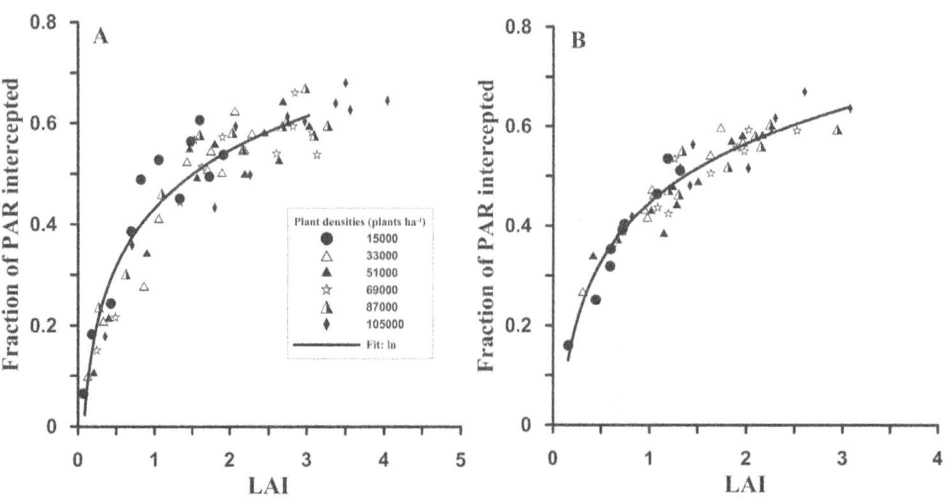

Figure 4. Relationship between the LAI and IPAR in 2011 (A) and 2012 (B).

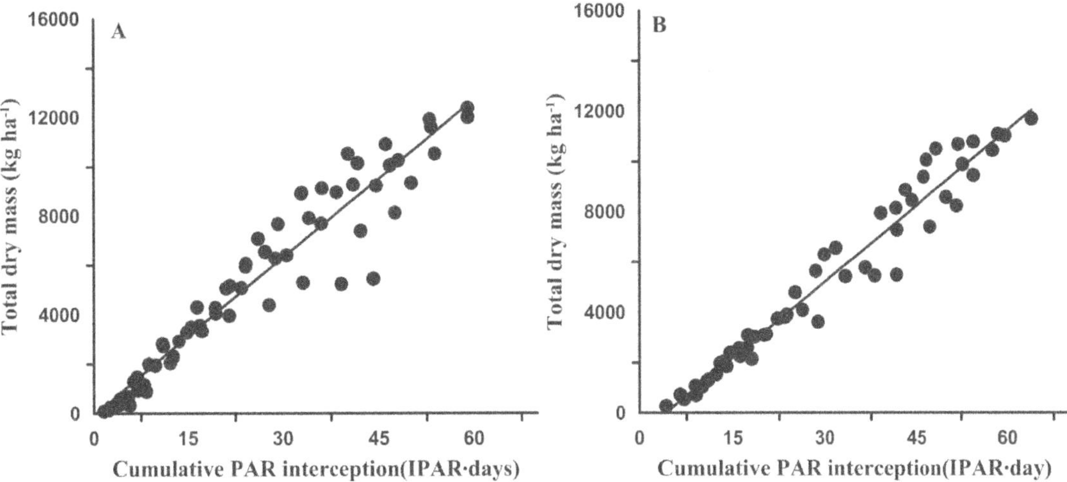

Figure 5. Relationship and fitted models between the cumulative IPAR and dry mass in 2011 (A) and 2012 (B).

Discussion

The results of this work demonstrate that both the TPAR and the vertical distribution of the PAR in the canopy are very important for crop photosynthesis, which is in agreement with previous research [57]. The present study used the grid method to detect existing spatial heterogeneities, and geo-statistics were applied to create contour maps for variables within the canopies. Baldocchi et al. [39] reported a spherical distribution in a canopy radiation transfer model. Subsequent studies by Campbell [58] and Campbell and Norman [59] conducted accurate experiments on a theoretical ellipsoidal distribution model, but validations of this model have been predominantly conducted in herbaceous crops. Canopy architecture influenced the spatial heterogeneity of the PAR in the 6 densities over the study periods, in agreement with the findings of Reta-Sanchez et al. [60] and Stewart et al. [61], who showed that the canopy architecture changes affected light penetration into the canopy. The PAR and canopy architecture determined for cotton in this experiment are in good agreement with the results of Sassenrath-Cole [62], who reported that the extreme cupping of cotton leaves may further increase the photosynthetically active area. Detailed canopy light distribution had been previously shown to improve the total canopy photosynthesis and yield [63].

1 Cotton PAR of the entire canopy

The TPAR and RPAR decreased rapidly for light sheltered by the cotton plant in the early developmental stage; however, it increased slowly in the later stage because of leaf senescence. Similarly, Cooper [64] reported a significant correlation between the crop growth rate and the canopy extinction coefficient. In the flowering stage, the horizontal differences between the TPAR were small in the 0 cm vertical position but markedly increased at 40 cm. The TPAR ranged from 11.2% to 38.3% at the 0 cm vertical position but from 13.0% to 62.6% at the 40 cm vertical position. This result might be related to the fact that the lower canopy layers predominantly receive diffuse radiation, which is in general more homogeneously distributed within the canopy.

2 Relationship between the intercepted PAR and leaf area index (LAI)

Morris [65] found that the vertically oriented leaves at the top of the canopy intercepted less light than did the horizontal leaves, which is consistent with present research on canopy architecture. In this study, the IPAR was highest at the tops of the canopies, which was in strong agreement with the results of both Sakamoto and Shaw [66] and Hatfield and Carlson [67], who found that approximately 90.0% of light interception occurred in the top and peripheral sections of the canopy. One possible explanation for this result is that the LAI was reduced due to the abscission of lower canopy leaves exposed to shade. When the LAI was at its maximum for all of the plant densities, the differences in the IPAR were unrelated to differences in the maximum LAI [35], which might be due to self-shading of the lower canopy leaves. Tharakan et al. [68] indicated that the LAI and canopy duration were very important for biomass production, which was supported by the findings of the present study.

3 Relationship between the intercepted PAR and biomass

The present research identified a close relationship between the biomass and IPAR within the canopy, and biomass was an appropriate index for assessing the IPAR, which was in agreement with the results from Russell et al. [69] and Kiniry et al. [70]. Additionally, Robinson et al. [71] showed that the leaf extension rate and cell number per leaf determined the final leaf size, which in turn influenced the IPAR and yield. However, the IPAR did not correspondingly increase with increased biomass, especially after canopy closure [37,72]; and this finding was consistent with the results reported by Ceotto et al. [73]. A linear relationship was found between the cumulative IPAR and the biomass, which was in agreement with the results from Christensen [74].

Conclusions

The results indicated the following: 1) the cotton canopy architecture affected the spatial distribution of the TPAR; 2) the distribution discrepancy was larger at the top of the canopy than at the bottom in the flowering stage; 3) the TPAR variation of the minimum and maximum densities was smaller than those of the other four densities; 4) the IPAR was affected by the LAI for different canopy structure; 5) and the biomass was highly correlated with the IPAR.

The amount and distribution of the leaf area in a crop canopy determined the way that the PAR was intercepted and consequently influenced the canopy photosynthesis and yield.

This study analyzed the PAR spatial distribution within the cotton canopy and the relationship between the LAI and IPAR to assess the importance of the IPAR in explaining biomass variation and identifying the optimal canopy structure from a suite of leaf measurements and plant density.

The results of this study will assist breeding researchers in the selection of cultivars with more erect leaves, especially at the top of the canopy, to improve light environments within canopies and canopy photosynthesis [75]. A desirable ideotype will include a more open structure with greater IPAR, leading to increased photosynthesis and yield. Modern developments in cotton breeding can use PAR spatial distribution information to produce more efficient genotypes for canopy photosynthesis thus increasing lint yield. The results of this study may be used to further map the PAR in spatially heterogeneous canopies using geo-statistical interpolation methods, thereby contributing to the development of an ideal plant shape and the breeding of high light use efficiency crops.

Acknowledgments

We greatly appreciate the work of the technicians from the experimental station at the Institute of Cotton Research at the Chinese Academy of Agricultural Sciences.

Author Contributions

Conceived and designed the experiments: YL. Performed the experiments: YH WD BY GW ZF LF JL SM. Analyzed the data: XZ. Contributed reagents/materials/analysis tools: YL. Wrote the paper: XZ LF.

References

1. Chen JM (1996) Canopy architecture and remote sensing of the fraction of photosynthetically active radiation absorbed by boreal conifer forests. IEEE T - Geosci Remote 34: 1353–1368.
2. McCree KJ (1981) Photosynthetically active radiation. Physiol Plant Ecol I 12: 41–55.
3. Maddonni GA, Otegui ME (1996) Leaf area, light interception, and crop development in maize. Field Crops Res 48: 81–87.
4. Tadahisa H (2009) Light interception by tomato plants (Solanum lycopersicum) grown on a sloped field. Agr Forest Meteorol 149: 756–762.
5. Maddonni GA, Chelle M, Drouet J (2001) Light interception of contrasting azimuth canopies under square and rectangular plant spatial distributions simulation and crop measurements. Field Crops Res 70: 1–13.
6. Ross J (1981) The radiation regime and architecture of plant stands. The Hague, the Netherlands: Kluwer Academic Publishers 391.
7. Monsi M, Saeki T (1953) Uber den Lichfaktor in denpflanzengesells chaften und seine bedeutung fur die stoffproduktion. Jap J Bot 14: 22–52.
8. McCree KJ (1966) A solarimeter for measuring photosynthetically active radiation. Agr Meteorol 3: 353–366.
9. Hall FG, Huemmrich KF, Goward SN (1990) Use of narrow-band spectra to estimate the fraction of absorbed photosynthetically active radiation. Remote Sens Environ 32: 47–54.
10. Frouin R, Pinker RT (1995) Estimating photosynthetically active radiation (PAR) at the earth's surface from satellite observations. Remote Sensing of Environment 51: 98–107.
11. Alados I, Foyo-Moreno I, Alados-Arboledas L (1996) Photosynthetically active radiation: measurements and modeling. Agr Forest Meteorol 78: 121–131.
12. Ritchie JT, Otter S (1985) Description and performance of CERES-Wheat: A user-oriented wheat yield model, In: Willis WO (editor), ARS Wheat Yield Project, ARS-38. U.S. Department of Agriculture, Agricultural Research Service, Washington D.C., USA, 159–175.
13. Jones CA, Kiniry JR, Dyke PT (1986) CERES-Maize: A simulation model of maize growth and development. College Station, TX, USA: Texas A & M University Press 194.
14. Boote KJ, Jones JW, Hoogenboom G, Pickering NB (1998) The CROPGRO model for grain legumes. In: Gordon Y. Tsuji GY, Hoogenboom G, Thornton PK, (editors). Understanding Options for Agricultural Production: Systems Approaches for Sustainable Agricultural Development (Volume 7) Dordrecht, the Netherlands: Springer Netherlands 99–128.
15. Jamieson PD, Porter JR, Goudriaan J, Ritchie JT, van Keulen H, et al. (1998) A comparison of the models AFRCWHEAT2, CERES-Wheat, Sirius, SUCROS2 and SWHEAT with measurements from wheat grown under drought. Field Crops Res 55: 23–44.
16. Stockle CO, Donatelli M, Nelson R (2003) CropSyst, a cropping systems simulation model. Eur J Agron 18: 289–307.
17. Rosenthal UD, Kanemasu T, Raney RJ, et al. (1977) Evaluation of an evaporation model for corn. Agron J 69: 461–464
18. Campbell GS (1986) Extinction Coefficients for Radiation in Plant Canopies Calculated Using An Ellipsoidal Inclination Angle Distribution. Agr Forest Meteorol 36: 317–321.
19. Campbell GS (1991) Application Note:Canopy leaf area index from sunfleck ceptometer PAR measurements. Agr Forest Meteorol 36: 107–128.
20. Kiniry J, Johnson MV, Mitchell R, Vogel K, Kaiser J, et al. (2011) Switchgrass leaf area index and light extinction coefficients. Agron J 103: 119–122.
21. Muchow RC, Spillman MF, Wood AW, Thomas MR (1994) Radiation interception and biomass accumulation in a sugarcane crop grown under irrigated tropical conditions. Aust J Agr Res 45: 37–49.
22. Alados I, Olmo FJ, Foyo-Moreno I, Alados-Arboledas L (2000) Estimation of photosynthetically active radiation under cloudy conditions. Agr Forest Meteorol 102: 39–50.
23. Singer JW, Meek DW, Sauer TJ, Prueger JH, Hatfield JL (2011) Variability of light interception and radiation use efficiency in maize and soybean. Field Crops Res 121: 147–152.
24. Tournebize R, Sinoquet H (1995) Light interception and partitioning in a shrub/grass mixture. Agr Forest Meteorol 72: 277–294.
25. Fila G, Sartoratoc I (2011) Using leaf mass per area as predictor of light interception and absorption in crop/weed monoculture or mixed stands. Agr Forest Meteorol 151(5): 575–584.
26. Campillo C, Prieto MH, Daza C, Monino MJ, Garcia MI (2008) Using digital images to characterize canopy coverage and light interception in a processing tomato crop. Hortscience 43: 1780–1786.
27. Vargas LA, Andersen MN, Jensen CR, Jørgensen U (2002) Estimation of leaf area index, light interception and biomass accumulation of Miscanthus sinensis 'Goliath' from radiation measurements. Biomass Bioenergy 22: 1–14.
28. Van der Zande D, Stuckens J, Verstraeten WW, Mereu S (2011) 3D modeling of light interception in heterogeneous forest canopies using ground-based LiDAR data. Int J Appl Earth Obs Geoinform 13: 792–800.
29. Gonias ED, Oosterhuis DM, Bibi AC (2011) Light interception and radiation use efficiency of okra and normal leaf cotton isolines. Environ Exp Botany 72: 217–222.
30. Naraghi M, Lotfi M (2010) Effect of different levels of shading on yield and fruit quality of cucumber (Cucumis sativus). Acta Hort. (ISHS) 871: 385–388.
31. Reta-Sanchez DG, Fowler JL (2002) Canopy light environment and yield of narrow-row cotton as affected by canopy architecture. Agron J 94: 1317–1323.
32. Baldissera TC, Frak E, Carvalho PCD, Louarn G (2014) Plant development controls leaf area expansion in alfalfa plants competing for light. Ann Bot 113: 145–157.
33. Yang YM, Liu XJ, Ouyang Z, Yang YH, Wang XL (2006) An optimal staggered canopy system for high yield cultivation of cotton and light distribution in the canopy - art. No. 62982G. Remote Sensing and Modeling of Ecosystems for Sustainability III 6298: G2982–G2982.
34. Yaseen M, Singh M, Singh UB, Singh S, Ram M (2013) Optimum planting time, method, plant density, size of planting material, and photo synthetically active radiation for safed musli (Chlorophytum borivilianum). Indust Crops Prod 43: 61–64.
35. Francescangeli N, Sangiacomo MA, Marti H (2006) Effects of plant density in broccoli on yield and radiation use efficiency. Sci Hortic-Amsterdam 110: 135–143.
36. Westgate ME, Forcella F, Reicosky DC (1997) Rapid canopy closure for maize production in the northern US corn belt: Radiation-use efficiency and grain yield. Field Crops Res 49: 249–258.
37. Mao LL, Zhang L, Zhao X, Liu S, van der Werf W, et al. (2014) Crop growth, light utilization and yield of relay intercropped cotton as affected by plant density and a plant growth regulator. Field Crops Res 155: 67–76.
38. Ratjen AM, Kage H (2013) Is mutual shading a decisive factor for differences in overall canopy specific leaf area of winter wheat crops? Field Crops Res 149: 338–346.
39. Baldocchi DD, Hutchison BA, Matt DR, McMillen RT (1985) Canopy radiative transfer models for spherical and known leaf inclination angle distributions: a test in an oak-hickory forest. J Appl Ecol 22: 539–555.
40. Cohen S, Fuchs M (1987) The distribution of leaf area, radiation, Photosynthesis and transpiration in a Shamouti orange hedgerow orchard. Agr Forest Meteorol 40: 123–144.
41. Campillo C, Fortes R, Del Henar Prieto M (2012) Solar radiation effect on crop production. In: Babatunde EB (editor). Solar Radiation. ISBN: 978-953-51-0384-4.
42. Delagrange S (2011) Light-and seasonal-induced plasticity in leaf morphology, N partitioning and photosynthetic capacity of two temperate deciduous species. Environ Exp Botany 70: 1–10.

43. Stewart DW, Costa C, Dwyer LM, Smith DL, Hamilton RL, et al. (2003) Canopy structure, light interception, and photosynthesis in maize. Agronomy J 95: 1465–1474.
44. Gonsamo A, Petri Pellikka P (2008) Methodology comparison for slope correction in canopy leaf area index estimation using hemispherical photography. Forest Ecol Manag 256: 749–759.
45. Leblanc SG, Chen JM, Fernandes R, Deering DW, Conley A (2005) Methodology comparison for canopy structure parameters extraction from digital hemispherical photography in boreal forests. Agr Forest Meteorol 129: 187–207.
46. Hipps LE, Asrar G, Kanemasu ET (1983) Assessing the interception of photosynthetically active radiation in winter wheat. Agr Meteorol 28: 253–259.
47. Fortin MJ, James PMA, MacKenzie A, Melles SJ, Rayfeld B (2012) Spatial statistics, spatial regression, and graph theory in ecology. Spatial Stat 1: 100–109.
48. Arbia G (2014) Pairwise likelihood inference for spatial regressions estimated on very large datasets. Spatial Stat 7: 21–39.
49. Griffith DA (2012) Spatial statistics: a quantitative geographer's perspective. Spatial Stat 1: 3–15.
50. Melo JD, Carreno EM, Calvino A, Padilha-Feltrin A (2014) Determining spatial resolution in spatial load forecasting using a grid-based model. Electr Pow Syst Res 111: 177–184.
51. Stockle CO (1992) Canopy photosynthesis and transpiration estimates using radiation interception models with different levels of detail. Ecol Model 60: 31–44.
52. Zhu X, Tang L, Zhang W, Cao MY, Cao WX, et al. (2012) Transfer characteristics of canopy photosynthetically active radiation in different rice cultivars under different cultural conditions. Sci Agr Sinica 45: 34–43.
53. Zhu K, Cui ZD, Jiang B, Yang GY, Chen ZJ, et al. (2013) A DEM-based residual kriging model for estimating groundwater levels within a large-scale domain: a study for the Fuyang River Basin. Clean Technol Environ Policy 15: 687–698.
54. Castillo-Santiago MA, Ghilardi A, Oyama K, Hernandez-Stefanoni JL, Torres I, et al. (2013) Estimating the spatial distribution of woody biomass suitable for charcoal making from remote sensing and geostatistics in central Mexico. Energy Sustain Develop 17: 177–188.
55. Kilicman A, Dehkordi LK, Kajani MT (2012) Numerical Solution of Nonlinear Volterra Integral Equations System Using Simpson's 3/8 Rule. Mathematical Problems in Engineering, Article ID 603463. doi: 10.1155/2012/603463.
56. O'Neal ME, Landis DA, Isaacs R (2002) An inexpensive, accurate method for measuring leaf area and defoliation through digital image analysis. Field Forage Crops 95: 1190–1194.
57. Sarlikioti V, de Visser PHB, Buck-Sorlin GH, Marcelis LFM (2011) How plant architecture affects light absorption and photosynthesis in tomato: towards an ideotype for plant architecture using a functional-structural plant model. Ann Bot-London 108: 1065–1073.
58. Campbell GS (1986) Extinction coefficients for radiation in plant canopies calculated using an ellipsoidal inclination angle distribution. Agr Forest Meteorol 36: 317–321.
59. Campbell GS, Norman JM (1989) The description and measurement of plant canopy structure. In: Russell G, Marshall B, Jarvis PG (editors). Plant Canopies: Their Growth, Form and Function. Cambridge: Cambridge University Press 1–19.
60. Reta-Sanchez DG, Fowler JL (2002) Canopy light environment and yield of narrow-row cotton as affected by canopy architecture. Agronomy J 94: 1317–1323.
61. Stewart DW, Costa C, Dwyer LM, Smith DL, Hamilton RI, et al. (2003) Canopy structure, light interception, and photosynthesis in maize. Agronomy J 95: 1465–1474.
62. Sassenrath-Cole GF (1995) Dependence of canopy light distribution on leaf and canopy structure for two cotton (Gossypium) species. Agr Forest Meteorol 77: 55–72.
63. Song Q, Zhang G, Zhu X (2013) Optimal crop canopy architecture to maximize canopy photosynthetic CO2 uptake under elevated CO2-a theoretical study using a mechanistic model of canopy photosythesis. Funct Plant Biol 40: 109–124.
64. Cooper JP (1970) Potential production and energy conversion in temperate and tropical grasses. Herbage Abstr 40: 1–15.
65. Morris JT (1989) Modelling light distribution within the canopy of the marsh grass Spartina alterniflora as a function of canopy biomass and solar angle. Agr Forest Meteorol 46: 349–361.
66. Sakamoto CM, Shaw RH (1967) Apparent photosynthesis in field soybean communities. Agron J 59: 73–75.
67. Hatfield JL, Carlson RE (1978) Photosynthetically active radiation, CO2 uptake, and stomatal diffusive resistance profiles within soybean canopies. Agron J 70: 592–596.
68. Tharakan PJ, Volk TA, Nowak CA, Ofezu GJ (2008) Assessment of canopy structure, light interception, and light-use efficiency of first year regrowth of shrub willow (Salix sp.). BioEnergy Res 1: 229–238.
69. Russell G, Jarvis P, Monteith J (1989) Absorption of radiation by canopies and stand growth. In: Russell G, Marshall B, Jarvis P, editors.Plant canopies: their growth, form and function. Cambridge: Cambridge University Press 21–44.
70. Kiniry JR, Jones CA, O'toole JC, Blanchet R, Cabelguenne M, et al. (1989) Radiation-use efficiency in biomass accumulation prior to grain filling for five grain-crop species. Field Crops Res 20: 51–64.
71. Robinson KM, Karp A, Taylor G (2004) Defining leaf traits linked to yield in short-rotation coppice Salix. Biomass Bioenergy 26: 417–431.
72. Purcell LC, Ball RA, Reaper JD, Vories ED (2002) Radiation use efficiency and biomass production in soybean at different plant population densities. Crop Sci 42: 172–177.
73. Ceotto E, Di Candilo M, Castelli F, Badeck F-W, Rizza F, et al. (2013) Comparing solar radiation interception and use efficiency for the energy crops giant reed (Arundo donax L.) and sweet sorghum (Sorghum bicolor L. Moench), Field Crops Res 149: 159–166.
74. Christensen S (1993) Deriving light interception and biomass from spectral reflectance ratio. Remote Sens Environ 43: 87–95.
75. Long SP, Zhu XG, Naidu SL, Ort D (2006) Can improvement in photosynthesis increase crop yields? Plant Cell Environ 29: 315–330.

10

Understanding the Relationship between Cotton Fiber Properties and Non-Cellulosic Cell Wall Polysaccharides

Dhivyaa Rajasundaram[1,2], Jean-Luc Runavot[3], Xiaoyuan Guo[4], William G. T. Willats[4], Frank Meulewaeter[3], Joachim Selbig[1,2]*

1 Institute of Biochemistry and Biology, University of Potsdam, Potsdam-Golm, 14476, Germany, **2** Max-Planck Institute of Molecular Plant Physiology, Potsdam-Golm, 14476, Germany, **3** Bayer CropScience NV-Innovation Center, Technologiepark 38, 9052 Gent, Belgium, **4** Department of Plant and Environmental Sciences, Faculty of Sciences, University of Copenhagen, Thorvaldsensvej, 40 1.1871, Fredriksberg C, Denmark

Abstract

A detailed knowledge of cell wall heterogeneity and complexity is crucial for understanding plant growth and development. One key challenge is to establish links between polysaccharide-rich cell walls and their phenotypic characteristics. It is of particular interest for some plant material, like cotton fibers, which are of both biological and industrial importance. To this end, we attempted to study cotton fiber characteristics together with glycan arrays using regression based approaches. Taking advantage of the comprehensive microarray polymer profiling technique (CoMPP), 32 cotton lines from different cotton species were studied. The glycan array was generated by sequential extraction of cell wall polysaccharides from mature cotton fibers and screening samples against eleven extensively characterized cell wall probes. Also, phenotypic characteristics of cotton fibers such as length, strength, elongation and micronaire were measured. The relationship between the two datasets was established in an integrative manner using linear regression methods. In the conducted analysis, we demonstrated the usefulness of regression based approaches in establishing a relationship between glycan measurements and phenotypic traits. In addition, the analysis also identified specific polysaccharides which may play a major role during fiber development for the final fiber characteristics. Three different regression methods identified a negative correlation between micronaire and the xyloglucan and homogalacturonan probes. Moreover, homogalacturonan and callose were shown to be significant predictors for fiber length. The role of these polysaccharides was already pointed out in previous cell wall elongation studies. Additional relationships were predicted for fiber strength and elongation which will need further experimental validation.

Editor: David D. Fang, USDA-ARS-SRRC, United States of America

Funding: This work was supported by the European Union Seventh Framework Programme (FP7 2007–2013) under the WallTraC project (grant agreement No. 263916). This paper reflects the authors' views only. The European Community is not liable for any use that may be made of the information contained herein. The funders had no role in study design, data collection and analysis, decision to publish, or preparation of the manuscript.

Competing Interests: Co-authors Dr. Jean-Luc Runavot and Dr. Frank Meulewaeter are employed by Bayer CropScience NV, Innovation Center.

* Email: jselbig@uni-potsdam.de

Introduction

Cell walls, the key determinant of overall plant growth and development are primarily composed of polysaccharides, namely cellulose, hemicellulose, and pectins, lignin, and structural proteins [1,2]. Cell wall biology has been an area of prominent research over many years with the use of novel technologies to probe these higher order structures in the native state. Since the early 1970's, comparative biochemical analyses revealed that all plant cell walls share several common features. However, they exhibit diversity with respect to their chemical composition [3–5]. Indeed, cell walls are structurally complex as they are constantly remodeled and reconstructed during plant growth and development. Also, walls are modulated according to functional requirements, thereby limiting our knowledge on cell wall design [6–8]. Biochemical analyses complemented by genetic analyses have identified genes and gene products associated with cell wall synthesis. However, an understanding of how these genes are expressed across cells of different tissues and their impact on cell wall design and maintenance is still lacking [9–11]. Furthermore, the glycan-rich cell walls influence the nutritional and processing properties of plant based products such as pulp for paper manufacture, textile fibers, timber products, pharmaceuticals, and materials for fuel and composite manufacture [12–15]. Therefore, understanding the plant cell walls is not only fundamental to plant sciences but also of industrial relevance.

Microarrays are widely used in plant research for the high throughput analysis of nucleotides, proteins and increasingly, carbohydrate [16–18]. Carbohydrate microarrays also referred to as glycan arrays enable hundreds of glycans to be analyzed in parallel. Glycans on the arrays can include oligosaccharides, polysaccharides, glycoproteins and glycolipids [19–21]. Glycan arrays have several biological and medical applications which include glycoproteomic methods to identify new glycoproteins and glycans [22,23], characterization of glycan probes [24], profiling carbohydrate-lectin interactions [25,26], glycosaminoglycans-growth factor and cytokine interactions [27,28], pathogen-induced antibody interaction [29,30], cancer-antibody induced interaction [31,32], carbohydrate-virus interactions [33], quantitative carbohydrate-protein interactions [34], and drug discovery [35,36].

Comprehensive microarray polymer profiling (CoMPP), a microarray based glycan screening method is mostly used for high throughput characterization of plant cell walls. In this technique, generation of microarrays by sequential extraction of cell wall polysaccharides and screening samples against a large number of well-defined cell wall probes such as antibodies, carbohydrate binding proteins and modules is done. This methodology was first described in *Arabidopsis thaliana* and *Physcomitrella patens* [37]. In the study of Singh et al, application of CoMPP to study cotton fibers showed that towards the end of elongation, there was a loss in certain cell wall polymer epitopes [38]. Despite the availability of glycan arrays from several experiments, computational analysis has mostly been restricted to collection of glycobiology information in databases, motif analysis of glycans, and oligosaccharide structure determination [39–41].

In our study, we used the glycan array technology to study cotton fibers, one of the most important raw materials for the textile industry. There are four different domesticated species producing cotton fibers namely *Gossypium hirsutum* ('Upland cotton'), *Gossypium barbadense* ('Pima'or 'Egyptian'cotton), *Gossypium arboreum* ('Tree cotton'), and *Gossypium herbaceum* [42]. The development of cotton fibers occurs in four major stages: initiation, elongation, secondary wall synthesis and maturation. Although much work has already been done on the cotton fiber transcriptome, the key question in cotton fiber research is to link the cell wall profile of different cotton types to the cotton fiber properties and to a better understanding of fiber development [43–46]. Here, we aim to study the relation between fiber properties and non-cellulosic polysaccharide composition using univariate and multivariate regression based approaches on a diverse set of cotton fibers. To this end, we analyzed two datasets for the same cotton fibers: a glycan array profile and the physical fiber properties as determined by HVI and AFIS. We elucidated the usefulness of regression based approaches to determine the functional relationship between the two datasets and we also selected a subset of variables which have a good prediction of the phenotypic traits.

Materials and Methods

Plant material and evaluation of phenotypic traits

In this study, we used 32 different cotton lines of which three are from *Gossypium arboreum*, three from *Gossypium barbadense*, two from *Gossypium herbaceum* and 24 from *Gossypium hirsutum*. The cotton lines used in this study are listed in Table S1, including the plant introduction number (PI number) from the USDA National Plant Germplasm System (http://www.ars-grin.gov/npgs/). Seeds were sown in soil compost and plants were grown at constant conditions in a greenhouse at 26–28°C during a 16 h photoperiod. Mature cotton fibers were collected by harvesting all fully open bolls from several plants. The impact of boll position and plant-to-plant variation was minimized by mixing the fiber from all harvested bolls. Two types of analyses were performed on these fibers, the first being the glycan array measurements (Table S2) and the second being fiber characteristics/phenotype measurements (Table S1). For each line, High Volume Instrument (HVI) and Advanced Fiber Information System (AFIS) measurements were performed on 40 g of mature cotton fiber by CIRAD (France) according to the standard methods ASTM D3818-92 and D5867-05. These measurements were done on 6 and 5 replicates for HVI and AFIS, respectively, except for micronaire where only 2 replicates were performed. Five fiber characteristics which include length from HVI and AFIS, strength, elongation and micronaire were selected for further analysis due to their importance for textile processing. Length HVI refers to the average fiber length of the longer 50% of fibers in a given sample. Length AFIS (W) L deduces length parameters from individual fiber measurements. Strength of the cotton fiber refers to the force required to break a bundle of fibers 1 tex in size (1 tex equals the weight in grams of 1000 meters of fibers). Elongation of the cotton fibers is the measurement of the elasticity of cotton fibers with a higher number indicating more elasticity. Micronaire is obtained by measuring the resistance of the fibers to airflow and depends on the fiber fineness and degree of maturation.

Comprehensive Microarray Polymer Profiling (CoMPP) of mature cotton fiber cell wall

CoMPP analysis was performed on mature cotton fibers as previously described by [38] with minor modifications. Mature cotton fiber samples were extracted sequentially in 50 mM cyclohexanediamine tetraacetic acid (CDTA) and 4 M Sodium hydroxide (NaOH) with 1% (v/v) sodium tetrahydridoborate ($NaBH_4$). These two solvents were used to extract pectins and non-cellulosic polysaccharides, respectively. For each line, 300 ul of solvent was added to 10 mg of sample and incubated with shaking for 2 h. After centrifugation, supernatant from each extraction was printed in four replicates and four dilutions (1:2, 1:6, 1:18 and 1:54 [v/v] dilutions). Cadoxen extraction was omitted because it is mainly used to extract cellulose which we do not aim to analyse in our study. The array was probed with eleven monoclonal antibodies (mAbs) recognizing different carbohydrate epitopes as listed out in Table 1. A heat map was generated to display the relative intensity of each signal to the maximum signal observed within each antibody detection (Table S2). CoMPP is a semi-quantitative technique and should not be taken to obtain absolute amounts. Practically speaking, we set the maximum value in the whole data sheet as 100 and the other values are divided by this maximum value and multiplied by 100 to obtain numbers comprised between 0 and 100. When the quantification is done, the arrays are manually checked to make sure that there are clear dots on it and not only background or noise. The negative control is an array incubated with 5% milk in PBS and probed with secondary antibody and then developed as the others.

Pre-processing of the data

For the statistical analysis, we use R version 3.1.0 on a 64 bit linux platform [56]. The numerical values from both datasets were of different physical quantities and on different scales of magnitude. Moreover, there is no external knowledge that variables with higher numeric variation should be considered more important. Standardization of the raw data was done by computing z- scores of the raw data. Z- scores were calculated for each data point by subtracting the mean and dividing by the standard deviation of all data points.

Linear methods to delineate the relationship between the two datasets

Multiple regression models the relationship between a single scalar response variable and a set of explanatory (or independent) variables. Here, we used multiple regression analysis to model which of the cell wall probes were associated to the fiber

Table 1. List of probes used in the glycan array.

Probes used in the analysis	Specificity of the probes	Reference	Source
BS-400-2	(1,3)-β-D-glucan (callose)	[47]	Purchased from Biosupplies (Australia)
JIM5	Partially methyl-esterified homogalacturonan (HG)	[48]	Paul Knox lab
LM19	Un-esterified homogalacturonan (HG)	[49]	Paul Knox lab
JIM13	Arabinogalactan (AGP)	[50]	Paul Knox lab
JIM20	Extensin glycoproteins	[51]	Paul Knox lab
LM11	Xylan	[52]	Paul Knox lab
LM15	XXXG xyloglucans (XG) epitope	[53]	Paul Knox lab
LM24	XXLG and XLLG xyloglucan (XG) epitopes	[24]	Paul Knox lab
LM25	XXLG and XLLG xyloglucan (XG) epitopes	[24]	Paul Knox lab
BS-400-4	Mannan	[54]	Purchased from Biosupplies (Australia)
LM21	Mannan	[55]	Paul Knox lab

characteristics. This allowed us to determine the overall fit (variance explained) of the model and the relative contribution of each of the cell wall probes to the total variance explained. The results from the analysis were reported in the coefficients and ANOVA tables. Summary of the fitted model object gave an account of the residuals, the estimates of the intercept, the slope (with the results of a t-test), the residual standard error, the R^2 statistic and the results of an F-test. The terms used in the output of regression analysis are defined as follows: residual standard error is the standard deviation of the data about the regression line. The squared multiple correlation coefficient (R^2) is the proportion of variability in the response that is fitted in the model and the F value is a test statistic to decide whether the model as a whole has statistically significant predictive capability. p values give the statistically significant predictive capability in the presence of other variables [57,58]. Based on this, five models were selected to determine which of the cell wall polysaccharides play an important role in determining that particular fiber characteristic.

In addition to the multiple regression analysis, relationships between multiple dependent and independent variables were investigated simultaneously using canonical correlation analysis (CCA). The two sets of data were represented by matrices X (dimension $n \times p$) and Y (dimension $n \times q$) and columns in X and Y denote the variables p (glycan measurements) and q (fiber characteristics) respectively. Classification of variables as dependent or independent is of little importance for the statistical estimation of the canonical functions as canonical correlation finds linear combinations of sets of multiple dependent and independent variables which are maximally correlated [59,60].

The first step in CCA was to derive one or more canonical function between the glycan and phenotypic measurements. Each function consisted of a pair of variates, one representing the cell wall probes and the other representing the fiber characteristics. The maximum number of canonical variates (functions) that could be extracted from the sets of variables equals the number of variables in the smallest data set, independent or dependent. As a result, the first pair of canonical variates was derived so as to have the highest intercorrelation possible between the glycan array and the fiber measurements. Technically, the second pair of canonical variates exhibits the maximum relationship between the two sets of variables (variates) not accounted for by the first pair of variates and successive pairs of canonical variates were based on residual variance. Therefore each of the pairs of variates is orthogonal and independent of all other variates derived from the same set of data. The strength of the relationship between the pairs of variates obtained from both datasets was determined by the canonical correlation. An estimate of shared variance between the canonical variates was provided by the squared canonical correlations, also called canonical roots or eigenvalues. The statistical significance of each canonical function was assessed using multivariate tests of significance namely Wilk's lambda, Hotelling's trace, Pillai's trace and Roy's greatest characteristic criterion (Roy's gcr). The statistically significant canonical functions were then interpreted using canonical loadings, cross-loadings and redundancy index [61–65]. We used the "mixOmics" package [66] in R to perform the canonical correlation analysis.

Sparse partial least square regression to predict the cell wall probes associated to fiber characteristics

Partial least squares (PLS), a well-known regression technique dealing with collinear matrices, clearly has an edge over other regression techniques [67]. Unlike CCA, the PLS latent variables are linear combinations of the variables based on the maximization of covariance but do not allow feature selection. There are many variants of PLS of which we focused on a sparse partial least squares approach (sPLS) which includes a built-in feature to select variables while integrating the data. We used the "mixOmics" package [66] in the regression mode. Specifically, we use a two block data setup, X be the nxp matrix and Y be the nxq matrix where n denotes the samples, variables p and q denote the glycan measurements and fiber characteristics respectively. Sparse PLS, based on lasso regression penalizes the loading vectors using singular value decomposition and has an additional advantage to perform better even when the covariates are highly correlated. We used sPLS in the regression mode and the aim was to model the relationship between the variables and also predict one group of variables from the other [68–71].

Results

Standardization of the raw data

In this study, we attempted to assess the relationship between the cell wall polysaccharides and the physical fiber properties of mature cotton fibers, the data of which are provided as Table S1 and S2. The glycan array values used for the regression analysis were the sums from the CDTA and the NaOH extractions as performing the analysis using the individual values gave the same correlations. For the fiber characteristics dataset, the values were in different units and scales such as mm (for length), g/tex (for strength), and percentage (for elongation). To make the fiber characteristics dataset compliant to the glycan array, the raw data were jointly standardized using z scores prior to the analysis.

Modelling the fiber properties using linear regression models

We investigated the linear relationship between the fiber properties and their corresponding array values by a series of regression analyses. Multiple regression models were built considering one fiber characteristic at a time as the dependent variable and multiple probes as the independent variables. Five such models were predicted for the phenotypic traits and the overall model prediction result (Table 2) shows that the model for length HVI, length AFIS and micronaire are statistically significant. The significant predictor variables of length HVI are BS-400-2, LM19 and the ones for length AFIS include BS-400-2, JIM5, JIM20, LM15, and LM19. LM15, LM19, LM24 and LM25 are the significant predictor variables for the model predicting cotton fiber micronaire and the overall model has a p value of 4.906e-06. The models for strength and elongation do not show any statistical significance.

Assessing the relationship between multiple probes and all of the fiber characteristics simultaneously using canonical correlation analysis

The multiple regression analysis can predict the value of a single (metric) dependent variable from a linear function of a set of independent variables. However, to explore the relationship of sets of multiple predictor variables (probe measurements) to sets of multiple response variables (phenotypic traits) CCA was used. As CCA uses information from all the variables in both the predictor and response sets, it serves as a more efficient approach than methods routinely used, such as multiple linear regression.

For the CCA analysis, the glycan array measurements (probed by 11 antibodies) are designated as the set of independent variables. The fiber characteristics namely length AFIS, length HVI, strength, elongation and micronaire were specified as the set of dependent variables (Figure 1). However, it is of little importance to classify the variables as independent or dependent as the technique aims to maximize the correlation between the two sets of variables. In Figure 1, the terms r_{x1} to r_{x11} represent the canonical loadings which reflect the variance that the eleven variables from the glycan array shares with the independent canonical variate U_1. Similarly the terms r_{y1} to r_{y5} represent the canonical loadings which reflect the variance that the five phenotypic variables share with the dependent canonical variate V_1. The canonical correlation between the independent and dependent canonical variates is measured by the canonical functions which are represented by R^2_{c1} to R^2_{c5}. The statistical problem involved identifying any latent relationships (relationships between composites of variables rather than the individual variables themselves) between the glycan and the fiber measurements.

The canonical correlation which is based on the linear relationship of the glycan array data and fiber characteristics was computed to derive five canonical functions. Each of these functions consists of a pair of variates, one for the glycan array data and the other for the fiber characteristics. Since the study includes 11 independent variables and 5 dependent variables, the maximum number of canonical functions which could be derived is five (Table 3).

In addition to tests of each canonical function separately, multivariate tests of these five functions simultaneously were also performed. The test statistics employed include Wilks' lambda, Pillai's criterion, Hotelling's trace, and Roy's gcr. Table 4 details the p-values from the multivariate test statistics, which all indicate that only the first canonical function, taken collectively, is statistically significant at 1% level.

From the results of these tests, we proceeded to interpret other aspects of the analysis based on the first canonical function. A redundancy index was calculated for the independent and dependent variates of the first function in Table 5. The redundancy index is calculated as the average loading squared times the canonical R^2. As can be seen, the redundancy index for the dependent (0.191) and independent variates (0.200) is quite low. The low values result from the relatively low shared variance in the dependent variates (0.214) and independent variates (0.225), not the canonical R^2. With such a small percentage, this is an example of a statistically significant canonical function that does not have practical significance because it does not explain a large proportion of the dependent variables' variance.

The interpretations involve examining the canonical functions to determine the relative importance of each of the original variables in deriving the canonical relationships (Table 6). The three methods for interpretation are (1) canonical weights (standardized coefficients), (2) canonical loadings (structure correlations), and (3) canonical cross-loadings.

Table 6 contains the standardized canonical weights for each canonical variate for both dependent and independent variables. As mentioned earlier, the magnitude of the weights represent their relative contribution to the variate. Based on the size of the weights, the order of contribution of independent variables to the

Table 2. Summary statistics of the five possible multiple regression models.

Fiber characteristics	Residual standard error	Multiple R-squared	Adjusted R-squared	F-statistic	p-value	Significant predictors
Length HVI	0.696	0.706	0.545	4.372 on 11 and 20 DF	0.002	BS-400-2, LM19
Length AFIS	0.632	0.720	0.566	4.677 on 11 and 20 DF	0.001	BS-400-2, JIM5, JIM20, LM15, LM19
Strength	0.940	0.378	0.036	1.107 on 11 and 20 DF	0.404	JIM20
Elongation	0.825	0.376	0.033	1.098 on 11 and 20 DF	0.410	–
Micronaire	0.469	0.851	0.769	10.4 on 11 and 20 DF	4.906e-06	LM15, LM19, LM24, LM25

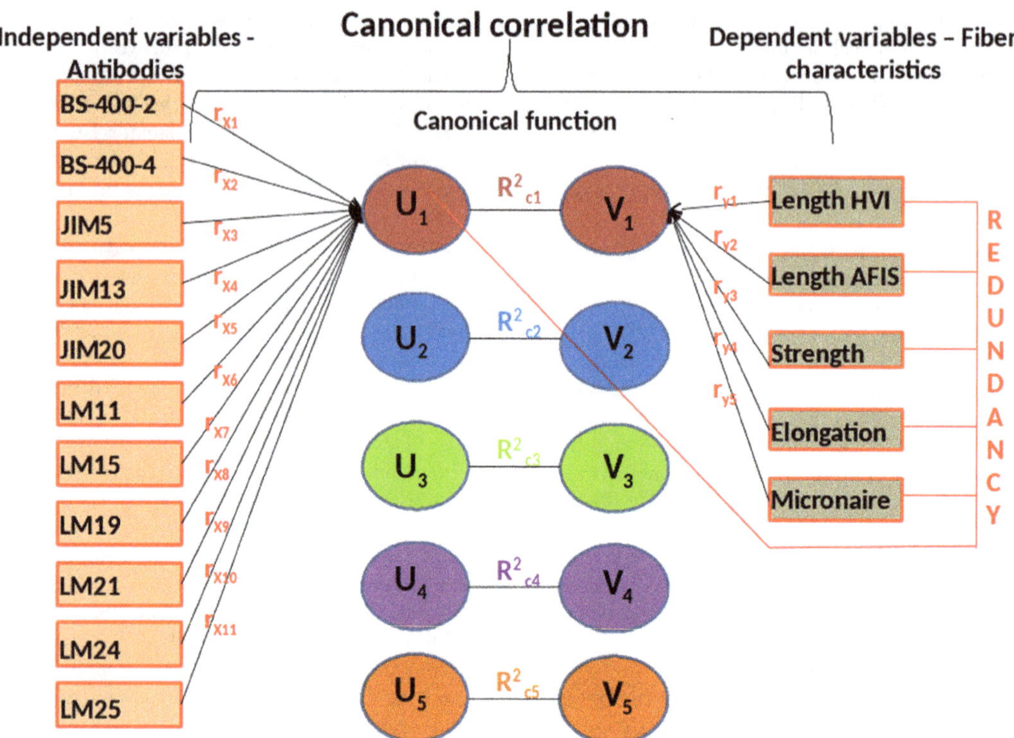

Figure 1. Canonical correlation analysis maximizes the correlation between the linear combination of the cell wall polysaccharides in the glycan array and the fiber properties. In this figure, given a linear combination of X variables: $U_1 = f_{1\times1} + f_{2\times2} + \ldots + f_pX_p$ and a linear combination of Y variables: $V_1 = g_1Y_1 + g_2Y_2 + \ldots + g_qY_q$, the first canonical correlation is the maximum correlation coefficient between U_1 and V_1, for all U_1 and V_1.

first variate is LM19, LM25, JIM5, LM15, BS-400-4, LM21, LM24, JIM13, and JIM20 and the dependent variable order on the first variate is micronaire followed by length AFIS, length HVI, strength and elongation. Because canonical weights are typically unstable, particularly in instances of multicollinearity, owing to their calculation solely to optimize the canonical correlation, the canonical loading and cross-loadings are considered more appropriate.

Table 6 also contains the canonical loadings for the dependent and independent variates for the first canonical functions. In the first dependent variates, all the five variables had different values of loadings resulting in low shared variance (0.214). This indicates a low degree of inter-correlation among the five dependent variables. Observing the independent variates, there is a different pattern and loading values ranged from 0.06 to 0.77. The variables with the highest loadings on the independent variate are LM25, LM19, LM15, and JIM5. We also observed some loadings with negative values which include those of BS-400-4, JIM20, and LM11.

In case of the cross loadings, micronaire has a value of −0.890 and interestingly has a negative loading. Length AFIS to some extent has a loading value of 0.387 while those of the other variables is low. By squaring these terms, we find the percentage of the variance for each of the variables explained by function 1. The results show that 79.21 percent of the variance in micronaire, 14.97 percent of the variance in length AFIS is explained by function 1 whereas strength, elongation and length HVI have very low values. Similarly for the independent variables' cross loadings, variables LM25, LM19, LM15, JIM5 have high correlations of 0.73, 0.67, 0.61, and 0.61 respectively. From this information, approximately 51.8% of the variance in LM25, 45.1% of the variance in LM19, 36.3% of the variance in LM15, and 35.7% of the variance in JIM5 is explained by the dependent canonical variates.

Table 3. Canonical Correlation analysis relating probe signals and fiber characteristics with the measure of overall model fit.

Canonical function	Canonical correlation	Canonical R^2	F statistics	p-value
1	0.945	0.883	2.85	1.57338e-05
2	0.868	0.753	1.88	1.072479e-02
3	0.803	0.645	1.30	2.035076e-01
4	0.523	0.273	0.59	8.715784e-01
5	0.342	0.116	0.34	9.040993e-01

Table 4. Multivariate tests of significance for the canonical functions.

Canonical function	Wilks' Lambda, using F-approximation (Rao's F):	Hotelling-Lawley Trace, using F-approximation:	Pillai-Bartlett Trace using F-approximation	Roy's largest root using F-approximation
1	1.57338e-05	2.666759e-07	0.00	8.732348e-12
2	1.072479e-02	1.221924e-03	0.042	
3	2.035076e-01	5.475660e-02	0.285	
4	8.715784e-01	8.365373e-01	0.801	
5	9.040993e-01	8.855773e-01	0.848	

The final step of interpretation is examining the signs of the cross-loadings. Examining the signs of the independent variables' cross loadings, those with high correlations have a positive direct relationship whereas BS-400-4, JIM20 and LM11 have an inverse relationship. The four highest cross-loadings of the first independent variate correspond to the variables with the highest canonical loadings as well. Observing the cross loadings of the dependent variables, we see that micronaire has the highest canonical loading and an inverse relationship. Also, elongation is observed to have an inverse relation but since it is of very low value, it was not taken into account.

sPLS approach to predict specific cell wall polysaccharides involved in fiber properties

sPLS was computed in the regression mode and the input for the analysis included the 11 cell wall probes along with the five fiber characteristics The number of dimensions H to be retained was estimated with the Qh^2 criterion, for which a value below the threshold 0.0975 indicates a significant contribution for the prediction purpose. The Qh^2 values calculated for each dimension of the sPLS showed that 2 dimensions were enough to capture the whole information. From Figure 2, we can interpret the results from the sPLS via the correlation circle plot where the predictor variables are in red and the response variables are represented in blue. A correlation circle plot is a graphic tool to represent variables of two different data-types and examine the relationships between the variables and variates. In this plot, variables namely cell wall probes and fiber measurements can be represented as vectors. The relationship between these two data-types is approximated by the inner product between the associated vectors which is defined as the product of the two vector lengths and their cosine angle. For better interpretation, two circles of radii 0.5 and 1 are represented to visualize the variables. The longer the distance to the origin, the stronger is the relationship between the variables.

Using the interpretation which is detailed, we find that BS-400-4, LM21, and JIM13 share a positive relationship with elongation characteristic of cotton fibers. We were also able to attribute the strength of the cotton fibers to JIM20, LM11 and LM24. Interestingly, LM19, JIM5, LM15, and LM25 were projected diametrically opposite to that of the micronaire in the correlation circle, thereby indicating a strongly negative relationship. Length HVI and length AFIS share a negative relation to BS-400-2. To estimate the significance of the predicted relationships, the root mean squared error prediction (RMSEP) values were computed for each response variable (fiber properties) and ranked according to the absolute value of their loadings in v_2. The lower the RMSEP value, the better the prediction of the model is. In this case, the model for micronaire was the best one (RMSEP of 0.71), followed by that of length AFIS (1.13), strength (1.14), elongation (1.15), and length HVI (1.21).

Figure 3 displays the graphical representation of the cotton lines in dimension 1 and 2. This plot shows that some of the lines are clustered together, with Acala SJ1, Germains Acala (GC 352 and GC 362), TAM-90C-19 S, and FM966 forming one cluster, Acala red okra, okra leaf, multiple marker, Tidewater, and TTU 202-1107B forming a second cluster and PIMAS7, Lankart 57, IV4F-91057, GA161, Ting tao tzu ching chung mien, Brymer brown, Malla guza, Selection of SHIH, China 10, Texas rust brown, Tex 1000 and 30834 (A1660) forming a third cluster. Strikingly, some of these clusters contain lines from different *Gossypium* species and lines from one species often belong to multiple clusters. The

Table 5. Redundancy analysis of dependent and independent variates for the first canonical function.

Standardized variance of the dependent variables explained by				
Their own Canonical variates (shared variance)			The opposite canonical variates (Redundancy)	
Percentage	Cumulative percentage	Canonical R^2	Percentage	Cumulative percentage
0.214	0.214	0.883	0.191	0.191
Standardized variance of the independent variables explained by				
Their own Canonical variates (shared variance)			The opposite canonical variates (Redundancy)	
Percentage	Cumulative percentage	Canonical R^2	Percentage	Cumulative percentage
0.225	0.225	0.883	0.200	0.200

Table 6. Canonical weights, loadings, and cross-loadings for the first canonical function.

	Canonical weights	Canonical loadings	Canonical cross-loadings
Dependent variables			
Length HVI	−0.636	0.127	0.120
Strength	−0.226	0.040	0.038
Elongation	−0.033	0.056	0.053
Micronaire	−0.843	−0.941	−0.890
Length AFIS	0.810	0.409	0.387
Independent variables			
BS-400-2	0.184	0.362	0.342
BS-400-4	−0.487	−0.119	−0.113
JIM5	−0.823	0. 0.632	0.598
JIM13	−0.290	0.288	0.272
JIM20	−0.204	−0.376	−0.355
LM11	−0.114	−0.360	−0.340
LM15	−0.719	0.638	0.603
LM19	1.243	0.712	0.672
LM21	0.356	0,275	0.260
LM24	−0.324	0.066	0.062
LM25	1.082	0.767	0.724

variation in fiber characteristics and composition is thus clearly not species-specific. However, one should be careful in interpreting the results from the individual lines as the study was designed to discover correlations between fiber properties and composition and not to study properties of individual lines.

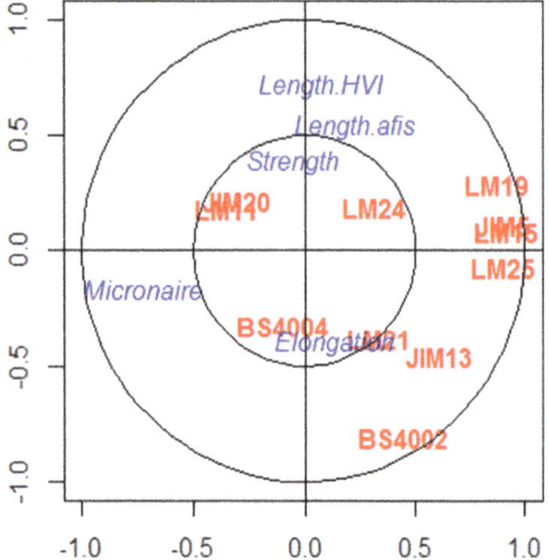

Figure 2. Graphical representation of the variables selected by sPLS on the first two dimensions predicts specific cell wall polysaccharides linked to the fiber properties. The coordinates of each variable are obtained by computing the correlation between the latent variable vectors and the original dataset. The selected variables are then projected onto correlation circles where highly correlated variables cluster together. These graphics help to identify association between the two datasets. The correlation between two variables is positive if the angle is sharp $\cos(\alpha)>0$, negative if the angle is obtuse $\cos(\theta)<0$, and null if the vectors are perpendicular $\cos(\beta)\sim 0$.

Discussion

Understanding the genetics and physiology of cotton fibers is of importance to the textile industry. There have been numerous studies, both profiling and sequencing based experiments to study cotton fiber development at the transcriptional level. The high degree of transcriptional complexity in the development of cotton fibers has been the focus of these studies [43,72–75]. We used the CoMPP technique in our analysis to study directly the glycan composition of cotton lines from different species. The work presented here demonstrates the potential of glycan microarrays in combination with multivariate statistical approaches for understanding the cell wall composition responsible for the fiber characteristics. Specifically, the use of regression based approaches in our study helps to predict models for each of the fiber trait under study.

We studied the association between glycan array measurements and their relation with fiber characteristics using linear approaches like multiple regression, CCA and sPLS. From the results of multiple regression (Table 2), we were able to predict three models for length HVI, length AFIS and micronaire of cotton fibers but not for strength and elongation characteristics. Moreover, to extend our understanding of the data to situations involving more than one fiber characteristic at a time, CCA was used as it simultaneously models effects of multiple independent variables on multiple dependent variables. As CCA uses information from all the variables in both the exposure and outcome variable sets and maximizes the estimation of the relationship between the two sets, CCA may offer a more efficient approach for assessing the relationship of the cell wall probes with fiber characteristics than methods routinely used such as multiple linear regression. CCA starts with simultaneous consideration of both glycan array measurements and the phenotype measures, limiting the inefficiencies that may accompany conventional multiple testing, and thus, reducing type-1 error. The resulting procedure gives a global view of association between indicators of both datasets. Thus,

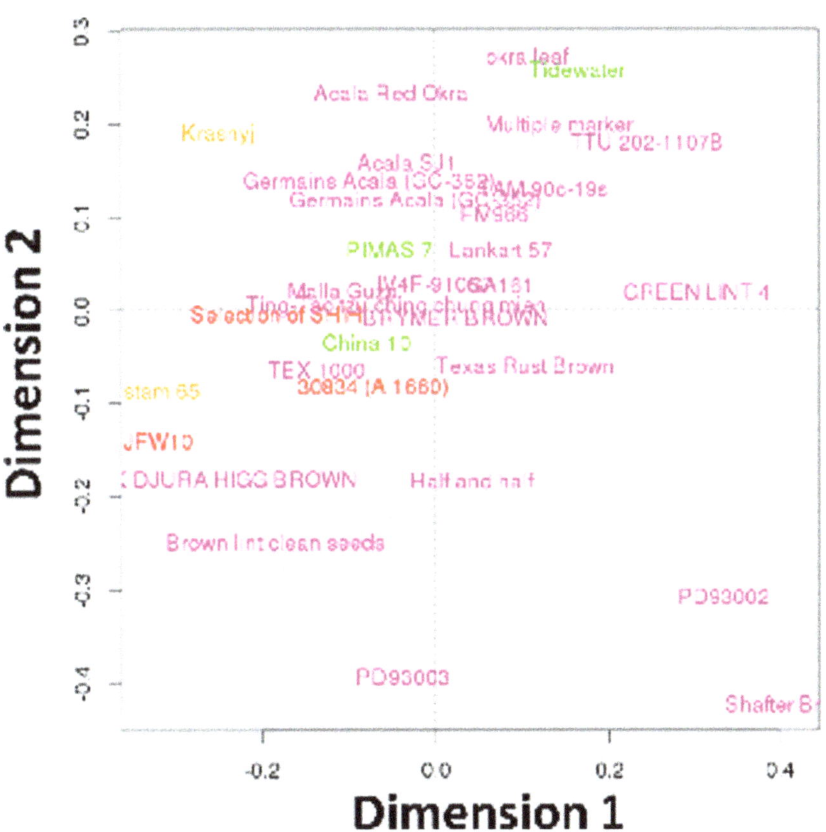

Figure 3. Graphical representation of the cotton lines on the first two sPLS dimensions shows the trend in clustering of specific cotton lines across different species. Four different species of cotton are shown in different colors. *Gossypium hirsutum* is colored in magenta, *Gossypium barbadense* in green, *Gossypium herbaceum* in orange and *Gossypium arboreum* in red.

CCA could be used as a comprehensive approach to extract information from data simultaneously. Another major advantage of using the CCA to multiple regression analysis is to deal with the issue of multicollinearity. In multiple regression, the interpretation is usually based on the significance of weights, which is highly influenced by multicollinearity. If two variables have a high correlation one of them will be completely eliminated even if both have a high correlation to the outcome. In our analysis this is illustrated by JIM5 and LM19 (both detecting homogalacturonan), with both showing a high correlation with micronaire in CCA but only LM19 being identified as a predictor of micronaire in the linear regression model. From the results of the CCA, we obtained an overall picture of associations between the glycan and phenotype measurements, with information about the relative contribution of the variables to that particular canonical variate through canonical loadings. The canonical analysis revealed that the canonical correlation was statistically significant at 1%. However, canonical correlation based methods are statistically difficult to assess as they do not fit into a regression framework. In this context penalized CCA adapted with elastic net (CCA-EN) could be used but the elastic net is similar to a lasso soft-thresholding penalization and the algorithm uses partial least squares and not canonical correlation computations [63]. From [63] it is evident that sPLS made a good compromise between all of these approaches and includes variable selection. Additionally, we used the sPLS approach to be able to predict specific cell wall polysaccharides linked with fiber characteristics. Moreover, sPLS maximizes the covariance between the latent variables whereas the canonical correlation based methods maximize the correlation.

There were both unique and common findings from the three types of regression analysis. The major and most significant finding in common to all these analyses is that micronaire is negatively correlated with the xyloglucan (XG) and homogalacturonan (HG) probes. One possible explanation for this observation is that cotton fiber with a high micronaire usually has a very thick secondary cell wall resulting in very high levels of cellulose and lower levels of the non-cellulosic components. However, we do not find a negative correlation of micronaire with other non-cellulosic compounds suggesting that increased cellulose levels of high micronaire fibers affect the XG and HG epitopes in a different way than the other non-cellulosic epitopes. For instance, it could specifically decrease extractability of the XG and HG epitopes. As micronaire measures a combination of fiber fineness and maturity, we wanted to understand whether the observed correlation is with maturity or fineness or a combination of both. We tested this using linear regression models once again and built models for fineness and maturity of the fibers. We observed that the regression models for fineness had an adjusted R^2 value of 0.803 with JIM5, LM19, and LM25 as significant predictors at a 1% threshold. The regression model for maturity was also significant at the 1% threshold but with no particular significant predictors thereby suggesting that the observed correlation is attributed to fiber fineness. This indicates that this correlation is linked to the thickness rather than the shape of the fiber, which is consistent with a link to the cellulose levels.

Since only the first canonical function of the CCA analysis is statistically significant and this function explains only for micronaire a large fraction of the variance, the results of the CCA analysis are not informative with respect to the other fiber properties. For these fiber properties, the correlation between fiber length and callose is the only one that was detected in both the linear regression and the sPLS analysis. Callose has been described to play a role in cotton fiber elongation. Indeed, it was reported that plasmodesmatal closure was positively correlated with the rapid fiber elongation and that callose was involved in the gating of these plasmodesmata [76]. However, this observation involves transient callose detection, only after 5 dpa and already significantly reduced at 20 dpa, what makes it unlikely to be detected in mature fibers. Other callose deposition was reported by [77]. This callose is supposed to be deposited in the secondary cell wall and remains in the fiber. From the results of the multiple regression models (Table 2), a positive correlation between several of the homogalacturonan probes and length property of the fibers is apparent. The link between pectins and the elongation of cell walls is already observed in several plant systems [78] and studies in flax stems, pea stems and maize coleoptiles revealed a negative correlation between pectin levels and cell elongation. In cotton fibers and trichomes, there exists a positive correlation between pectic sheath and elongation [79] and recent studies by [80] have established that pectic polysaccharides and xyloglucan containing uronic acids were the major polysaccharides extracted during elongation. Hence, our results are in agreement with various studies which state that pectin biosynthesis promotes fiber elongation [81] and that the degree of esterification is a key factor in controlling the elongation [38,82]. The correlation between length and HG was not detected in the sPLS analysis most likely because the stronger (negative) correlation of HG with micronaire.

Furthermore, relationships between fiber strength or elongation and specific carbohydrate epitopes could be deduced from the results of the sPLS analysis (Figure 2). For instance, fiber strength was associated both with the xylan (LM11) and the extensin (JIM20) epitope. A role of xylan in fiber strength would be consistent with the function of heteroxylan in other cell types which is commonly related to the strengthening of cell walls as revealed by defects in cellulose deposition in xylan mutants [83]. A role of extensin in fiber strength is less expected and would need experimental validation. In the linear regression analysis, extensin was identified as a significant predictor for length AFIS but not for length HVI. A role for extensin in determining cotton fiber length would be more consistent with its role in other plant cell types [84]. Finally, AGP glycan (JIM13) and mannan (BS-400-4 and LM21) epitopes were found to predict cotton fiber elongation from the sPLS model. Interestingly, studies have indicated that AGPs are important players during fiber development. Immunofluorescence assays by JIM 13 showed distinct patterns in developing fiber cells indicating that polysaccharide chains of AGPs are involved in initiation and elongation stages of cotton fibers [85–87]. However, it is not clear how these AGPs would affect the elongation property of the mature fiber. These unexpected correlations present thus interesting hypotheses for further structure-function relationship studies of the cotton fiber.

Overall, CoMPP assays of cell wall polysaccharides from cotton fibers suggest that it will be a powerful tool in detecting and quantifying the differences between large sets of cotton lines thus gathering lot of information which is necessary for a proper statistical approach. With the use of predictive statistical approaches to integrate different kinds of datasets, this analysis has thus discovered some correlations that are in line with already known biological functions and others for which the biological relevance still has to be tested. Also, it confirmed the relevance of this type of analysis to enable a detailed understanding of the data from CoMPP assays of cell wall polysaccharides. However, the use of mature cotton fibers in this analysis only allows detecting relevant correlations for components that are still present at maturity. In addition, many changes in polysaccharide composition occur between the fiber elongation stage and maturity. One would thus expect to identify only a fraction of the relationships between polysaccharide composition and fiber properties by analysis of mature fibers, especially for fiber properties such as length that are determined in the early stages of development. Hence it would be interesting to perform a similar kind of analysis using the polysaccharide composition of developing fibers to see whether additional relationships with fiber properties can be determined. The panel of cotton lines used in this study was selected to have maximal diversity in fiber properties and composition. Applying this type of analysis to commercially important cotton lines would allow to understand whether differences in polysaccharide composition affect properties of commercial cotton in the same way as observed in this study and to get insight into the developmental polysaccharides that are essential to obtain high quality cotton fibers. With the sequencing of the *G. hirsutum* genome, cotton fiber research is an exciting field and the work presented here will provide a base for future studies, with potential to translate this study on the developing fibers.

Supporting Information

Table S1 Fiber characteristics/phenotype measurements for the 32 cotton lines used in the study. The plant introduction number (PI number) from the USDA national plant germplasm is also included for each cotton line.

Table S2 Comprehensive microarray polymer profiling (CoMPP) heat map of CDTA and NaOH extractions of mature cotton fibers from 32 cotton lines. References for probe specificity are listed in Table 1.

Acknowledgments

We would like to thank Prof. JP Paul Knox, University of Leeds for the cell wall epitopes used in the glycan array experiment.

Author Contributions

Conceived and designed the experiments: DR JLR XG WW FM JS. Performed the experiments: JLR XG. Analyzed the data: DR. Contributed reagents/materials/analysis tools: DR JLR XG. Contributed to the writing of the manuscript: DR JLR. Read and approved the manuscript: DR JLR XG WW FM JS.

References

1. Heredia A, Jiménez A, Guillén R (1995) Composition of plant cell walls. Z Für Lebensm-Unters -Forsch 200: 24–31.
2. Keegstra K (2010) Plant Cell Walls. Plant Physiol 154: 483–486. doi:10.1104/pp.110.161240.
3. Somerville C, Bauer S, Brininstool G, Facette M, Hamann T, et al. (2004) Toward a Systems Approach to Understanding Plant Cell Walls. Science 306: 2206–2211. doi:10.1126/science.1102765.

4. Minorsky PV (2002) The wall becomes surmountable. Plant Physiol 128: 345–353. doi:10.1104/pp.900022.
5. Carpita NC, Gibeaut DM (1993) Structural models of primary cell walls in flowering plants: consistency of molecular structure with the physical properties of the walls during growth. Plant J Cell Mol Biol 3: 1–30.
6. Roberts K (2001) How the Cell Wall Acquired a Cellular Context. Plant Physiol 125: 127–130. doi:10.1104/pp.125.1.127.
7. McCann M, Rose J (2010) Blueprints for Building Plant Cell Walls. Plant Physiol 153: 365–365. doi:10.1104/pp.110.900324.
8. Pilling E, Höfte H (2003) Feedback from the wall. Curr Opin Plant Biol 6: 611–616.
9. Somerville C (2006) Cellulose synthesis in higher plants. Annu Rev Cell Dev Biol 22: 53–78. doi:10.1146/annurev.cellbio.22.022206.160206.
10. Mutwil M, Debolt S, Persson S (2008) Cellulose synthesis: a complex complex. Curr Opin Plant Biol 11: 252–257. doi:10.1016/j.pbi.2008.03.007.
11. Ellis M, Egelund J, Schultz CJ, Bacic A (2010) Arabinogalactan-Proteins: Key Regulators at the Cell Surface? Plant Physiol 153: 403–419. doi:10.1104/pp.110.156000.
12. Chapple C, Carpita N (1998) Plant cell walls as targets for biotechnology. Curr Opin Plant Biol 1: 179–185. doi:10.1016/S1369-5266(98)80022-8.
13. Thakur BR, Singh RK, Handa AK (1997) Chemistry and uses of pectin-a review. Crit Rev Food Sci Nutr 37: 47–73. doi:10.1080/10408399709527767.
14. Sticklen MB (2008) Plant genetic engineering for biofuel production: towards affordable cellulosic ethanol. Nat Rev Genet 9: 433–443. doi:10.1038/nrg2336.
15. Morris G, Kök S, Harding S, Adams G (2010) Polysaccharide drug delivery systems based on pectin and chitosan. Biotechnol Genet Eng Rev 27: 257–284.
16. Schena M (1996) Genome analysis with gene expression microarrays. BioEssays News Rev Mol Cell Dev Biol 18: 427–431. doi:10.1002/bies.950180513.
17. Ekins R, Chu FW (1999) Microarrays: their origins and applications. Trends Biotechnol 17: 217–218.
18. Wang D (2003) Carbohydrate microarrays. Proteomics 3: 2167–2175. doi:10.1002/pmic.200300601.
19. Park S, Lee M-R, Shin I (2008) Carbohydrate microarrays as powerful tools in studies of carbohydrate-mediated biological processes. Chem Commun Camb Engl: 4389–4399. doi:10.1039/b806699j.
20. Shin I, Park S, Lee M (2005) Carbohydrate microarrays: an advanced technology for functional studies of glycans. Chem Weinh Bergstr Ger 11: 2894–2901. doi:10.1002/chem.200401030.
21. Wang R, Liu S, Shah D, Wang D (2005) A practical protocol for carbohydrate microarrays. In: Zanders ED, editor. Methods Mol Biol Clifton NJ 310: 241–252.
22. Hsu T-L, Hanson SR, Kishikawa K, Wang S-K, Sawa M, et al. (2007) Alkynyl sugar analogs for the labeling and visualization of glycoconjugates in cells. Proc Natl Acad Sci U S A 104: 2614–2619. doi:10.1073/pnas.0611307104.
23. Hanson SR, Hsu T-L, Weerapana E, Kishikawa K, Simon GM, et al. (2007) Tailored glycoproteomics and glycan site mapping using saccharide-selective bioorthogonal probes. J Am Chem Soc 129: 7266–7267. doi:10.1021/ja0724083.
24. Pedersen HL, Fangel JU, McCleary B, Ruzanski C, Rydahl MG, et al. (2012) Versatile high-resolution oligosaccharide microarrays for plant glycobiology and cell wall research. J Biol Chem: jbc.M112.396598. doi:10.1074/jbc.M112.396598.
25. Uchiyama N, Kuno A, Koseki-Kuno S, Ebe Y, Horio K, et al. (2006) Development of a lectin microarray based on an evanescent-field fluorescence principle. Methods Enzymol 415: 341–351. doi:10.1016/S0076-6879(06)15021-1.
26. Gupta G, Surolia A, Sampathkumar S-G (2010) Lectin microarrays for glycomic analysis. Omics J Integr Biol 14: 419–436. doi:10.1089/omi.2009.0150.
27. Gama CI, Tully SE, Sotogaku N, Clark PM, Rawat M, et al. (2006) Sulfation patterns of glycosaminoglycans encode molecular recognition and activity. Nat Chem Biol 2: 467–473. doi:10.1038/nchembio810.
28. De Paz JL, Noti C, Seeberger PH (2006) Microarrays of Synthetic Heparin Oligosaccharides. J Am Chem Soc 128: 2766–2767. doi:10.1021/ja057584v.
29. Ratner DM, Seeberger PH (2007) Carbohydrate microarrays as tools in HIV glycobiology. Curr Pharm Des 13: 173–183.
30. Wang L-X, Ni J, Singh S, Li H (2004) Binding of high-mannose-type oligosaccharides and synthetic oligomannose clusters to human antibody 2G12: implications for HIV-1 vaccine design. Chem Biol 11: 127–134. doi:10.1016/j.chembiol.2003.12.020.
31. Huang C-Y, Thayer DA, Chang AY, Best MD, Hoffmann J, et al. (2006) Carbohydrate microarray for profiling the antibodies interacting with Globo H tumor antigen. Proc Natl Acad Sci U S A 103: 15–20. doi:10.1073/pnas.0509693102.
32. Lawrie CH, Marafioti T, Hatton CSR, Dirnhofer S, Roncador G, et al. (2006) Cancer-associated carbohydrate identification in Hodgkin's lymphoma by carbohydrate array profiling. Int J Cancer J Int Cancer 118: 3161–3166. doi:10.1002/ijc.21762.
33. Blixt O, Head S, Mondala T, Scanlan C, Huflejt ME, et al. (2004) Printed covalent glycan array for ligand profiling of diverse glycan binding proteins. Proc Natl Acad Sci U S A 101: 17033–17038. doi:10.1073/pnas.0407902101.
34. Liang P-H, Wang S-K, Wong C-H (2007) Quantitative analysis of carbohydrate-protein interactions using glycan microarrays: determination of surface and solution dissociation constants. J Am Chem Soc 129: 11177–11184. doi:10.1021/ja072931h.
35. Bryan MC, Wong C-H (2004) Aminoglycoside array for the high-throughput analysis of small molecule-RNA interactions. Tetrahedron Lett 45: 3639–3642. doi:10.1016/j.tetlet.2004.03.035.
36. Disney MD, Barrett OJ (2007) An aminoglycoside microarray platform for directly monitoring and studying antibiotic resistance. Biochemistry (Mosc) 46: 11223–11230. doi:10.1021/bi701071h.
37. Moller I, Marcus SE, Haeger A, Verhertbruggen Y, Verhoef R, et al. (2008) High-throughput screening of monoclonal antibodies against plant cell wall glycans by hierarchical clustering of their carbohydrate microarray binding profiles. Glycoconj J 25: 37–48. doi:10.1007/s10719-007-9059-7.
38. Singh B, Avci U, Eichler Inwood SE, Grimson MJ, Landgraf J, et al. (2009) A specialized outer layer of the primary cell wall joins elongating cotton fibers into tissue-like bundles. Plant Physiol 150: 684–699. doi:10.1104/pp.109.135459.
39. Aoki-Kinoshita KF, Kanehisa M (2006) Bioinformatics approaches in glycomics and drug discovery. Curr Opin Mol Ther 8: 514–520.
40. Von der Lieth C-W, Bohne-Lang A, Lohmann KK, Frank M (2004) Bioinformatics for glycomics: status, methods, requirements and perspectives. Brief Bioinform 5: 164–178.
41. Marchal I, Golfier G, Dugas O, Majed M (2003) Bioinformatics in glycobiology. Biochimie 85: 75–81.
42. Wendel JF, Brubaker C, Alvarez I, Cronn R, Stewart JM (2009) Evolution and Natural History of the Cotton Genus. In: Paterson AH, editor. Genetics and Genomics of Cotton. Plant Genetics and Genomics: Crops and Models. Springer US. 3–22.
43. Wang QQ, Liu F, Chen XS, Ma XJ, Zeng HQ, et al. (2010) Transcriptome profiling of early developing cotton fiber by deep-sequencing reveals significantly differential expression of genes in a fuzzless/lintless mutant. Genomics 96: 369–376. doi:10.1016/j.ygeno.2010.08.009.
44. Al-Ghazi Y, Bourot S, Arioli T, Dennis ES, Llewellyn DJ (2009) Transcript Profiling During Fiber Development Identifies Pathways in Secondary Metabolism and Cell Wall Structure That May Contribute to Cotton Fiber Quality. Plant Cell Physiol 50: 1364–1381. doi:10.1093/pcp/pcp084.
45. Gou J-Y, Wang L-J, Chen S-P, Hu W-L, Chen X-Y (2007) Gene expression and metabolite profiles of cotton fiber during cell elongation and secondary cell wall synthesis. Cell Res 17: 422–434. doi:10.1038/sj.cr.7310150.
46. Avci U, Pattathil S, Singh B, Brown VL, Hahn MG, et al. (2013) Cotton fiber cell walls of Gossypium hirsutum and Gossypium barbadense have differences related to loosely-bound xyloglucan. PloS One 8: e56315. doi:10.1371/journal.pone.0056315.
47. Meikle PJ, Bonig I, Hoogenraad NJ, Clarke AE, Stone BA (1991) The location of (1→3)-β-glucans in the walls of pollen tubes of Nicotiana alata using a (1→3)-β-glucan-specific monoclonal antibody. Planta 185: 1–8. doi:10.1007/BF00194507.
48. Willats WG, Limberg G, Buchholt HC, van Alebeek GJ, Benen J, et al. (2000) Analysis of pectic epitopes recognised by hybridoma and phage display monoclonal antibodies using defined oligosaccharides, polysaccharides, and enzymatic degradation. Carbohydr Res 327: 309–320.
49. Verhertbruggen Y, Marcus SE, Haeger A, Ordaz-Ortiz JJ, Knox JP (2009) An extended set of monoclonal antibodies to pectic homogalacturonan. Carbohydr Res 344: 1858–1862. doi:10.1016/j.carres.2008.11.010.
50. Yates EA, Valdor JF, Haslam SM, Morris HR, Dell A, et al. (1996) Characterization of carbohydrate structural features recognized by anti-arabinogalactan-protein monoclonal antibodies. Glycobiology 6: 131–139.
51. Smallwood M, Beven A, Donovan N, Neill SJ, Peart J, et al. (1994) Localization of cell wall proteins in relation to the developmental anatomy of the carrot root apex. Plant J 5: 237–246. doi:10.1046/j.1365-313X.1994.05020237.x.
52. McCartney L, Marcus SE, Knox JP (2005) Monoclonal antibodies to plant cell wall xylans and arabinoxylans. J Histochem Cytochem Off J Histochem Soc 53: 543–546. doi:10.1369/jhc.4B6578.2005.
53. Marcus SE, Verhertbruggen Y, Hervé C, Ordaz-Ortiz JJ, Farkas V, et al. (2008) Pectic homogalacturonan masks abundant sets of xyloglucan epitopes in plant cell walls. BMC Plant Biol 8: 60. doi:10.1186/1471-2229-8-60.
54. Pettolino FA, Hoogenraad NJ, Ferguson C, Bacic A, Johnson E, et al. (2001) A (1->4)-beta-mannan-specific monoclonal antibody and its use in the immuno-cytochemical location of galactomannans. Planta 214: 235–242.
55. Marcus SE, Blake AW, Benians TAS, Lee KJD, Poyser C, et al. (2010) Restricted access of proteins to mannan polysaccharides in intact plant cell walls. Plant J Cell Mol Biol 64: 191–203. doi:10.1111/j.1365-313X.2010.04319.x.
56. R Core Team 2013 (2013) A language and environment for statistical computing. R Foundation for Statistical Computing, Vienna, Austria. URL http://www.R-project.org/.
57. Tabachnick BG, Fidell LS (2012) Using Multivariate Statistics. 6 edition. Boston: Pearson. 1024 p.
58. Schneider A, Hommel G, Blettner M (2010) Linear Regression Analysis. Dtsch Ärztebl Int 107: 776–782. doi:10.3238/arztebl.2010.0776.
59. Hair JF (2010) Multivariate data analysis. Upper Saddle River, NJ: Prentice Hall.
60. Lutz JG, Eckert TL (1994) The Relationship between Canonical Correlation Analysis and Multivariate Multiple Regression. Educ Psychol Meas 54: 666–675. doi:10.1177/0013164494054003009.
61. Tenenhaus A, Philippe C, Guillemot V, Cao K-AL, Grill J, et al. (2014) Variable selection for generalized canonical correlation analysis. Biostatistics. doi:10.1093/biostatistics/kxu001.

62. Witten DM, Tibshirani RJ (2009) Extensions of Sparse Canonical Correlation Analysis with Applications to Genomic Data. Stat Appl Genet Mol Biol 8.
63. Le Cao K-A, Martin PG, Robert-Granie C, Besse P (2009) Sparse canonical methods for biological data integration: application to a cross-platform study. BMC Bioinformatics 10: 34. doi:10.1186/1471-2105-10-34.
64. Rencher AC (2003) Canonical Correlation. Methods of Multivariate Analysis. John Wiley & Sons, Inc. 361–379.
65. Thompson B (1984) Canonical correlation analysis uses and interpretation. Beverly Hills, Calif.: Sage Publications.
66. Dejean S, Gonzalez I, Lê Cao K-A with contributions from Monget P, Coquery J, Yao F, Liquet B and Rohart F (2013) mixOmics: Omics Data Integration Project. R package version 5.0-1. Available: http://CRAN.R-project.org/package=mixOmics.
67. Boulesteix A-L, Strimmer K (2007) Partial least squares: a versatile tool for the analysis of high-dimensional genomic data. Brief Bioinform 8: 32–44. doi:10.1093/bib/bbl016.
68. Lê Cao K-A, Rossouw D, Robert-Granié C, Besse P (2008) A sparse PLS for variable selection when integrating omics data. Stat Appl Genet Mol Biol 7: Article 35. doi:10.2202/1544-6115.1390.
69. GonzáLez I, DéJean S, Martin PGP, GonçAlves O, Besse P, et al. (2009) Highlighting relationships between heterogeneous biological data through graphical displays based on regularized canonical correlation analysis. J Biol Syst 17: 173–199. doi:10.1142/S0218339009002831.
70. González I, Cao K-AL, Davis MJ, Déjean S (2012) Visualising associations between paired "omics" data sets. BioData Min 5: 19. doi:10.1186/1756-0381-5-19.
71. Chun H, Keles S (2010) Sparse partial least squares regression for simultaneous dimension reduction and variable selection. J R Stat Soc Ser B Stat Methodol 72: 3–25. doi:10.1111/j.1467-9868.2009.00723.x.
72. Bowman MJ, Park W, Bauer PJ, Udall JA, Page JT, et al. (2013) RNA-Seq Transcriptome Profiling of Upland Cotton (Gossypium hirsutum L.) Root Tissue under Water-Deficit Stress. PloS One 8: e82634. doi:10.1371/journal.pone.0082634.
73. Lacape J-M, Claverie M, Vidal RO, Carazzolle MF, Guimarães Pereira GA, et al. (2012) Deep Sequencing Reveals Differences in the Transcriptional Landscapes of Fibers from Two Cultivated Species of Cotton. PLoS ONE 7: e48855. doi:10.1371/journal.pone.0048855.
74. Rambani A, Page JT, Udall JA (2014) Polyploidy and the petal transcriptome of Gossypium. BMC Plant Biol 14: 1–14. doi:10.1186/1471-2229-14-3.
75. Gilbert MK, Turley RB, Kim HJ, Li P, Thyssen G, et al. (2013) Transcript profiling by microarray and marker analysis of the short cotton (Gossypium hirsutum L.) fiber mutant Ligon lintless-1 (Li 1). BMC Genomics 14: 403. doi:10.1186/1471-2164-14-403.
76. Ruan Y-L, Xu S-M, White R, Furbank RT (2004) Genotypic and Developmental Evidence for the Role of Plasmodesmatal Regulation in Cotton Fiber Elongation Mediated by Callose Turnover. Plant Physiol 136: 4104–4113. doi:10.1104/pp.104.051540.
77. Salnikov VV, Grimson MJ, Seagull RW, Haigler CH (2003) Localization of sucrose synthase and callose in freeze-substituted secondary-wall-stage cotton fibers. Protoplasma 221: 175–184. doi:10.1007/s00709-002-0079-7.
78. Goldberg R, Morvan C, Jauneau A, Jarvis MC (1996) Methyl-esterification, de-esterification and gelation of pectins in the primary cell wall. In: J Visser and A.G.J Voragen, editor. Progress in Biotechnology. Pectins and Pectinases Proceedings of an International Symposium. Elsevier, Vol. Volume 14. 151–172.
79. Vaughn KC, Turley RB (1999) The primary walls of cotton fibers contain an ensheathing pectin layer. Protoplasma 209: 226–237. doi:10.1007/BF01453451.
80. Tokumoto H, Wakabayashi K, Kamisaka S, Hoson T (2002) Changes in the sugar composition and molecular mass distribution of matrix polysaccharides during cotton fiber development. Plant Cell Physiol 43: 411–418.
81. Haigler CH, Betancur L, Stiff MR, Tuttle JR (2012) Cotton fiber: a powerful single-cell model for cell wall and cellulose research. Front Plant Sci 3. Available: http://www.ncbi.nlm.nih.gov/pmc/articles/PMC3356883/. Accessed 24 Jun 2014.
82. Wang H, Guo Y, Lv F, Zhu H, Wu S, et al. (2010) The essential role of GhPEL gene, encoding a pectate lyase, in cell wall loosening by depolymerization of the de-esterified pectin during fiber elongation in cotton. Plant Mol Biol 72: 397–406. doi:10.1007/s11103-009-9578-7.
83. Hao Z, Mohnen D (2014) A review of xylan and lignin biosynthesis: foundation for studying Arabidopsis irregular xylem mutants with pleiotropic phenotypes. Crit Rev Biochem Mol Biol 49: 212–241. doi:10.3109/10409238.2014.889651.
84. Sadava D, Chrispeels MJ (1973) Hydroxyproline-rich cell wall protein (extensin): Role in the cessation of elongation in excised pea epicotyls. Dev Biol 30: 49–55. doi:10.1016/0012-1606(73)90047-X.
85. Huang G-Q, Gong S-Y, Xu W-L, Li W, Li P, et al. (2013) A fasciclin-like arabinogalactan protein, GhFLA1, is involved in fiber initiation and elongation of cotton. Plant Physiol 161: 1278–1290. doi:10.1104/pp.112.203760.
86. Qin L-X, Rao Y, Li L, Huang J-F, Xu W-L, et al. (2013) Cotton GalT1 Encoding a Putative Glycosyltransferase Is Involved in Regulation of Cell Wall Pectin Biosynthesis during Plant Development. PLoS ONE 8: e59115. doi:10.1371/journal.pone.0059115.
87. Bowling AJ, Vaughn KC, Turley RB (2011) Polysaccharide and glycoprotein distribution in the epidermis of cotton ovules during early fiber initiation and growth. Protoplasma 248: 579–590. doi:10.1007/s00709-010-0212-y.

11

Alternative Splicing and Highly Variable Cadherin Transcripts Associated with Field-Evolved Resistance of Pink Bollworm to Bt Cotton in India

Jeffrey A. Fabrick[1]*, Jeyakumar Ponnuraj[2], Amar Singh[3], Raj K. Tanwar[3], Gopalan C. Unnithan[4], Alex J. Yelich[4], Xianchun Li[4], Yves Carrière[4], Bruce E. Tabashnik[4]

1 U.S. Department of Agriculture, Agricultural Research Service, U.S. Arid Land Agricultural Research Center, Maricopa, Arizona, United States of America, **2** National Institute of Plant Health Management, Rajendranagar, Hyderabad, Andhra Pradesh, India, **3** National Centre for Integrated Pest Management, Indian Agricultural Research Institute, New Delhi, Delhi, India, **4** Department of Entomology, University of Arizona, Tucson, Arizona, United States of America

Abstract

Evolution of resistance by insect pests can reduce the benefits of insecticidal proteins from *Bacillus thuringiensis* (Bt) that are used extensively in sprays and transgenic crops. Despite considerable knowledge of the genes conferring insect resistance to Bt toxins in laboratory-selected strains and in field populations exposed to Bt sprays, understanding of the genetic basis of field-evolved resistance to Bt crops remains limited. In particular, previous work has not identified the genes conferring resistance in any cases where field-evolved resistance has reduced the efficacy of a Bt crop. Here we report that mutations in a gene encoding a cadherin protein that binds Bt toxin Cry1Ac are associated with field-evolved resistance of pink bollworm (*Pectinophora gossypiella*) in India to Cry1Ac produced by transgenic cotton. We conducted laboratory bioassays that confirmed previously reported resistance to Cry1Ac in pink bollworm from the state of Gujarat, where Bt cotton producing Cry1Ac has been grown extensively. Analysis of DNA from 436 pink bollworm from seven populations in India detected none of the four cadherin resistance alleles previously reported to be linked with resistance to Cry1Ac in laboratory-selected strains of pink bollworm from Arizona. However, DNA sequencing of pink bollworm derived from resistant and susceptible field populations in India revealed eight novel, severely disrupted cadherin alleles associated with resistance to Cry1Ac. For these eight alleles, analysis of complementary DNA (cDNA) revealed a total of 19 transcript isoforms, each containing a premature stop codon, a deletion of at least 99 base pairs, or both. Seven of the eight disrupted alleles each produced two or more different transcript isoforms, which implicates alternative splicing of messenger RNA (mRNA). This represents the first example of alternative splicing associated with field-evolved resistance that reduced the efficacy of a Bt crop.

Editor: Mario Soberón, Instituto de Biotecnología, Universidad Nacional Autónoma de México, Mexico

Funding: This work was funded by U.S. Department of Agriculture base funding (Project Number 5347-22620-021-00D) to J.A.F., research grants from the U.S. Department of Agriculture Agriculture and Food Research Initiative (2008-35302-0390) and Biotechnology Risk Assessment Research Grants Program (2011-33522-30729) to B.E.T., and the Department of Science and Technology (India) Better Opportunities for Young Scientists in Chosen Areas of Science and Technology (BOYSCAST) fellowship to J.P. The funders had no role in study design, data collection and analysis, decision to publish, or preparation of the manuscript.

Competing Interests: J.A.F. is coauthor of a patent "Cadherin Receptor Peptide for Potentiating Bt Biopesticides" (patent numbers: US20090175974A1, US8354371, WO2009067487A2, WO2009067487A3). B.E.T. is coauthor of a patent on modified Bt toxins, "Suppression of Resistance in Insects to Bacillus thuringiensis Cry Toxins, Using Toxins that do not Require the Cadherin Receptor" (patent numbers: CA2690188A1, CN101730712A, EP2184293A2, EP2184293A4, EP2184293B1, WO2008150150A2, WO2008150150A3). Pioneer, Dow AgroSciences, Monsanto and Bayer CropScience did not provide funding to support this work, but may be affected financially by publication of this paper and have funded other work by J.A.F., Y.C., and B.E.T. There are no further patents, products in development or marketed products to declare.

* E-mail: jeff.fabrick@ars.usda.gov

Introduction

Insecticidal crystalline proteins from the bacterium *Bacillus thuringiensis* (Bt) kill some major insect pests, but are harmless to most non-target organisms including people [1–3]. To provide a new tool for pest management, scientists genetically engineered crops to produce Bt proteins for insect control [3]. The area planted to transgenic Bt crops increased from 1 million hectares in 1996 to more than 75 million hectares worldwide in 2013 [4]. These Bt crops can decrease reliance on conventional insecticides, suppress some key pests, and increase yields and farmers' profits [5–10]. However, the evolution of resistance to Bt crops by insect pests can diminish such benefits [11–13].

Although several mechanisms of resistance to Bt toxins occur, the most common type entails mutations that reduce binding of Bt toxins to larval midgut proteins [2,14–17]. Identification of the genes conferring pest resistance to Bt toxins has been limited to laboratory-selected strains, with three notable exceptions: mutations in an ABCC2 transporter gene are linked with resistance to Cry1Ac in a field-selected strain of *Plutella xylostella* and a greenhouse-selected strain of *Trichoplusia ni* that were derived from populations exposed to sprays containing Cry1Ac [18], and in *Helicoverpa armigera*, mutations in a gene encoding a cadherin protein that binds Cry1Ac are linked with resistance to Cry1Ac in a laboratory-selected strain and in field-selected populations from northern China that were exposed intensively to Bt cotton producing Cry1Ac [19–23]. Relative to susceptible populations, the percentage of individuals resistant to Cry1Ac was significantly higher in field populations from northern China, yet it was less than 5% as of 2010 and reduced efficacy of Bt cotton producing Cry1Ac has not been reported there [22,24].

By contrast with the knowledge of genes responsible for many examples of laboratory-selected resistance and the three examples of field- and greenhouse-selected resistance described above, the genes conferring resistance to Bt toxins have not been identified for any of the first five cases in which reduced efficacy of Bt crops is associated with field-evolved resistance [13,25–29]. Here we examined the genetic basis of resistance for one of these five cases: field-evolved resistance to Bt cotton producing Cry1Ac in India by pink bollworm (*Pectinophora gossypiella*), which is a global pest of cotton [13,29–31].

In India, which grew more hectares of Bt cotton than any other country in the world in 2012 and 2013 [4,32], Bt cotton hybrids producing Cry1Ac were commercialized in 2002 [33]. However, Bt cotton was planted illegally before 2002 in the state of Gujarat, which leads India in cotton production and typically produces a third of the nation's cotton [33–35]. The estimated mean percentage of all cotton hectares planted with Bt cotton from 2003 to 2007 was 75% (range = 54 to 90%) in Gujarat, compared with 30% (range = 2 to 73%) in Maharashtra, India's second leading cotton-producing state [33–34].

Pink bollworm resistance to Cry1Ac was documented with diet bioassays showing that mean survival at a diagnostic toxin concentration was 72% for a population sampled in 2008 from the district of Amreli in Gujarat, compared with 0 to 4% for populations from four sites outside of Gujarat including Akola in Maharashtra [29]. Monsanto (2010) also reported "unusual survival of pink bollworm" on Bt cotton producing Cry1Ac during 2009 and "confirmed" pink bollworm resistance to Cry1Ac in four districts of Gujarat: Amreli, Bhavnagar, Junagarh and Rajkot [30]. Farmers in India have switched to cotton hybrids producing two Bt toxins (Cry1Ac and Cry2Ab), which are effective against pink bollworm larvae resistant to Cry1Ac [29–30,36–37]. These two-toxin hybrids were planted on 10.4 million hectares in 2013, representing 94% of India's cotton [4].

We hypothesized that field-evolved resistance to Cry1Ac of pink bollworm in India is associated with mutations in a cadherin gene called *PgCad1*, because resistance to Cry1Ac is linked with mutations in this gene for five laboratory-selected strains of pink bollworm from Arizona in the southwestern United States [38–42]. Unlike the situation in India, pink bollworm field populations in Arizona have remained susceptible to Cry1Ac despite more than 16 years of extensive exposure to Bt cotton producing this toxin [9,43–44]. From 1996–2005, the main factors that delayed pink bollworm resistance in Arizona appear to be abundant refuges of non-Bt cotton, recessive inheritance of resistance, fitness costs associated with resistance and incomplete resistance [43,44]. Since 2006, an eradication program using mass releases of sterile pink bollworm moths and other tactics in combination with up to 98% adoption of Bt cotton statewide has dramatically suppressed this pest in Arizona [9,44]. In contrast, lack of compliance with the refuge strategy apparently promoted rapid evolution of pink bollworm resistance to Cry1Ac in India [44–46]. Despite the absence of field-evolved resistance of pink bollworm to Bt cotton in the United States, our previous work identified four recessive cadherin alleles (*r1*, *r2*, *r3*, and *r4*) of *PgCad1* linked with resistance to Cry1Ac in laboratory-selected strains from Arizona [38–42].

In this study of pink bollworm from India, we detected none of the four cadherin resistance alleles from Arizona, but we discovered eight novel, severely disrupted cadherin alleles associated with resistance to Cry1Ac. Analysis of messenger RNA (mRNA) from these eight alleles revealed 19 transcript isoforms. Each of these 19 transcript isoforms has a premature stop codon, a deletion of at least 99 base pairs (bp), or both. For seven of the eight disrupted cadherin alleles, we detected two or more mRNA transcripts produced by a single allele, which indicates alternative splicing of pre-cursor mRNA (pre-mRNA) [47–48].

Results

2.1 Larval Survival in Diet Bioassays with Cry1Ac

We used diet incorporation bioassays with a diagnostic concentration (10 micrograms Cry1Ac per ml diet) [49] to evaluate resistance to Cry1Ac of the first-generation progeny of pink bollworm collected from the field during the 2010–2011 growing season from Anand in Gujarat (AGJ) and from Akola in Maharashtra (AMH) (Fig. 1). We obtained F1 larvae from AGJ parents collected from Bt Cry1Ac cotton whereas the AMH parents were from non-Bt cotton. Larval survival adjusted for control mortality was 65% for AGJ (n = 17 treated and 10 control larvae) and 0% for AMH (n = 43 treated and 60 control larvae) (Fisher's exact test, $P<0.0001$). These results indicate that a substantial proportion of the AGJ population was resistant to Cry1Ac, whereas the AMH population was predominantly susceptible.

2.2 DNA Screening of Populations from India for Cadherin Resistance Alleles from Arizona

We used established PCR methods to screen the genomic DNA (gDNA) of pink bollworm from India for three cadherin alleles that are linked with laboratory-selected resistance to Cry1Ac in pink

Figure 1. Sampling locations for pink bollworm field populations in India. We screened DNA of 425 pink bollworm collected from all seven sites for cadherin resistance alleles *r1*, *r2*, and *r3* (triangles). We sequenced cadherin cDNA and gDNA of 11 larvae from three sites: Akola (AMH), Anand (AGJ), and Khandwa (KMP) (circles) and conducted bioassays with 130 larvae from two sites: AMH and AGJ (squares). Based on cadherin DNA sequences (circles) and bioassay data (squares) from this study, red indicates evidence of resistance for AGJ and KMP; blue indicates evidence of susceptibility for AMH. Resistance was reported previously from four districts of Gujarat including Rajkot [29–30].

bollworm from Arizona (*r1*, *r2* and *r3*) [38,50–51]. We found none of these three cadherin alleles in 425 pink bollworm collected during 2010 and 2011 from seven sites in India (Fig. 1, Table S1). The sample from India screened for *r1*, *r2* and *r3* included 46 individuals from two resistant populations in Gujarat: 19 from AGJ, where resistance was detected in our bioassay (described above); and 27 from Rajkot, where resistance was reported previously [30]. In addition, the screened samples included 38 individuals from Khandwa in the state of Madhya Pradesh (KMP) that were collected as fourth instars on Bt cotton and were expected to be predominantly resistant. These results indicate that cadherin resistance alleles *r1*, *r2* and *r3* from Arizona were not common in India, even in samples expected to have a high proportion of individuals resistant to Cry1Ac.

2.3 Cadherin DNA and Transcripts from Resistant and Susceptible Larvae

To determine if resistance to Cry1Ac in pink bollworm from India was associated with cadherin mutations different from those identified in Arizona, we sequenced cadherin gDNA and cDNA of larvae preserved in RNAlater from three sources: AMH, AGJ, and KMP. Based on 0% survival of AMH larvae at a diagnostic concentration of Cry1Ac, we inferred that the AMH larvae were susceptible (as described above). We analyzed DNA from three AGJ larvae that we identified as resistant because they became fourth instars while feeding on diet containing a diagnostic concentration of Cry1Ac. We also analyzed DNA from five individuals from KMP that we expected to be predominantly resistant because they were collected as second and third instars from bolls in Bt cotton fields.

Sequencing revealed no severe disruptions in the cDNA of cadherin from the three susceptible larvae from AMH (Fig. 2, Fig. S1), whereas severe disruptions occur in all three of the cadherin alleles from the resistant AGJ larvae, and in 5 of the 6 alleles from the KMP larvae that were collected from Bt cotton (Table 1, Fig. 2). In the eight larvae analyzed from AGJ and KMP, we found eight novel, severely disrupted cadherin alleles (*r5–r12*) with a total of 19 different cDNA sequences (Table 1, Fig. 2). Seven of these eight alleles have at least two transcript isoforms, which implicates alternative splicing of these alleles (Table 1, Fig. 2).

As expected for susceptible pink bollworm [38], cadherin cDNA isolated from three susceptible AMH larvae had 5,208 bp encoding a predicted protein of 1,735 amino acids (Fig. S1, Fig. 2). The predicted open reading frame (ORF) for the consensus AMH cDNA has 99% homology with the translated sequence from the *PgCad1 s* allele (AY198374.1) from the susceptible APHIS-S strain of pink bollworm from Arizona [38]. As with the *s* allele from Arizona, the translated protein encoded by cDNA from AMH includes a putative membrane signal sequence, 11 extracellular cadherin repeats (CR1-CR11), a membrane-proximal region, a transmembrane domain, and a cytoplasmic domain (Fig. 2).

Eight of the nine complete cDNA sequences we obtained from three susceptible AMH larvae have no insertions or deletions (indels) (Fig. S1). In the exceptional sequence from one AMH individual, we found a single deletion of 3 bp corresponding to nucleotides 72–74 of the *s* allele from Arizona encoding alanine in the membrane signal sequence (sequence AMH-3_16, Figs. S1 and S2). This deletion was also detected in one larva from AGJ (AGJ-1, Table 1) and two larvae from KMP (KMP-7 and KMP-8, see details below). We also identified 195 putative single nucleotide polymorphisms (SNPs) in the full-length cDNA sequences from AMH (Fig. S1). Of the 96 putative SNPs encoding amino acid changes, 52 are conservative substitutions (Fig. S2). Several missense mutations (e.g., Leu/His1274, Asp/Gly1371, Glu/Gly1381and Arg/Gly1469) occur in CR10-CR11, the region involved in binding Cry1Ac in pink bollworm [52]. However, we found no insertions, deletions, or missense mutations in the specific portions of these domains that bind Cry1Ac in pink bollworm [52].

In contrast with the conserved cadherin cDNA sequences from susceptible AMH larvae, the cadherin cDNA sequences from three resistant AGJ larvae are highly variable and severely disrupted (Table 1, Figs. 2, 3, and S3). In three AGJ larvae, we found three novel cadherin alleles (*r5*, *r6*, and *r7*; Table 1 and Fig. 2). Two of these three alleles have multiple isoforms (e.g., *r5A*, *r5B*, and *r5C* of allele *r5*) yielding a total of six isoforms (Table 1, Figs. 2 and 3). Five of these six isoforms have premature stop codons; the sixth isoform (*r7B*) has a 99-bp deletion encoding a cadherin protein that lacks the entire CR10 (Table 1, Fig. 2).

The cadherin cDNA sequences are also highly variable and severely disrupted in four of the five larvae from KMP (Table 1, Figs. 2, 4 and S5), which were collected from Bt cotton and expected to be resistant (the fifth larva is described below). These four KMP larvae carried a total of five different disrupted cadherin alleles (*r8–r12*). Two of these four larvae each had two different disrupted alleles (alleles *r8* and *r9* in individual KMP-4 and alleles *r11* and *r12* in individual KMP-6, Table 1). Each of the five mutant cadherin alleles in KMP has two to four isoforms, yielding 13 isoforms in four larvae (Table 1 and Figs. 2 and 4). In each of the five mutant KMP alleles, we identified one or more indels of 1 to 1,157 bp, with 10 of the 13 isoforms bearing indels that introduce premature stop codons (Table 1, Fig. 2). In addition, cDNA from isoform *r9A* has a single base substitution (guanine 2,289 to adenine) that introduces a premature stop codon. Of the three disrupted KMP isoforms lacking a premature stop codon (*r9B*, *r10A*, and *r10B*), *r10A* and *r10B* shared deletions of 126 and 105 bp; *r10B* also had a third deletion of 303 bp (Table 1). The *r9B* isoform has the largest deletion identified: 1,157 bp corresponding to the portion of the cadherin protein from CR9 to the membrane-proximal region (Table 1, Fig. 2).

Unlike the cDNA sequences from the four KMP larvae described above, none of the five cDNA sequences obtained from five different clones isolated from one KMP larva (KMP-8) are severely disrupted by indels or substitutions (Figs. 2, S5, and S6). Of the two deletions in KMP-8 (Fig. S5), one is the same 3-bp deletion found in one sequence from the susceptible larva AMH-3 (Fig. S1, Fig. S2). The second is the 3-bp deletion corresponding to bases 1,008–1,010 in the *s* allele from Arizona encoding glutamate in CR2 (sequence KMP-8_35; Figs. S5 and S6). Both of these deletions result from alternative mRNA splicing, as they both occur at exon-intron splice junctions and are not present in gDNA. The consensus ORF from KMP-8 has 5,205 bp encoding 1,734 amino acids and shares 99% identity with the *PgCad1 s* allele (AY198374.1) (Fig. S5, Fig. S6). Although the cDNAs from AMH-1, AMH-2, AMH-3, and KMP-8 are not severely disrupted, the 14 cDNA sequences from these four individuals have 27 informative SNPs corresponding to seven unique *s* alleles (Fig. S7). Even with these 27 informative SNPs, no more than two alleles are evident from each single diploid individual. Each of these seven *s* alleles from India shares >99% identity with the PgCad1 *s* allele from Arizona (AY198374.1).

In total, fifteen of the 19 transcript isoforms of the eight severely disrupted alleles have deletions corresponding to the complete loss of one or more exons (Table 1). This includes *r6A*, the only transcript we detected for allele *r6*, which contains a premature stop codon in its cDNA and lacks exons 8–13 (Table 1, Fig. 3, Fig. S3). Although we were not able to obtain gDNA for allele *r6*, the

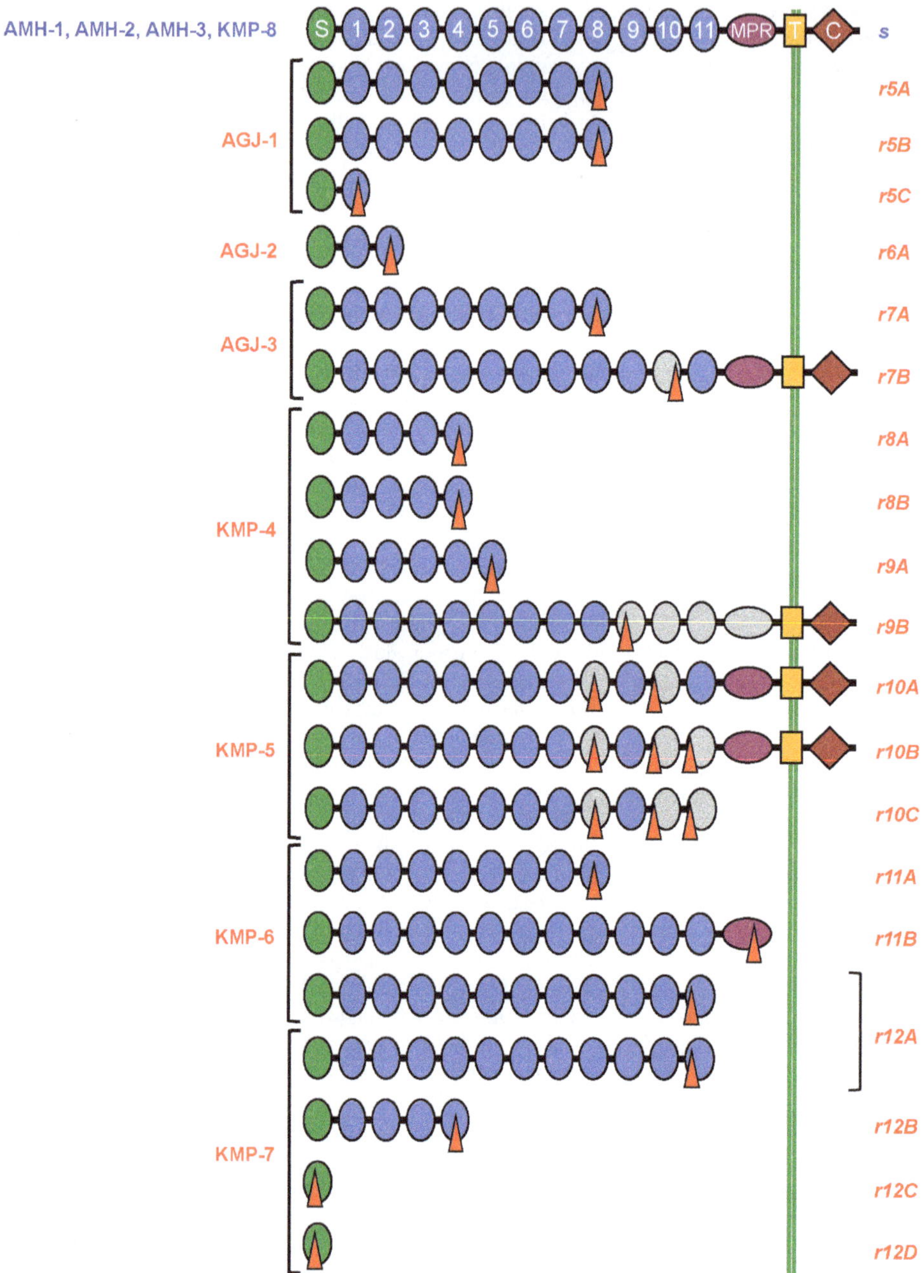

Figure 2. Predicted cadherin proteins in pink bollworm from three populations in India. We isolated and sequenced full-length *PgCad1* cDNA clones from 11 individuals: three from Akola, Maharashtra (AMH-1 to AMH-3), three from Anand, Gujarat (AGJ-1 to AGJ-3), and five from Khandwa, Madhya Pradesh (KMP-4 to KMP-8). Predicted proteins are shown for cDNA of the *PgCad1* susceptible (*s*) allele and 19 isoforms (*r5A, r5B,* etc.) of mutant alleles *r5–r12*. The amino-terminal membrane signal sequence (S), cadherin repeats (1–11), membrane-proximal region (MPR), transmembrane region (T), and cytoplasmic domain (C) are shown for the *s* allele. Red triangles indicate mutations predicted to cause loss of at least 33 amino acids (see Table 1). Truncated structures indicate proteins predicted from cDNA with premature stop codons. Gray indicates missing regions of proteins caused by deletions. The 3-bp deletion (corresponding to bp 72–74 in the *s* allele) that occurred in one sequence from AMH-3 and four sequences from KMP-8 as well as in two sequences from AGJ-1 and one sequence from KMP-7 is not shown.

deletion of exons 8–13 in the *r6A* transcript occurs exactly at the exon-intron junctions (Fig. S3 and Fig. S4). Thus, we suspect that mis-splicing, which entails a mistake in splicing [47], causes the disruption in transcript *r6A* in this allele. Mis-splicing is also implicated in the 3-bp deletion found in cDNA but not gDNA from larvae in each of the three populations studied (AMH-3, AGJ-1, KMP-7, and KMP-8) (Table 1 and Figs. S1, S3, and S5).

In addition to a 20-bp insertion that occurs only in the *r5C* isoform and reflects alternative splicing, the gDNA of *r5* and all three isoforms of *r5* have an insertion of 3,120 bp that causes the loss of exons 21–24 (Table 1, Fig. 3, Fig. S8). Thus, this 3,120-bp insertion reflects mis-splicing rather than alternative splicing. A CENSOR search of Repbase [53] reveals that this insert is similar to several transposable elements (Table 2). Several smaller insertions that introduce premature stop codons also occur in

Table 1. Nineteen transcript isoforms of eight disrupted cadherin alleles in seven pink bollworm larvae from two populations in India: Anand, Gujarat (AGJ) and Khandwa, Madhya Pradesh (KMP).

Indivi-dual(s)	Allele	Iso-form	Deletion size(s) (bp)[a]	Inser-tion size (bp)[a]	Cadherin region[b]	Pre-mature stop codon(s)	Complete exon(s) missing
AGJ-1	*r5*	*r5A*	**478**[c]	-	CR8-9	Yes	21–24
AGJ-1	*r5*	*r5B*	3[d], **478**[c]	-	CR8-9	Yes	21–24
AGJ-1	*r5*	*r5C*	478[c]	**20**	Signal-CR1	Yes	21–24
AGJ-2	*r6*	*r6A*	**1051**[e]	-	CR2-5	Yes	8–13
AGJ-3	*r7*	*r7A*	**247**	-	CR8-9	Yes	21–22
AGJ-3	*r7*	*r7B*	99	-	CR10	No	27
KMP-4	*r8*	*r8A*	**170**	4	CR4-5	Yes	13
KMP-4	*r8*	*r8B*	-	**4**	CR4	Yes	No
KMP-4	*r9*	*r9A*	165[f]	-	CR5, CD	Yes	32
KMP-4	*r9*	*r9B*	1157	-	CR9-MPR	No	23–31
KMP-5	*r10*	*r10A*	126, 105	-	CR8, 10	No	21, 25
KMP-5	*r10*	*r10B*	126, 105, 303	-	CR8, 10, 11	No	21, 25, 28–29
KMP-5	*r10*	*r10C*	126, 105, **193**	-	CR8, 10, 11	Yes	21, 25, 28
KMP-6	*r11*	*r11A*	**23**[g]	127	CR8, MPR	Yes	No
KMP-6	*r11*	*r11B*	-	**125**	MPR	Yes	No
KMP-6 KMP-7	*r12*	*r12A*	-	**1**[h]	CR11	Yes	No
KMP-7	*r12*	*r12B*	3[d], **118**[i]	1[h]	CR4, 11	Yes	11
KMP-7	*r12*	*r12C*	**11**[j], 148[k]	1[h]	Signal, CR11	Yes	5
KMP-7	*r12*	*r12D*	**11**[j], 230[l]	1[h]	Signal, CR4, 11	Yes	11–12

[a]Mutations shown in bold cause premature stop codons.
[b]Region of cadherin protein where major mutations occur (see Figure 2).
[c]The 478-bp deletion found in *r5A*, *r5B* and *r5C* is caused by insertion of 3,120 bp similar to transposons (Table 2), causing the loss of exons 21–24 from gDNA and cDNA.
[d]The 3-bp deletion in *r5B* and *r12B* is caused by mis-splicing, occurs at exon-intron splice junction 1, and is found in both *r* and *s PgCad1* alleles.
[e]gDNA from AGJ-2 was not available to compare with cDNA, but the absence of exons 8–13 occurs exactly at the exon-intron junctions, suggesting that mis-splicing occurred.
[f]*r9A* includes A-to-G (I) RNA editing at base position 2,289 and results in the introduction of a premature stop codon (see Fig. 4). The 165-bp deletion causes the loss of exon 32.
[g]The 23-bp deletion corresponds to the final 23 nucleotides of exon 20 in cDNA clone KMP-6_3.
[h]The single base insertion introduces a premature stop codon and truncates the mRNA transcript in CR11.
[i]The 118-bp deletion causes the loss of exon 11 resulting in the introduction of a premature stop codon and truncates the mRNA transcript in CR4.
[j]The 11-bp deletion occurs in the membrane signal sequence of *r12C* and *r12D* transcripts resulting in the introduction of a premature stop codon.
[k]The 148-bp deletion causes the loss of exon 5 in mRNA transcript between the membrane signal sequence and CR1.
[l]The 230-bp deletion causes the loss of exons 11–12 in mRNA transcript found in CR4.

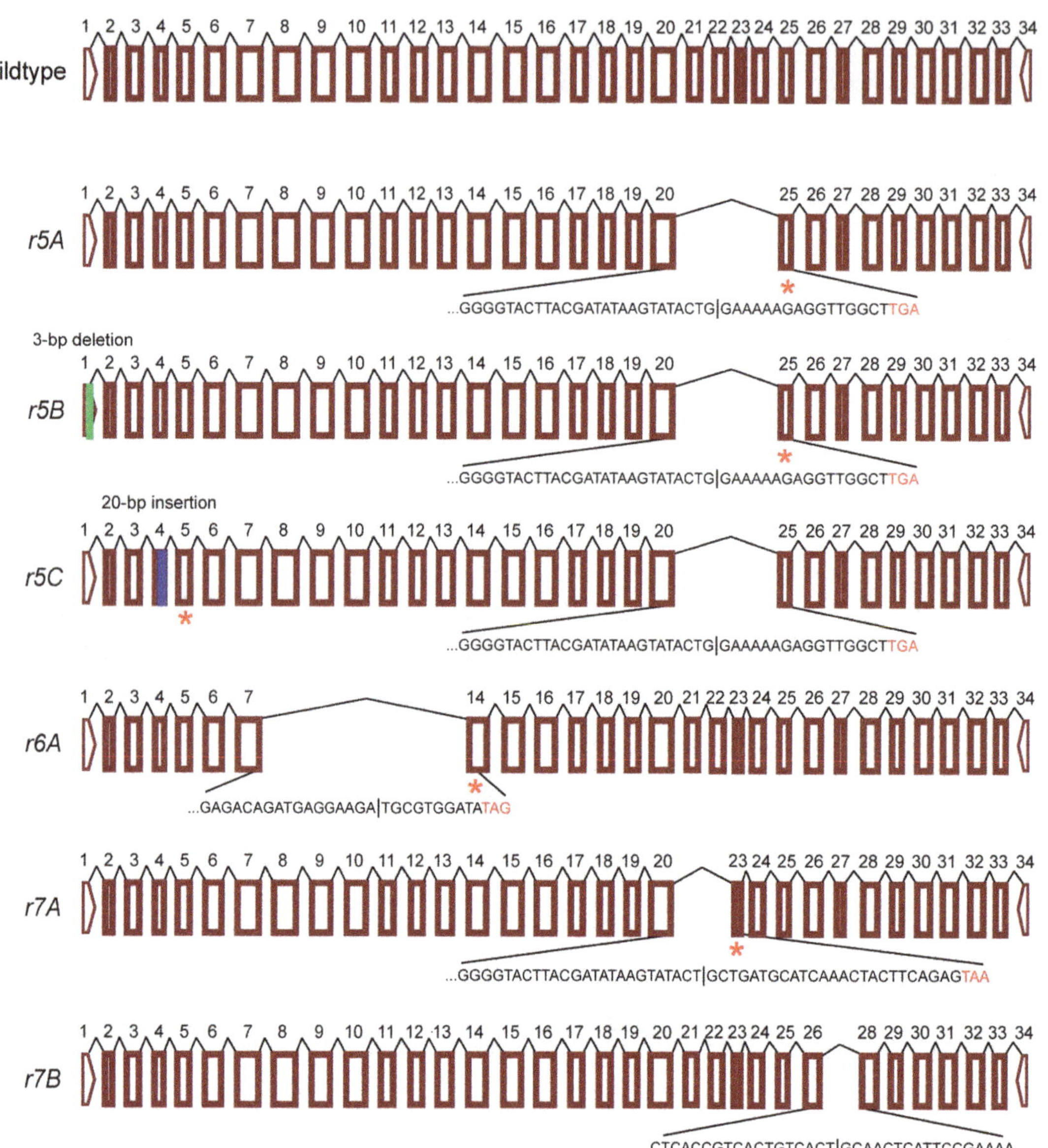

Figure 3. Cadherin mRNA transcripts of a susceptible allele and three severely disrupted alleles found in three resistant pink bollworm larvae from Anand, Gujarat (AGJ). Exons are numbered (1–34). Sequences are shown for exons missing from transcripts. Blue boxes show insertions, green boxes show deletions, and stars show premature stop codons. The six transcript isoforms shown are *r5A-r7B* (GenBank accession KJ480757-KJ480762).

both cDNA and gDNA and do not reflect altered splicing (four bp in *r8*, 125 to 127 bp in *r11*, and one bp in *r12*; Table 1, Table S2, Fig. 4, Fig. S8).

Discussion

The bioassay results here with pink bollworm derived from the field in India during 2010 and 2011 show 65% of individuals resistant to Cry1Ac in the Anand population from Gujarat (AGJ) compared with 0% in the Akola population from Maharashtra (AMH). These results confirm previous reports from 2008 and 2009 indicating pink bollworm resistance to Cry1Ac in Gujarat, where Bt cotton was adopted rapidly, but not in Akola, where adoption was much slower [29–30,33].

Whereas previous results show that resistance to Cry1Ac in laboratory-selected strains of pink bollworm from Arizona is linked with mutations in a gene encoding a cadherin protein that binds Cry1Ac in the larval midgut [38–42], the data here show an association between field-evolved resistance to Cry1Ac in India and different mutations in the same gene. In the susceptible AMH population, none of the cadherin DNA sequences from three larvae were severely disrupted. By contrast, all of the cadherin DNA sequences were severely disrupted in the three resistant larvae from AGJ that survived exposure to a diagnostic concentration of Cry1Ac.

Among five individuals from Khandwa in Madhya Pradesh (KMP) collected as second or third instars from Bt cotton and expected to be predominantly resistant, four had only severely disrupted cadherin alleles and the fifth had no disrupted cadherin alleles. We cannot exclude the hypothesis that the fifth larva from KMP was susceptible, because we did not determine the

4-bp insertion
1–12, 14–34
r8A
...ACCTCCTACTCTGAGGCT|GATGCGTGGATATAG...
4-bp insertion
1–34
r8B
1-bp substitution
1–31, 33 34
r9A
...ACCCGAGCT|GACACATCTG...
1–22, 32 33 34
r9B
...CTGACACAGGCGCTTCCTG|TGCCTTATTTTGCTCA...
1–20, 22 23 24, 26–34
r10A
...AGTATACTG|GAACGCGCA...
...TAGCATTCT|TTGGTGGTG...
1–20, 22 23 24, 26 27, 30–34
r10B
...AGTATACTG|GAACGCGCA...
...TAGCATTCT|TTGGTGGTG...
...ACCGTTCGT|ATCGCAGAA...
1–20, 22 23 24, 26 27, 29–34
r10C
...AGTATACTG|GAACGCGCA...
...TAGCATTCT|TTGGTGGTG...
...ACCGTTCGT|ATGAGAAGA...TGA
127-bp insertion
1–34
r11A
...GGATATTGGG|GCGTTCGACCACGGTATTCCTCAGCAGATATCTCATGA
23-bp deletion
1–34
r11B
1–34
r12A
3-bp deletion
1-bp insertion
1–10, 12–34
r12B
...CAGCGTTGT|GTTCGCGCT...TAG
11-bp deletion
1-bp insertion
1–4, 6–34
r12C
...AACCAATAG|GAATTGGGG...
11-bp deletion
1-bp insertion
1–10, 13–34
r12D
...CAGCGTTGT|CCGCGCGTA...

Figure 4. Cadherin mRNA transcripts from five severely disrupted alleles found in four pink bollworm larvae collected on Bt cotton in Khandwa, Madhya Pradesh (KMP). Transcript isoforms of alleles *r8–r12* from individuals KMP-4, KMP-5, KMP-6, and KMP-7. Exons are numbered. Sequences are shown for exons missing from transcripts. Blue boxes show insertions, green boxes show deletions, black boxes show substitutions, and stars show premature stop codons. The 13 transcript isoforms shown are *r8A-r12D* (GenBank accession KJ480763-KJ480775).

concentration of Cry1Ac in the bolls on which the field-collected larvae fed and cannot rule out the possibility that the fifth larva fed on plant tissues with a reduced concentration of Cry1Ac. We also cannot exclude an alternative hypothesis that the fifth KMP larva was resistant, with the resistance conferred by a gene other than cadherin. Although cadherin mutations are sufficient to cause resistance to Cry1Ac in pink bollworm, mutations at other loci also can confer resistance to this toxin in pink bollworm and other Lepidoptera [18,54–60].

In eight larvae from the field-selected populations AGJ and KMP, we discovered eight novel, severely disrupted cadherin alleles (*r5–r12*) with a total of 19 novel cDNA isoforms (Table 1 and Fig. 2). Among the 19 isoforms, 15 have premature stop codons and the other four have one or more deletions of at least 99 bp in the sequence encoding the Cry1Ac-binding region (Table 1, Fig. 2). The premature stop codons are expected to yield truncated cadherin proteins that are not anchored in the midgut membrane and cannot mediate toxicity of Cry1Ac. The predicted omission of at least 33 amino acids from the Cry1Ac-binding region of cadherin protein could also reduce binding of Cry1Ac and thus confer resistance to this toxin. In contrast with these severely disrupted alleles from India, among the four pink bollworm cadherin resistance alleles from Arizona, only *r2* has a deletion (202 bp) that introduces a premature stop codon [38,42] and each of the other three (*r1, r3,* and *r4*) has only a single deletion (24, 126 and 15 bp, respectively) that does not occur in the sequence encoding the Cry1Ac-binding region [38,42]. Given that the relatively minor disruptions in three of four cadherin alleles of pink bollworm from Arizona are genetically linked with resistance to Cry1Ac, we conclude that the severe disruptions in the eight cadherin alleles in pink bollworm from India probably confer resistance to Cry1Ac.

Although mutations in the same cadherin gene are associated with pink bollworm resistance to Cry1Ac in laboratory-selected strains from Arizona and field-selected populations from India, we did not find any of the four cadherin resistances alleles from Arizona in the 436 pink bollworm from India that we analyzed. These include 425 individuals from seven populations screened for alleles *r1, r2* and *r3* and 11 individuals from AMH, AGJ and KMP from which we sequenced cadherin cDNA. The difference in cadherin resistance alleles between Arizona and India could reflect the respective geographic origins from which the pink bollworm were derived, as well as laboratory versus field selection. With highly variable cadherin in the AGJ and KMP populations from India, we also found no resistance alleles in common between these two field-selected populations separated by ca. 400 km, and only one resistance allele that occurred in two individuals within a population from India (*r12* in KMP-6 and KMP-7, Table 1). Given the high diversity of cadherin resistance alleles within each population, it is surprising that all three AGJ individuals and three of the five KMP individuals were homozygous for disrupted alleles at the cadherin locus (Table 1). This pattern may reflect assortative mating, because random mating would generate a higher frequency of individuals carrying two different resistance alleles.

To our knowledge, the two or more transcript isoforms associated with seven of the eight severely disrupted cadherin alleles from India (Table 1) represent the first examples of alternative splicing associated with resistance to a Bt toxin. Although alternative splicing generated five cadherin isoforms in a Cry1Ac-resistant strain of *T. ni* [61], resistance in this strain is genetically linked with the ABCC2 gene, and is not associated with variation in either the transcripts or gDNA for cadherin [18,55]. However, mutations in cadherin gDNA of pink bollworm and *H. armigera* that cause mis-splicing and produce a single altered transcript linked with resistance to Cry1Ac have been reported

Table 2. Similarity between transposons and the insertion in intron 20 of the *r5 PgCad1* allele.

Position in insertion (bp)[a]	Repbase transposon name[b]	Position in transposon (bp)[c]	Transposon class	Orientation[d]	Sim[e]	BLAST score[f]
524–619	LYDIA_LTR	205–300	LTR/Gypsy	comp.	0.71	229
1,580–1,737	TED	1–162	LTR/Gypsy	comp.	0.75	609
2,332–2,432	CoeSINE4	81–178	NonLTR/SINE/SINE2	comp.	0.78	306
2,449–2,489	HaSE3	112–152	NonLTR/SINE/SINE3	comp.	0.83	237
2,587–2,651	HATN3_DR	274–338	DNA/hAT	comp.	0.73	280
3,146–3,197	Transib-4_DBp	2,848–2,899	DNA/Transib	direct	0.83	213
3,568–3,660	ISL2EU-3_HM	1,655–1,746	DNA/ISL2EU	direct	0.74	207

[a]Nucleotide position in the 3,827-bp fragment from pink bollworm cadherin (which includes the 3,120-bp insertion in the *r5* allele) cloned from AGJ-1 gDNA using primers 20PgCad5 + 81PgCad3 (See Figure S8).
[b]LYDIA_LTR, long terminal repeat retrotransposon from LYDIA, a gypsy-like endogenous retrovirus from *Lymantria dispar*; TED, internal part of retrotransposon TED inserted in *Autographa californica* nuclear polyhedrosis virus; CoeSINE4, coelacanth SINE non-long terminal repeat retrotransposon from *Latimeria chalumnae*; HaSE3, SINE non-long terminal repeat retrotransposon from *Helicoverpa armigera*; HATN3_DR, nonautonomous DNA transposon from *Danio rerio*; Transib-4_DBp, Transib-type DNA transposon from the *Drosophila bipectinata* genome; ISL2EU-3_HM, autonomous ISL2EU DNA transposon from *Hydra magnipapillata*.
[c]Nucleotide position in the transposon sequence.
[d]Orientation of the insertion sequence relative to the corresponding sequence in the transposon; comp. indicates complementary.
[e]Similarity between the fragment sequence and the corresponding sequence in the transposon; calculated as the number of exact matches/(alignment length - total length gaps in the fragment sequence - total length of gaps in the transposon sequence + total number of gaps).
[f]Alignment score from BLAST.

[62–63]. In *H. armigera*, four different indels in gDNA yield the same altered cDNA transcript that lacks exon 32 [63]. For the previously characterized pink bollworm cadherin resistance allele *r3*, insertion of a non-LTR chicken-repeat retrotransposon (*CR1-1_Pg*) causes splicing out of exon 21 from mRNA [62]. Here we found that loss of exons 21-24 in all three isoforms of the pink bollworm *r5* allele is caused by a 3,120-bp insertion that has sequences similar to several transposons (Table 2).

Because we found eight different cadherin resistance alleles and 19 variant isoforms in only eight pink bollworm larvae from two field-selected populations in India, we expect that larger sample sizes from these and other field-selected populations in India would reveal even more genetic variation at the pink bollworm cadherin locus. To put the diversity of pink bollworm cadherin from India in perspective, we note that only 22 cadherin resistance alleles have been reported previously based on more than a decade of work by several research teams analyzing thousands of individuals representing three major cotton pests. These previously reported cadherin resistance alleles consist of the four in pink bollworm from Arizona [38,42], one in *H. virescens* from the southeastern United States [64], and 17 in *H. armigera* from northern China and western India [19–23,63,65–66]. Mis-splicing was reported for one cadherin resistance allele from pink bollworm [62] and another from *H. armigera* [63], as noted above, but not for the other previously reported cadherin alleles. Genetic variation in cadherin that is not associated with resistance has also been reported in other pests [59,67–69].

Whereas severe disruptions occurred in all three of the cadherin alleles from the resistant AGJ larvae and in 5 of the 6 alleles from the KMP larvae collected from Bt cotton, we found no severe disruptions in the cDNA of cadherin from the three susceptible larvae from AMH. Likewise, our previous work with pink bollworm from Arizona revealed four disrupted cadherin alleles linked with resistance to Cry1Ac in laboratory-selected strains and no such disruptions in susceptible insects [38–42]. These results suggest that in the AGJ and KMP populations, the high genetic variation in cadherin and the high frequency of disrupted cadherin alleles reflect selection of these populations in the field for resistance to Bt cotton producing Cry1Ac. We hypothesize that fitness costs, which have been identified for cadherin resistance alleles of pink bollworm from Arizona [40–41,70–74], keep the frequency of such alleles low in the absence of selection for resistance.

Similar to the results with pink bollworm, the only other comparison reported between the molecular genetic basis of laboratory- and field-selected resistance to a Bt toxin in a transgenic crop shows cadherin resistance alleles linked with resistance to Cry1Ac selected in both environments for *H. armigera* from northern China [22]. In northern China, the *r1* cadherin resistance allele of *H. armigera*, which includes a premature stop codon and was first detected in a laboratory-selected strain derived in 2001 [19], was also found in three independently isolated resistant strains initiated in 2009 from the field-selected Anyang population in Henan province [22]. In that case, the collection sites for the laboratory- and field-selected populations are separated by only 300 km.

Also similar to the results with pink bollworm in India, previous studies identified 15 cadherin resistance alleles from four populations of *H. armigera* in China [20–23]. Only two of these 15 alleles were found in more than one individual within a population (*r1* from Anyang and *r8* from Jiangpu) [20] and only one allele (*r15*) was detected in more than one population [23,63].

The diversity of cadherin mutations associated with resistance to Cry1Ac in field-selected populations of pink bollworm in India and *H. armigera* in China implies that it would not be efficient to monitor resistance in these populations by screening cadherin DNA for specific resistance alleles, as was done previously in the United States for pink bollworm and *H. virescens* [9,51,75]. An alternative approach that would detect any resistance alleles at the cadherin locus, as well as non-recessive resistance alleles at any locus, is the F1 screen in which field-collected adults are allowed to mate in single pairs with adults from a strain that is homozygous for a recessive cadherin mutation [20–22,76–77]. In general, laboratory-selected strains that are homozygous for recessive resistance alleles at any locus can be used in this way to screen field populations for recessive resistance alleles at the same locus, even if the gene is not identified and the alleles differ between the lab- and field-selected populations [76–77].

Generation of resistance alleles by alternative splicing, as seen in seven of eight cadherin resistance alleles from India (Table 1), can reduce the feasibility of resistance monitoring with DNA screening not only by increasing the diversity of transcripts, but also by making it necessary to analyze mRNA, which requires better sample preservation and more steps than screening gDNA. Alternative splicing may also accelerate evolution of resistance by generating a greater diversity of mutations that include altered proteins conferring higher levels of resistance, lower fitness costs associated with resistance, or both.

Mutations affecting splicing of mRNA are pervasive in eukaryotes [48] and are associated with some cases of resistance to neurotoxic insecticides [78–82]. Whereas previous work identified mis-splicing of cadherin mRNA linked with resistance to Cry1Ac in pink bollworm [38,62] and *H. armigera* [63], our results suggest that alternative splicing at this genetic locus is important in field-evolved resistance of pink bollworm to Cry1Ac produced by Bt cotton in India. The general significance of this genetic mechanism in pest resistance to Bt crops remains to be determined.

Materials and Methods

4.1 Pink Bollworm Field Collections

We studied pink bollworm collected at seven sites from five states in India (Fig. 1 and Table S1). No permission or permit was required for these collections. Pink bollworm is a crop pest that is not an endangered or protected species.

4.2 Diet Bioassays

We conducted diet bioassays at the National Centre for Integrated Pest Management laboratory in New Delhi to determine susceptibility to Cry1Ac of first-generation (F_1) progeny of field-collected pink bollworm from Anand in Gujarat (AGJ) and Akola in Maharashtra (AMH). We obtained 37 live AGJ larvae from 650 bolls of Bt cotton that produces Cry1Ac (Bollgard) collected on 17 January 2011. We obtained ca. 100 live AMH larvae from ca. 1000 non-Bt cotton bolls collected at the Panjabrao Deshmukh Agricultural University Cotton Research Station in Akola on 30 November 2010. Field-collected larvae from each site were reared to pupation on untreated wheat germ diet [49] and allowed to emerge as adults and mate. We obtained eggs and tested the resulting F_1 neonates individually in 30-mL plastic cups with ca. 5 g diet containing either 0 (control) or 10 micrograms Cry1Ac per mL diet [49,83]. The source of Cry1Ac was MVPII (Dow Agrosciences, San Diego, CA); a liquid formulation containing protoxin encapsulated in *Pseudomonas fluorescens* [84]. After 21 d at 25°C and a photoperiod of 12 light:12 dark, we scored live third instars, fourth instars, pupae, and adults as survivors. We used Fisher's exact test (http://

graphpad.com/quickcalcs/contingency1/) to determine if survival differed significantly between the AGJ and AMH.

4.3 DNA Screening of Populations from India for Cadherin Resistance Alleles from Arizona

For DNA-based detection of three cadherin resistance alleles (*r1*, *r2*, and *r3*) previously identified from laboratory-selected strains of pink bollworm from Arizona [38,40,50–51], we collected larvae from cotton bolls and adults from pheromone traps at seven sites in five states of India (Table S1 and Fig. 1). As detailed in Table S1, some of the field-collected larvae were reared to the pupal or adult stage on diet in the laboratory. Larvae, pupae, and adults were frozen in ethanol (>95%) for subsequent analysis.

We extracted gDNA from each individual using the PUREGENE DNA Isolation Kit (Qiagen, Valencia, CA). We screened the gDNA of 425 field-collected insects from India for *r1*, *r2*, and *r3* using the protocol and PCR primers described by Morin et al. (2003, 2004) [38,50] and Tabashnik et al. (2005) [40]. PCR products were separated on 1% agarose gels and visually inspected for the presence of DNA bands of appropriate size. Individuals were counted as screened only if their cadherin gDNA was of good quality, as indicated by successful amplification of one or both bands from conserved portions of the pink bollworm cadherin gene: the ~700 bp "intron control" band and the ~1,600 bp "X" band from the *r3x* reaction [42,50]. Furthermore, we used gDNA previously extracted from laboratory-selected resistant strains containing known *r* alleles (*r1* from AZP-R and BX-H [38,40]; *r2* from AZP-R [38,40]; and *r3* from BX-R [37,40]) as positive controls for genotyping [50]. For 58 insects, we screened the gDNA separately for each individual. For 367 insects, we tested gDNA in 39 pools with 3 to 10 insects per pool (mean = 9.4 per pool). The tests for each of the three known cadherin *r* alleles included a positive control for each individual or pool, as well as additional positive and negative controls.

4.4 Cloning and Sequencing of Pink Bollworm Cadherin cDNA and Gene

We cloned and sequenced cadherin DNA of 11 fourth instar larvae of pink bollworm that were preserved in RNAlater (Ambion-Life Technologies, Carlsbad, CA) from three sites in India: three from AGJ, three from AMH, and five from Khandwa in Madhya Pradesh (KMP) (Fig. 1). The three AGJ larvae used for cloning and sequencing were a subset of the resistant F_1 larvae that survived exposure to a diagnostic toxin concentration (10 micrograms Cry1Ac per mL diet) in the bioassay described above. The AMH larvae used for cloning and sequencing were collected on 30-November-2010 from non-Bt cotton bolls at the Panjabrao Deshmukh Agricultural University Cotton Research Station in Akola and immediately preserved in RNAlater. KMP larvae were collected during the first week of December 2010 from bolls of Bollgard cotton (Rasi variety) grown by farmers.

4.4.1 cDNA Cloning. Each of the 11 larvae (three from AGJ, three from AMH, and five from KMP) were removed from RNAlater and cut in half. The posterior halves were used to extract RNA, while the anterior halves were returned to RNAlater for later genomic DNA extractions. Total RNA was extracted using TRIzol (Invitrogen-Life Technologies, Carlsbad, CA) according to the manufacturer's instructions. RNA concentration was determined using a NanoDrop ND1000 spectrophotometer (Thermo Scientific, Wilmington, DE) and total RNA quality was assessed with an Agilent BioAnalyzer 2100 with RNA Nano 6000 LabChip Kit (Agilent Technologies, Santa Clara, CA). cDNA was produced using random hexamer primers and ThermoScript RT-PCR System (Invitrogen-Life Technologies) according to the manufacturer's recommendations. From each individual, we used primers 52PgCad5 and 25PgCad3 with high-fidelity SuperTaq Plus DNA Polymerase (Ambion-Life Technologies) to amplify full-length *PgCad1* cDNA (Table S3). PCR products were A-tailed with 1 unit of Takara ExTaq (Takara Bio USA, Madison, WI) and precipitated in ammonium acetate and ethanol. PCR products were resuspended and separated on 0.8% agarose gels stained with Crystal Violet (Invitrogen-Life Technologies). DNA bands were gel-purified and ligated into pCR-XL-TOPO using TOPO XL Gel Purification and PCR cloning kits (Invitrogen-Life Technologies). Plasmids were propagated in OneShot TOP10 electro-competent *Escherichia coli* (Invitrogen-Life Technologies) and purified using QIAprep Spin MiniPrep kit in QIAcube robot (Qiagen, Valencia, CA). Inserts were sequenced using M13 reverse vector primer, 52PgCad5, 89PgCad5, 57PgCad3, 70PgCad5, 72PgCad5, 73PgCad3, 75PgCad3, 76PgCad5, 77PgCad3, 78PgCad5, 79PgCad3, 20PgCad5, 81PgCad3, 85PgCad3, 87PgCad3, 25PgCad3, and T7 vector primer as appropriate. The nucleotide sequences reported in this paper are deposited in the GenBank public database.

4.4.2 gDNA Cloning. We used a PUREGENE DNA Isolation Kit (Qiagen, Valencia, CA) to extract gDNA from the anterior half of 9 of the 11 larvae described above: AMH-1, AMH-3, AGJ-1, AGJ-3, KMP-4, KMP-5, KMP-6, KMP-7, and KMP-8. gDNA was not extracted from individuals AMH-2 and AGJ-2 because we used all of their tissue for cDNA preparation. *PgCad1*-specific primers (Table S3), designed using Primer3Plus [85], were used with SuperTaq Plus DNA Polymerase to PCR-amplify partial genomic fragments corresponding to mutations found in cDNA from each of the eight individuals. PCR products were gel-purified, ligated into pCR-XL-TOPO or pCR2.1-TOPO (Invitrogen-Life Technologies), and plasmids were propagated in *E. coli* as indicated above. Additional gene-specific primers were used to completely sequence genomic clones (Table S3). The Arizona State University DNA Core Lab (Tempe, AZ) performed the DNA sequencing.

4.5 DNA Sequence Analysis

DNA sequences were trimmed, edited, and assembled in Vector NTI (LifeTechnologies). Multiple sequence alignments for DNA and predicted translated proteins were performed using CLUSTAL Omega (1.2.0) [86]. Repbase (http://www.girinst.org/) was searched using CENSOR [53]. Protein translations were obtained using ExPASy Translate tool (http://web.expasy.org/translate/).

Supporting Information

Figure S1 Alignment of cadherin cDNA sequences of pink bollworm from Akola, Maharashtra (AMH) with the susceptible allele *PgCad1 s* (AY198374.1). Eight of the nine cDNA clones from three individuals (AMH-1, AMH-2, AMH-3) have no insertions or deletions. One cDNA clone (AMH-3_16) has a single 3-bp deletion at base positions 72–74. Stars show nucleotides conserved in all of the sequences. The deletion is highlighted in gray.

Figure S2 Alignment of predicted amino acid sequences of pink bollworm cadherin from Akola, Maharashtra (AMH) with *PgCad1 s* (AY198374.1). Stars show amino acids conserved in all of the sequences. The symbols ":" and "." indicate conservative amino acid substitutions scoring >0.5 and ≤0.5 in the Gonnet PAM 250 matrix, respectively. Red boxes show amino acids corresponding to lepidopteran cadherin Cry1Ac toxin binding regions.

Figure S3 Alignment of cadherin cDNA sequences of pink bollworm from Anand, Gujarat (AGJ) with the susceptible allele *PgCad1 s* (AY198374.1). Thirteen clones from three individuals (AGJ-1, AGJ-2, AGJ-3) had six isoforms of three alleles [*r5A* (KJ480757), *r5B* (KJ480758), *r5C* (KJ480759), *r6A* (KJ480760), *r7A* (KJ480761), and *r7B* (KJ480762)]. Stars show nucleotides conserved in all of the sequences. Deletions are highlighted in gray and the insertion is highlighted in yellow. Codons highlighted in red indicate the positions of premature stop codons.

Figure S4 Alignment of predicted amino acid sequences of pink bollworm cadherin from Anand, Gujarat (AGJ) with *PgCad1 s* (AY198374.1.1). Stars show amino acids conserved in all of the sequences. The symbols ":" and "." indicate conservative amino acid substitutions scoring >0.5 and ≤0.5 in the Gonnet PAM 250 matrix, respectively. Red boxes show amino acids corresponding to lepidopteran cadherin Cry1Ac toxin binding regions.

Figure S5 Alignment of cadherin cDNA sequences of pink bollworm from Khandwa, Madhya Pradesh (KMP) with the susceptible allele *PgCad1 s* (AY198374.1). Twenty-three clones from five individuals (KMP-4, KMP-5, KMP-6, KMP-7, KMP-8) had thirteen isoforms of five *r* alleles [*r8A* (KJ480763), *r8B* (KJ480764), *r9A* (KJ480765), *r9B* (KJ480766), *r10A* (KJ480767), *r10B* (KJ480768), *r10C* (KJ480769), *r11A* (KJ480770), *r11B* (KJ480771), *r12A* (KJ480772), *r12B* (KJ480773), *r12C* (KJ480774), and *r12D* (KJ480775)] and two *s* alleles [clones KMP-8_5, KMP-8_24, and KMP-8_46 for *s6A* (KJ480754), clone KMP-8_35 for *s6B* (KJ480755), and clone KMP-8_3 for *s7* (KJ480756)]. Stars show nucleotides conserved in all of the sequences. Deletions are highlighted in gray and insertions are highlighted in yellow. Codons highlighted in red indicate the positions of premature stop codons.

Figure S6 Alignment of predicted amino acid sequences of pink bollworm cadherin from Khandwa, Madhya Pradesh (KMP) with *PgCad1 s* (AY198374.1). Stars show amino acids conserved in all of the sequences. The symbols ":" and "." indicate conservative amino acid substitutions scoring > 0.5 and ≤0.5 in the Gonnet PAM 250 matrix, respectively. Red boxes show amino acids corresponding to lepidopteran cadherin Cry1Ac toxin binding regions.

Figure S7 Alignment of cadherin cDNA sequences corresponding to susceptible alleles from Akola, Maharashtra (AMH) and Khandwa, Madhya Pradesh (KMP) with the susceptible allele *PgCad1 s* (AY198374.1). Fourteen clones from four individuals (AMH-1, AMH-2, AMH-3, KMP-8) have 27 allelic sites [single nucleotide polymorphisms that occur more than once and are not from C-to-U or A-to-I (G) RNA editing]. A total of seven *s* alleles are present from four individuals, including AMH-1 with two alleles, *s1* (clone AMH-1_2, KJ480749) and *s2* (clones AMH-1_7 and 11, KJ480750), AMH-2 with *s3* (clones AMH-2_1, 4, and 5, KJ480751), AMH-3 with *s4* (clones AMH-3_1 and 13, KJ480752) and *s5* (clone AMH-3_16, KJ480753), and two alleles from KMP-8 (*s6A* from clones KMP-8_5, 24, and 46, KJ480754; *s6B* from clone KMP-8_35, KJ480755; and *s7* from KMP-8_3, KJ480756). Allelic bases are shown in red boxes. Stars show nucleotides conserved in all of the sequences. Deletions from mis-spliced mRNA are highlighted in gray.

Figure S8 Partial genomic DNA sequencing of seven novel disrupted cadherin alleles in pink bollworm larvae from Anand (AGJ) in Gujarat and Khandwa (KMP) in Madhya Pradesh. Four mutations (found in isoforms *r5A*, *r8B*, *r11B*, and *r12A*) have altered gDNA, whereas 16 mutations are due to post-transcription modifications. Green-highlighted sequences show location of sense and antisense primers (from Table S3). Exon coding regions are shown as normal text and introns are highlighted in gray. Exon/intron splice junction nucleotides are highlighted in light blue. Yellow-highlighted sequence indicates insertions. Pink-highlighted sequence indicates gaps in sequencing. The 20 gDNA fragments shown are r5A_20-81, r5B_227-228, r5C_89-10, r7A_20-165, r7B_164-163, r8A_186-166, r8B_219-220, r9A_171-25, r9B_58-87, r10A-r10C_20-21, r10A-r10C_169-170, r10B_86-167, r10C_24-85, r11A_20-49, r11B_171-172, r12A_221-222, r12B_168-187, r12C-r12D_227-228, r12C_89-10, and r12D_186-73 (GenBank accession KJ724990-KJ725008). Note that r12A_221-222 does not have an accession number because it does not meet the minimum number of bases required by GenBank.

Table S1 Pink bollworm from India screened for cadherin alleles *r1-r3* from Arizona.

Table S2 gDNA sequencing of eight novel disrupted cadherin alleles in pink bollworm larvae from Anand (AGJ) in Gujarat and Khandwa (KMP) in Madhya Pradesh.

Table S3 Nucleotide primers used to amplify and sequence *PgCad1* from India pink bollworm.

Acknowledgments

We thank Yidong Wu and Dale Spurgeon for comments on an earlier draft. We thank the Department of Science and Technology (India) for sponsoring the fellowship from BOYSCAST (Better Opportunities for Young Scientists in Chosen Areas of Science and Technology) to J. Ponnuraj. This is a cooperative investigation between USDA-ARS and the University of Arizona. Mention of trade names or commercial products in this article is solely for the purpose of providing specific information and does not imply recommendation or endorsement by the U.S. Department of Agriculture. USDA is an equal opportunity provider and employer.

Author Contributions

Conceived and designed the experiments: JAF BET JP XL. Performed the experiments: JAF JP AS RKT GCU AJY. Analyzed the data: JAF BET XL JP. Contributed reagents/materials/analysis tools: JAF BET JP AS RKT. Wrote the paper: JAF BET XL YC.

References

1. Mendelsohn M, Kough J, Vaituzis Z, Matthews K (2003) Are Bt crops safe? Nat Biotechnol 21: 1003–1009.
2. Bravo A, Likitvivatanavong S, Gill SS, Soberón M (2011) *Bacillus thuringiensis*: A story of a successful bioinsecticide. Insect Biochem Mol Biol 41: 423–431.
3. Sanahuja G, Banakar R, Twyman RM, Capell T, Christou P (2011) *Bacillus thuringiensis*: A century of research, development and commercial applications. Plant Biotechnol J 9: 283–300.
4. James C (2013) Global status of commercialized biotech/GM crops: 2013. ISAAA Briefs No. 46. ISAAA: Ithaca, NY.
5. Wu KM, Lu YH, Feng HQ, Jiang YY, Zhao JZ (2008) Suppression of cotton bollworm in multiple crops in China in areas with Bt toxin-containing cotton. Science 321: 1676–1678.
6. Carpenter JE (2010) Peer-reviewed surveys indicate positive impact of commercialized GM crops. Nat Biotechnol 28: 319–321.
7. Hutchison WD, Burkness EC, Mitchell PD, Moon RD, Leslie TW, et al. (2010) Areawide suppression of European corn borer with Bt maize reaps savings to non-Bt maize growers. Science 330: 222–225.
8. National Research Council (2010) The impact of genetically engineered crops on farm sustainability in the United States. National Academies Press, Washington D.C..
9. Tabashnik BE, Sisterson MS, Ellsworth PC, Dennehy TJ, Antilla L, et al. (2010) Suppressing resistance to Bt cotton with sterile insect releases. Nat Biotechnol 28: 1304–1307.
10. Lu Y, Wu K, Jiang Y, Guo Y, Desneux N (2012) Widespread adoption of Bt cotton and insecticide decrease promotes biocontrol services. Nature 487: 362–365.
11. Tabashnik BE (1994) Evolution of resistance to *Bacillus thuringiensis*. Annu Rev Entomol 39: 47–79.
12. Tabashnik BE, Gassmann AJ, Crowder DW, Carrière Y (2008) Insect resistance to Bt crops: Evidence versus theory. Nat Biotech 26: 199–202.
13. Tabashnik BE, Brevault T, Carrière Y (2013) Insect resistance to Bt crops: Lessons from the first billion acres. Nat Biotech 31: 510–521.
14. Tabashnik BE, Liu YB, Malvar T, Masson L, Ferré J (1998) Insect resistance to *Bacillus thuringiensis*: Uniform or diverse? Phil Trans Roy Soc London B 353: 1751–1756.
15. Ferré J, Van Rie J (2002) Biochemistry and genetics of insect resistance to *Bacillus thuringiensis*. Annu Rev Entomol 47: 501–533.
16. Caccia S, Hernández-Rodríguez CS, Mahon RJ, Downes S, James W, et al. (2010) Binding site alteration is responsible for field-isolated resistance to *Bacillus thuringiensis* Cry2A insecticidal proteins in two *Helicoverpa* species. PLoS ONE 5: e9975.
17. Jurat-Fuentes JL, Karumbaiah L, Jakka SRK, Ning C, Liu C, et al. (2011) Reduced levels of membrane-bound alkaline phosphatase are common to lepidopteran strains resistant to Cry toxins from *Bacillus thuringiensis*. PLoS ONE 6: e17606.
18. Baxter SW, Badenes-Pérez FR, Morrison A, Vogel H, Crickmore N, et al. (2011) Parallel evolution of Bt toxin resistance in Lepidoptera. Genetics 189: 675–679.
19. Xu X, Yu L, Wu Y (2005) Disruption of a cadherin gene associated with resistance to Cry1Ac delta-endotoxin of *Bacillus thuringiensis* in *Helicoverpa armigera*. Appl Environ Microbiol 71: 948–954.
20. Yang Y, Chen H, Wu Y, Yang Y, Wu S (2007) Mutated cadherin alleles from a field population of *Helicoverpa armigera* confer resistance to *Bacillus thuringiensis* toxin Cry1Ac. Appl Environ Microbiol 73: 6939–6944.
21. Zhao J, Jin L, Yang Y, Wu Y (2010) Diverse cadherin mutations conferring resistance to *Bacillus thuringiensis* toxin Cry1Ac in *Helicoverpa armigera*. Insect Biochem Mol Biol 40: 113–118.
22. Zhang H, Tian W, Zhao J, Jin L, Yang J, et al. (2012) Diverse genetic basis of field-evolved resistance to Bt cotton in cotton bollworm from China. Proc Natl Acad Sci USA 109: 10275–10280.
23. Zhang H, Wu S, Yang Y, Tabashnik BE, Wu Y (2012) Non-recessive Bt toxin resistance conferred by an intracellular cadherin mutation in field-selected populations of cotton bollworm. PLoS ONE 7: e53418.
24. Zhang H, Yin W, Zhao J, Jin L, Yang Y, et al. (2011) Early warning of cotton bollworm resistance associated with intensive planting of Bt cotton in China. PLoS ONE 6: e22874.
25. Luttrell RG, Ali I, Allen KC, Young SY, Szalanski A, et al. (2004) Resistance to Bt in Arkansas populations of cotton bollworm. *In* Proceedings of the 2004 Beltwide Cotton Conferences, San Antonio, TX, January 5–9, 2004 (ed. Richter, DA) 1373–1383 (National Cotton Council of America, Memphis, TN, 2004).
26. van Rensburg JBJ (2007) First report of field resistance by stem borer, *Busseola fusca* (Fuller) to Bt-transgenic maize. S African J Plant Soil 24: 147–151.
27. Storer NP, Babcock JM, Schlenz M, Meade T, Thompson GD, et al. (2010) Discovery and characterization of field resistance to Bt maize: *Spodoptera frugiperda* (Lepidoptera: Noctuidae) in Puerto Rico. J Econ Entomol 103: 1031–1038.
28. Gassmann AJ, Petzold-Maxwell JL, Keweshan RS, Dunbar MW (2011) Field-evolved resistance to Bt maize by Western corn rootworm. PLoS ONE 6: e22629.
29. Dhurua S, Gujar GT (2011) Field-evolved resistance to Bt toxin Cry1Ac in the pink bollworm, *Pectinophora gossypiella* (Saunders) (Lepidoptera: Gelechiidae), from India. Pest Manag Sci 67: 898–903.
30. Monsanto (2010) Cotton in India. http://www.monsanto.com/newsviews/Pages/india-pink-bollworm.aspx (subsequently revised to "Pink bollworm resistance to GM cotton in India," revised page accessed 29 April 2014).
31. Henneberry TJ, Naranjo SE (1998) Integrated management approaches for pink bollworm in the southwestern United States. Integr Pest Manage Rev 3: 31–52.
32. James C (2012) Global status of commercialized biotech/GM crops: 2012. ISAAA Briefs No. 44. ISAAA: Ithaca, NY.
33. Lalitha N, Ramaswami B, Viswanathan PK (2009) India's experience with Bt cotton: Case studies from Gujarat and Maharashtra. *In* Biotechnology and Agricultural Development: Transgenic Cotton, Rural Institutions and Resource-Poor Farmers (ed. Tripp, R.) 135–167 (Routledge, New York, 2009).
34. Desh Gujarat (2013) State wise figures of cotton production in India in last four years, even with lower production Gujarat continues to top. Accessed 22 March 2013. http://deshgujarat.com/2013/03/22/state-wise-figures-of-cotton-production-in-india-in-last-four-years-even-with-lower-production-gujarat-continues-to-top/
35. Herring R.J. (2007) Stealth seeds: Bioproperty, biosafety, biopolitics. J Dev Stud 43: 130–157.
36. Tabashnik BE, Dennehy TJ, Sims MA, Larkin K, Head GP, et al. (2002) Control of resistant pink bollworm (*Pectinophora gossypiella*) by transgenic cotton that produces *Bacillus thuringiensis* toxin Cry2Ab. Appl Environ Microbiol 68: 3790–3794.
37. Tabashnik BE, Unnithan GC, Masson L, Crowder DW, Li X, et al. (2009) Asymmetrical cross-resistance between *Bacillus thuringiensis* toxins Cry1Ac and Cry2Ab in pink bollworm. Proc Natl Acad Sci USA 106: 11889–11894.
38. Morin S, Biggs RW, Sisterson MS, Shriver L, Ellers-Kirk C, et al. (2003) Three cadherin alleles associated with resistance to *Bacillus thuringiensis* in pink bollworm. Proc Natl Acad Sci USA 100: 5004–5009.
39. Tabashnik BE, Liu Y-B, Unnithan DC, Carrière Y, Dennehy TJ, et al. (2004) Shared genetic basis of resistance to Bt toxin Cry1Ac in independent strains of pink bollworm. J Econ Entomol 97: 721–726.
40. Tabashnik BE, Biggs RW, Higginson DM, Henderson S, Unnithan DC, et al. (2005) Association between resistance to Bt cotton and cadherin genotype in pink bollworm. J Econ Entomol 98: 635–644.
41. Carrière Y, Ellers-Kirk C, Biggs RW, Nyboer MK, Unnithan GC, et al. (2006) Cadherin-based resistance to *Bacillus thuringiensis* cotton in hybrid strains of pink bollworm: Fitness costs and incomplete resistance. J Econ Entomol 99: 1925–1935.
42. Fabrick JA, Tabashnik BE (2012) Similar genetic basis of resistance to Bt toxin Cry1Ac in boll-selected and diet-selected strains of pink bollworm. PLoS ONE 7: e35658.
43. Tabashnik BE, Dennehy TJ, Carrière Y (2005) Delayed resistance to transgenic cotton in pink bollworm. Proc Natl Acad Sci USA 102: 15389–15393.
44. Tabashnik B, Morin S, Unnithan G, Yelich A, Ellers-Kirk C, et al. (2012) Sustained susceptibility of pink bollworm to Bt cotton in the United States. GM Crops and Food: Biotechnology in Agriculture and the Food Chain 3: 1–7.
45. Stone GD (2014) Biotechnology and the political ecology of information in India. Human Organization 63: 127–140.
46. Choudhary B, Gaur K (2010) Bt cotton in India: A country profile. ISAAA Series of Biotech Crop Profiles (ISAAA, Ithaca, NY, 2010).
47. Roy SW, Irimia M (2009) Splicing in the eukaryotic ancestor: Form, function and dysfunction. Trends Ecol Evol 24: 447–455.
48. Nilsen TW, Gravely BR (2010) Expansion of the eukaryotic proteome by alternative splicing. Nature 463: 457–463.
49. Liu Y-B, Tabashnik BE, Meyer SK, Carrière Y, Bartlett AC (2001) Genetics of pink bollworm resistance to *Bacillus thuringiensis* toxin Cry1Ac. J Econ Entomol 94: 248–252.
50. Morin S, Henderson S, Fabrick JA, Carrière Y, Dennehy TJ, et al. (2004) DNA-based detection of Bt resistance alleles in pink bollworm. Insect Biochem Mol Biol 34: 1225–1233.
51. Tabashnik BE, Fabrick JA, Henderson S, Biggs RW, Yafuso CM, et al. (2006) DNA screening reveals pink bollworm resistance to Bt cotton remains rare after a decade of exposure. J Econ Entomol 99: 1525–1530.
52. Fabrick JA, Tabashnik BE (2007) Binding of *Bacillus thuringiensis* toxin Cry1Ac to multiple sites of cadherin in pink bollworm. Insect Biochem Mol Biol 37: 97–106.
53. Kohany O, Gentles AJ, Hankus L, Jurka J (2006) Annotation, submission and screening of repetitive elements in Repbase: RepbaseSubmitter and Censor. BMC Bioinformatics 7: 474.
54. Gahan LJ, Pauchet Y, Vogel H, Heckel DG (2010) An ABC transporter mutation is correlated with insect resistance to *Bacillus thuringiensis* Cry1Ac toxin. PLoS Genet 6: e1001248.
55. Zhang X, Tiewsiri K, Kain W, Huang L, Wang P (2012) Resistance of *Trichoplusia* ni to *Bacillus thuringiensis* toxin Cry1Ac is independent of alteration of the cadherin-like receptor for Cry toxins. PLoS ONE 7: e35991.
56. Hernández-Martínez P, Hernández-Rodríguez CS, Krishnan V, Crickmore N, Escriche B, et al. (2012) Lack of Cry1Fa binding to the midgut brush border membrane in a resistant colony of *Plutella xylostella* moths with a mutation in the ABCC2 locus. Appl Environ Microbiol 78: 6759–6761.
57. Atsumi S, Miyamoto K, Yamamoto K, Narukawa J, Kawai S, et al. (2012) Single amino acid mutation in an ATP-binding cassette transporter gene causes

resistance to Bt toxin Cry1Ab in the silkworm, *Bombyx mori*. Proc Natl Acad Sci USA 109: E1591–8.
58. Tanaka S, Miyamoto K, Noda H, Jurat-Fuentes JL, Yoshizawa Y, et al. (2013) The ATP-binding cassette transporter subfamily C member 2 in *Bombyx mori* larvae is a functional receptor for Cry toxins from *Bacillus thuringiensis*. FEBS Journal 280: 1782–1794.
59. Coates BS, Sumerford DV, Siegfried BD, Hellmich RL, Abel CA (2013) Unlinked genetic loci control the reduced transcription of aminopeptidase N 1 and 3 in the European corn borer and determine tolerance to *Bacillus thuringiensis* Cry1Ab toxin. Insect Biochem Mol Biol 43: 1152–1160.
60. Tabashnik BE, Biggs RW, Fabrick JA, Gassmann AJ, Dennehy TJ, et al. (2006) High-level resistance to *Bacillus thuringiensis* toxin Cry1Ac and cadherin genotype in pink bollworm. J Econ Entomol 99: 2125–2131.
61. Zhang X, Kain W, Wang P (2013) Sequence variation and differential splicing of the midgut cadherin gene in *Trichoplusia ni*. Insect Biochem Mol Biol 43: 712–723.
62. Fabrick JA, Mathew LG, Tabashnik BE, Li X (2011) Insertion of an intact CR1 retrotransposon in a cadherin gene linked with Bt resistance in the pink bollworm, *Pectinophora gossypiella*. Insect Mol Biol 20: 651–665.
63. Zhang H, Tang M, Yang F, Yang Y, Wu Y (2013) DNA-based screening for an intracellular cadherin mutation conferring non-recessive Cry1Ac resistance in field populations of *Helicoverpa armigera*. Pestic Biochem Physiol 107: 148–152.
64. Gahan LJ, Gould F, Heckel DG (2001) Identification of a gene associated with Bt resistance in *Heliothis virescens*. Science 293: 857–860.
65. Jin L, Wei Y, Zhang L, Yang Y, Tabashnik BE, et al. (2013) Dominant resistance to Bt cotton and minor cross-resistance to Bt toxin Cry2Ab in cotton bollworm from China. Evol Appl 6: 1222–1235.
66. Nair R, Kalia V, Aggarwal KK, Gujar GT (2013) Variation in the cadherin gene sequence of Cry1Ac susceptible and resistant *Helicoverpa armigera* (Lepidoptera: Noctuidae) and the identification of mutant alleles in resistant strains. Curr Sci 104: 215–223.
67. Coates BS, Sumerford DV, Hellmich RL, Lewis LC (2005) Sequence variation in the cadherin gene of *Ostrinia nubilalis* a tool for field monitoring. Insect Biochem Mol Biol 35: 129–139.
68. Coates BS, Sumerford DV, Lewis LC (2008) Segregation of European corn borer, *Ostrinia nubilalis*, aminopeptidase 1, cadherin, and bre5-like alleles, from a colony resistant to *Bacillus thuringiensis* Cry1Ab toxins, are not associated with F_2 larval weights when fed a diet containing Cry1Ab. J Insect Sci 8: 1–8.
69. Bel Y, Siqueira HA, Siegfried BD, Ferré J, Escriche B (2009) Variability in the cadherin gene in an *Ostrinia nubilalis* strain selected for Cry1Ab resistance. Insect Biochem Mol Biol 39: 218–23.
70. Carrière Y, Tabashnik BE (2001) Reversing insect adaptation to transgenic insecticidal plants. Proc Roy Soc Lond B 268: 1475–1480.
71. Williams JL, Ellers-Kirk C, Orth RG, Gassmann AJ, Head G, et al. (2011) Fitness cost of resistance to Bt cotton linked with increased gossypol content in pink bollworm larvae. PLoS ONE 6: e21863.
72. Carrière Y, Ellers-Kirk C, Liu YB, Sims MA, Patin AL, et al. (2001) Fitness costs and maternal effects associated with resistance to transgenic cotton in the pink bollworm (Lepidoptera: Gelechiidae). J Econ Entomol 94: 1571–1576.
73. Carrière Y, Ellers-Kirk C, Patin AL, Sims M, Meyer S, et al. (2001) Overwintering cost associated with resistance to transgenic cotton in the pink bollworm. J Econ Entomol 94: 935–941.
74. Higginson DM, Nyboer ME, Biggs RW, Morin S, Tabashnik BE, et al. (2005) Evolutionary trade-offs of insect resistance to Bt crops: Fitness cost affecting paternity. Evolution 59: 915–920.
75. Gahan LJ, Gould F, López JD, Micinski S, Heckel DG (2007) A polymerase chain reaction screen of field populations of *Heliothis virescens* for a retrotransposon insertion conferring resistance to *Bacillus thuringiensis* toxin. J Econ Entomol 100: 187–194.
76. Gould F, Anderson A, Jones A, Sumerford D, Heckel DG, et al. (1997) Initial frequency of alleles for resistance to *Bacillus thuringiensis* toxins in field populations of *Heliothis virescens*. Proc Natl Acad Sci USA 94: 3519–3523.
77. Mahon RJ, Downes S, James W, Parker T (2010) Why do F1 screens estimate higher frequencies of Cry2Ab resistance in *Helicoverpa armigera* (Lepidoptera: Noctuidae) than do F2 screens? J Econ Entomol 103: 472–481.
78. ffrench-Constant RH, Rocheleau TA (1993) *Drosophila* gamma-aminobutyric acid receptor gene Rdl shows extensive alternative splicing. J Neurochem 60: 2323–2326.
79. Hemingway J, Hawkes N, Prapanthadara L, Jayawardenal KG, Ranson H (1998) The role of gene splicing, gene amplification and regulation in mosquito insecticide resistance. Philos Trans R Soc Lond B Biol Sci 353: 1695–1699.
80. Dong K (2007) Insect sodium channels and insecticide resistance. Invert Neurosci 7: 17–30.
81. Rinkevich FD, Chen M, Shelton AM, Scott JG (2010) Transcripts of the nicotinic acetylcholine receptor subunit gene Pxyla6 with premature stop codons are associated with spinosad resistance in diamondback moth, *Plutella xylostella*. Invertebr Neurosci 10: 25–33.
82. Baxter SW, Chen M, Dawson A, Zhao JZ, Vogel H, et al. (2010) Mis-spliced transcripts of nicotinic acetylcholine receptor alpha6 are associated with field evolved spinosad resistance in *Plutella xylostella* (L.). PLoS Genet 6: e1000802.
83. Tabashnik BE, Liu Y-B, de Maagd RA, Dennehy TJ (2000) Cross-resistance of pink bollworm (*Pectinophora gossypiella*) to *Bacillus thuringiensis* toxins. Appl Environ Microbiol 66: 4582–4584.
84. Tabashnik BE, Liu Y-B, Dennehy TJ, Sims MA, Sisterson MS, et al. (2002) Inheritance of resistance to Bt toxin Cry1Ac in a field-derived strain of pink bollworm (Lepidoptera: Gelechiidae). J Econ Entomol 95: 1018–1026.
85. Untergasser A, Nijveen H, Rao X, Bisseling T, Geurts R, et al. (2007) Primer3Plus, an enhanced web interface to Primer3. Nucleic Acids Res 35: W71–W74.
86. Sievers F, Wilm A, Dineen D, Gibson TJ, Karplus K, et al. (2011) Fast, scalable generation of high-quality protein multiple sequence alignments using Clustal Omega. Mol Systems Biol 7: 539.

12

Transcriptomic Analysis of Fiber Strength in Upland Cotton Chromosome Introgression Lines Carrying Different *Gossypium barbadense* Chromosomal Segments

Lei Fang, Ruiping Tian, Jiedan Chen, Sen Wang, Xinghe Li, Peng Wang, Tianzhen Zhang*

National Key Laboratory of Crop Genetics and Germplasm Enhancement, Cotton Hybrid R & D Engineering Center (the Ministry of Education), Nanjing Agricultural University, Nanjing, China

Abstract

Fiber strength is the key trait that determines fiber quality in cotton, and it is closely related to secondary cell wall synthesis. To understand the mechanism underlying fiber strength, we compared fiber transcriptomes from different *G. barbadense* chromosome introgression lines (CSILs) that had higher fiber strengths than their recipient, *G. hirsutum* acc. TM-1. A total of 18,288 differentially expressed genes (DEGs) were detected between CSIL-35431 and CSIL-31010, two CSILs with stronger fiber and TM-1 during secondary cell wall synthesis. Functional classification and enrichment analysis revealed that these DEGs were enriched for secondary cell wall biogenesis, glucuronoxylan biosynthesis, cellulose biosynthesis, sugar-mediated signaling pathways, and fatty acid biosynthesis. Pathway analysis showed that these DEGs participated in starch and sucrose metabolism (328 genes), glycolysis/gluconeogenesis (122 genes), phenylpropanoid biosynthesis (101 genes), and oxidative phosphorylation (87 genes), etc. Moreover, the expression of MYB- and NAC-type transcription factor genes were also dramatically different between the CSILs and TM-1. Being different to those of CSIL-31134, CSIL-35431 and CSIL-31010, there were many genes for fatty acid degradation and biosynthesis, and also for carbohydrate metabolism that were down-regulated in CSIL-35368. Metabolic pathway analysis in the CSILs showed that different pathways were changed, and some changes at the same developmental stage in some pathways. Our results extended our understanding that carbonhydrate metabolic pathway and secondary cell wall biosynthesis can affect the fiber strength and suggested more genes and/or pathways be related to complex fiber strength formation process.

Editor: Jinfa Zhang, New Mexico State University, United States of America

Funding: This work was financially supported in part by grants from the National Science Foundation of China (31330058) and the Priority Academic Program Development of Jiangsu Higher Education Institutions. The funders had no role in study design, data collection and analysis, decision to publish, or preparation of the manuscript.

* E-mail: cotton@njau.edu.cn

These authors contributed equally to this work.

Introduction

The cotton fiber is a terminally differentiated single cell derived from the epidermal cell of the developing ovule. After initiation, the fiber cell undergoes 1000- to 3000-fold elongation during its development. The development of cotton fibers involves four partially overlapping stages: initiation (−3 to +3 days post-anthesis; DPA), elongation and primary cell wall formation (3–23 DPA), secondary cell wall formation (16–40 DPA) and maturation (40–50 DPA) [1–6]. The most rapid period of fiber cell elongation begins around 10–16 DPA and continues to ~20 DPA. Primary and secondary cell wall synthesis overlaps during the period of 16–25 DPA. During the secondary cell wall formation stage, the speed of cell elongation slows down and even stops.

Fiber strength is an important indicator of cotton fiber quality, and depends on formation of the secondary cell wall. Cellulose synthesis plays a predominant role in fiber cells, and cellulose accounts for >95% of the dry weight of the mature cotton fiber [3,7]. Genome and EST sequencing have revealed that there are at least ten different CesA genes for cellulose synthase in *Arabidopsis*; CesA-like genes have also been reported in rice and barley [8–10]. In cotton (*Gossypium raimondii*), at least 15 cellulose synthase (CESA) sequences are required for cellulose synthesis [11]. A recent investigation in *Arabidopsis thaliana* using microarrays led to the identification of genes that are highly co-expressed with cellulose synthase genes and two mutants, irx8 and irx13, that have irregular xylem phenotypes, were also identified [12]. Sucrose synthase (Susy) is the enzyme that catalyzes the hydrolysis of sucrose to UDP-glucose that is then used as a substrate for cellulose synthesis. In cotton, the expression of Susy is higher at 16–32 DPA, and this enzyme plays a major role in partitioning carbon toward cellulose synthesis in the fiber [13]. SusC is another new sucrose synthase gene with a high level of expression during secondary cell wall synthesis [14]. Peroxide, mainly as H_2O_2, promotes cellulose synthesis as a signal of secondary cell wall synthesis [15,16].

At present, many ovule- and fiber-specific cDNA libraries have been constructed and sequenced, and more than 268,000 expressed sequence tags (ESTs) from *Gossypium* are deposited in the NCBI database (http://www.ncbi.nlm.nih.gov). For genetic characterization of rapid cell elongation in cotton fibers, approximately 14,000 unique genes were assembled from 46,603 expressed sequence tags (ESTs) from developmentally-staged fiber cDNAs of a cultivated diploid species (*G. arboreum* L.). Eighty-one genes that were significantly up-regulated during secondary cell wall synthesis were found to be involved in cell wall biogenesis and energy/carbohydrate metabolism, which is consistent with the stage of cellulose synthesis during secondary cell wall modification in developing fibers [17]. Transcriptome profiling of the cotton fiber early in development by high-throughput tag-sequencing (Tag-seq) analysis using the Solexa Genome Analyzer reveals significant differential expression of genes in a fuzzless/lintless mutant [18]. High-throughput, genome-wide transcriptomic analysis of cotton under drought stress revealed a significant down-regulation of genes and pathways involved in fiber elongation, and an up-regulation of defense response genes [19]. More research have been processed in fiber initiation and elongation stage [20–24]. Saturated very-long-chain fatty acids (VLCFAs; C20:0–C30:0) exogenously applied in ovule culture medium significantly promoted fiber cell elongation in cotton (*G. hirsutum L.*) by activating ethylene biosynthesis [25,26]. Previous investigations into cotton fiber development mainly focused on the elongation stage, and the number of genes reported from the later stages is quite small. Most of the genes up-regulated during secondary cell wall synthesis were related to cellulose synthesis, cell wall biosynthesis, and carbohydrate metabolism [17,22,27].

Chromosome segment introgression lines (CSILs) consist of a battery of near-isogenic lines that have been developed to cover the entire genomes of some crops, including tomato, rice, wheat, and cotton [28–31]. With the exception of a single, homozygous chromosome segment transferred from a donor parent, the remaining genome of each CSIL is the same as the recipient parent [31]. We used *G. barbadense* CSILs in the background of the standard genetic line of *G. hirsutum*, cv. TM-1, in order to understand the molecular mechanism behind superior quality fiber formation. Multi-point tests showed that three CSILs produced stronger fibers when compared to the recipient parent TM-1, but one CSIL produced weaker fibers. Using Solexa Genome sequencing, we analyzed transcriptome profiles from the CSILs and TM-1. We found that many genes were either up- or down-regulated at the stage of secondary cell wall synthesis, and that many metabolic pathways were altered in the CSILs.

Materials and Methods

Plant materials

G. hirsutum cv. TM-1, the genetic standard for Upland cotton, was obtained from the Southern Plains Agricultural Research Center, USDA-ARS, College Station/Texas, USA [32]. *G. barbadense* cv. Hai7124, an extra-long staple cotton that is widely grown in China, is descended from a selected individual in a study of inheritance of resistance to *Verticillium dahlia* [33,34]. In this study, we identified three CSILs with stronger fiber or high fiber strength that carried different *G. barbadense* chromosome segment(s) in the recurrent parent TM-1. The detailed method of developing CSILs has been described previously [31]. We selected three CSILs, CSIL-35431, CSIL-31134, and CSIL-31010, in which the average fiber strength were 35.1, 34.73 and 34.28 cN/tex, respectively, significantly higher than TM-1, and also CSIL-35368 which had poorer fiber strength than TM-1(28.71 cN/tex) (Table S1). The introgressed *G. barbadense* chromosomal segments were different in the four lines [35]. Fiber samples were collected at 15, 20, and 25 DPA, frozen in liquid nitrogen, and stored at −70°C.

RNA isolation and evaluation

Total RNA was extracted from frozen tissue using an improved CTAB extraction protocol [36]. RNAs were evaluated for quality using RNA Pico Chips on an Agilent 2100 Bioanalyzer (Agilent Technologies, Santa Clara, CA, USA). All RNA samples were quantified and qualified with an RNA Integrity Number (RIN) > 8, and 28S/18S rRNA band intensity (2:1).

Library construction and sequencing

Digital gene expression libraries were constructed using the Illumina Gene Expression Sample Preparation Kit according to the manufacturer's instructions. We constructed and sequenced 14 libraries derived from immature fibers at 15, 20, and 25 DPA using the Solexa Genome Sequencing Analyzer system provided by BGI (Beijing Genomics Institute at Shenzhen, China), which gave 21 bp tags. The process was described in detail previously [18].

Data processing, statistical evaluation, and selection of differentially expressed genes (DEGs)

Raw data reads were filtered by the Illumina pipeline to produce clean data. All low-quality data, such as short tags (< 21 nt) and singletons, were removed. A database of 21-base-long sequences was produced beginning with CATG using 37,505 reference genes from the diploid species *G. raimondii* (http://www.phytozome.net). The remaining high quality sequences were then mapped to this database; only a single mismatch was allowed, and more than one match was excluded. Gene expression levels were the summation of tags aligned to the different positions of the same gene. Expression levels are expressed as TPM, transcripts per million. To identify DEGs during fiber elongation, we compared pairs of DEG profiles from different libraries. Three fiber development periods for the four CSILs were compared with the same period for TM-1, and 11 comparisons were obtained. P- and Q-values were also calculated for every comparison [37]. DEGs were defined as FDR≤0.001 with an absolute value of $|\log_2 \text{Ratio}| \geq 1$ to judge the significance of differences in transcript abundance.

Digital tag profiling analysis

DEG clustering in CSILs at different developmental stages were performed with Cluster3.0 (http://bonsai.hgc.jp/~mdehoon/software/cluster/software.htm). We also performed clustering with the 'Self-organizing tree algorithm' (SOTA, Multiple Array Viewer software, MeV 4.9.0) [38].

GO enrichment and KEGG (Kyoto Encyclopedia of Genes and Genomes) pathway analysis was done using BLAST2GO (http://www.blast2go.com/b2ghome). Mapman was also used to analyze metabolic pathway base on KEGG database [39].

Quantitative RT-PCR

Quantitative RT-PCR assays were performed on a 7500 Real-Time PCR system (Applied Biosystems, San Francisco, CA, USA). Reactions were performed in a final volume of 15 μL and contained 2 μL of diluted cDNA, 7.5 μL of 2× SYBR mix (Roche, Basel, Switzerland), and 200 nM of the forward and reverse primers. Primer lengths were designed to be between 18 and 24 nt using Beacon Designer 7, and PCR amplicon lengths were designed to be between 100 bp and 150 bp (Table S2). The thermal cycling conditions were 40 cycles of 95°C for 15 s, 60°C for 30 s, and 72°C for 30 s. All reactions were run in triplicate, and the cotton *histone3* gene (ACC NO. AF024716) was used as an internal control for normalization of expression levels (F: 5′-GGTGGTGTGAAGAAGCCTCAT-3′, and R: 5′-AATTTCACGAACAAGCCTCTGGAA-3′). The relative gene expression levels were presented as $2^{-\Delta CT}$.

Results

Statistical analysis of transcriptome data

The total number of sequence tags per library ranged from 7.0 to 8.5 million, and the number of distinct sequence tags was between 1.8 and 2.2 million. Approximately 50% of the clean tags were mapped to reference genes, and 60% of the reference genes were mapped with unambiguous tag (Table 1 and Table S3).

To see whether the fiber transcriptomes at different developmental stages were different, the 23,237 genes which were expressed in at least three libraries at one stage (15 DPA, 20 DPA, or 25 DPA) were classified into six groups using the Multiple Array Viewer using TPM value (Figure 1A). Genes in Group 3 had higher expression levels at 15 DPA and 20 DPA than at the later stage (25 DPA). Genes in Group 4 had higher expression levels at 15 DPA than at either 20 DPA or 25 DPA. Genes in Group 5 showed the opposite expression pattern, with higher expression levels at 20 DPA and 25 DPA compared to 15 DPA. The other groups also showed distinct expression patterns (Figure 1A).

Classification by gene function revealed that Group 3 is enriched in genes involved in protein catabolism, cell division, and cellulose biosynthesis, Group 4 is enriched in genes for cell morphogenesis, fatty acid biosynthesis, lipid transport, and wax biosynthesis, and Group 5 has more genes involved in glucose catabolism, response to chitin, and sucrose metabolism (Figure 1B). The unbalanced pattern of the expressed-gene functional distribution could possibly reflect some physiological events involved in secondary cell wall biosynthesis.

Cluster analysis of differentially expressed genes (DEGs) between and/or among CSILs

We specifically looked for DEGs in secondary cell wall fibers from 15 to 25 DPA, because previous studies have reported that the different sets of transcripts responsible for fiber secondary cell wall formation may be enriched at these stages of development [17,22,27]. Three fiber development periods for the four CSILs were compared with TM-1 at the same period. DEGs were defined as FDR≤0.001 with an absolute value of $|\log_2\text{Ratio}|\geq 1$. Analysis of the data indicated that many genes showed differential expression in the 11 comparison groups. The number of DEGs were about 6,000–8,000 in CSILs from 15 DPA to 25 DPA (Figure 2A). But the number of DEGs in CSIL-31010 at 20 DPA, CSIL-31010 at 25 DPA, and CSIL-31134 at 15 DPA, were 4,600, 10,106 and 2,060, respectively. We also found that the DEGs that were up-regulated or down-regulated were different in CSILs. There were ~1,500–3,500 DEGs in common from 15 DPA to 25 DPA between CSIL-35431, CSIL-31010, and CSIL-35368 (Figure 2B).

To understand the mechanisms behind the changes in fiber strength observed in the CSILs, we also analyzed the common DEGs among CSIL-35431, CSIL-31010 and CSIL-31134 (Table S4). A total of 727 and 1796 common DEGs were selected at 15 and 20 DPA in three stronger fiber CSILs, respectively (Figure 3). More functional enrichment were shown at 15 DPA, including

Table 1. The distribution of total and distinct tags.

Summary	TM-1			CSIL-35431			CSIL-31010		
	15DPA	20DPA	25DPA	15DPA	20DPA	25DPA	15DPA	20DPA	25DPA
Raw Data	8374304	7231305	14267931	7461562	7389820	7298224	7471212	7007447	7203054
Distinct Raw Tag	389162	394430	494345	375999	371748	317147	283130	337525	386302
Clean Tag	8144920	7013147	14002534	7264426	7213791	7124575	7369012	6840101	7002766
Distinct Clean Tag	169700	177887	251173	182629	199257	148121	182184	172775	188639
Unique Clean Tag Mapping to Gene	3983215	3206328	6829327	3961228	3411249	3463314	3671014	3381955	3552566
Total % of clean tag	48.90%	45.72%	48.77%	54.53%	47.29%	48.61%	49.82%	49.44%	50.73%
Unique Distinct Clean Tag Mapping to Gene	45380	40475	56172	53187	48453	41065	46720	45762	48147
Total % of distinct tag	26.74%	22.75%	22.36%	29.12%	24.32%	27.72%	25.64%	26.49%	25.52%
Unambiguous Tag-mapped Genes	21498	20901	22594	21781	22325	20590	21900	22457	22811
Percentage of reference genes	57.32%	55.73%	60.24%	58.07%	59.53%	54.90%	58.39%	59.88%	60.82%

Clean tags: tags after filtering dirty tags (low quality tags) from raw data.
Distinct tags: different kinds of tags.
Unambiguous tags: the clean tags after removing tags mapped to reference sequences from multiple genes.

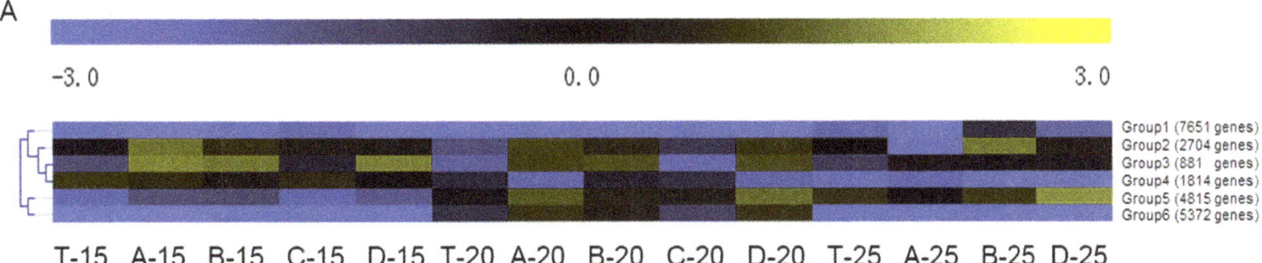

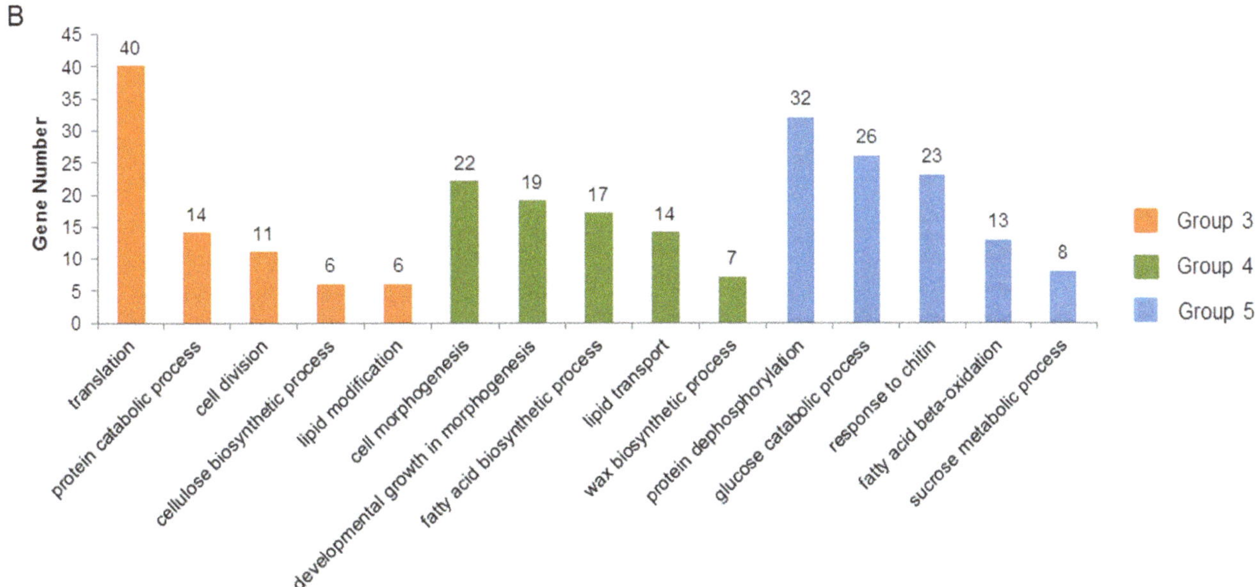

Figure 1. Statistical analysis of transcriptome data. (A) SOTA clustering of the different genes using Log2(TPM). T, TM-1; A, CSIL-35431; B, CSIL-31010; C, CSIL-31134; D, CSIL-35368. 15, 15 DPA; 20, 20 DPA; 25, 25 DPA. (B) Distribution of functions of genes in different clusters. Yellow square indicated group 3, green square indicated group 4 and blue square indicated group 5. X-axis indicated different enriched process and Y-axis indicated number of hit-found genes in these processes.

major CHO metabolism (carbohydrate), cell wall biosynthesis, amino acid metabolism and secondary metabolism (Figure 3E). Among these genes, 321 and 998 common upregulated DEGs between the same CSILs at 15 and 20 DPA were indentified, respectively (Figure 3). These common DEGs or processes maybe directly related to the fiber strength. However, these DEGs maybe function as downstream genes altered by the introgressed segments since these CSILs were inserted different *G. barbadense* segments in recipient TM-1.

To visualize the expression patterns of DEGs, we performed cluster analysis of 18,288 genes that were differentally expressed between CSIL-35431 and CSIL-31010 (Figure 4). These DEGs could be grouped into six clusters, designated G1–G6, based on their expression patterns. From 15 DPA to 20 DPA, the stages of fast fiber elongation and secondary cell wall deposition overlap, with the latter reaching a peak at around 20–25 DPA. We focused on clusters G1, G4, and G6 to conduct data analysis in order to identify genes that were either up-regulated or down-regulated during the secondary cell wall synthesis stage. Compared to the TM-1 control, 3,658 genes in cluster G1 were highly expressed at 15 and 20 DPA, 4,487 genes in G4 were highly expressed at 15 DPA, 20 DPA, and 25 DPA, 3,033 genes in G6 were highly expressed only at 25 DPA, and the other three groups showed various different expression patterns. Clustering results for 19,742 DEGs from the four CSILs showed five groups, indicating that the gene expression pattern in CSIL-31134 was distinct from the others at 15 DPA and 20 DPA, and that CSIL-35368 was similar to CSIL35431 and CSIL-31010 (Figure S1).

Functional annotation by GO enrichment and KEGG analysis

To understand the mechanisms behind the changes in fiber strength observed in the CSILs, we analyzed DEG enrichment in the major functional GO categories of biological process, molecular function, and cellular component between CSIL-35431 and CSIL-31010. Based on the clustering results shown in Figure 4, G1 was enriched in genes for secondary cell wall biogenesis, glucuronoxylan biosynthesis, microtubule-based movement, and cellulose biosynthesis, G4 was enriched in genes for protein phosphorylation, response to chitin, and sugar-mediated signaling pathways, and G6 was enriched in fatty acid biosynthesis genes (Table 2). These data suggest that in the developmental stage of secondary cell wall deposition, DEGs were enriched for carbohydrate synthesis and cell wall formation.

We applied the same GO analysis to the common DEGs at 15 DPA and 20 DPA in CSIL-35431 and CSIL-31010, respectively. These DEGs were enriched in genes for similar functional

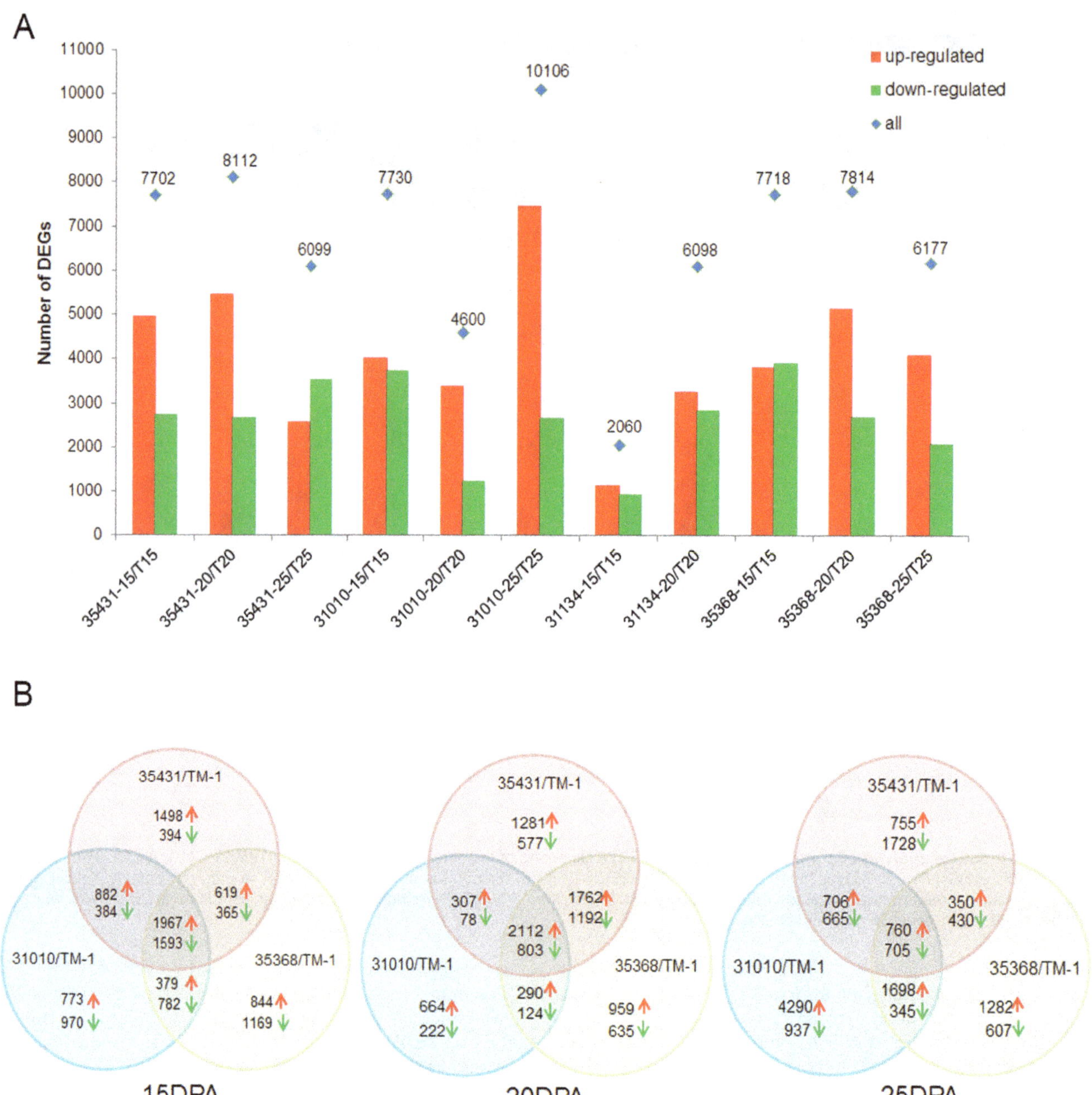

Figure 2. Statistical of DEGs between CSILs and TM-1 at 15, 20 and 25 DPA. (A) Up-regulated and down-regulated genes in different comparison. Red bar, up-regulated genes compared to TM-1; green bar, stand for down-regulated genes compared to TM-1, Blue square, total DEGs. CSILs included CSIL-35431, CSIL-3010, CSIL-31134, CSIL-35368 and TM-1. 15, 15DPA; 20, 20DPA; 25, 25DPA. (B) Common and special DEGs at 15 DPA, 20 DPA and 25 DPA.

categories, such as cellular metabolic processes and carbohydrate metabolism, etc. We also found genes for some processes that were enriched only in CSIL35431 or CSIL-31010 (Figure S2).

Further GO analysis for CSIL-35368 and CSIL-31134 indicated that the DEGs in CSIL-35368 at 15 and 20 DPA were enriched in genes for lignin biosynthesis, secondary cell wall biogenesis, and response to chitin, which was similar to the enrichment found in CSIL-35431 and CSIL-31010. But at 15 and 20 DPA in the stronger fiber line CSIL-31134, GO enrichments were different from the other three lines, mainly in genes for ATP synthesis, proton transport, copper ion export, and oxidoreductase activity, but not in cell wall biosynthesis (Table S5).

Based on the results of GO analysis, we know that the secondary cell wall related biological process were impacted in the CSILs, but it is still not very clear how secondary cell wall biosynthesis was affected in the CSILs. Therefore, we performed pathway analysis on 18,288 DEGs in CSIL-35431 and CSIL-31010. The most highly enriched pathways found are listed in Table 3. KEGG analysis showed that the genes were enriched in pathways for starch and sucrose metabolism (328 genes), glycolysis/gluconeogenesis (122 genes), phenylpropanoid biosynthesis (101 genes), and

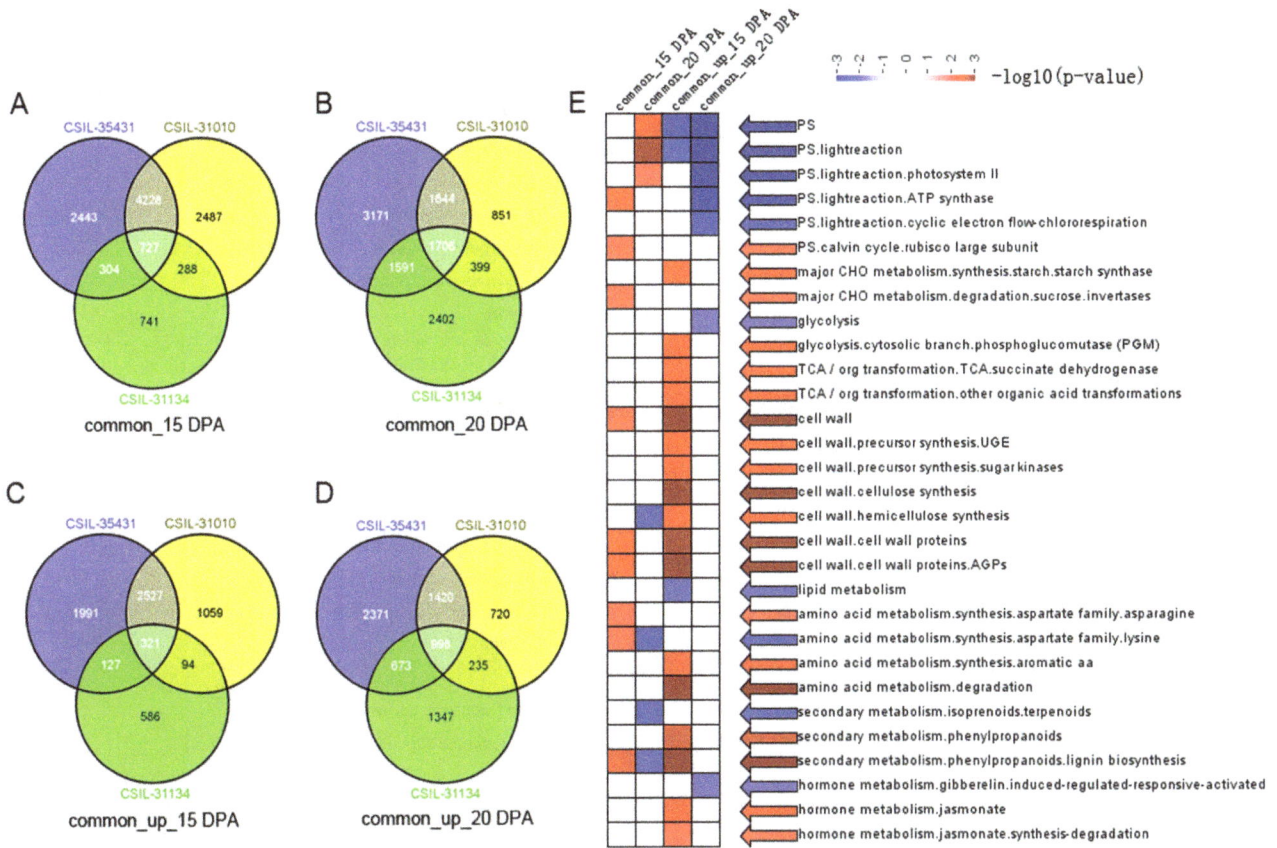

Figure 3. Analysis of common and common upregulated DEGs among three stronger fiber CSILs. (A, B, C, D) Common and common upregulated DEGs among three stronger fiber CSILs at 15 and 20 DPA. Common_up, common regulated DEGs. (B) Functional enrichment analysis of these DEGs using mapman software (Summary statistic type, wlcoxon). Colors from blue to red indicated that functions were enriched more significantly with smaller p-values.

oxidative phosphorylation (87 genes) (Table 3 and Figure S3). The regulation of some enzymes that catalyze sucrose, starch, and cellulose biosynthesis may have a direct or indirect impact on fiber quality. This could be especially true for sucrose and pectin metabolism, and many genes in these pathways were up-regulated. We also found that genes involved in phenylpropanoid and flavonoid biosynthetic processes were enriched in the CSILs.

Based on the cluster analysis of the weaker fiber line CSIL-35368, we hypothesized that changes in other biochemical pathways led to reduced fiber strength (Figure S1). Considering only those that were down-regulated in CSIL-35368, we found

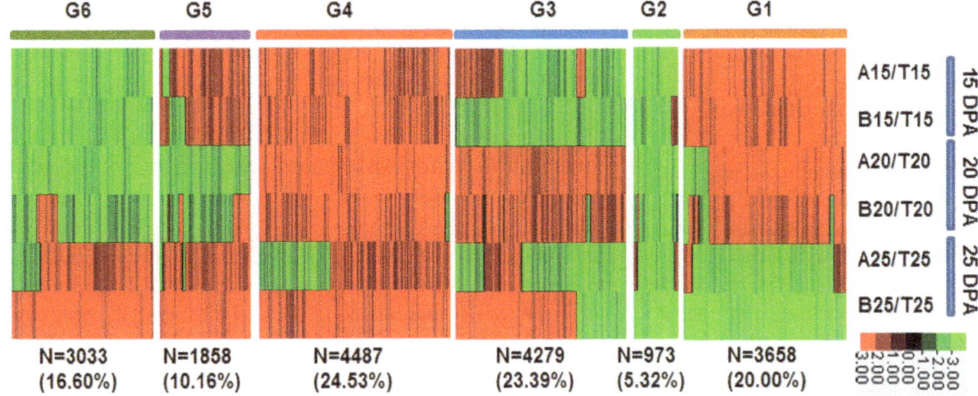

Figure 4. Heat map analysis of the expression of DEGs between CSILs and TM-1. A, B and T indicated CSIL-35431, CSIL-31010 and TM-1, respectively. 15, 15DPA; 20, 20DPA; 25, 25DPA. Red color indicated up-regulated genes and green color indicated down-regulated genes. N = number of DEGs in different group.

Table 2. Enrichment analysis of gene ontologies from 15 to 25 DPA.

Cluster	GO-ID	GO Ontology (Biological process)
G1	GO:0010417	glucuronoxylan biosynthetic process
15DPA up-regulated	GO:0009834	secondary cell wall biogenesis
20DPA up-regulated	GO:0007018	microtubule-based movement
25DPA down-regulated	GO:0030244	cellulose biosynthetic process
	GO:0009753	response to jasmonic acid stimulus
G2	GO:0015031	protein transport
15DPA down-regulated	GO:0015991	ATP hydrolysis
20DPA down-regulated	GO:0032544	plastid translation
25DPA down-regulated	GO:0016075	rRNA catabolic process
	GO:0006511	ubiquitin catabolic process
G3	GO:0009734	auxin mediated signaling pathway
15DPA down-regulated	GO:0030259	lipid glycosylation
20DPA up-regulated	GO:0018106	peptidyl-histidine phosphorylation
25DPA down-regulated	GO:0009722	detection of cytokinin stimulus
G4	GO:0006468	protein phosphorylation
15DPA up-regulated	GO:0010200	response to chitin
20DPA up-regulated	GO:0006096	glycolysis
25DPA up-regulated	GO:0010182	sugar mediated signaling pathway
	GO:0009966	regulation of signal transduction
G5	GO:0015884	protein folding
15DPA up-regulated	GO:0006886	intracellular protein transport
20DPA down-regulated	GO:0006122	mitochondrial electron transport
25DPA up-regulated	GO:0015914	phospholipid transport
G6	GO:0007267	cell-cell signaling
15DPA down-regulated	GO:0010025	wax biosynthetic process
20DPA down-regulated	GO:0006633	fatty acid biosynthetic process
25DPA up-regulated	GO:0006723	hydrocarbon biosynthetic process

G1–G6 according to Figure 3.

genes that participated in fatty acid degradation and biosynthesis, and also in carbohydrate metabolic pathways (Figure 2B and Figure S4).

Eight genes previously reported in the carbohydrate pathway were selected for quantitative RT-PCR. The expression patterns of these genes were consistent with the DEG data in TM-1 (Figure 5) and in the CSILs as well (Figure 6B and Figure 7B).

Carbohydrate metabolism in the secondary cell wall synthesis stage

Following the start of secondary cell wall formation, protein and carbohydrate metabolism genes involved in cell wall biosynthesis will be up-regulated [27]. We selected 72 DEGs associated with carbohydrate metabolism to investigate the mechanism of fiber development. These genes were related to pectin, sucrose, galactan, glucan, xyloglucan, and cellulose biosynthesis. We were interested in genes that are up-regulated in fiber cells at 15 DPA and 20 DPA, at the start of secondary cell wall formation. A heat map showing the different expression levels for these genes including cellulose synthase, sucrose synthase, pectin lyase, and other polysaccharides degradation in CSIL-35431 and CSIL-31010 is shown in Figure 6A. We found that the cellulose synthase genes were up-regulated in the CSILs at 15 DPA-25 DPA. It has been reported that cellulose biosynthesis predominates, and that many other metabolic pathways are down-regulated during secondary cell wall synthesis [27]. Moreover, we confirmed the expression patterns of cellulose synthase genes, annotated with the *Arabidopsis* genes *AtCESA4*, *AtCESA7* and *AtCESA8*, using quantitative RT-PCR (Figure 4B). Proteins encoded by *AtCESA4*, *7*, and *8* are specifically required to form a functional cellulose synthase complex (CSC) that is essential for secondary cell wall formation [40–42].

Transcription factors associated with secondary cell wall synthesis

Recent molecular and genetic studies have identified transcription factors that are involved in regulating secondary cell wall synthesis in *Arabidopsis* [43–45]. In our study, 97 MYB-type and 68 NAC-type transcription factors showed changes in expression between the CSILs and TM-1 (Table S6, Table S7). It was interesting that some NACs and MYBs were up-regulated in CSIL-35431 and CSIL-31010 during the secondary cell wall synthesis stage, especially at 15 DPA and 20 DPA. Defined as $|\log_2\text{Ratio}|\geq 2$, 59 MYB and 47 NAC transcription factors were selected for heat-map analysis (Figure 7A). Among these transcription factors, genes homologous with *ATMYB2*, *AT-*

Table 3. KEGG analysis of DEGs in CSIL-35431 and CSIL-31010.

Pathway	DEGs with pathway annotation (2576)	References genes with pathway annotation(4601)	Ratio	DEGs distribution in each groups					
				G1	G2	G3	G4	G5	G6
Starch and sucrose metabolism	328	563	58.26%	87	15	54	89	31	52
Purine metabolism	251	459	54.68%	35	20	53	65	30	48
Phenylalanine metabolism	140	261	53.64%	29	5	26	47	12	21
Amino sugar and nucleotide sugar metabolism	137	211	64.93%	47	5	23	36	12	14
Pyrimidine metabolism	128	227	56.39%	15	13	26	37	15	22
Glycolysis/Gluconeogenesis	122	188	64.89%	28	4	23	35	13	19
T cell receptor signaling pathway	115	212	54.25%	28	4	23	34	11	15
Pentose and glucuronate interconversions	109	227	48.02%	24	8	20	27	7	23
Glycerolipid metabolism	102	174	58.62%	18	4	29	21	15	15
Pyruvate metabolism	102	182	56.04%	16	8	18	29	12	19
Phenylpropanoid biosynthesis	101	220	45.91%	26	6	17	28	8	16
Galactose metabolism	94	160	58.75%	23	2	20	24	7	18
Cysteine and methionine metabolism	94	145	64.83%	21	2	21	28	10	12
Glycerophospholipid metabolism	94	167	56.29%	14	8	29	21	10	12
Arginine and proline metabolism	93	144	64.58%	16	5	22	19	12	19
Oxidative phosphorylation	87	184	47.28%	19	13	12	11	13	19
Carbon fixation in photosynthetic organisms	84	151	55.63%	19	3	11	34	8	9
Fatty acid degradation	83	133	62.41%	17	5	18	21	7	15

G1–G6 according to Figure 3.

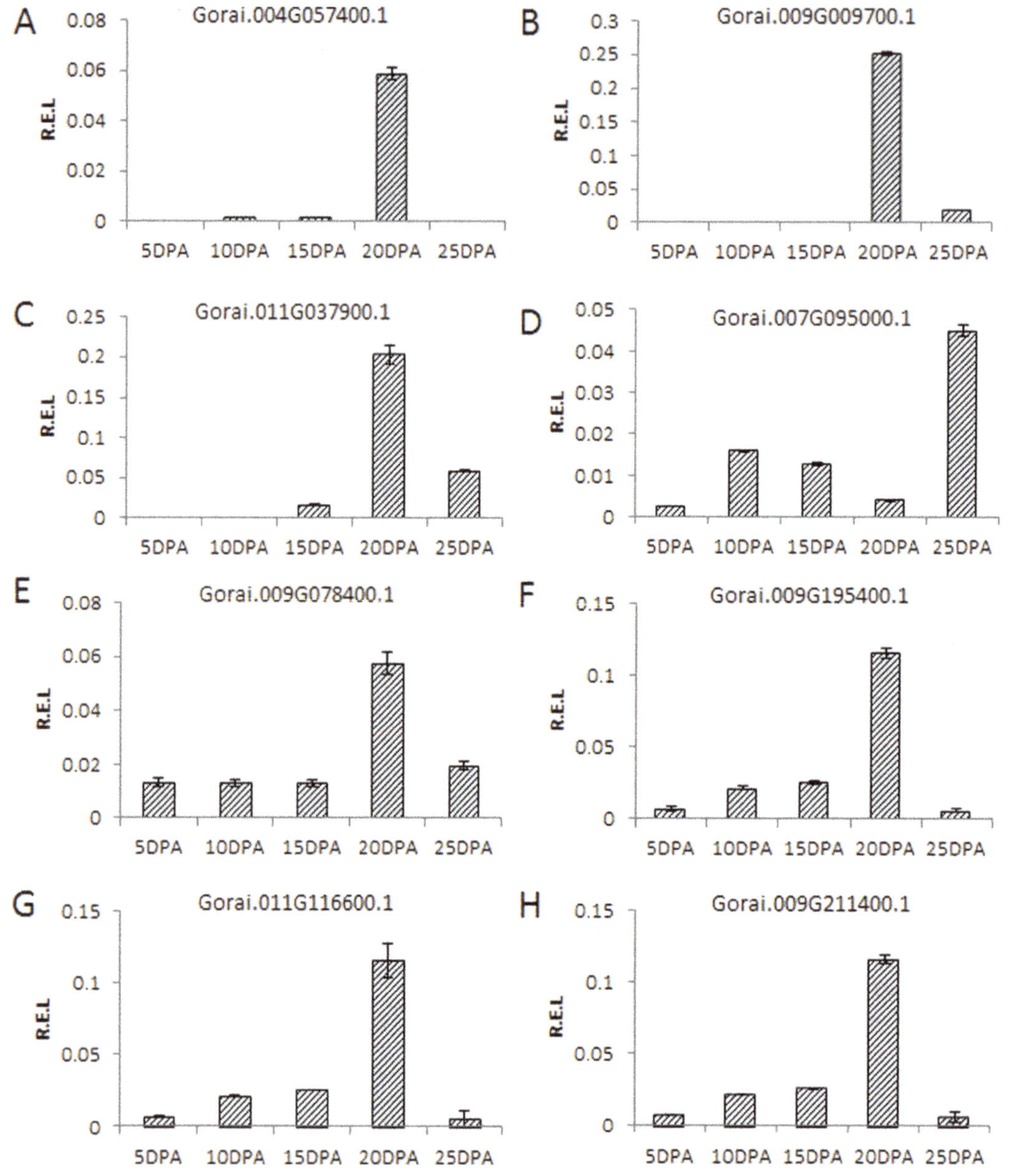

Figure 5. Quantitative RT–PCR validation of tag-mapped genes in TM-1. These genes have been reported before, including 3 CesA genes (A,B,C) (homologous with AtCESA4, AtCESA7, AtCESA8, respectively), xyloglucan endotransglucosylase (D), beta -galactosidase (E), glycosyl hydrolase 9B7 (F), xylan alpha-glucuronosyltransferase 1, GUX1 (G), xylan alpha-glucuronosyltransferase 2, GUX2 (H).

MYB43, *ATMYB73*, *ATNAC52*, and *ATNAC61* were expressed at higher levels in the CSILs. We confirmed that three transcription factors were up-regulated in CSILs from 15 DPA to 25 DPA (Figure 7B). In the MYB family, it has been reported that the expression of genes for MYB85, MYB52, MYB54, MYB69, MYB42, and MYB43 are developmentally associated with cells undergoing secondary wall thickening [45].

Different metabolic pathways associated with altered fiber strength

In order to investigate the mechanisms underlying changes in fiber strength, we analyzed several metabolic pathways including cell wall, lipids, minor CHO (carbohydrate) metabolism, and two secondary metabolite pathways. It is interesting that DEGs involved in cell wall proteins, cell wall pectin esterase, cell wall modification, cell wall cellulose synthesis, cell wall degradation/pectate lyases, lipid metabolism/FA synthesis, and lipid degradation showed distinct expression patterns or differential up/down-regulation at 20 DPA (Figure 8A). We found that up-regulated DEGs were similar to down-regulated DEGs both in CSIL-35431 and CSIL-35368. However, most of DEGs in CSIL-31010 were up-regulated at 20 DPA, while the opposite was true for DEGs in CSIL-31134, especially those genes involved in cell wall modification. In CSIL-31134, we also found a few genes in these metabolic pathways that were changed at 15 DPA except in cell wall modification, and in CSIL-31010, we found DEGs enriched in these metabolic pathways at 25 DPA (Figure S5). From the secondary metabolism results, we identified a few DEGs involved in flavonoid biosynthesis in CSIL-35431 and CSIL-31010 at 15

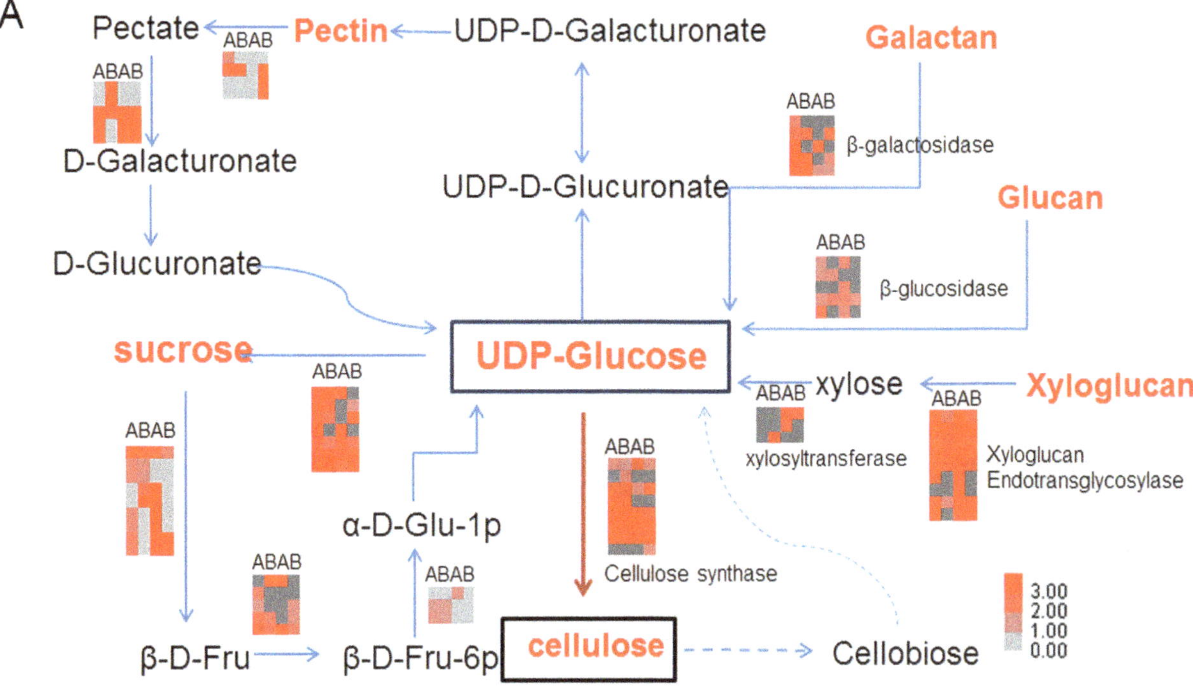

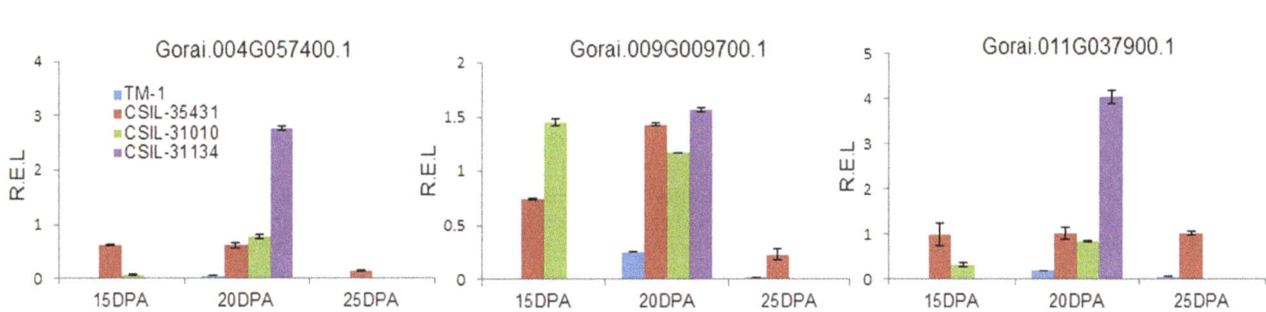

Figure 6. Carbohydrate pathways that are differentially regulated during the secondary cell wall synthesis stage. (A) Carbohydrate pathways. Genes up-regulated in CSIL-315431 and CSIL-31010 were selected to do heat map. ABAB indicated DEGs in CSIL-35431 at 15 DPA, CSIL-35431 at 20DPA, CSIL-3010 at 15DPA and CSIL-31010 at 20DPA, from left to right. Every square stand for one gene and every line stand for the same gene. Genes with red color expressed higher in CSILs than TM-1 and gray color stand for no difference. β-D-Fru, β-D-Fructose; α-D-Glu-1p, α-D-Glucose-1-phosphate; β-D-Fru-6p, β-D-Fructose-6-phosphate. (B) Quantitative RT–PCR validation of four CesA genes in CSILs and TM-1, Gorai.004G057400.1, Gorai.009G009700.1 and Gorai.011G037900.1 homologous with *AtCESA4*, *AtCESA7* and *AtCESA8*, respectively.

DPA. In contrast, more genes were up-regulated or down-regulated in CSIL-35368 at 15 DPA. It was obvious that DEGs from the phenylpropanoid pathways at 25 DPA were different from one another, and the expression pattern of DEGs in CSIL-31010 changed dramatically. Moreover, there were few genes that were up-regulated or down-regulated in CSIL-35368 at 25 DPA (Figure 7B). We assume that metabolic pathways in the CSILs at different developmental stages were changed in various ways as a result of the introgressed chromosmal segments from *G. barbadense*.

Discussion

G. hirsutum produces a high yield of cotton with moderate fiber strength. *G. barbadense* is characterized by a low yield, but with increased fiber fineness and strength. As a breeding target, we tried to combine the high yield of *G. hirsutum* with the superior fiber qualities of *G. barbadense*, and we also wanted to elucidate the molecular mechanism behind the formation of superior quality fibers. Fiber strength is an important indicator of the cotton fiber quality, and depends on the formation of the secondary cell wall. Genome-wide transcriptome profiling is effective at revealing significant genes and pathways involved in secondary cell wall formation. Transcriptome analysis showed that gene expression patterns and functional distribution were different during secondary cell wall biosynthesis.

Carbohydrate metabolism plays an important role in secondary cell wall synthesis

It is well known that the mature cotton fiber is composed of nearly pure cellulose, and that such a high level of cellulose synthesis requires an abundant supply of UDP-glucose [46,47]. This means that a large amount of cellulose is required during the

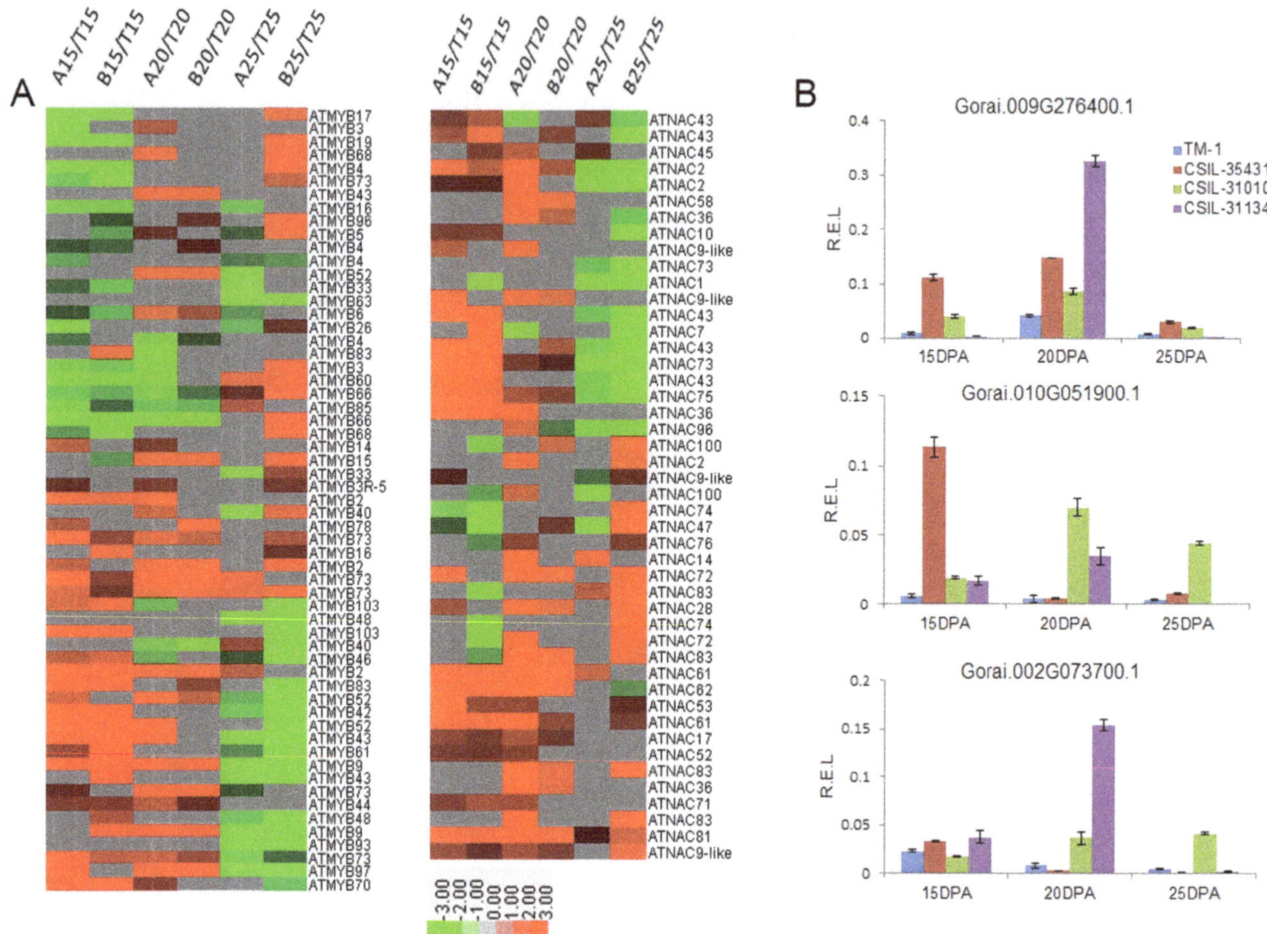

Figure 7. NAC and MYB family genes involved in the regulation of secondary wall biosynthesis. (A) 59 MYB family genes and 47 NAC family genes showed different expression level between CSILs and TM-1 at 15DPA, 20DPA and 25DPA. |Ratio|>2 and FDR<0.001. A, B, T indicated CSIL-35431, CSIL-31010 and TM-1. 15, 15DPA; 20, 20DPA; 25, 25DPA. (B) Quantitative RT–PCR validation of three transcription factors.

secondary cell wall synthesis stage. Functional classification and enrichment analysis showed that following the initiation of secondary cell wall synthesis, DEGs were enriched for secondary cell wall biogenesis, glucuronoxylan biological processes, and other carbohydrate metabolic pathways in the CSILs (Table 2). Focusing on carbohydrate metabolic pathways, it is obvious that the key intermediate in the multiple pathways is UDP-glucose, a substrate for cellulose synthesis. Our results showed that several CesA genes are expressed at higher levels during secondary cell wall synthesis than they are at earlier stages (Figure 6B). Ten *AtCESA* genes have been reported in *Arabidopsis*, and *AtCESA4*, *7*, and *8* are specifically required to form the cellulose synthase complex (CSC) that is essential for secondary cell wall synthesis [40–42]. Similarly, three CESA isoforms have been identified during secondary cell wall synthesis in rice, maize, and *Populus* [10,48,49]. Also, many genes that participate in the degradation of poly- and oligo-saccharides were found to be up-regulated at 15 and 20 DPA, in order to produce more UDP-glucose for cellulose biosynthesis. Similarly, it has also been reported that during the secondary cell wall synthesis stage, certain metabolic pathways, including hydrolysis of fatty acids and non-cellulose poly- and oligo-saccharides, would be up-regulated [27]. Sucrose synthase (SuSy) has long been studied as a cytoplasmic enzyme in plant cells, where it serves to degrade sucrose and provide carbon for respiration and synthesis of cell wall polysaccharides and starch [50]. It has also been shown that genes associated with secondary cell wall biosynthesis are involved in sugar metabolism [51].

Multiple mechanisms affect fiber strength development

Except for carbohydrate metabolism, recent research has shown that transcription factors also affect fiber development during secondary cell wall biosynthesis. Several NAC- and MYB-type transcription factors were up-regulated in the CSILs compared to TM-1 from 15 DPA to 25 DPA, and these included cotton homologs of *AtMYB2*, *AtMYB43*, and *AtNAC52* etc. (Figure 7A). The NAC-mediated transcriptional regulation of secondary wall biosynthesis is a conserved mechanism throughout vascular plants [44,52]. *SND2*, a NAC transcription factor gene, regulates genes involved in secondary cell wall development in *Arabidopsis* fibers and increases fiber cell area in *Eucalyptus* [53]. A MYB75-associated protein complex is likely to be involved in modulating secondary cell wall biosynthesis in both the *Arabidopsis* inflorescence and stem [54]. It has also been found that the rice and maize MYB transcription factors, OsMYB46 and ZmMYB46, are functional orthologs of *Arabidopsis* MYB46/MYB83 and, when overexpressed in *Arabidopsis*, are able to activate the entire secondary wall biosynthetic program [55].

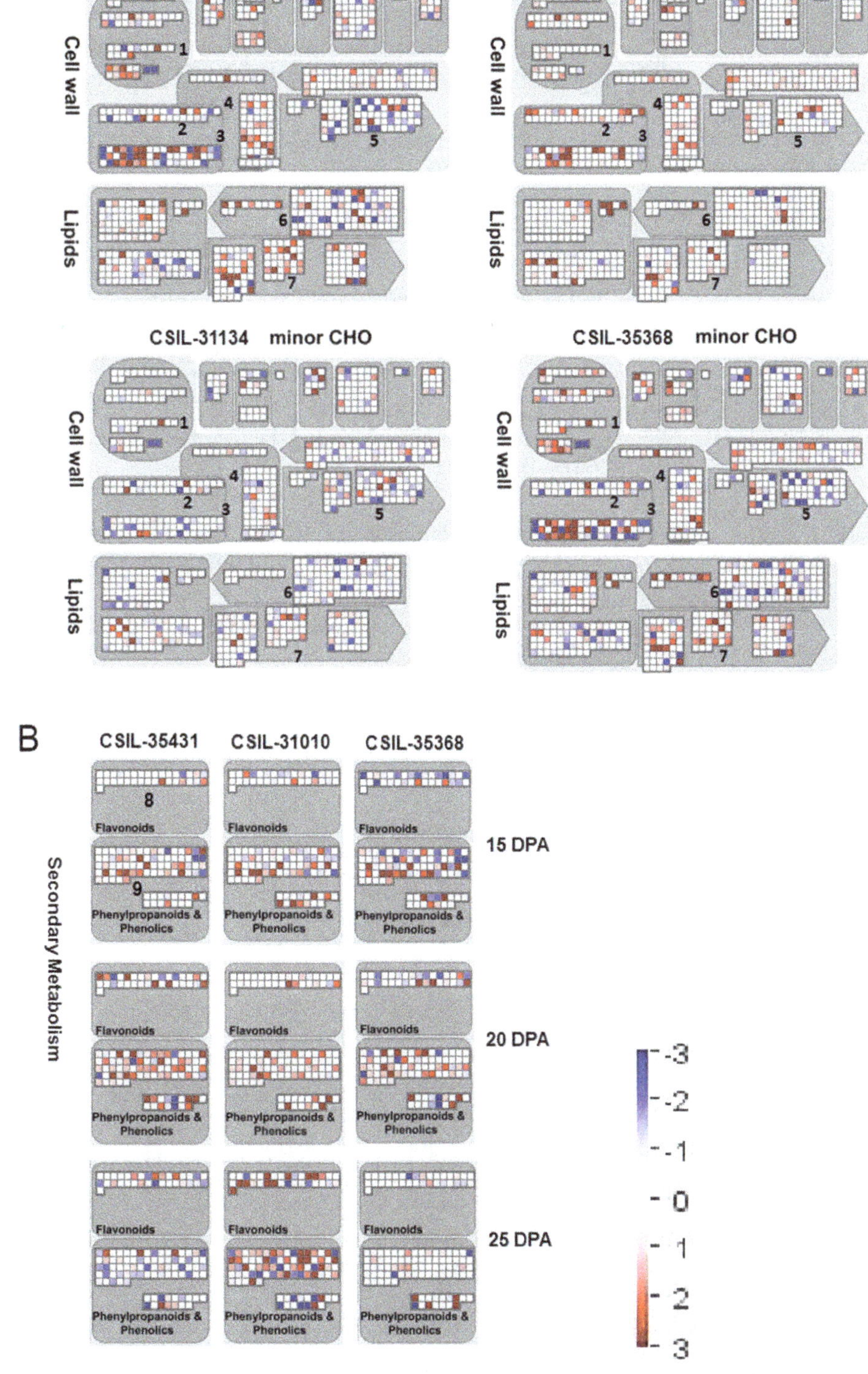

Figure 8. Metabolism analysis of DEGs in CSILs during the secondary cell wall biosynthesis stage. (A) Motabolism overview in four CSILs at 20 DPA. (B) Secondary motabolism analysis in three CSILs at 15 DPA, 20 DPA and 25 DPA. 1, cell wall protein; 2, cell wall pectin esterases; 3, cell wall modification; 4, cell wall cellulose synthesis; 5, cell wall degradation/pectate lyases; 6, lipid metabolism/FA synthesis; 7, lipid degradation; 8, flavonoids; 9, phenylpropanoids/lignin biosynthesis. Blue square, down-regulated genes; Red square, up-regulated genes.

Several metabolic pathways were examined to determine the mechanism behind changes in fiber strength; these included cell wall, lipids, minor CHO metabolism, and two secondary metabolic pathways. Although results of the GO and KEGG analyses showed that CSIL-35431, CSIL-31010, and CSIL-35368 had similar patterns, fiber strength in these three lines were different. Our results support the hypothesis that different metabolic pathways can affect fiber strength, and the same pathway in the CSILs can be altered differentially at various times in development. DEGs in CSIL-31010 were up-regulated at 20 DPA, while the opposite was found for DEGs in CSIL-31134, especially those genes involved in cell wall modification. The expression levels of genes involved in flavonoid biosynthesis in the weak fiber line CSIL-35368 were changed dramatically at 15 DPA, but there were few genes changed at 25 DPA; this patter was the opposite of that in CSIL-35431 and CSIL-31010, lines with high quality fiber. We hypothesize that phenylpropanoid and flavonoid metabolism generally affected the fiber strength of CSIL-35368. Genes for phenylpropanoid and flavonoid biosynthesis showed significant enrichment and temporal differences in gene expression patterns which are associated with xylem formation [56]. It has been reported that expression levels of phenylpropanoid genes showed high correlations with specific fiber properties, supporting a role in determining fiber strength [57].

In conclusion, upland cotton CSILs carrying distinct *G. barbadense* chromosomal segments provide valuable material for research into fiber development. The *G. barbadense* chromosome segments resulted in different patterns of differentially expressed genes, and altered different metabolic pathways, mainly in carbohydrate metabolism. In addition, several transcription factor genes were found to be specifically up-regulated in the CSILs. Metabolic pathways involved in cell wall, lipid, phenylpropanoid, and flavonoid biosynthesis play a significant role during secondary cell wall formation, and are associated with the development of cotton fiber strength.

Supporting Information

Figure S1 Heat map of the expression of DEGs between 4 CSILs at 15–25 DPA.

Figure S2 Enrichment analysis of common DEGs at 15DPA and 20DPA in CSIL-35431 and CSIL-31010.

Figure S3 Heat map of DEGs participated in four metabolic pathways from 15 DPA to 25 DPA.

Figure S4 Pathway analysis of genes only down-regulated in CSIL-35368 from 15 DPA to 25 DPA.

Figure S5 Metabolism analysis of DEGs in CSILs at 15 DPA and 25 DPA.

Table S1 Average fiber quality of 4 CSILs and TM-1.

Table S2 Primer for quantitative RT-PCR.

Table S3 Categorization and abundance of tags.

Table S4 List of common DEGs among CSIL-35431, CSIL-31134 and CSIL-31010.

Table S5 Enrichment analysis of gene ontologies in CSIL-35368 and CSIL-31010 at 15 DPA and 20 DPA.

Table S6 Different expression level of 97 MYB transcription factors.

Table S7 Different expression level of 68 NAC transcription factors.

Author Contributions

Conceived and designed the experiments: TZ. Performed the experiments: LF RT SW XL PW. Analyzed the data: LF JC. Wrote the paper: LF TZ.

References

1. Basara AS, Malik CP (1984) Development of cotton fiber. Inter Rev Cyto. pp. 65–113.
2. Haigler TA, Jernstedt JA (1999) Molecular genetics of developing cotton fibers. In: AM Basra (Ed), Cotton Fibers. Hawthorne Press, New York, 231–267.
3. Kim HJ, Triplett BA (2001) Cotton fiber growth in planta and in vitro. Models for plant cell elongation and cell wall biogenesis. Plant Physiol 127: 1361–1366.
4. Lee JJ, Hassan OS, Gao W, Wei NE, Kohel RJ, et al. (2006) Developmental and gene expression analyses of a cotton naked seed mutant. Planta 223: 418–432.
5. Lee JJ, Woodward AW, Chen ZJ (2007) Gene expression changes and early events in cotton fibre development. Ann Bot 100: 1391–1401.
6. Wilkins TA, Arpat AB (2005) The cotton fiber transcriptome. Physiol Plant 124: 295–300.
7. Meinert MC, Delmer DP (1977) Changes in biochemical composition of the cell wall of the cotton fiber during development. Plant Physiol 59: 1088–1097.
8. Bolton JJ, Soliman KM, Wilkins TA, Jenkins JN (2009) Aberrant Expression of Critical Genes during Secondary Cell Wall Biogenesis in a Cotton Mutant, Ligon Lintless-1 (Li-1). Comp Funct Genom: 659301.
9. Richmond TA, Somerville CR (2000) The cellulose synthase superfamily. Plant Physiol 124: 495–498.
10. Tanaka K, Murata K, Yamazaki M, Onosato K, Miyao A, et al. (2003) Three distinct rice cellulose synthase catalytic subunit genes required for cellulose synthesis in the secondary wall. Plant Physiol 133: 73–83.
11. Paterson AH, Wendel JF, Gundlach H, Guo H, Jenkins J, et al. (2012) Repeated polyploidization of Gossypium genomes and the evolution of spinnable cotton fibres. Nature 492: 423–427.
12. Persson S, Wei H, Milne J, Page GP, Somerville CR (2005) Identification of genes required for cellulose synthesis by regression analysis of public microarray data sets. Proc Natl Acad Sci USA 102: 8633–8638.
13. Ruan YL, Chourey PS, Delmer DP, Perez-Grau L (1997) The Differential Expression of Sucrose Synthase in Relation to Diverse Patterns of Carbon Partitioning in Developing Cotton Seed. Plant Physiol 115: 375–385.
14. Brill E, van Thournout M, White RG, Llewellyn D, Campbell PM, et al. (2011) A novel isoform of sucrose synthase is targeted to the cell wall during secondary cell wall synthesis in cotton fiber. Plant Physiol 157: 40–54.
15. Potikha TS, Collins CC, Johnson DI, Delmer DP, Levine A (1999) The involvement of hydrogen peroxide in the differentiation of secondary walls in cotton fibers. Plant Physiol 119: 849–858.
16. Yang YM, Xu CN, Wang BM, Jia JZ (2001) Effects of plant growth regulators on secondary wall thickening of cotton fibres. Plant Growth Regul 35: 233–237.
17. Arpat AB, Waugh M, Sullivan JP, Gonzales M, Frisch D, et al. (2004) Functional genomics of cell elongation in developing cotton fibers. Plant Mol Biol 54: 911–929.
18. Wang QQ, Liu F, Chen XS, Ma XJ, Zeng HQ, et al. (2010) Transcriptome profiling of early developing cotton fiber by deep-sequencing reveals significantly differential expression of genes in a fuzzless/lintless mutant. Genomics 96: 369–376.
19. Padmalatha KV, Dhandapani G, Kanakachari M, Kumar S, Dass A, et al. (2012) Genome-wide transcriptomic analysis of cotton under drought stress reveal significant down-regulation of genes and pathways involved in fibre elongation and up-regulation of defense responsive genes. Plant Mol Biol 78: 223–246.
20. Chaudhary B, Hovav R, Rapp R, Verma N, Udall JA, et al. (2008) Global analysis of gene expression in cotton fibers from wild and domesticated *Gossypium barbadense*. Evol Dev 10: 567–582.
21. Hovav R, Udall JA, Chaudhary B, Hovav E, Flagel L, et al. (2008) The evolution of spinnable cotton fiber entailed prolonged development and a novel metabolism. PLoS Genet 4: e25.
22. Hovav R, Udall JA, Hovav E, Rapp R, Flagel L, et al. (2008) A majority of cotton genes are expressed in single-celled fiber. Planta 227: 319–329.
23. Ji SJ, Lu YC, Feng JX, Wei G, Li J, et al. (2003) Isolation and analyses of genes preferentially expressed during early cotton fiber development by subtractive PCR and cDNA array. Nucleic Acids Res 31: 2534–2543.

24. Udall JA, Flagel LE, Cheung F, Woodward AW, Hovav R, et al. (2007) Spotted cotton oligonucleotide microarrays for gene expression analysis. BMC Genomics 8: 81.
25. Qin YM, Hu CY, Pang Y, Kastaniotis AJ, Hiltunen JK, et al. (2007) Saturated very-long-chain fatty acids promote cotton fiber and *Arabidopsis* cell elongation by activating ethylene biosynthesis. Plant Cell 19: 3692–3704.
26. Shi YH, Zhu SW, Mao XZ, Feng JX, Qin YM, et al. (2006) Transcriptome profiling, molecular biological, and physiological studies reveal a major role for ethylene in cotton fiber cell elongation. Plant Cell 18: 651–664.
27. Gou JY, Wang LJ, Chen SP, Hu WL, Chen XY (2007) Gene expression and metabolite profiles of cotton fiber during cell elongation and secondary cell wall synthesis. Cell Res 17: 422–434.
28. Eshed Y, Zamir D (1995) An introgression line population of *Lycopersicon pennellii* in the cultivated tomato enables the identification and fine mapping of yield-associated QTL. Genetics 141: 1147–1162.
29. Liu S, Zhou R, Dong Y, Li P, Jia J (2006) Development, utilization of introgression lines using a synthetic wheat as donor. Theor Appl Genet 112: 1360–1373.
30. Takai T, Nonoue Y, Yamamoto SI, Yamanouchi U, Matsubara K, et al. (2007) Development of chromosome segment substitution lines derived from backcross between indica donor rice cultivar '*Nona bokra*' and japonica recipient cultivar '*Koshihikari*'. Breeding Sci 57: 257–261.
31. Wang P, Ding YZ, Lu QX, Guo WZ, Zhang TZ (2008) Development of *Gossypium barbadense* chromosome segment substitution lines in the genetic standard line TM-1 of *Gossypium hirsutum*. Chi Sci Bull 53: 1512–1517.
32. Kohel R, Richmond T, Lewis C (1970) Texas marker-1. Description of a genetic standard for *Gossypium hirsutum* L. Crop Sci 10: 670–671.
33. Pan J, Zhang T, Kuai B (1994) Studies on the inheritance of resistance to *Verticillium* dahliae in cotton. J Nanj Agric Univ 17.
34. Yang C, Guo W, Li G, Gao F, Lin S, et al. (2008) QTLs mapping for *Verticillium* wilt resistance at seedling and maturity stages in *Gossypium barbadense* L. Plant Sci 174: 290–298.
35. Wang P, Zhu Y, Song X, Cao Z, Ding Y, et al. (2012) Inheritance of long staple fiber quality traits of *Gossypium barbadense* in *G. hirsutum* background using CSILs. Theor Appl Genet 124: 1415–1428.
36. Jiang JX, Zhang TZ (2003) Extraction of total RNA in cotton tissues with CTAB-acidic phenolic method. Cott Sci 15: 166–167.
37. Benjamini Y, Yekutieli D (2001) The control of the false discovery rate in multiple testing under dependency. Ann Stat: 1165–1188.
38. Herrero J, Valencia A, Dopazo J (2001) A hierarchical unsupervised growing neural network for clustering gene expression patterns. Bioinformatics 17(2):126–136
39. Kanehisa M, Araki M, Goto S, Hattori M, Hirakawa M, et al. (2008) KEGG for linking genomes to life and the environment. Nucleic Acids Res 36: D480–D484.
40. Taylor NG, Howells RM, Huttly AK, Vickers K, Turner SR (2003) Interactions among three distinct CesA proteins essential for cellulose synthesis. Proc Natl Acad Sci USA 100: 1450–1455.
41. Taylor NG, Laurie S, Turner SR (2000) Multiple cellulose synthase catalytic subunits are required for cellulose synthesis in *Arabidopsis*. Plant Cell 12: 2529–2540.
42. Taylor NG, Scheible WR, Cutler S, Somerville CR, Turner SR (1999) The irregular xylem3 locus of Arabidopsis encodes a cellulose synthase required for secondary cell wall synthesis. Plant Cell 11: 769–780.
43. Olsen AN, Ernst HA, Leggio LL, Skriver K (2005) NAC transcription factors: structurally distinct, functionally diverse. Trends Plant Sci 10: 79–87.
44. Zhong R, Lee C, Ye ZH (2010) Functional characterization of poplar wood-associated NAC domain transcription factors. Plant Physiol 152: 1044–1055.
45. Zhong R, Lee C, Zhou J, McCarthy RL, Ye ZH (2008) A battery of transcription factors involved in the regulation of secondary cell wall biosynthesis in *Arabidopsis*. Plant Cell 20: 2763–2782.
46. Delmer DP, Amor Y (1995) Cellulose biosynthesis. Plant Cell 7: 987–1000.
47. Delmer DP, Haigler CH (2002) The regulation of metabolic flux to cellulose, a major sink for carbon in plants. Metab Eng 4: 22–28.
48. Appenzeller L, Doblin M, Barreiro R, Wang HY, Niu XM, et al. (2004) Cellulose synthesis in maize: isolation and expression analysis of the cellulose synthase (CesA) gene family. Cellulose 11: 287–299.
49. Song DL, Shen JH, Li LG (2010) Characterization of cellulose synthase complexes in *Populus* xylem differentiation. New Phytol 187: 777–790.
50. Amor Y, Haigler CH, Johnson S, Wainscott M, Delmer DP (1995) A membrane-associated form of sucrose synthase and its potential role in synthesis of cellulose and callose in plants. Proc Natl Acad Sci USA 92: 9353–9357.
51. Hinchliffe DJ, Meredith WR, Yeater KM, Kim HJ, Woodward AW, et al. (2010) Near-isogenic cotton germplasm lines that differ in fiber-bundle strength have temporal differences in fiber gene expression patterns as revealed by comparative high-throughput profiling. Theor Appl Genet 120: 1347–1366.
52. Zhong R, Lee C, Ye ZH (2010) Evolutionary conservation of the transcriptional network regulating secondary cell wall biosynthesis. Trends Plant Sci 15: 625–632.
53. Hussey SG, Mizrachi E, Spokevicius AV, Bossinger G, Berger DK, et al. (2011) SND2, a NAC transcription factor gene, regulates genes involved in secondary cell wall development in *Arabidopsis* fibres and increases fibre cell area in Eucalyptus. BMC Plant Biol 11: 173.
54. Bhargava A, Ahad A, Wang S, Mansfield SD, Haughn GW, et al. (2013) The interacting MYB75 and KNAT7 transcription factors modulate secondary cell wall deposition both in stems and seed coat in *Arabidopsis*. Planta 237: 1199–1211.
55. Zhong R, Lee C, McCarthy RL, Reeves CK, Jones EG, et al. (2011) Transcriptional activation of secondary wall biosynthesis by rice and maize NAC and MYB transcription factors. Plant Cell Physiol 52: 1856–1871.
56. Brown DM, Zeef LA, Ellis J, Goodacre R, Turner SR (2005) Identification of novel genes in *Arabidopsis* involved in secondary cell wall formation using expression profiling and reverse genetics. Plant Cell 17: 2281–2295.
57. Al-Ghazi Y, Bourot S, Arioli T, Dennis ES, Llewellyn DJ (2009) Transcript profiling during fiber development identifies pathways in secondary metabolism and cell wall structure that may contribute to cotton fiber quality. Plant Cell Physiol 50: 1364–1381.

13

Morphological and Physiological Responses of Cotton (*Gossypium hirsutum* L.) Plants to Salinity

Lei Zhang, Huijuan Ma, Tingting Chen, Jun Pen, Shuxun Yu*, Xinhua Zhao*

State Key Laboratory of Cotton Biology, Institute of Cotton Research of CAAS, Anyang, P. R. China

Abstract

Salinization usually plays a primary role in soil degradation, which consequently reduces agricultural productivity. In this study, the effects of salinity on growth parameters, ion, chlorophyll, and proline content, photosynthesis, antioxidant enzyme activities, and lipid peroxidation of two cotton cultivars, [CCRI-79 (salt tolerant) and Simian 3 (salt sensitive)], were evaluated. Salinity was investigated at 0 mM, 80 mM, 160 mM, and 240 mM NaCl for 7 days. Salinity induced morphological and physiological changes, including a reduction in the dry weight of leaves and roots, root length, root volume, average root diameter, chlorophyll and proline contents, net photosynthesis and stomatal conductance. In addition, salinity caused ion imbalance in plants as shown by higher Na^+ and Cl^- contents and lower K^+, Ca^{2+}, and Mg^{2+} concentrations. Ion imbalance was more pronounced in CCRI-79 than in Simian3. In the leaves and roots of the salt-tolerant cultivar CCRI-79, increasing levels of salinity increased the activities of superoxide dismutase (SOD), ascorbate peroxidase (APX), and glutathione reductase (GR), but reduced catalase (CAT) activity. The activities of SOD, CAT, APX, and GR in the leaves and roots of CCRI-79 were higher than those in Simian 3. CAT and APX showed the greatest H_2O_2 scavenging activity in both leaves and roots. Moreover, CAT and APX activities in conjunction with SOD seem to play an essential protective role in the scavenging process. These results indicate that CCRI-79 has a more effective protection mechanism and mitigated oxidative stress and lipid peroxidation by maintaining higher antioxidant activities than those in Simian 3. Overall, the chlorophyll a, chlorophyll b, and Chl (a+b) contents, net photosynthetic rate and stomatal conductance, SOD, CAT, APX, and GR activities showed the most significant variation between the two cotton cultivars.

Editor: Wagner L. Araujo, Universidade Federal de Vicosa, Brazil

Funding: This work was supported by the National Natural Science Foundation of China (31301262) and China Postdoctoral Science Foundation (Grant no. 2013M540169). The funders had no role in study design, data collection and analysis, decision to publish, or preparation of the manuscript.

Competing Interests: The authors have declared that no competing interests exist.

* Email: yu@cricaas.com.cn (SY); zhaoxinhua1968@126.com (XZ)

Introduction

The proportion of agricultural land that is negatively affected by high salinity is increasing worldwide, owing to natural causes and agricultural practices [1]. This problem has been aggravated by the development of more recent agricultural practices such as irrigation. Approximately 20% of the world's cultivated lands and more than half of all irrigated lands are affected by salinity [2]. High salt concentrations in the soil cause various events that negatively impact agricultural production, such as delays in plant growth and development, inhibition of enzymatic activities and reductions in photosynthetic rates [3]. Therefore, before attempts can be made to introduce genetic and environmental factors to alleviated salt stress, it is critical to elucidate the morphological and physiological responses of particular crops and cultivars to salinity.

In general, salt stress causes an imbalance of the cellular ions resulting in ion toxicity, osmotic stress and production of reactive oxygen species (ROS) [4], thus affecting plant growth, morphology, and survival [5]. High concentrations of NaCl outside the roots reduce the water potential and make it more difficult for the root to extract water. On the other hand, high concentrations of Na^+ and Cl^- ions inside plant cells are inhibitory to many enzyme processes. Salt-tolerant plants can not only regulate ion and water movements more efficiently but should also have a better antioxidant system for effective removal of ROS [6]. Salt stress causes excessive generation of ROS such as superoxide anions (O_2^-), hydrogen peroxide (H_2O_2), and hydroxyl radicals (OH•) [7]. To mitigate the oxidative damage initiated by ROS formed under salt stress, plants possess a complex antioxidant system, including non-enzymatic antioxidants such as ascorbic acid, glutathione (GSH), tocopherols, and carotenoids; antioxidant enzymes such as superoxide dismutase (SOD, EC 1.15.1.1), catalase (CAT, EC 1.11.1.6), glutathione peroxidase (EC 1.11.1.9), and peroxidases (POD, EC 1.17.1.7); and enzymes of the so-called ascorbate-glutathione cycle, including ascorbate peroxidase (APX, EC 1.11.1.11) and glutathione reductase (GR, EC 1.6.4.2). These components of the antioxidant system act in concert to alleviate the cellular damage accumulated under conditions of oxidative stress [8], [9].

SOD is generally considered as the first line of the antioxidant defense system, as it catalyzes the dismutation of O_2^- into H_2O_2 and O_2 in the cytosol, chloroplasts, and mitochondria [10]. POD is mainly located in the apoplastic space and vacuoles, where it plays an important role in catalyzing the conversion of H_2O_2 to H_2O and O_2 [11]. H_2O_2 is scavenged by CAT and APX. CAT dismutates H_2O_2 to H_2O and O_2, whereas APX, together with monodehydroascorbate reductase, dehydroascorbate reductase, and GR, converts H_2O_2 to H_2O via the ascorbate-glutathione pathway. APX is the first enzyme in this pathway, which eliminates H_2O_2 by using ascorbate as an electron donor in an oxidation-reduction reaction [12]. GSH is the final enzyme in this pathway and functions to protect plants from oxidative stress by maintaining GSH in the reduced state [13]. Under salt stress, malondialdehyde (MDA), the decomposition product of the polyunsaturated fatty acids of biomembranes, tends to accumulate [14]. Accordingly, cell membrane stability has been widely used to differentiate between stress-tolerant and -susceptible cultivars of some crops [15], in some cases, higher membrane stability could be correlated with abiotic stress tolerance.

In most plants, higher levels of the activity of the above-mentioned antioxidant enzyme are considered as a salt tolerance mechanisms [9]. Indeed, previous studies have shown that within the same species, salt-tolerant cultivars generally have enhanced or higher constitutive antioxidant enzyme activity under salt stress when compared with sensitive-cultivars. This has been demonstrated in numerous plant species such as cotton [16], rice [17], and pea [18]. Moreover, the response of plant antioxidant enzymes to salinity has been shown to vary among plant species, tissues, and subcellular localizations [19]. Several studies have demonstrated that salt-tolerant species show increased antioxidant enzyme activities and antioxidant contents in response to salt stress, whereas salt-sensitive species fail to do so [20]. Thus, the evidence accumulated to data indicates that intrinsic antioxidant resistance mechanisms of plants may provide a strategy to enhance salt tolerance. However, to achieve efficient selection of genetically-transformed salt-tolerant plants, the mechanisms underlying the effects of salt on the morphology, physiology, growth, and antioxidative responses of plants must first be identified [21].

Salt may affect plant growth indirectly by decreasing the rate of photosynthesis. Indeed, under saline conditions, substantial reduction in photosynthesis has been associated with a decrease in total chlorophyll content and distortion in chlorophyll ultra-structures [22]. Although the factors that limit photosynthesis in salt-stressed plants have been investigated for a number of species, the mechanistic pattern of inhibition remains unclear [23].

In addition, the relationships between ion accumulation, morphological and physiological changes, salt stress, and resulting plant injury are poorly understood. A number of conflicting views have been proposed in the literature over the toxic ions and enzymatic protection and activity involved in the response to salinity [24], [25]. Oxygen radicals are generated during plant metabolism that need to be scavenged by antioxidant systems to maintain normal growth; therefore, determining any adverse effects on these antioxidant systems due to salinity is an important consideration for appropriate cultivar selection.

Cotton is one of the most economically important crops in China. Although it is classified as a salt-tolerant crop, this tolerance is not only limited but also varies according to the growth and developmental stages of the plant [26]. Several studies have been conducted to assess the effect of salinity on the germination, vegetative growth, or yield of cotton [27–29]. However, the interactions between growths rates, ionic content, enzymatic activity, and oxidation reactions are likely to be complex and perhaps vary significantly between cultivars; therefore, such interactions deserve more detailed investigation.

In this study, the effect of NaCl on the growth behavior of two cotton cultivars that differ in their tolerance to salt was investigated. Changes in growth, ion concentration, pigment contents, and photosynthesis were assessed and linked to differences in the antioxidant system observed during salt stress. Since no detailed investigations have been conducted on this responses to data, the information presented here will not only provide criteria for improving salt-tolerant cotton specifically but also for selecting other salt-tolerant species and cultivars.

Materials and Methods

Experimental design

Seeds of two cotton cultivars, CCRI-79 (salt tolerant) and Simian 3 (salt sensitive), were obtained from the National Medium-Term Gene Bank of Cotton in China and soaked in sterile deionized water at 28°C for 6 h. They were then transferred to two sheets of sterile filter paper moistened with deionized water and placed in plastic trays for germination at 28°C for 72 h in the dark. The seeds were then sown into pots filled with perlite and grown under controlled conditions (light/dark regime of 16/8 h at 23°C, relative humidity of 60%–70%, photosynthetic photon flux density of 350 $\mu mol \cdot m^{-2} \cdot s^{-1}$). Germinated seeds were sown into holes in Styrofoam boards that were placed in deionized water and grown hydroponically in a growth room for 3 weeks under fluorescent and incandescent lights.

After 3 weeks, healthy and uniform seedlings were transplanted to 4-L plastic pots (10 plants per pot) filled with an aerated Hoagland nutrition solution (pH 5.2). The nutrition solution was aerated constantly and replaced twice a week throughout the experiment. Plants were cultured under non-saline conditions for 10 d to ensure full establishment before starting the salinity treatments. Salt stress treatment was initiated by providing the plants with full-strength Hoagland's solutions containing 0, 80, 160, or 240 mM NaCl. To avoid osmotic shock, salt concentrations were increased daily by 40 mM NaCl, until reaching the required concentration. After 1 week, the plants were harvested, cleaned and their fresh weights were measured.

Leaf photosynthesis measurements

Net photosynthetic rate (Pn) and stomatal conductance (gs) of leaves were measured in three plants per cultivar per treatment with a photosynthesis system (Li-6400, Li-COR Inc., NE, USA) under 1500 $\mu mol \cdot m^2 \cdot s^2$ light intensity, 65%±5% relative humidity, 32°C±2°C leaf temperature, and 380 $\mu mol \cdot mol^{-1}$ CO_2 concentrations at 9:30–11:00 AM.

Growth parameter measurements

From each treatment group, 10 plants were randomly selected and separated into leaf and root fractions. The leaf area of the youngest fully developed leaf of each plant was measured. Root samples were placed in a rectangular glass dish with a thin layer of water (4–5 mm depth) to allow all roots to spread appropriately. Entire roots were scanned with an EPSON Transparency unit (Epson Perfection V700 Photo; Indonesia), and then analyzed with WinRHIZO software version 5.0 (Regent Instruments, Inc.; Canada) to calculate the total root length, total root surface area, total root volume, and average root diameter. Leaves and roots were washed with deionized water and dried at 70°C for 48 h to determine dry weight before being ground to determine the ion contents.

In addition, another six plants per replicate of each of the four treatments were harvested. Fresh roots of seedlings were separated and frozen immediately in liquid nitrogen before being stored at −80°C pending further analysis.

Chlorophyll and carotenoid content measurements

To determine chlorophyll a (Chl a), chlorophyll b (Chl b), and carotenoids levles, 3–5 discs (0.8-cm diameter) were cut from the upper-most fully expanded leaves randomly selected from five plants per replicate. Discs were homogenized with 2 mL of acetone (80%) and washed twice with an additional 2 mL of acetone. The absorbance of the pooled extracts was measured using a spectrophotometer at 480 nm, 645 nm, and 663 nm. Contents of Chl a, Chl b, and carotenoids in the extracts was determined using MacKinney equations [30]:

$$Chla(mg\ g^{-1}DW) = (12.72 \times A663) - (2.58 \times A645)$$

$$Chlb(mg\ g^{-1}DW) = (22.87 \times A645) - (4.67 \times A663)$$

$$Carotenoids(mg\ g^{-1}DW) = (0.114 \times A663) + A480 - (0.638 \times A645)$$

Ion analyses

The concentration of calcium (Ca), magnesium (Mg), potassium (K), and sodium (Na) were analyzed in subsamples of dried plant materials, which were finely ground in a mill grinder. Approximately 0.5 g of finely ground plant samples were placed into digesting tubes, to which 10 mL of concentrated nitric acid and 3 mL perchlorate acid were added. All the samples were soaked for 12 h and then burned at 300°C for 3 h. The residue was transferred to a 50-mL volumetric flask, which was topped up to 50 mL with distilled water. The cation content was then measured using an atomic absorption spectrophotometer (TAS-986; Persee; China) [31]. For the determination of Cl^- content, leaf samples (0.1 g) were extracted in 10 mL of distilled water by heating at 80°C for 3 h [32]. The Cl^- content in the extracts was analyzed by ion chromatography (DX-300; Sunnyvale, CA, USA) [33].

Determination of lipid peroxidation

Frozen leaf and root segments (0.5 g) were homogenized in a 0.1% (w/v) trichloroacetic acid (TCA) solution. The homogenate was centrifuged at 15,000× *g* for 10 min and 1 mL of the supernatant was added to 4 mL 0.5% (w/v) 2-Thiobarbituric acid (TBA) in 20% (w/v) TCA. The mixture was incubated at 90°C for 30 min, and the reaction was stopped by placing the reaction tubes in an ice water bath. Samples were centrifuged at 10,000× *g* for 5 min and the absorbance of the supernatant was read at 532 nm. The value for non-specific absorption at 600 nm was subtracted from the measured values. The concentration of MDA was calculated using an extinction coefficient of 155 $mM^{-1}\cdot cm^{-1}$.

H_2O_2 determination

H_2O_2 content was estimated according to the methods of Bernt and Bergmeyer [34]. Approximately 0.5 g of root and leaf samples from control and treatment groups were homogenized with liquid nitrogen, and the powders were suspended in 1.5 mL of 100 mM potassium phosphate buffer (pH 6.8). The suspensions were then centrifuged at 18,000× *g* for 20 min at 4°C. The enzymatic reaction was initiated with 0.25 mL supernatant and 1.25 mL peroxidase reagent, consisting of 83 mM potassium phosphate buffer (pH 7.0), 0.005% (w/v) *O*-dianizidine, and 40 μg peroxidase/mL at 30°C. The reaction was stopped after 10 min by adding 0.25 mL of 1 N perchloric acid and the reaction mixture was centrifuged at 5000× *g* for 5 min. The absorbance of the supernatant was read at 436 nm, and the amount of H_2O_2 was determined using an extinction coefficient of 39.4 $mM^{-1}\cdot cm^{-1}$.

Extraction and assay of antioxidant enzyme assay

For enzyme extractions, frozen root and leaf samples (0.3 g) were ground into a fine powder by using a mortar that was placed in an ice bath and a pestle that was pre-cooled with liquid nitrogen, and homogenized in 50 mM potassium phosphate buffer (pH 7.8) containing 1 mM ascorbate and 2% (w/v) polyvinylpolypyrrolidone. Homogenates were then centrifuged at 20,000× *g* for 30 min at 4°C.

SOD activity was determined according to the methods of Foster and Hess [35]. The reaction was performed in a total volume of 1 mL containing 50 mM potassium phosphate buffer (pH 7.8, containing 0.1 mM ethylenediaminetetraacetic acid [EDTA]), 0.1 mM cytochrome c, 0.1 mM xanthine, enzyme extract, and 0.3 U/mL of xanthine oxidase. The reaction was initiated by the addition of xanthine oxidase and absorbance was measured at 560 nm. One unit of SOD activity was defined as the amount of enzyme that inhibits the rate of cytochrome c reduction by 50%.

Total CAT activity was measured according to the method reported by Beers and Sizer [36], with minor modifications. The reaction mixture (1.5 mL) consisted of 100 mM phosphate buffer (pH 7.0), 0.1 mM EDTA, 20 mM H_2O_2, and 50 μL enzyme extract. The reaction was initiated by the addition of the enzyme extract. The decrease in H_2O_2 was monitored at 240 nm and was quantified by its molar extinction coefficient (36 $M^{-1}\cdot cm^{-1}$).

Peroxidase activity was analyzed in 2.9 mL of 0.05 M phosphate buffer, containing 1.0 mL of 0.05 M guaiacol and 1.0 mL of 2% H_2O_2 [37]. Increases in absorbance at 470 nm were recorded after adding 2.0 mL of 20% chloroacetic acid.

APX activity was determined according to Nakano and Asada [38] by following the decline in absorbance at 290 nm as ascorbate was oxidized. The oxidation rate of ascorbate was estimated between 1 and 60 s after starting the reaction with the addition of H_2O_2. The 1-mL reaction mixture contained 50 mM HEPES-NaOH (pH 7.6), 0.22 mM ascorbate, 1.0 mM H_2O_2, and an enzyme extract. Corrections were made for the low, non-enzymatic oxidation of ascorbate in the absence of H_2O_2.

GR activity was measured as described by Foyer et al. [39]. The assay medium contained 1 mM EDTA in 50 mM potassium phosphate buffer (pH 7.8), 0.1 mM NADPH, enzyme extract, and 0.1 mM glutathiol (GSSG) in a total volume of 1 mL. The reaction was initiated by adding GSSG and the NADPH oxidation rate was monitored at 340 nm. GR activity was calculated using an extinction coefficient of 6.2 $mM^{-1}\cdot cm^{-1}$ for NADPH, and one unit of enzyme was defined as the amount of enzyme required to oxidize 1 μmol NADPH per minute. The specific enzyme activity for all enzymes was expressed as units/mg protein.

Statistical analysis

The experiments were set up as a completely randomized design, including two cotton cultivars and four salinity levels. All data obtained were subjected to ANOVA, and the mean difference was compared by the LSD test at 95% or 99% levels of probability. In all figures, the spread of values is shown as error bars representing standard errors of the means.

Results and Discussion

Growth parameters

Salinity exposure can lead to various physiological and biochemical changes within plant cells causing numerous changes in their structure and function [40]. In the present study, salt-induced changes in the growth and antioxidant profile of cotton plants were evaluated. Increasing NaCl concentration, up to 240 mM, gradually reduced leaf and root growth (Fig. 1). In general, both cultivars showed decreased growth rates of the roots and leaves with increasing salt concentration, but there was no significant variation in the leaf dry weight of Simian 3 when

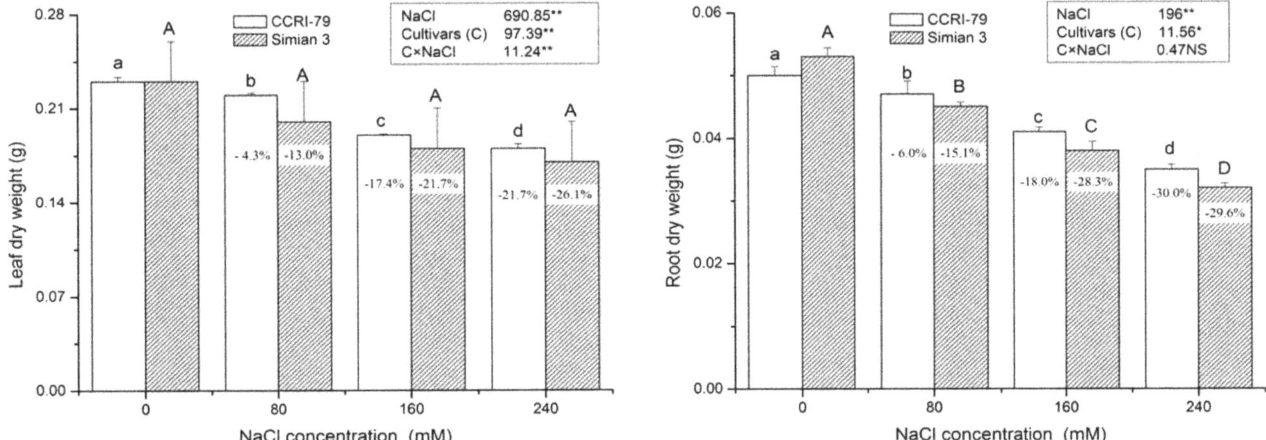

Figure 1. Effect of NaCl salinity on the leaf and root dry weight of two cotton cultivars. Vertical bars represent ± standard error (n = 3). Bars labeled with the different lowercase letters on open square bars or uppercase letters on closed square bars are significant difference (P<0.05). *, ** Significant at P<0.05 and P<0.01, respectively. NS, not significant. Figures in column indicate± increase/decrease under salinity stress over control.

subjected to salt stress. However, the percentage reduction in leaf and root dry weights due to salinity over control was lower in CCRI-79 as compared with Simian 3, indicating that CCRI-79 is a more salt-tolerant cultivar. Inhibition of growth due to NaCl stress in CCRI-79 is comparable to the observations of Takemura et al. [41]. This reduction in growth may be due to osmotic injury or specific ion toxicity caused by the uptake of salt [42]. However, the differential response of growth to salinity observed between CCRI-79 and Simian 3 could be due to genotypic differences, which have also been reported by Qidar and Shams [26]. An increase in the tissue maintenance process (through respiration) is believed to be the primary cause of growth decline during salinity stress, and could represent a mechanism of adaptation to salinity [25]. The sacrifice of leaf photosynthetic tissue during salt adaptation may serve to conserve energy that can then be redirected to maintaining leaf multiplication and growth, indicating successful use of the tissue maintenance process. Thus, the fact that the leaf dry weight of the salt-sensitive cultivar Simian 3 was maintained indicates that it is not reallocating its energy reserves when faced with high-salinity conditions as compared to the more salt-tolerant cultivar CCRI-79.

When plants are grown under conditions of salt stress, the immediate response is a cessation in the expansion of the leaf surface [43]. Several authors have reported the phenomenon of leaf expansion in response to salinity in halophytes as well as in glycophytes [44]. Similarly, in our study, leaf area was highest in the control group (0 mM NaCl), whereas it decreased continuously as salinity increased (Table 1). One possible reason for this decline might be related to salt osmotic effects, which affect cell turgor and expansion [45].

Compared with the no-salt control treatment, salt stress significantly (P<0.01) reduced the length, surface area, volume,

Table 1. Morphological parameters of cotton at different NaCl concentrations and results of ANOVA (F-ratios).

Cultivar	NaCl (mM)	Leaf area (cm^2)	Root length (cm)	Surface area (cm^2)	Volume (cm^3)	Average diameter (mm)
CCRI-79	0	36.19±0.91a	370.86±3.45a	52.07±0.94a	0.57±0.014a	0.54±0.016a
	80	35.18±1.06a	320.33±2.53b	44.04±1.29b	0.48±0.016b	0.45±0.014b
	160	34.43±0.46a	192.72±4.38c	32.45±1.67c	0.42±0.009c	0.43±0.009b
	240	34.07±0.12a	161.28±4.88d	28.03±0.71d	0.34±0.014d	0.35±0.007c
Simian 3	0	35.13±0.85a	302.58±3.02a	49.07±0.78a	0.62±0.041a	0.52±0.005a
	80	33.99±1.07a	260.86±8.67b	35.71±1.28b	0.39±0.043b	0.49±0.018a
	160	33.46±1.36a	188.32±14.26c	31.12±0.98c	0.38±0.023b	0.40±0.014b
	240	32.98±0.63a	138.58±1.05d	24.19±0.60d	0.30±0.008c	0.31±0.004c
NaCl		4.31NS	673.99**	338.36**	75.49**	182.17**
Cultivars(C)		5.73*	134.92**	50.99**	4.60 NS	3.12 NS
C×NaCl		0.01 NS	20.54**	6.70*	5.15*	9.41**

Values are the mean of three replicates ± S.E. Means followed by a different letter within a column for each cotton cultivar are significantly different at P<0.05 according to the Student's LSD test.
*, ** Significant at P<0.05 and P<0.01, respectively.
NS, not significant.

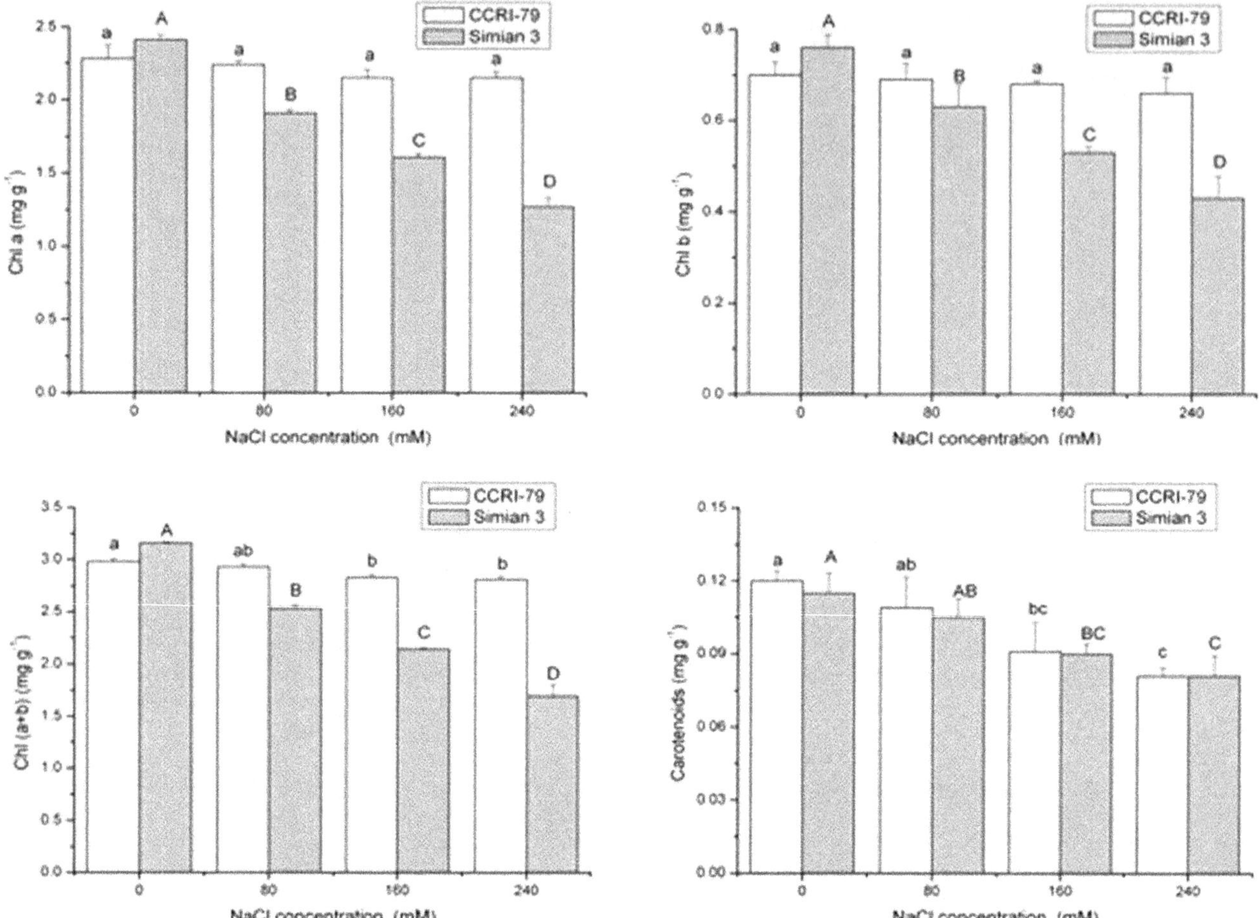

Figure 2. Concentrations of chlorophylls and carotenoids in cotton grown at different NaCl concentrations. Vertical bars represent ± standard error (n = 3). Bars labeled with the different lowercase letters on open square bars or uppercase letters on closed square bars are significant difference (P<0.05).

and average diameter of the roots in both cotton cultivars (Table 1). Significant differences were observed when NaCl concentration was increased from 80 to 240 mM. There are several reasons for the reduced root length, including cell growth restriction due to low water potential of the external medium, interference caused by ions, or the toxicity of accumulated ions [46]. Our findings are consistent with the results of Siroka et al. [47], who reported that salinity suppressed the development of maize roots cell. The inhibition of root growth in terms of root length, surface area, volume, and average diameter can be attributed to the inhibition of mitosis, reduced synthesis of cell wall components, damage to the Golgi apparatus, and changes in polysaccharide metabolism [48]. However, this decrease was more

Table 2. *F* values of ANOVA of the effects of NaCl, cultivars, and their interaction for chlorophylls, carotenoids content, net photosynthetic rate (Pn), and stomatal conductance (gs).

Item	NaCl	Cultivars(C)	C×NaCl
Chl a	108.52**	245.65**	66.29**
Chl b	45.49**	61.62**	26.86**
Chl (a+b)	186.62**	380.12**	113.08**
Carotenoids	13.62**	0.28 NS	0.10 NS
Pn	98.15**	200.15**	51.5**
gs	56.31**	77.17**	23.44**

*, ** Significant at P<0.05 and P<0.01, respectively.
NS, not significant.

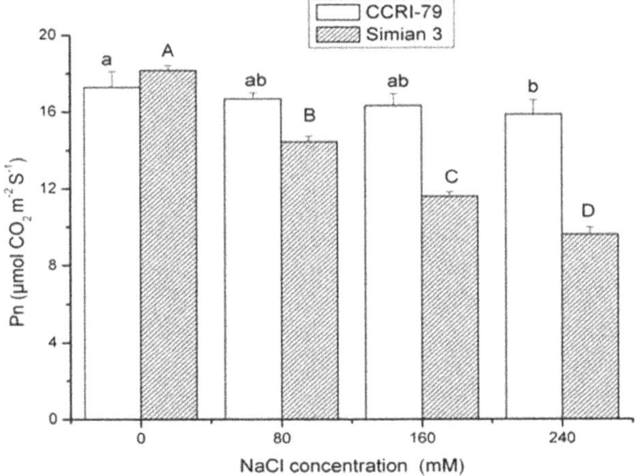

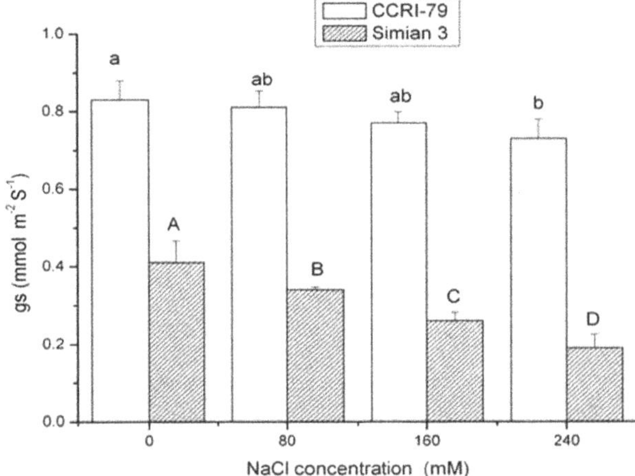

Figure 3. Net photosynthetic rate (Pn) and stomatal conductance (gs) of two cotton cultivars as affected by different NaCl concentrations. Vertical bars represent ± standard error (n = 3). Bars labeled with the different lowercase letters on open square bars or uppercase letters on closed square bars are significant difference ($P<0.05$).

predominant in Simian 3 than in CCRI-79, indicating that CCRI-79 was more tolerant to salinity than Simian 3.

Chlorophyll and carotenoid contents and photosynthesis

In general, the reduction in growth and productivity when plants are grown under conditions of salt stress is accompanied by as strong reduction in the rate of photosynthesis owing to severe impairments in photosynthetic activities and the photosynthetic apparatus, the degree of which depends on the varieties of species considered [49]. As shown in Fig. 2, the contents of Chl a, Chl b, and Chl (a+b) in the plants significantly ($P<0.01$) decreased as salinity increased in Simian 3, but only marginal changes were observed in CCRI-79 (Table 2). Maintenance of chlorophyll content has been reported in other salt-tolerant crops such as durum wheat and legume species [50]. One possible mechanisms of salt tolerance in these species may be the possession of a salt exclusion and/or sequestration trait that prevents leaf injury, thus maintaining chlorophyll content. Since chlorophyll content is directly correlated with growth and development of the plant [51], the decrease in chlorophyll content in Simian 3 suggested substantial damage to the photosynthetic mechanism, as reported previously in salt-treated rice and sorghum plants [52], [53]. The inhibitory effects of salt on chlorophylls could be due to the suppression of specific enzymes responsible for the synthesis of chlorophyll. Our findings regarding total chlorophyll content are comparable with the observations of Meloni et al. [54], who demonstrated that Guazuncho, another salt-sensitive cotton cultivar, showed a 35% reduction in total chlorophyll content after 21 days of salinity treatment.

Carotenoids are reported to play an important role in ROS scavenging, thereby protecting membranes from salt stress [55]. However, we did not observe changes in carotenoid contents in response to salinity treatments in either cultivar, suggesting that these pigments are not involved in the response to salt stress in cotton. Although the rate of change was slower in carotenoids than in chlorophylls, carotenoid content also showed a decreasing trend with increasing salt stress, indicating that this trait could also serve as a useful indicator of NaCl stress in cotton.

Since plant growth is dependent on photosynthesis, environmental stresses affecting growth will also affect photosynthesis [56]. In the present study, the net photosynthetic rate (Pn) and stomatal conductance (gs) of both cotton cultivars were inhibited by salinity due to NaCl (Fig. 3, Table 2). However, the net photosynthetic rate and stomatal conductance were significantly lower for Simian 3 than for CCRI-79 under conditions of salt stress. Compared with the control treatment, the net photosynthetic rate and stomatal conductance of Simian 3 significantly decreased with increasing salinity, whereas there was no significant difference in either trait in CCRI-79.

Photosynthesis was reduced in both cotton cultivars in response to salt stress, which was likely caused by the reduction in stomatal conductance. Indeed, parallel decreases in stomatal conductance and net photosynthesis due to NaCl salinity have previously been reported for cotton [57]. Our results suggest that stomatal closure limited the leaves' photosynthetic capacity in the NaCl-treated plants of both cultivars, Although only Simian 3 showed a significant decline in the leaf chlorophyll contents due to NaCl stress for 7 days. Similarly, Delfine et al. reported no changes in the chlorophyll content in spinach plants (*Spinacia oleracea* L.) exposed to salt stress for 20 days [58]. Our results suggest that the greater reduction in stomatal conductance accompanied by decreased leaf chlorophyll content could have contributed to the higher reduction in the leaf photosynthetic rate of Simian 3 when compared with that of CCRI-79.

Further examination showed that the decrease in the Chl a, Chl b, Chl (a+b), net photosynthetic rate, and stomatal conductance levels with increasing salt concentrations occurred more rapidly compared with the rate of decrease in carotenoid content; this trend was more conspicuous in Simian 3 than in CCRI-79.

Ion concentrations

High external salt concentration causes an ion imbalance or disturbance in ion homeostasis [43]. In our experiments, the leaves and roots of both cultivars had higher levels of Na^+ and Cl^- ions under salt stress due to nonspecific ion uptake and/or membrane leakage. However, as the NaCl concentration increased, the levels of Na^+ and Cl^- also increased further, suggesting that these cultivars may not differ in terms of Na^+ uptake and its transportation to leaves, and thus the increase observed can not be explained by an ionic exclusion mechanism (Fig. 4). In addition, Na^+ ion concentrations in leaves were higher than those in the roots of both cultivars at all salinity levels, which indicated

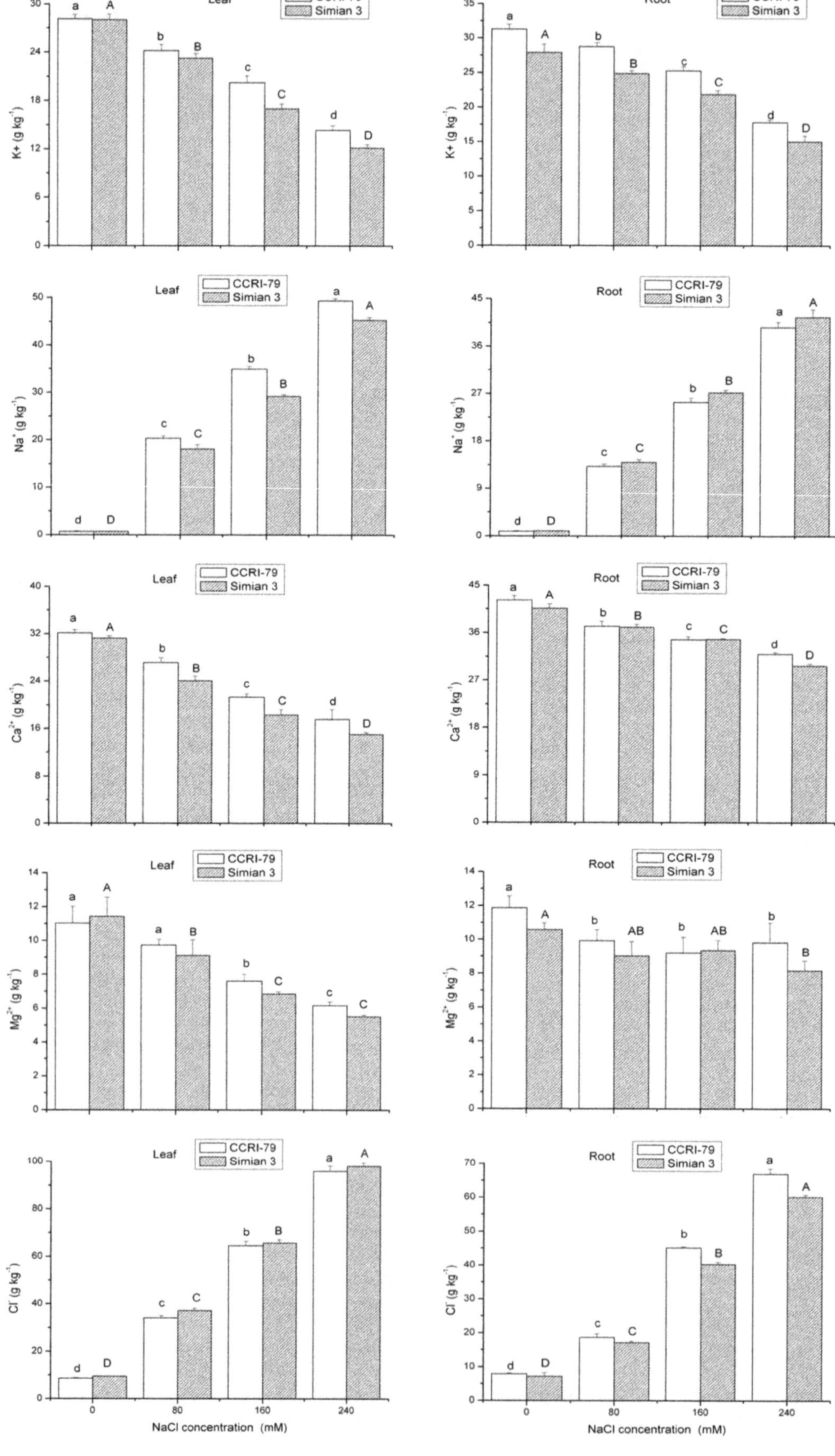
Leaf
Root
CCRI-79
Simian 3
K+ (g kg⁻¹)
Na⁺ (g kg⁻¹)
Ca²⁺ (g kg⁻¹)
Mg²⁺ (g kg⁻¹)
Cl⁻ (g kg⁻¹)
NaCl concentration (mM)
0
80
160
240

Figure 4. Effect of NaCl salinity on the concentrations of K^+, Na^+, Ca^{2+}, Mg^{2+} and Cl^- in the roots and leaves of cotton. Vertical bars represent ± standard error (n = 3). Bars labeled with the different lowercase letters on open square bars or uppercase letters on closed square bars are significant difference ($P<0.05$).

the inability of these cultivars to prevent Na^+ ion transportation from the roots to leaves. Chlorine ions showed a similar distribution pattern to Na^+ ions, despite being at higher salinity levels. Na^+ and Cl^- are highly water-soluble and are readily taken up by plants and transported into leaves; these ions most likely act as osmotica, but only moderate concentrations can be tolerated before growth and photosynthesis are reduced.

High Na^+ and Cl^- absorption competes with the uptake of other nutrient ions such as K^+, Mg^{2+}, and Ca^{2+}, leading to a deficiency of these ions and an imbalance among cations [59]. During the same period, K^+ and Ca^{2+} content in the leaves and roots of both cultivars were significantly reduced, and a further decrease in these ions was observed with increasing NaCl concentration (Fig. 4, Table 3). Salt tolerance involves not only adaptation to Na^+ influx but also acquisition of K^+, the uptake of which is adversely affected by high external Na^+ concentration due to the chemical similarity of these two ions [60], [61]. Indeed, selective uptake of K^+ as opposed to Na^+ is considered to be one of the key physiological mechanisms contributing to salt tolerance in many plant species [62]. K^+ efflux has already been used as an indicator of cellular toxicity for a range of toxic compounds, and losses of K^+ and Ca^{2+} have already been documented during salinity stress [63], [64]. A large and permanent efflux of K^+ and Ca^{2+} usually indicates damage to the limiting membranes.

On the other hand, salt treatments induced a significant decrease in Mg^{2+} concentrations in cotton leaves in the present study. The significantly lower levels of Mg^{2+} in the leaves under salinity conditions are probably related to the lower levels of chlorophylls present in the NaCl-treated leaves. However, in the roots of NaCl-treated plants, Mg^{2+} concentrations were close to or lower than those observed in control plants, even at the highest salt dose (240 mM NaCl). This was in contrast to the Ca^{2+} content, which was significantly lower in salt-treated plants. Our findings are similar to those of Khan [65], who reported that NaCl treatment induced a decline in Ca^{2+} and Mg^{2+} levels in *Ceriops tagal* plants.

In Addition, there were no significant differences between the two cotton cultivars in the variation of ion content (including K^+, Na^+, Mg^{2+}, Ca^{2+} and Cl^-) under NaCl stress. This similarity in ionic levels may be a consequence of the shoot culture method used, which would not involve a selective mechanisms of ion transport. For example, there was likely to be no regulation between the xylem parenchyma and xylem interface, in which ion selection and reabsorption from the medium may be regulated. Previous results reported for other species are consistent with these observations [66]. However, the restriction of entry of ions into metabolically active areas of cells in the more-tolerant CCRI-79 cultivar can not be ruled out as a mechanism to maintain ionic equilibrium when ions are highly concentrated in the external environment [67].

Lipid peroxidation and proline content

Salt stress is known to result in extensive lipid peroxidation, which has often been used as an indicator of salt-induced oxidative damage in membranes [68]. The MDA content increased with increasing salinity in the leaves and roots of both cotton cultivars (Fig. 5, Table 4), indicating cell membrane damage in both cotton cultivars. However, as the salinity increased, the accumulation of MDA was higher in Simian 3 as compared to CCRI-79, indicating a higher degree of lipid peroxidation in Simian 3 due to salt stress. Lipid peroxidation could be a result of light-dependent formation of singlet oxygen during stress conditions [69]. The low values of MDA content obtained with CCRI-79 might account for the lower lipid peroxidation levels observed and the reduced effect on membrane permeability. Low levels of lipid peroxidation may have contribute to the observed tolerance of CCRI-79 plants exposed to high levels of salinity. Similar results for lipid peroxidation have been reported by other researchers in barley [70].

Many plants accumulate proline as a nontoxic and protective osmoylte under saline conditions [55]. In this study, the levels of proline continued to increase in both of the cultivars as the NaCl concentration increased (Fig. 5). The proline concentration of

Table 3. *F* values of ANOVA of the effect of NaCl, cultivars, and their interaction for K^+, Na^+, Ca^{2+}, Mg^{2+}, and Cl^- contents.

Tissue	Item	Cultivars (C)	NaCl	C×NaCl
Leaf	K^+	22.47**	349.67**	4.05 NS
	Na^+	131.7**	5755.89**	22.16**
	Ca^{2+}	27.27**	220.03**	1.28 NS
	Mg^{2+}	2.09 NS	69.77**	0.94 NS
	Cl^-	7.03*	3175.89**	0.58 NS
Root	K^+	72.66**	204.40**	0.34 NS
	Na^+	21.19**	4546.89**	3.12 NS
	Ca^{2+}	15.30**	311.67**	4.96*
	Mg^{2+}	5.19NS	6.21*	0.91 NS
	Cl^-	65.95**	3497.79**	11.22**

*, ** Significant at $P<0.05$ and $P<0.01$, respectively.
NS, not significant.

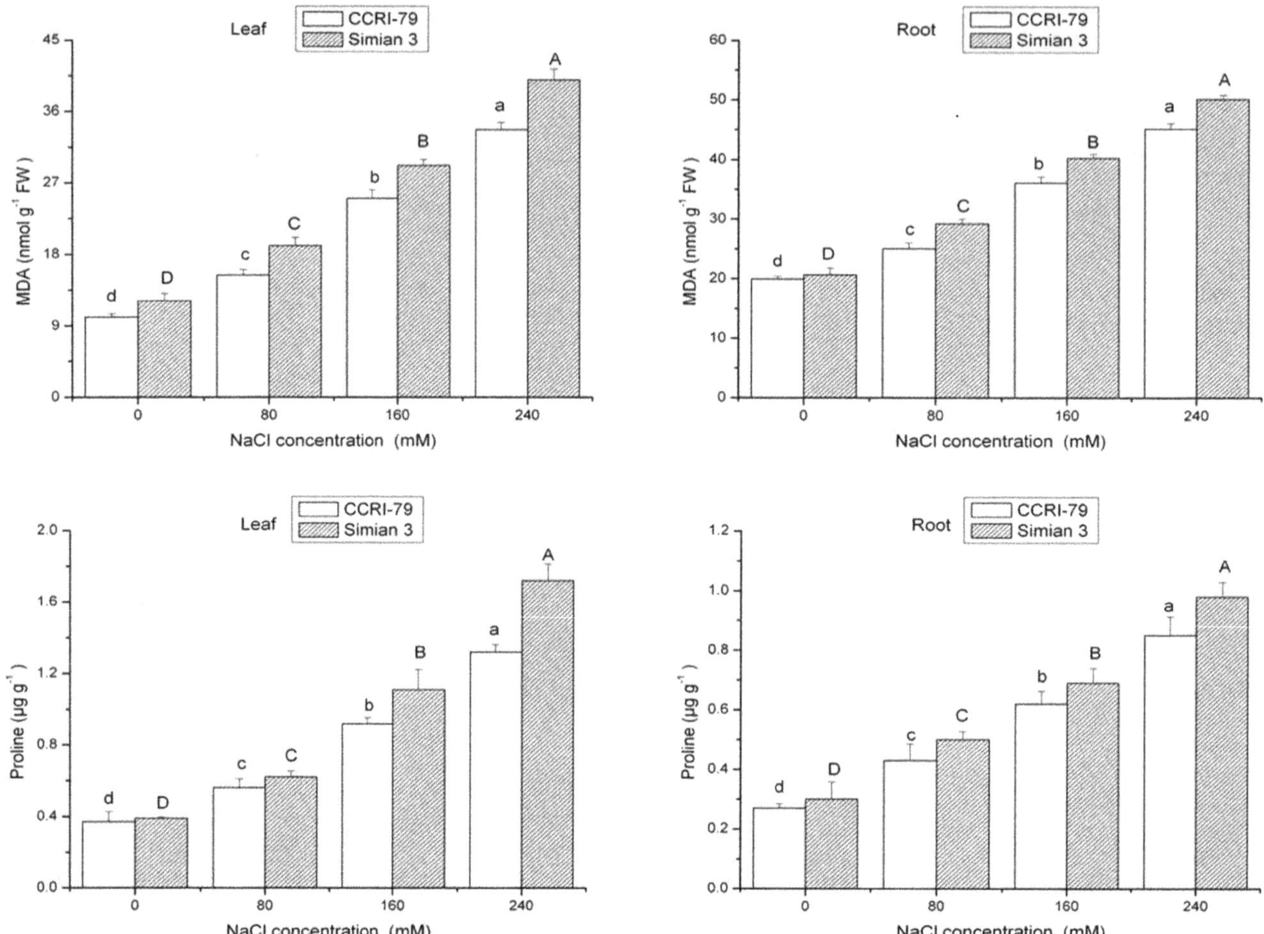

Figure 5. Effect of NaCl salinity on the concentrations of malondialdehyde (MDA) and proline in the roots and leaves of cotton. Vertical bars represent ± standard error (n = 3). Bars labeled with the different lowercase letters on open square bars or uppercase letters on closed square bars are significant difference (P<0.05).

CCRI-79 plants was lower than that of Simian 3 plants, especially at the highest salinity level, which could be attributable to the greater salt resistance of the CCRI-79 cultivar, i.e., less injury was induced by the salt [71]. Similar trends were observed by Rabie and Almadini in broad bean plants [72]. Moreover, in our study, proline accumulation in the leaves was higher than that in the roots, which is similar to the findings of Sharma and Dietz [73], indicating that proline plays a more important protective role in the leaves of cotton seedlings than in the roots under salinity stress.

H_2O_2 content

Stress conditions enhance H_2O_2 production in different compartments of plants cells through both enzymatic and non-enzymatic processes [74]. In our study, both genotypes had similar levels of H_2O_2 levels at 0 mM NaCl treatment. However, salinity treatments caused a marked increase in H_2O_2 content, and the Simian 3 cultivar had a higher H_2O_2 content than did CCRI-79 after NaCl treatment (Fig. 6, Table 4), which could be mainly due to the decreased H_2O_2-scavenging activity in the salt-sensitive cultivar. These results are comparable to those reported in a previous study [75], where accumulation of H_2O_2 in the roots of the salt-tolerant rice cultivar FL478 was significantly higher than that of the salt-sensitive rice cultivar IR-29 in response to moderate salt stress applied for 12 days. Simultaneously, the H_2O_2 content were markedly lower in the leaves than in the roots, regardless of NaCl dose, which contrasts with the findings of Lee et al. [76] in rice. This discrepancy in the effects of salt on H_2O_2 content between studies may be related to technical difficulties, and therefore these results should be evaluated with caution [77]. Furthermore, H_2O_2 has been shown to induce cytosolic APX [78]; therefore, accumulation of H_2O_2 under high salinity conditions may be a signal to initiate an adaptive response to the stress [79]. Although differences in salt tolerance among different cultivars are not necessarily related to differences in the ability to detoxify ROS, many comparative studies using salt-tolerant and salt-sensitive genotypes have shown a correlation between salt tolerance and increased activity of antioxidant enzymes [80].

Activities of antioxidant enzymes

Environmental stresses that limit photosynthesis can increase oxygen-induced cellular damage due to increased ROS generation [81]. Therefore, salt stress resistance may depend, at least in part, on the enhancement of the antioxidative defense system, which involves antioxidant compounds and several antioxidant enzymes. In the present study, the responses of SOD, POD, CAT, APX, and GR enzyme activities suggested that oxidative stress is an important component of salt stress in cotton plants.

Table 4. *F* values of ANOVA of the effect of NaCl, cultivars, and their interaction for malondialdehyde (MDA), proline, H_2O_2 content, and antioxidant enzyme activities.

Tissue	Item	Cultivars (C)	NaCl	C×NaCl
Leaf	MDA	78.15**	608.62**	3.65 NS
	Proline	25.01**	230.23**	6.58*
	H_2O_2	74.86**	1357.74**	11.54**
	SOD	315.65**	213.58**	63.01**
	CAT	874.75**	445.12**	43.46**
	POD	0.47 NS	5.52*	0.49 NS
	APX	1.36 NS	5.93*	0.08 NS
	GR	28.88**	194.18**	15.77**
Root	MDA	60.16**	708.42**	4.47*
	Proline	9.10*	119.51**	0.72 NS
	H_2O_2	174.18**	4622.30**	40.92**
	SOD	8559.93**	1510.48**	1357.86**
	CAT	111.07**	1.92 NS	28.60**
	POD	50.42**	322.16**	2.08 NS
	APX	37.10**	208.76**	5.83*
	GR	17.47**	82.69**	1.45 NS

*, ** significant at P = 0.05 and P = 0.01 levels, respectively.
NS, not significant.
H_2O_2, hydrogen peroxide; SOD, superoxide dismutase; CAT, catalase; POD, peroxidase; APX, ascorbate peroxidase; GR, glutathione reductase.

Because SOD can catalyze the dismutation of superoxide to molecular oxygen and H_2O_2, this enzyme is considered the most effective intracellular enzymatic antioxidant. Indeed, it has been suggested that SOD plays an important role in plant stress tolerance and provides the first line of defense against the toxic effects of elevated levels of ROS [82]. In this study, salt stress increased SOD activity in the leaves of both cultivars and in the roots of CCRI-79 only (Fig. 7, Table 4). However, increased SOD activity in both the leaves and roots was more conspicuous in the salt tolerant cultivar CCRI-79 than in the salt-sensitive cultivar Simian 3, suggesting that the salt-tolerant genotype has a more efficient $O_2^{\bullet-}$ radical-scavenging ability. Similar results have also been shown in both the leaves and roots of cotton and pea plants [14], [83]. In plants, high induction of SOD activity can lead to H_2O_2 accumulation as well as lipid peroxidation [84], which could contribute to the increased H_2O_2 content observed in the roots than in the leaves of cotton seedlings exposed to salinity.

POD is the primary enzyme that detoxifies H_2O_2 in the chloroplasts and cytosol of plant cells [85]. CAT plays an important role in the antioxidant system because it converts H_2O_2 into oxygen and water [86]. These two enzymes constitute the main H_2O_2-scavenging systems in cells. The present data showed that the roots had higher POD activity compared to the leaves in both cultivars; however, the enzyme activity in the roots and leaves responded differently to incremental levels of salinity. In the roots of both cultivars, there was a significant decline in POD activity with an increase in salinity levels, whereas there was no significant difference in the leaf of Simian 3 across salt treatments. However, POD activity in the leaf of CCRI-79 showed no significant difference when subjected to 0 to 160 mM NaCl concentrations. Conversely, the 240 mM NaCl concentration induced a significant decrease in the leaf POD activity of CCRI-79 when compared to the control (0 mM NaCl concentration). Amor et al. reported that H_2O_2 accumulation under salinity stress was related to a decrease in CAT activity in *C. maritima* [87]. In our study, CAT activity in the leaves and roots of Simian 3 declined with an increase in NaCl concentration (Fig. 7, Table 4). However, although the CAT activity in CCRI-79 increased with salinity in the roots, it decreased with increasing salinity in the leaves. CAT activity in the leaves and roots of CCRI-79 was significantly higher than that in Simian 3 leaves and roots for all salinity treatments. This indicates that the scavenging of H_2O_2 is more effective in CCRI-79 than in Simian 3. The roots of both the CCRI-79 and Simian 3 cultivars showed higher CAT activities than did the leaves. The stimulation of POD and CAT suggests that these enzymes are important in the detoxification of H_2O_2 in plant seedlings under salinity stress [88]. In contrast, POD and CAT activities were found to be reduce in response to excess salinity in the leaves and roots of various other plants species [52], [89]. Thus, the effect of salinity on antioxidant enzyme activities varies among plant species, organs, and even treatment concentrations.

H_2O_2 scavenging is also accomplished by the glutathione-ascorbate cycle, a series of coupled redox reactions involving three enzymes, APX, GR and monodehydroascorbate reductase [90]. APX plays a key role in protecting the plant against oxidative stress by scavenging H_2O_2 in different cell compartments. It also has a higher affinity for H_2O_2 than POD and CAT and, as such, may play a more crucial role in the management of ROS during stress [82]. As shown in Fig. 7, there was a significant difference in the effect of salt on APX activity between the leaves and roots of the two cultivars. As salinity increased, APX activity in the root increased in CCRI-79 but decreased in Simian 3. However, there was no significant difference in the leaf APX activity of plants subjected to 0 to 160 mM NaCl concentrations in both cultivars. Conversely, the 240 mM NaCl concentration induced a significant

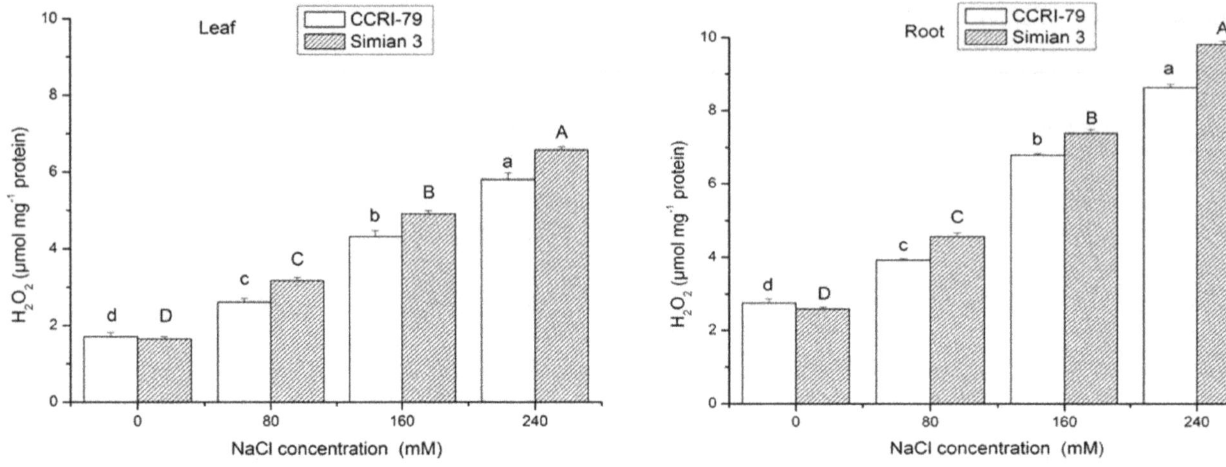

Figure 6. Effect of NaCl salinity on the concentrations of H_2O_2 in the roots and leaves of cotton. Vertical bars represent ± standard error (n = 3). Bars labeled with the different lowercase letters on open square bars or uppercase letters on closed square bars are significant difference (P< 0.05).

decrease in the leaf APX activity of both cultivars when compared to the control (0 mM NaCl concentration). We also found that higher APX activities were accompanied by higher CAT activities in CCRI-79, implying that CCRI-79 has a more effective H_2O_2-scavenging mechanism than Simian 3.

The role of GSH and GR in H_2O_2 scavenging in plant cells has been well established in the Halliwell-Asada enzyme pathway. GR catalyzes the rate-limiting and last step of the ascorbate-glutathione pathway [91]. In our study, GR activities increased in the leaves of both cotton cultivars and in the roots of CCRI-79 only in response to salt stress. On the other hand, salinity significantly reduced GR activity in the roots of Simian 3. Several studies investigating salt-tolerant and salt-sensitive cultivars have suggested that the salt tolerance trait is related to increased GR activity in salt-tolerant cultivars [83], [92], which is similar to our results. The elevated levels of GR activity may increase the rate of NADPH oxidation to $NADP^+$, thereby ensuring the availability of $NADP^+$ to accept electrons from the photosynthetic electron transport chain. Under such conditions, the flow of electrons to O_2^-, and therefore the formation of O_2^- can be minimized. However, in the roots of Simian 3, the reduction in GR activity suggests a decrease in the GSH turnover rate. Considering that salinity also reduced APX activity in the roots of Simian 3, these results suggests that salt-sensitive cultivars exhibit a less-active ascorbate-glutathione cycle in the roots, which may be a key enzyme for the development of salt-tolerance in cotton plants.

Furthermore, the variation in SOD, CAT, APX, and GR activities differed significantly between CCRI-79 and Simian 3 under the NaCl concentrations tested. The increased salinity resistance of CCRI-79 was associated with its ability to maintain higher activity of these antioxidant enzymes, which resulted in lower H_2O_2 production, lipid peroxidation, and higher membrane stability. This provides further evidence that the H_2O_2-scavenging mechanisms were more effective in CCRI-79. By contrast, the relatively lower CAT, APX, and GR activities in salt-stressed Simian 3 compared to control plants indicated that H_2O_2 scavenging was less effective in this cultivar. This excess of H_2O_2 may be the main contributor to the extensive lipid peroxidation and growth inhibition observed in Simian 3.

When compared with the other scavenging enzymes tested, CAT and APX had much higher H_2O_2-scavenging activities in the leaves and roots of both cotton cultivars in our study. The importance of these two enzymes in H_2O_2 scavenging has been demonstrated in several previous studies, in which increased CAT and APX activities were correlated with tolerance to salt and other environmental stresses [93]. These enzymes were also shown to be important in salt tolerance of barley and mulberry [94], [95]. Therefore, the results suggest that the coordination of CAT and APX with SOD activity could comprise an additional constituent in the enzymatic antioxidant mechanism of cotton plants against oxidative stress.

Conclusions

In this study, we compared the response of two cotton cultivars that differ with respect to salt tolerance to increasing NaCl concentrations. Overall, salinity significantly reduced the leaf and root dry weights, root volume, root length, root surface area, root average diameter, chlorophyll content and photosynthesis in the cotton plants of both cultivars. In contrast, antioxidant enzyme activity and proline and MDA contents increased in response to salinity. The salt-tolerant cultivar CCRI-79 showed evidence of possessing a more efficient antioxidant defense system against oxidative stress and lipid peroxidation by maintaining higher SOD, CAT, APX, and GR activities than those in Simian 3 during salt stress. The differences in the antioxidant enzyme activity of the leaves and roots may, at least in part, explain the greater tolerance to salt stress exhibited by CCRI-79 compared to that exhibited by Simian 3. Besides differences in antioxidant enzyme activities, the two cotton cultivars also showed marked variation in Chl a, Chl b, and Chl (a+b) contents, net photosynthetic rate, and stomatal conductance in response to NaCl stress. Therefore, acquisition of tolerance to salt may not only involve improved resistance to oxidative stress owing to enzymes that primarily function to protect membranes and tissues from such damage, but might also involve improvement in the biosynthetic pathway of photosynthetic pigments to maintain higher rates of photosynthesis in the face of stress. However, it should be noted that salt stress was only assessed at concentrations of 0, 80, 160, and 240 mM NaCl; therefore, further studies should be conducted to verify and screen the selection criteria for salt-tolerant species and cultivars. Nonetheless, the data presented

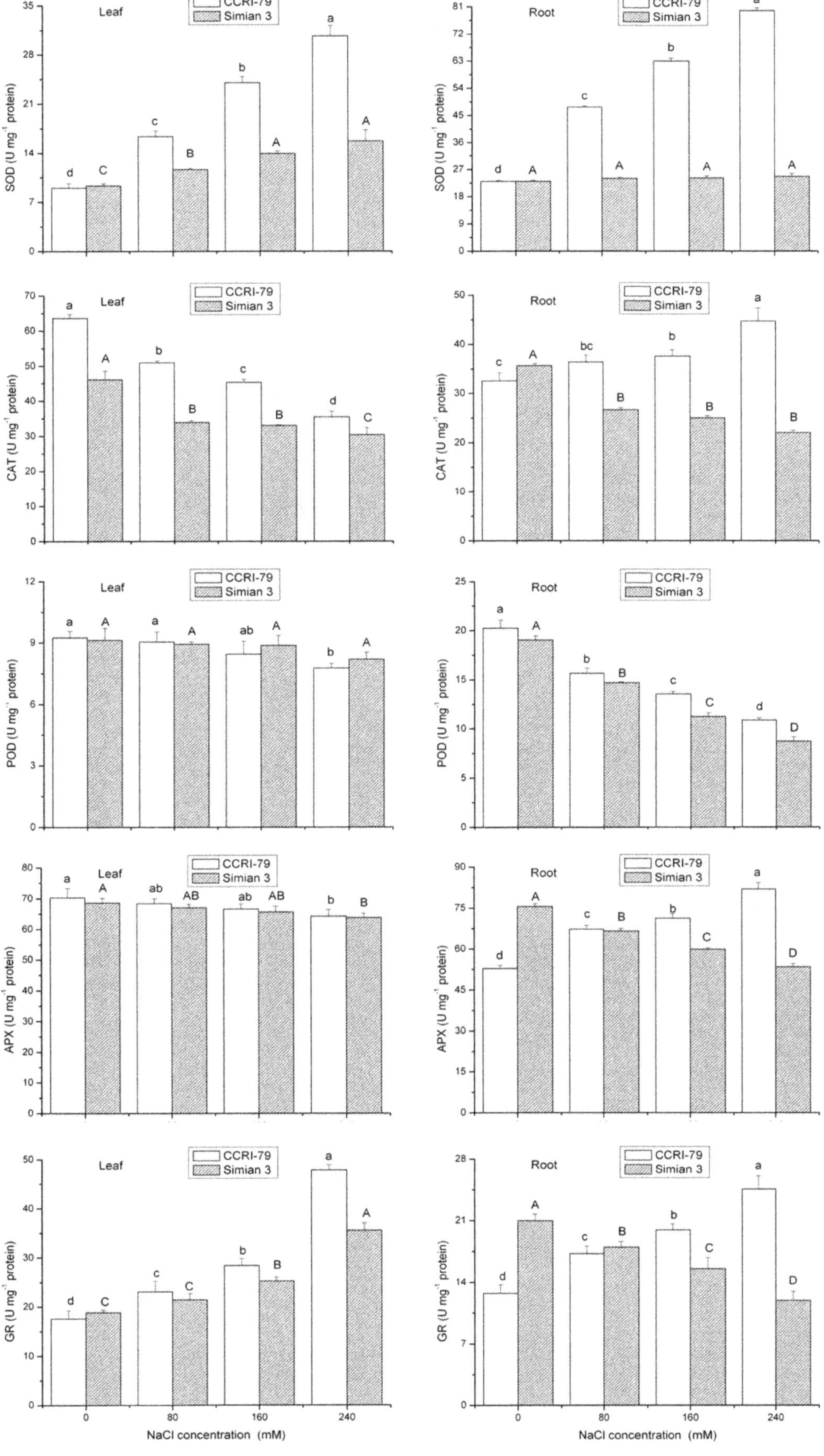

Leaf
Root
CCRI-79
Simian 3
SOD (U mg^{-1} protein)
CAT (U mg^{-1} protein)
POD (U mg^{-1} protein)
APX (U mg^{-1} protein)
GR (U mg^{-1} protein)
NaCl concentration (mM)
0
80
160
240

Figure 7. Effect of NaCl salinity on the antioxidant enzyme activities in the roots and leaves of cotton. Vertical bars represent ± standard error (n = 3). Bars labeled with the different lowercase letters on open square bars or uppercase letters on closed square bars are significant difference (P<0.05). SOD, superoxide dismutase; CAT, catalase; POD, peroxidase; APX, ascorbate peroxidase; GR, glutathione reductase.

herein provide novel information on the mechanisms and traits involved in salt tolerance, which could be exploited for cultivar selection and breeding to increase crop production in the face of increased salinity stress.

Acknowledgments

The authors thank Dr. Zhiguo Zhou for critically reading this manuscript and technical assistance.

Author Contributions

Conceived and designed the experiments: SXY XHZ. Performed the experiments: LZ JP HJM TTC. Analyzed the data: LZ. Contributed reagents/materials/analysis tools: LZ JP TTC. Wrote the paper: LZ. Final approval of the version to be published: XHZ.

References

1. Munns R, Tester M (2008) Mechanisms of salinity tolerance. Annu Rev Plant Biol 59: 651–681.
2. Arzani A (2008) Improving salinity tolerance in crop plants: a biotechnological view. In Vitro Cell Dev Biol Anim 44: 373–383.
3. Gaber MA (2010) Antioxidative defense under salt stress. Plant Signaling &Behaviour 5: 369–374.
4. Khan MA, Ungar IA, Showalter AM (2000) Effects of salinity on growth, water relations and ion accumulation in the subtropical perennial halophyte, *Atriplex griffithii var. stocksil*. Ann Bot 85: 225–232.
5. Locy RD, Chang CC, Neilson BL, Singh NK (1996) Photosynthesis in salt-adapted heterotrophic tobacco cells and regenerated plants. Plant Physiol 110: 321–328.
6. Rout NP, Shaw BP (2001) Salt tolerance in aquatic macrophytes: possible involvement of the antioxidative enzymes. Plant Sci 160: 415–423.
7. Zheng C, Jiang D, Liu F, Dai T, Jing Q, et al. (2009) Effects of salt and water logging stresses and their combination on leaf photosynthesis, chloroplast ATP synthesis, and antioxidant capacity in wheat. Plant Sci 176: 575–582.
8. Foyer CH, Halliwell B (2000) Oxygen processing in photosynthesis: regulation and signaling. New Phytol 146: 359–388.
9. Ashraf M (2009) Biotechnological approach of improving plant salt tolerance using antioxidants as markers. Biotechnol Adv 27: 84–93.
10. Sigaud-Kutner TSC, Pinto E, Okamoto OK, Latorre LR, Colepicolo P (2002) Changes in superoxide dismutases activity and photosynthetic pigment content during growth of marine phytoplankters in batch-cultures. Plant Physiol 114: 566–571.
11. Gratao PL, Polle A, Lea PJ, Azevedo RA (2005) Making the life of heavy metal-stressed plants a little easier. Funct Plant Biol 32: 481–494.
12. Noctor G, Foyer CH (1998) Ascorbate and glutathione: keeping active oxygen under control. Annu Rev Plant Phys 49: 249–279.
13. Blokhina O, Virolainen E, Fagestedt KV (2002) Antioxidants, oxidative damage, and oxygen deprivation stress: A review. Ann Bot 91:179–194.
14. Gosset DR, Millhollon EP, Lucas MC (1994) Antioxidant response to NaCl stress in salt-tolerant and salt-sensitive cultivar of cotton. Crop Sci. 34: 706–714.
15. Blum A, Ebercon A (1981) Cell membrane stability as a measure of drought and heat tolerance in wheat. Crop Sci 21: 43–47.
16. Gosset DR, Millhollon EP, Lucas MC (1994) Antioxidant response to NaCl stress in salt-tolerant and salt-sensitive cultivars of cotton. Crop Sci 34: 706–714.
17. Dionisio-Sese ML, Tobita S (1998) Antioxidant response of rice seedlings to salinity stress. Plant Sci 135: 1–9.
18. Hernandez JA, Jimenez A, Mullineaux P, Sevilla F (2000) Tolerance of pea (*Pisum sativum* L.) to long-term salt stress is associated with induction of antioxidant defenses. Plant Cell Environ 23: 853–862.
19. Mittova V, Tal M, Volokita M, Guy M (2003) Up-regulation of the leaf mitochondrail and peroxisomal antioxidative systems in response to salt-induced oxidative stress in the wild salt-tolerant tomato species *Lycopersicon pennellii*. Plant Cell Environ 26: 845–856.
20. Meneguzzo S, Navario-Izzo F, Izzo R (1999) Antioxidative responses of leaves and roots of wheat to increasing NaCl concentrations. J Plant Physiol 155: 274–280.
21. Xiong L, Zhu JK (2002) Molecular and genetic aspects of plant responses to osmotic stress. Plant Cell Environ 25: 131–139.
22. Meng HB, Jiang SS, Hua SJ, Lin XY, Li YL, et al. (2011) Comparison between a tetraploid turnip and its diploid progenitor (*Brassica rapa* L.): the adaptation to salinity stress. ASC 10: 363–375.
23. Steduto P, Albrizio R, Giorio P (2000) Gas exchange response and stomatal and non-stomatal limitations to carbon assimilation of sunflower under salinity. Environ Exp Bot 44: 243–255.
24. Fedina IS, Tsoner T, Guleva EI (1993) The effect of pretreatment with proline on the response of *Pisum sativum* to salt stress. Photosynthetica 29: 521–527.
25. Flowers TJ, Yeo AR (1995) Breeding for salinity resistance in crop plants: Where next? Aust J Plant Physiol 22: 875–884.
26. Qidar M, Shams M (1997) Some agronomic and physiological aspects of salt tolerance in cotton (*Gossypium hirsutum* L.). Journal of Agronomy and Crop Science 179: 101–106.
27. Guo WX, Mass SJ, Bronson KF (2012) Relationship between cotton yield and soil electrical conductivity, topography, and Landsat imagery. Precision Agronomy 13(2): 678–692.
28. Zhang L, Zhang GW, Wang YH, Zhou ZG, Meng YL, et al. (2013) Effect of soil salinity on physiological characteristics of functional leaves of cotton plants. J Plant Res 126(2): 293–304.
29. Ahmads S, Khan N, Iqbal MZ (2002) Salt tolerance of cotton (*Gossypium hirsutum* L.). Asian Journal of Plant Science 1(6): 715–719.
30. Sestak K, Catsky J, Jarvis PG (1971) Plant photosynthetic production. Dr. W Junk N.V publishers The Hague.
31. Zheng Y, Wang Z, Sun X, Jia A, Jiang G, et al. (2008) Higher salinity tolerance cultivars of winter wheat relieved senescence at reproductive stage. Environ Exp Bot 62(2): 129–138.
32. Ashraf M, Orooj A (2006) Salt stress effects on growth, ion accumulation and seed oil concentration in an arid zone traditional medicinal plant ajwain (*Trachyspermum ammi* [L.] Sprague). J Arid Environ 64(2): 209–220.
33. Liu J, Guo WQ, Shi DC (2010) Seed germination, seedling survival, and physiological response of sunflowers under saline and alkaline conditions. Photosynthetica 48(2): 278–286.
34. Bernt E, Bergmeyer HU (1974) Inorganic peroxidases. In: Bergmeyer HU (ed) methods of enzymatic analysis, vol 4. Academic press, New York: 2246–2248.
35. Foster JG, Hess JL (1980) Responses of superoxide dismutase and glutathione reductase activities in cotton leaf tissue exposed to an atmosphere enriched in oxygen. Plant Physiol 66: 482–487.
36. Beers RF, Sizer IW (1952) A spectrophotometric method of measuring the breakdown of hydrogen peroxide by catalase. J Biol Chem 195: 133–140.
37. Tan W, Liu J, Dai T, Jing Q, Cao W, et al. (2008) Alterations in photosynthesis and antioxidant enzyme activity in winter wheat subjected to post-anthesis waterlogging. Photosynthetica 46: 21–27.
38. Nakano Y, Asada K (1981) Hydrogen peroxide is scavenged by ascorbate-specific peroxidase in spinach chloroplasts. PCPhy 22(5): 867–880.
39. Foyer CH, Halliwell B (1976) The presence of glutathione and glutathione reductase in chloroplasts: a proposed role in ascorbic acid metabolism. Planta 133: 21–25.
40. Takemura T, Hanagata N, Sugihara K, Baba S, Karube I, et al. (2000) Physiological and biochemical responses to salt stress in the mangrove, *Bruguiera gymnorrhiza*. Aquat Bot 68: 15–28.
41. Garratt LC, Janagoudar BS, Lowe KC, Anthony P, Power JB, et al. (2002) Salinity tolerance and antioxidant status in cotton cultures. Free Radical Biol Med 33(4): 502–511.
42. Meloni DA, Oliva MA, Ruiz HA, Martinez CA (2001) Contribution of proline and inorganic solutes to osmotic adjustment in cotton under salt stress. J Plant Nutr 24: 599–612.
43. Parida AK, Das AB (2005) Salt tolerance and salinity effects on plants: a review. Ecotoxicol Environ Saf 60: 324–349.
44. Curtis PS, Läuchli A (1986) The role of the leaf area development and photosynthetic capacity in determining growth of Kenaf under moderate salt stress. Aust J Plant Physiol 18: 553–565.
45. Thiel G, Lynch J, Läuchli A (1988) Short-term effects of salinity stress on the turgor and elongation of growing barley leaves. J Plant Physiol 132: 38–44.
46. Cuartero J, Fernandez-Munoz R (1999) Tomato and salinity. HortScience 78: 83–125.
47. Siroka B, Huttova J, Tamas L, Simonoviva M, Mistrik I (2004) Effect of cadmium on hydrolytic enzymes in maize root and celeopptile. Biologia 59: 513–517.
48. Berkelaar E, Beverley H (2000) The relationship between morphology and cadmium accumulation in seedlings of two durum wheat cultivars. Canadian Journal of Botany 78: 381–387.

49. Singh AK, Dubey RS (1995) Changes in chlorophyll a and b contents and activities of photosystems I and II in rice seedlings induced by NaCl. Photosynthetica 31: 489–499.
50. Munns R, James RA, Lauchli A (2006) Approaches to increasing the salt tolerance of wheat and other cereals. J Exp Bot 57: 1025–1043.
51. Zhang Q, Alfarra MR, Worsnop D, Allan JD, Coe H, et al. (2005) Deconvolution and quantification of hydrocarbon-like and oxygenated organic aerosols based on aerosol mass spectrometry. Environment Science and Technology 39: 4938–4952.
52. Lee MH, Cho EJ, Wi SG, Bae H, Kim JE, et al. (2013) Divergences in morphological changes and antioxidant responses in salt-tolerant and salt-sensitive rice seedlings after salt stress. Plant Physiol Biochem 70: 325–335.
53. de Lacerda CF, Cambraia J, Oliva MA, Ruiz HA, Prisco JT (2003) Solute accumulation and distribution during shoot and leaf development in two sorghum genotypes under salt stress. Environ Exp Bot 49(2): 107–120.
54. Meloni DA, Oliva MA, Martinez CA, Cambraia J (2003) Photosynthesis and activity of superoxide dismutase, peroxidase and glutathione reductase in cotton under salt stress. Environ Exp Bot 49(1): 69–76.
55. Asharf M, Foolad MR (2005) Pre-sowing seed treatment- a shotgun approach to improve generation, plant growth, and crop yield under saline and non-saline conditions. Advances in Agronomy 88: 223–271.
56. Dubey RS (1997) Photosynthesis in plants under stressful conditions. In: Pessarakli, M. (Ed.), Handbook of Photosynthesis. Marcel Dekker, New York,: 859–975.
57. Brugnoli E, Lauteri M (1991) Effects of salinity on stomatal conductance, photosynthetic capacity and capacity and carbon isotope discrimination of salt-tolerant (*Gossypium hirsutum* L.) and salt-sensitive bean (*Phaseilus vulgaris* L.) C3 non-halophytes. Plant Physiol 95: 658–635.
58. Delfine S, Alvino A, Villani MC, Loreto F (1999) Restrictions to carbon dioxide conductance and photosynthesis in spainach leaves recovering from salt stress. Plant Physiol 119: 1101–1106.
59. Parida A, Das AB, Mittra B (2004) Effects of salt on growth, ion accumulation, photosynthesis and leaf anatomy of the mangrove, *Bruguiera parviflora*. Trees: Structure and Function 18: 167–174.
60. Kaya C, Tuna AL, Asharf M, Altunlu H (2007) Improved salt tolerance of molon (*Cucumis melo* L.) by the addition of proline and potassium nitrate. Environ Exp Bot 60: 397–403.
61. Borsani O, Valpuesta V, Botella MA (2003) Developing salt tolerant plants in a new century: a molecular biology approach. Plant Cell Tissue and Organ Culture 73: 101–115.
62. Asgari HR, Cornelis W, Damme PV (2012) Salt stress on wheat (*Triticum aestivum* L.) growth and leaf ion concentrations. International Journal of Plant Production 6: 195–208.
63. Taleisnik E, Grunberg K (1994) Ion balance in tomato cultivars differing in salt tolerance. I. Soidium and potassium accumulation and fluxes under moderate salinity. Plant Physiol 92: 528–534.
64. Graifenberg A, Giustiniani L, Temperini O, Paola ML (1995) Allocation of Na, Cl, K and Ca within plant tissues in globe artichoke (*Cynara scolimus* L.) under saline-sodic conditions. Scientia Horticulturae 63: 1–10.
65. Khan MA (2001) Experimental assessment of salinity tolerance of *Ceriops tagal* seedlings and saplings from the *Indus delta*. Pakistan Journal of Acquatic Botany 70: 259–268.
66. Munns R, Schachtman DP, Condon AG (1995) The significance of a two-phase growth response to salinity in wheat and barely. Aust J Plant Physiol 22: 561–569.
67. Niu X, Bressan RA, Hasegawa PM, Pardo JM (1995) Ion homeostasis in NaCl stress environments. Plant Physiol 109: 735–742.
68. Hernandez JA, Jimenez A, Mullineaux P, Sevilla F (2002) Tolerance of pea (*Pisum sativum* L.) to long-term salt stress is associated with induction of antioxidant defenses. Plant Cell Environ 23: 853–862.
69. Boo YC, Jung J (1999) Water deficit-induced oxidative stress and antioxidative defense in rice plants. Plant Physiol 155: 255–261.
70. Seckin B, Turkan I, Sekmen AH, Ozfidan C (2010) The role of antioxidant defense systems at differential salt tolerance of *Hordeum marinum* Hds. (sea barlygrass) and *Hordeum vulgare* L. (cultivated barley). Environ Exp Bot 69: 76–85.
71. Ruiz-Lozano JM, Azcon R (1997) Effect of calcium application on the tolerance of mycorrhizal lettuce plants to polyethylene glycol-induced water stress. Symbiosis 23: 9–21.
72. Rabie GH, Almadini AM (2005) Role of bioinoculants in development of salt-tolerance of *Vicia faba* plants under salinity stress. Africa Journal of Biotechnology 4: 210–220.
73. Sharma SS, Dietz KJ (2006) The significance of amino acids and amino acid derived molecules in plant responses and adaptation to heavy metal stress. J Exp Bot 57: 711–726.
74. Foyer CH, Lopez-Delgado H, Dat JF, Scott IM (1997) Hydrogen peroxide and glutathione-associated mechanism of acclimatory stress tolerance and signaling. Plant Physiol 100: 241–254.
75. Senadheera P, Tirimanne TLS, Maathuis FJM (2012) Long term salinity stress reveals variety specific differences in root oxidative stress response. Rice Sci 19: 36–43.
76. Lee DH, Kim YS, Lee CB (2001) The inductive responses of the antioxidant enzymes by salt stress in the rice (*Oryza sativa* L.). J Plant Physiol 158(6): 737–745.
77. Queval G, Hager J, Gakiere B, Noctor G (2008) Why are literature data for H_2O_2 contents so variable? A discussion of potential difficulties in the quantitative assay of leaf extracts. J Exp Bot 59: 135–146.
78. Morita S, Kaminaka H, Masumura K, Tanaka K (1999) Induction of rice cytosolic ascorbate peroxidase mRNA by oxidative stress: the involvement of hydrogen peroxide in oxidative stress signaling. Plant Cell Physiol 40: 417–422.
79. Breusegem FV, Vranova E, Dat JF, iNZE D (2001) The role of active oxygen species in plant signal transduction. Plant Sci 161: 405–414.
80. Sekmen AH, Turkan I, Takio S (2007) Differential responses of antioxidative enzymes and lipid peroxidation to salt stress in salt-tolerant Plantago maritima and salt-sensitive Plantago media. Plant Physiol 131: 399–411.
81. Mittler R (2002) Oxidative stress, antioxidants and stress tolerance. Trends in Plant Science 7(9): 405–410.
82. Gill SS, Tuteja N (2010) Reactive oxygen species and antioxidant machinery in abiotic stress tolerance in crop plants. Plant Physiol Biochem 48: 909–930.
83. Henandez JA, Jimenez A, Mullineaux P, Sevilla F (2000) Tolerance of pea to long-term salt stress in associated with induction of antioxidant defenses. Plant Cell Environ 23: 853–862.
84. Laspina NV, Groppa MD, Tomaro ML, Benavides MP (2005) Nitric oxide protects sunflower leaves against Cd-induced oxidative stress. Plant Sci 169: 323–330.
85. Zhang Q, Zhang JZ, Chow WS, Sun LL, Chen JW, et al. (2011) The influence of low temperature on phtosynthesis and antioxidant enzymes in sensitive banana and tolerant plantain (*Musa sp.*) cultivars. Photosynthetica 49: 201–208.
86. Asada K (2006) Production and scavenging of reactive oxygen species in chloroplasts and their functions. Plant Physiol 141: 391–396.
87. Amor NB, Jimenez A, Megdiche W, Lundqvist M, Sevilla M, et al. (2007) Kinetics of the antioxidant response to salinity to the halophyte *Cakile maritime*. J Integrative Plant Biol 126: 446–457.
88. Ben Amor N, Ben Hamed K, Debez A, Grignon C, Abdelly C (2005) Physiological and antioxidant responses of the perennial halophyte *Crithmum maritimum* to salinity. Plant Sci 168(4): 889–899.
89. Xu R, Yamada M, Fujiyama H (2013) Lipid peroxidation and antioxidative enzymes of two turgrass species under salinity stress. Pedosphere 23(2): 213–222.
90. Nakano Y, Asada K (1981) Hydrogen peroxide is scavenged by ascorbate-specific peroxidase in spinsth chloroplasts. PCPhy 22: 867–880.
91. Bower C, Montagu MV, Inze D (1992) Superoxide dismutase and stress tolerance. Annu Rev Plant Physiol Plant Mol Biol 43: 81–116.
92. Demiral T, Turkan I (2005) Comparative lipid peroxidation, antioxidant defense systems and proline content in roots of two rice cultivars differing in salt tolerance. Environ Exp Bot 53: 247–257.
93. Locato V, Gadaleta C, Gara I, Pinto MC (2008) Production of reactive species and modulation of antoxidant network in response to heat shock: a critical balance for cell fate. PCPhy 31: 1606–1619.
94. Liang Y, Chen Q, Liu Q, Zhang W, Ding R (2003) Exogenous silicon (Si) increases antioxidant enzyme activity and reduces lipid peroxidation in roots of salt-stressed barley (*Hordeum vulgare* L.). J Plant Physiol 160: 1157–1164.
95. Sudhakar C, Lakshmi A, Giridarakumar S (2001) Changes in the antioxidant enzyme efficiency in two high yielding genotypes of mulberry (*Morus alba* L.) under NaCl salinity. Plant Sci 161: 613–619.

14

Transcriptome and Biochemical Analyses Revealed a Detailed Proanthocyanidin Biosynthesis Pathway in Brown Cotton Fiber

Yue-Hua Xiao*, Qian Yan, Hui Ding, Ming Luo, Lei Hou, Mi Zhang, Dan Yao, Hou-Sheng Liu, Xin Li, Jia Zhao, Yan Pei

Biotechnology Research Center, Southwest University, Beibei, Chongqing, China

Abstract

Brown cotton fiber is the major raw material for colored cotton industry. Previous studies have showed that the brown pigments in cotton fiber belong to proanthocyanidins (PAs). To clarify the details of PA biosynthesis pathway in brown cotton fiber, gene expression profiles in developing brown and white fibers were compared via digital gene expression profiling and qRT-PCR. Compared to white cotton fiber, all steps from phenylalanine to PA monomers (flavan-3-ols) were significantly up-regulated in brown fiber. Liquid chromatography mass spectrometry analyses showed that most of free flavan-3-ols in brown fiber were in 2, 3-trans form (gallocatechin and catechin), and the main units of polymeric PAs were trihydroxylated on B ring. Consistent with monomeric composition, the transcript levels of flavonoid 3′, 5′-hydroxylase and leucoanthocyanidin reductase in cotton fiber were much higher than their competing enzymes acting on the same substrates (dihydroflavonol 4-reductase and anthocyanidin synthase, respectively). Taken together, our data revealed a detailed PA biosynthesis pathway wholly activated in brown cotton fiber, and demonstrated that flavonoid 3′, 5′-hydroxylase and leucoanthocyanidin reductase represented the primary flow of PA biosynthesis in cotton fiber.

Editor: Brett Neilan, University of New South Wales, Australia

Funding: This work was partially supported by the National Natural Science Foundation of China (31130039 to P.Y., 30971713 and 31271769 to X.Y.H) and by the Genetically Modified Organisms Breeding Major Project of China (2013ZX08005005-001 to X.Y.H). The funders had no role in study design, data collection and analysis, decision to publish, or preparation of the manuscript.

Competing Interests: The authors have declared that no competing interests exist.

* E-mail: xiaoyuehua@swu.edu.cn

Introduction

Naturally colored cotton is an important raw material for ecological textiles. With naturally colored cotton, textile manufacturers can eliminate dyeing during processing, and significantly reduce processing costs, environmental pollutions and chemical residues in fabrics [1,2]. In addition, naturally colored cotton may have lower flammability and higher ultraviolet protection value compared to traditional white cotton [3,4]. Brown cotton fibers with different shades are most widely used in the modern colored cotton industry. A wealth of information has been obtained about the chemical properties and biosynthesis pathway of brown pigments in cotton fiber, suggesting that these pigments belonged to proanthocyanidins (PAs) [5–9]. However, exact chemical properties of PA pigments and details of the PA biosynthesis pathway in brown cotton fiber are still to be elucidated.

PAs, also known as condensed tannins, are widely distributed in plants with various functions such as pigments in seed coat and protectants against herbivores and microbes [10,11]. PA is also an important factor affecting mouthfeel, contributing the bitter flavor and astringency to our daily foods and beverages [12]. In addition, PA's antioxidant and anti-inflammatory properties make it a potential chemopreventive and chemotherapeutic agent for some human diseases, including cancers [12,13]. Chemically, PAs are oligomers or polymers of polyhydroxy flavan-3-ol units. PA polymers are synthesized presumably by adding flavan-3, 4-diol (leucoanthocyanidin) molecules to an initiating flavan-3-ol unit or the terminal unit of a flavan-3-ol chain (Figure 1) [10,11,14]. Both flavan-3-ols and flavan-3, 4-diols are synthesized through plant flavonoid pathway. The details of this pathway vary with tissues and species and determine PA compositions in different plants [15].

In attempt to clarify the details of PA biosynthesis pathway in brown cotton fiber, we performed a digital gene expression (DGE) analysis to compare the gene expression profiles in brown and white fibers. A total of 24 PA synthase genes were identified to be significantly up-regulated in brown fiber. Furthermore, we determined the chemical properties of PAs in brown fiber by using liquid chromatography-mass spectrometry (LC-MS) method. It was found that the majority of PA initiating units consisted of gallocatechin and the main extension units in brown fiber were trihydroxylated on B ring. These results demonstrated a detailed PA pathway involved in the brown pigmentation in cotton fiber, which may be essential to manipulate the biosynthesis of pigment and other flavonoids in cotton fiber.

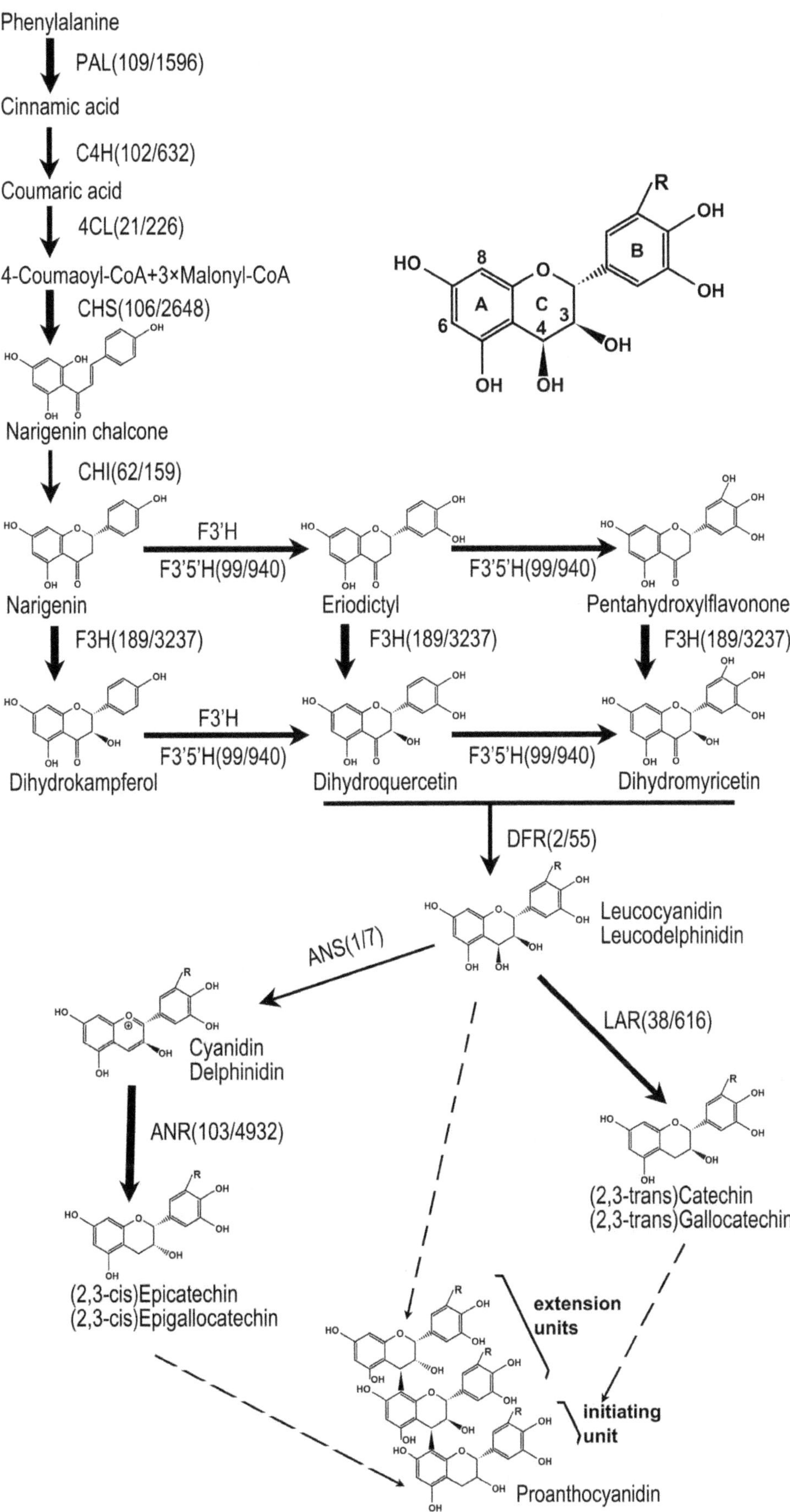

Phenylalanine
PAL(109/1596)
Cinnamic acid
C4H(102/632)
Coumaric acid
4CL(21/226)
4-Coumaoyl-CoA+3×Malonyl-CoA
CHS(106/2648)
Narigenin chalcone
CHI(62/159)
F3'H
F3'5'H(99/940)
Narigenin
Eriodictyl
Pentahydroxylflavonone
F3H(189/3237)
Dihydrokampferol
Dihydroquercetin
Dihydromyricetin
DFR(2/55)
Leucocyanidin
Leucodelphinidin
ANS(1/7)
LAR(38/616)
Cyanidin
Delphinidin
ANR(103/4932)
(2,3-trans)Catechin
(2,3-trans)Gallocatechin
(2,3-cis)Epicatechin
(2,3-cis)Epigallocatechin
extension units
initiating unit
Proanthocyanidin

Figure 1. The detailed PA biosynthesis pathway in brown cotton fiber. A typical PA monomer is depicted in the up right (R = H or OH). Solid arrows represent reactions from substrates to products with corresponding synthases indicated. The arrow line thickness roughly reflects the expression levels of corresponding synthases and putative flow rates in these steps. Expression levels are classified into 4 categories according to transcript abundance detected in the DGE analysis, i.e. low, moderate, high and very high expression (2~20, 20~200, 200~1000 and over 1000 TPM, respectively). Transcript levels (in TPM) of various PA synthases in white/brown fibers are indicated. Dashed arrows indicate the monomeric origins of oligomeric or polymeric PAs. PA synthases are abbreviated as in Table 1.

Materials and Methods

RNA Extraction and Solexa Sequencing

A recombination inbred line (RIL) population derived from the cross between white fiber cotton cultivar Yumian No.1 and brown fiber line T586 was constructed as described [16]. Total RNAs were extracted from fibers of 14 days post anthesis (DPA) using a modified CTAB method [17]. Equal amounts of total RNAs from each 10 white and brown fiber RILs were mixed to form two RNA bulks (WCF and BCF, respectively). DGE analyses of RNA bulks were performed in BGI-Shenzhen using standard procedure (Shenzhen, Guangdong, China) [18,19]. Briefly, biotin-labeled oligo d(T) primer was employed to initiate the synthesis of double-stranded cDNA. After *Nla*III restriction, the 3′ cDNA ends were separated by magnetic method and linked to an adaptor containing a *Mme*I recognition site. After *Mme*I digestion and ligation to the second adaptor, the tags were sequenced on an Illumina Genome Analyzer (Illumina, Inc., San Diego, CA) using Solexa technology. The raw sequence data were deposited in NIH Sequence Read Archive under the accession number SRP033354. Sequenced tags were annotated by aligning with *Gossypium hirsutum* unigenes and singleton ESTs (ftp://ftp.ncbi.nih.gov/repository/UniGene/Gossypium-hirsutum/Ghi.seq.uniq.gz, 2010) [20]. Differentially expressed genes (DEGs) in WCF and BCF bulks were identified according to the frequency of corresponding tags by setting a cutoff on a false discovery rate (FDR) <0.001 and a

Table 1. The real-time PCR primers used in this study.

Genes in D genome	PA synthases[1]	Primers	Sequences (5′ to 3′)
Gorai.002G248000	PAL	PAL-F	AGCTTGGAACTGGGTTGTTG
		PAL-R	AGCACCATTCCAACCCTTTA
Gorai.013G271700	C4H	C4H-F	TTTGGGTCGTTTGGTACAGA
		C4H-R	AAAATTGCCTTGGCTTAGCA
Gorai.003G052100	4CL	4CL-F	AAGGTGCACTTTGTTCATGC
		4CL-R	CGTTGCAATTTAAAAGCCAAAT
Gorai.011G161200	CHS	CHS-F	CAGGAGAAGGACTGGAGTGG
		CHS-R	AGCAGCAACACTATGGAGCA
Gorai.013G023400	CHI	CHI-F	ATGGAGTTTCTCCTCCAGCA
		CHI-R	GGTTTTTCACTGTCGACTCCA
Gorai.008G062900	F3H	F3H-F	CTGAAGAAGCTGGCCAAAGA
		F3H-R	TGCAAGGATTTCCTCCAATG
Gorai.008G198200	F3′H	F3′H-F	GCTGATGTTAGGGGCAATGA
		F3′H-R	CTCACCATGAAACGACAACG
Gorai.001G134900	F3′5′H	F3′5′H-F	AAACATGGATGAGGCCTTTG
		F3′5′H-R	GCAAGGGATGTGCTTAGGAA
Gorai.009G200600	DFR	DFR-F	CATGTTCGTAGGAGCTGTCG
		DFR-R	GGTAGGCACTCAATTGTTGAAA
Gorai.008G186500	LAR	LAR-F	GAATGAGCCATTCCGAACAT
		LAR-R	GCTTCGACTACTGGCTTTGG
Gorai.004G205900	ANS	ANS-F	GCCACCGAAGGATAAGATCA
		ANS-R	TGGGTCTTCCTGAACAGCTT
Gorai.009G175500	ANR	ANR-F	TGGGATCGAGGAAATCTACG
		ANR-R	ACCATAATCATTGGGGAAGC
	Histone 3	His-F	GAAGCTGCAGAGGCATACC
		His-R	CTACCACTACCATCATGGC
	Elongation initiation	Eif5-F	GGTTGCCATTGTGCAAGGA
	factor 5	Eif5-R	CCGTAGGTGAGCGTTAATCAGA

[1]4CL, 4-coumarate:CoA ligase; ANR, anthocyanidin reductase; ANS, anthocyanidin synthase; C4H, cinnamate 4-hydroxylase; CHI, chalcone isomerase; CHS, chalcone synthase; DFR, dihydroflavonol 4-reductase; F3H, flavone 3-hydroxylase; F3′H, flavonoid 3′-hydroxylase; F3′5′H, flavonoid 3′5′-hydroxylase; LAR, leucoanthocyanidin reductase; PAL, phenylalanine ammonia-lyase.

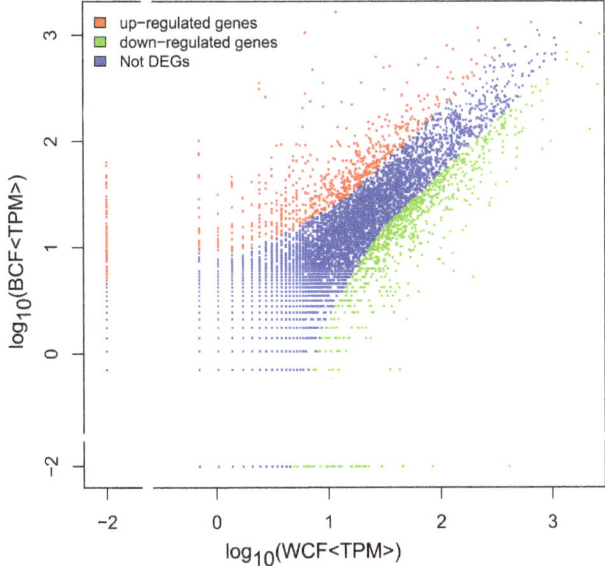

Figure 2. Distribution of gene expression levels in white and brown fibers. Gene expression level was standardized to transcripts per million (TPM). When no transcript was detected in a certain RNA bulk, the expression level was set to 0.01 TPM. WCF and BCF, white and brown cotton fiber RNA bulk, respectively. Differentially expressed genes (DEGs) include up-regulated and down-regulated genes with expression level ratio (BCF/WCF) >2 and <0.5, respectively, and the false discovery rate (FDR) <0.001.

frequency ratio of BCF to WCF >2 or <0.5. Interested unigenes and ESTs were mapped to cotton D genome [21] by blast searching in phytozome (http://www.phytozome.net/search.php) [22].

Quantitative RT-PCR

Quantitative RT-PCR (qRT-PCR) was employed to detect the expression levels of predominant PA synthase genes in each 10 brown and white fiber RILs. The investigated genes and corresponding primers were listed in table 1. The histone 3 and elongation initiation factor 5 genes from cotton were amplified as RNA standard [23,24]. PCRs were performed on a CFX96^{TM} real-time PCR detection system with SYB Green supermix (Bio-Rad, CA, USA). The thermocycling parameters were as follows: 95°C, 2 min, 40 cycles of 95°C, 10 s and 57°C, 20 s. A standard melting curve was added to monitor the specificity of PCR products. The reactions were duplicated for 3 times and data were analyzed using the software Bio-Rad CFX Manager 2.0 provided by manufacturer.

Extraction and Purification of Cotton Fiber PAs

PAs were extracted from developing brown cotton fibers and purified as described [25]. Around 10 g fibers harvested from bolls of 20 DPA were ground to fine powder in liquid N_2, and extracted in 50 ml 80% aqueous acetone twice at 4°C for 2 h. Solutions were centrifuged at 10 000 rpm for 5 min. Supernatants were combined and acetone was evaporated under vacuum at 35°C. Two-microliter aliquots of the remaining aqueous solution were applied to a packed Supelco Discovery DPA-6S polyamide cartridge (500 mg, Sigma-Aldrich Chemie Inc.). After washing with 5 ml 30% methanol, PAs were eluted with 2 ml 90% N, N-dimethylformamide. Finally, eluates were evaporated to dry under vacuum, re-suspended in methanol, and subjected to LC-MS analyses.

LC-MS Analysis

To detect monomeric flavan-3-ols in the purified PAs, LC-MS analyses were performed on LC-MS 2010A system (Shimadzu, Japan) with an Xtimate C18 column (2.1×150 mm, 5 µm, Welch Materials, Inc., Shanghai, China). Solvent A was water: formic acid (99:1, v/v), and solvent B was acetonitrile: formic acid (99:1, v/v). Around 5 µg PAs were injected and eluted with a gradient of solvent B to A at a flow rate of 0.2 ml/min. After 3-min isocratic elution in 90% solvent A and 10% solvent B, solvent B concentrations increased from 10% to 30% (v/v) in 12 min, followed 15-min washing with 100% of B and 3-min re-equilibration in 10% of B. Positive ions were monitored in a selected ion monitoring mode. Monomeric flavan-3-ols were identified according to the typical ions and retention times of authentic standards. Catechin and epicathechin with $[290+H]^+$ ions were eluted at 8.64 and 14.12 min, while gallocatechin and epigallocatechin with $[306+H]^+$ ions were detected at 3.92 and 6.62 min, respectively. Flavan-3-ol amounts were calculated by reference to the standard curves of authentic standards.

To elucidate the chemical properties of PA extension unit, purified PAs were hydrolyzed by acid butanol method [26]. PAs were mixed with equal volume (50 µl) of concentrated HCl: butanol (1:9, v/v) and incubated at 100°C for 1 h. The hydrolytes were evaporated to dry under vacuum, re-suspended in 100 µl aqueous methanol (15%, v/v), and subjected to LC-MS analysis. The column and solvents were the same as those used to detect monomeric flavan-3-ols, while the gradient profile was as follows: 0 to 3 min, 10% of solvent B; 3 to 20 min, 10% to 40% of B; 20 to 40 min 100% of B; and 20 to 23 min, 10% of B for re-equilibration. Typical ions and retention times for delphinidin, cyanidin and pelargonidin (Indofine, Hislsborough, NJ, USA) were $[304+H]^+$ at 14.81 min, $[288+H]^+$ at 16.73 min, and $[272+H]^+$ at 18.33 min, respectively. Anthocyanidins were quantified according to the areas of typical peaks by reference to the standard curves of corresponding standards.

Table 2. Over-represented GO cellular components in DEGs of brown and white fibers.

Gene ontology term	Clustered gene and frequency in DEGs (989)	Clustered gene and frequency in reference genes (25492)	P-value	GO ID
membrane-bounded vesicle	226, 22.9%	4474, 17.6%	0.003	GO:0031988
cytoplasmic membrane- bounded vesicle	225, 22.8%	4461, 17.5%	0.004	GO:0016023
vesicle	226, 22.9%	4503, 17.7%	0.005	GO:0016020
membrane	328, 33.2%	6939, 27.2%	0.005	GO:0031982
cytoplasmic vesicle	225, 22.8%	4489, 17.6%	0.006	GO:0031410

Table 3. Over-represented KEGG pathways in DEGs of brown and white fibers.

Pathway	DEGs with pathway annotation (1337)	Reference genes with pathway annotation (45320)	P value	Pathway ID
Oxidative phosphorylation	57, 4.26%	1123, 2.48%	0.000	ko00190
Ubiquitin mediated proteolysis	49, 3.66%	985, 2.17%	0.000	ko04120

Results

Digital Gene Expression (DGE) Analysis of Brown and White Cotton Fibers

Accumulation of flavonoids and pigments in fibers may retard the fiber development and reduce the final fiber quality and yield [9,27,28]. To dissect the molecular basis of pigment biosynthesis and its effect on fiber development, gene expression profiles in brown and white cotton fibers (BCF and WCF) were compared by DGE analysis. More than 3 million tags were generated from the developing fibers, including 2 860 036 and 2 913 186 clean tags in BCF and WCF libraries, respectively. Among these, 35664 (31.26%) distinct tags in BCF and 29048 (30.27%) in WCF were unambiguously mapped to a certain *G. hirsutum* unigene or singleton EST (ftp://ftp.ncbi.nih. gov/repository/UniGene/Gossypium-hirsutum/Ghi.seq.uniq.gz, 2010). Gene expression levels in BCF and WCF bulks were compared according to the corresponding tag frequencies. A total of 2079 differentially expressed genes (FDR<0.001) were identified, among which 1165 genes were up-regulated (expression level ratio BCF/WCF>2) and 914 were down-regulated (BCF/WCF<0.5) in brown fibers (Figure 2).

Gene ontology (GO) analysis suggested that the cellular components of vehicles (including membrane-bounded vesicle and cytoplasmic vesicle) and membrane were most significantly over-represented in differentially expressed genes (DEGs) between white and brown fibers (Table 2). The most significantly over-represented Kyoto Encyclopedia of Genes and Genomes (KEGG) pathways in DEGs were oxidative phosphorylation and ubiquitin mediated proteolysis (Table 3). In addition, some DEGs were related to the physiological processes involved in regulation of fiber development, such as reactive oxygen [29] and ethylene [30] (Table 4). These data indicated that the brown fiber gene (*Lc1*) and/or pigment accumulation in cotton fiber may affect multiple aspects in fiber development other than fiber coloration.

Expressions of PA Synthase Genes in White and Brown Fibers

In higher plants, PAs are synthesized through the flavonoid pathway from phenylalanine to flavan-3-ols (Figure 1) [10,11,14,15]. By DGE analysis of brown and white cotton fibers, a total of 34 PA synthase genes were identified (Table 5). Among these genes, 24 were significantly up-regulated in brown fiber, but none was significantly down-regulated. These up-regulated genes encoded 3 phenylalanine ammonia-lyases (PAL), 2 cinnamate 4-hydroxylases (C4H), 1 4-coumarate:CoA ligase (4CL), 6 chalcone synthases (CHS), 1 chalcone isomerase (CHI), 2 flavanone 3-hydroxylases (F3H), 2 flavonoid 3′, 5′-hydroxylases (F3′5′H), 2 dihydroflavonol 4-reductases (DFR), 2 leucoanthocyanidin reductases (LAR), 1 anthocyanidin synthase (ANS) and 2 anthocyanidin

Table 4. ROS- and ethylene- related DEGs of brown and white fibers.

Genes	WCB (TPM)	BCF (TPM)	P-Value	FDR	Homologous proteins
ROS related					
Ghi.16954	0.01	8.39	4.73E-08	4.93E-07	cytosolic ascorbate peroxidase
Ghi.9579	1.03	16.78	7.56E-05	4.16e-04	Cu/Zn superoxide dismutase
Ghi.16311	4.12	25.87	5.37E-07	4.79E-06	Cu/Zn superoxide dismutase
Ghi.9542	89.94	42.31	1.24E-12	2.23E-11	Mn superoxide dismutase
Ghi.9864	106.76	241.95	4.26E-14	1.04E-12	Peroxidase precursor
Ghi.16267	5.15	0.01	3.53E-05	2.11e-04	Peroxidase precursor
Ghi.16323	55.95	13.99	1.58E-18	5.42E-17	peroxisomal targeting signal 2 receptor
Ethylene related					
Ghi.16253	32.95	68.18	1.91E-09	2.44E-08	1-aminocyclopropane-1- carboxylate synthase
Ghi.16693	6.52	0.01	2.29E-06	1.80E-05	1-aminocyclopropane-1- carboxylate oxidase
Ghi.4454	46.34	18.53	2.89E-09	3.59E-08	ethylene-insensitive3 protein
Ghi.16250	239.26	95.8	1.28E-41	7.92E-40	ethylene-responsive element binding protein ERF2
Ghi.22814	0.01	5.24	2.63E-05	1.64 E-04	ethylene-responsive element binding protein ERF2
Ghi.9501	5.49	0.01	1.78E-05	1.15E-049	Ethylene-responsive transcription factor
Ghi.2915	0.01	5.24	2.63E-05	1.64 E-04	Ethylene-responsive transcription factor
Ghi.13085	22.31	46.5	5.85E-07	5.16E-06	ethylene-responsive element-binding factor

Table 5. PA synthase genes identified in cotton fiber and their expressions in BCF and WCF.

PA synthase[1]	Unigene/EST	Gene in D genome	WCF (TPM)	BCF (TPM)	FDR	Expression Changes[2]
PAL	Ghi.2173	Gorai.002G248000	104.7	1564.32	1.31E-11	U
	Ghi.3957	Gorai.009G416300	1.37	15.38	2.77E-04	U
	Ghi.24914	Gorai.011G238400	2.74	14.69	4.30E-05	U
	ES851993	Gorai.007G373600	0.01	1.4	0.03	N
C4H	Ghi.4426	Gorai.013G271700	99.55	460.48	6.82E-12	U
	Ghi.9347	Gorai.011G207000	2.06	171.68	1.12E-04	U
4CL	Ghi.9697	Gorai.003G052100	12.68	261.53	4.64E-07	U
	Ghi.1797	Gorai.011G053700	6.52	2.8	0.09	N
	Ghi.21520	Gorai.009G005900	1.55	0.71	0.54	N
	Ghi.24129	Gorai.013G269100	0.01	0.7	0.34	N
CHS	Ghi.1443	Gorai.011G161200	22.31	1726.6	4.65E-12	U
	Ghi.10134	Gorai.005G035100	74.15	690.55	6.24E-12	U
	Ghi.24663	Gorai.011G161200	1.37	172.03	2.78E-04	U
	Ghi.2516	Gorai.009G339300	3.09	24.13	2.32E-05	U
	Ghi.6103	Gorai.006G000200	3.43	18.18	1.43E-05	U
	BQ410047	Gorai.009G339300	1.71	16.78	2.46E-04	U
CHI	Ghi.10760	Gorai.013G023400	30.18	122.73	0	U
	Ghi.7860	Gorai.012G014600	31.91	36.36	0.47	N
F3H	Ghi.17176	Gorai.008G062900	189.14	3230.73	5.54E-12	U
	Ghi.7987	Gorai.007G194700	0.01	4.55	5.71E-04	U
	CO123602	Gorai.008G062900	0.34	1.75	0.10	N
F3′5′H	Ghi.14535	Gorai.001G134900	96.11	781.81	0	U
	BM360843	Gorai.001G134900	3.26	158.39	5.80E-05	U
DFR	Ghi.9795	Gorai.009G200600	1.55	31.12	3.17E-04	U
	Ghi.9683	Gorai.010G008900	0.01	23.08	0	U
	BQ401793	Gorai.009G200600	0.01	1.05	0.07	N
LAR	Ghi.18163	Gorai.008G186500	38.45	597.2	0	U
	Ghi.17301	Gorai.008G285400	0.01	15.73	4.77E-13	U
	Ghi.17277	Gorai.008G186500	0.01	2.8	0.01	N
	Ghi.2177	Gorai.008G285400	0.01	0.7	0.37	N
ANS	Ghi.1234	Gorai.004G205900	0.01	7.34	3.56E-06	U
	DW488687	Gorai.011G289300	0.69	0.01	0.44	N
ANR	Ghi.8422	Gorai.009G175500	101.63	4898.19	1.15E-11	U
	Ghi.964	Gorai.009G175500	1.37	33.57	2.79E-04	U

[1]PA synthases are abbreviated as in Table 1.
[2]U, significantly up-regulated in BCF; N, not significantly changed.

reductases (ANR). Thus, for every step from PAL to ANR in the PA biosynthesis pathway (Figure 1), there was at least one encoding gene significantly up-regulated in brown fiber (Table 5).

To verify the result of DGE analysis, expression levels of the predominant PA synthase genes identified for all steps from phenylalanine to flavan-3-ols (Figure 1, Table 1 and 5) and a flavonoid 3′-hydroxylase homologous gene (F3′H, Gorai.008G198200 from *G. raimondii*) were detected in brown and white fibers via qRT-PCR. Compared to white fiber, all of the 12 PA synthase genes were up-regulated in brown fiber (16~260 fold, Figure 3A). Furthermore, the high level expressions of PA synthase genes were co-segregated with brown fiber in individual RILs, as exemplified by LAR and F3′5′H genes (Figure 3B). These results were consistent with DGE profiles and confirmed that the brown fiber gene (*Lc1*) wholly activated PA biosynthesis pathway in cotton fiber.

DGE analysis also revealed that transcript levels of different flavonoid synthases varied dramatically in cotton fibers (Table 5 and Figure 4). In white fiber, the synthases related to common steps in flavonoid biosynthesis (PAL, C4H, 4CL, CHS, CHI, F3H and F3′5′H) and PA specific steps (LAR and ANR) showed moderate expression levels (20~200 TPM), while DFR and ANS had very low expression (<2 TPM). In brown fiber, the relative transcription pattern of different PA synthases was similar to that in white fiber (Figure 4). Although the whole PA pathway was significantly up-regulated in brown fiber, DFR and ANS had moderate (55.26 TPM) and low (7.34 TPM) expression levels, respectively, in comparison to much higher expression levels of

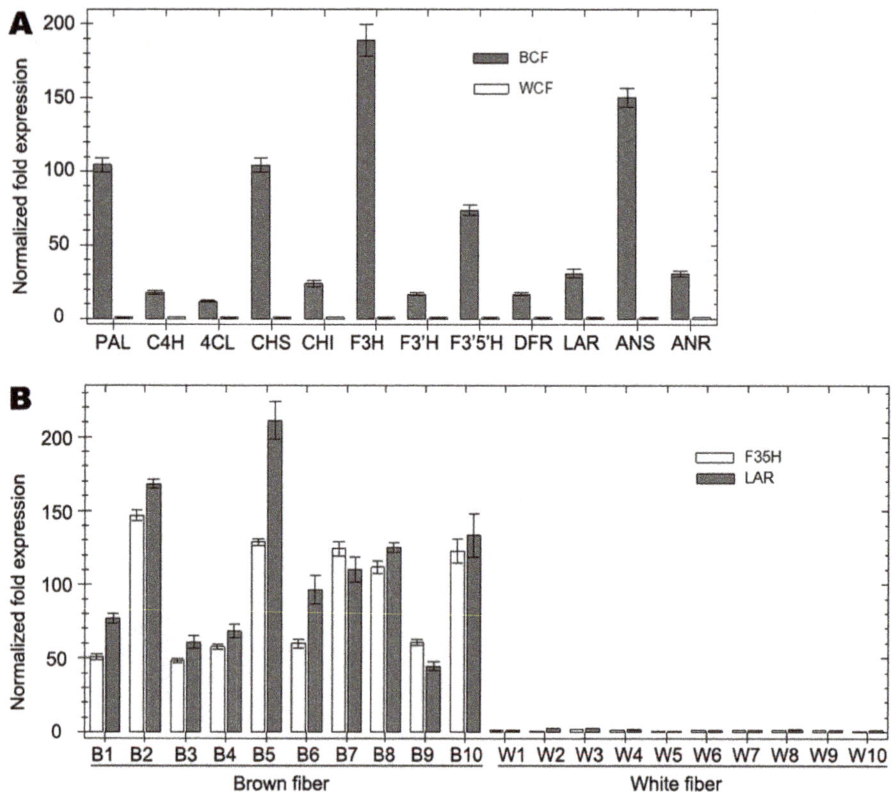

Figure 3. Detection of PA synthase genes in RILs by qRT-PCR. A, Comparison of expression levels of PA structural genes in white and brown fiber bulks (WCF and BCF, respectively). B, expression levels of LAR and F3′5′H genes in fibers of each 10 white and brown fiber RILs. Relative expression levels are normalized to cotton histone 3 and elongation initiation factor 5 genes. WCF and W1 are set as control in A and B, respectively.

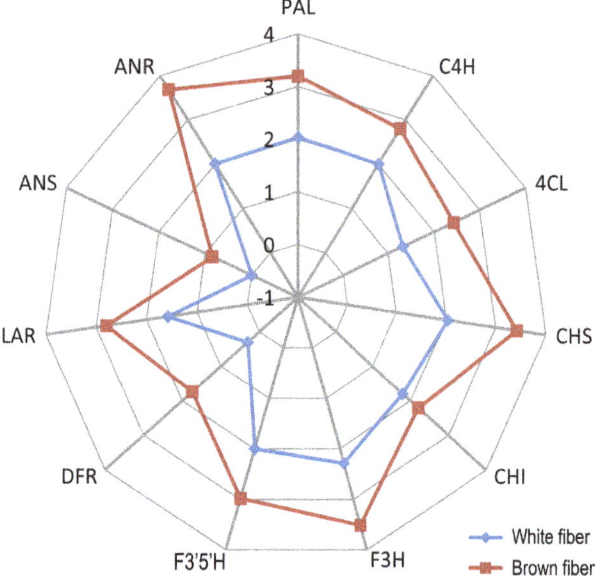

Figure 4. The transcription patterns of different PA synthases in brown and white fibers. Each point on the polygons represents a PA synthase and the scale corresponds to the transcript levels ($\log_{10}$TPM). PA synthases are abbreviated as in Table 1.

other PA synthases (e.g. 940.2 TPM for F3′5′H and 616.43 TPM for LAR). These data suggested that the relative expression profile of different flavonoid synthase genes in cotton fiber was strictly regulated at transcription level.

Monomeric Composition of PAs in Brown Cotton Fiber

Transcriptional analysis revealed a PA biosynthesis pathway wholly activated in brown cotton fiber. To further clarify the details in this pathway, we employed LC-MS method to determine the monomeric composition of PAs in brown fiber. Four flavan-3-ols (gallocatechin, epigallocatechin, catechin and epicatechin) were identified in the PAs from brown fiber by LC-MS analysis, with mol percentages of 85.4±1.4%, 3.0±0.2%, 10.8±1.2% and 0.8±0.1%, respectively (Figure 5A–C). Among these monomers, the 2, 3-trans-flavan-3-ols (gallocatechin and catechin) account for 96.2%, suggesting that most of free flavan-3-ols and then initiating units of polymeric PAs in brown cotton fiber are synthesized via LAR branch, rather than ANS/ANR (Figure 1).

The anthocyanidin composition in the acid hydrolysate reflected the extension unit composition of corresponding PAs [26]. To determine composition of PA extension units in brown fiber, we detected the anthocyanidins released by acid-butanol reactions. As shown in Figure 5D–G, three kinds of anthocyanidins (pelargonidin, cyanidin and delphinidin) were detected with the mol percentage of 9.4±0.7%, 12.8±1.4% and 77.8±1.8%, respectively. High percentage of delphinidin in PA acid hydrolysate indicated that the main PA extension units in brown cotton fiber were favan-3-ols trihydroxylated on B ring. Therefore, both initiating and extension units of PAs in brown cotton fiber consist

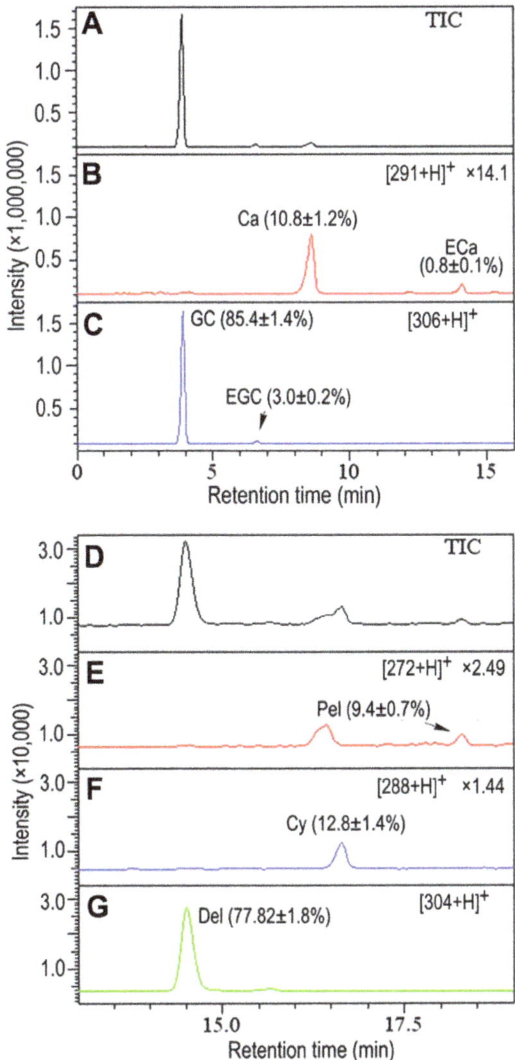

Figure 5. LC-MS analyses of PAs in brown cotton fiber. A, B and C, LC-MS profiles of monomeric flavan-3-ols in purified PAs showing total intensity count (TIC) and the intensity counts of $[290+H]^+$ and $[306+H]^+$, respectively. D, E, F and G, LC-MS profiles of anthocyanins in the PA acid hydrolysate showing TIC and the intensity counts of $[272+H]^+$, $[288+H]^+$ and $[304+H]^+$, respectively. The mol percentages of flavan-3-ols or anthocyanidins are shown in average±SD for 3 duplicate tests. The profiles in B, E and F are magnified by folds as indicated. Ca, catechin; Cy, cyanidin; Del, delphinidin; ECa, epicatechin; EGC, epigallocatechin; GC, gallocatechin and Pel, pelargonidin.

mainly of flavan-3-ols trihydroxylated on B ring, indicating that F3′5′H plays a primary role in PA biosynthesis in brown cotton fiber.

Taken together, transcriptome and biochemical analyses collectively demonstrated a detailed PA biosynthesis pathway wholly up-regulated in brown cotton fiber, in which F3′5′H and LAR represented the primary flow for PA biosynthesis (Figure 1).

Discussion

The exact chemical property of pigments is an important clue for exploitation of naturally colored cottons. Early extraction experiment suggested that pigments in naturally colored cotton belonged to flavonoids [1]. Hua *et al* revealed much higher PAL activity in brown cotton fiber compared to white fiber [9]. Expression analyses showed that several flavonoid synthase genes, such as CHI, F3H, DFR, ANS, ANR, C4H, CHS, F3′H and F3′5′H, were significantly up-regulated in brown fiber [6,8]. Recently, Li and coworkers identified 15 flavonoid-related proteins (including PAL, CHS, F3H, DFR and ANR) with high abundance in brown cotton fiber via comparative proteomic analysis of BCF and WCF near isogenic lines [5]. In addition, the concentrations of PAs and PA precursors in brown fiber were much higher than in white fiber [5,6,8,27]. These studies consistently indicated that the pigments in brown fiber belonged to PAs. In the present study, we aimed to dissect the details of PA biosynthesis pathway in brown cotton fiber. By DGE and qRT-PCR analyses, we found that all the investigated PA synthases (including PAL, C4H, 4CL, CHS, CHI, F3H, F3′H, F3′5′H, DFR, ANS, ANR and LAR) were significantly up-regulated in brown fiber, suggesting that the brown fiber gene (*Lc1*) activated the whole PA biosynthesis pathway in cotton fiber. Furthermore, biochemical analyses demonstrated that the main PA units were trihydroxylated on B ring and most of free flavan-3-ols were in 2, 3-trans form. These results demonstrated that F3′5′H and LAR represented the major flow for PA biosynthesis in brown cotton fiber. By dissecting the details of PA biosynthesis pathway in brown cotton fiber, our results paved the way to manipulate the biosynthesis of pigment and other flavonoids in cotton fiber via biotechnology techniques.

PAs are the predominant coloring compounds in seed coats, and may function as barrier to fungus infection of embryos [15]. It has been found for a long time that the PAs in cotton seed coats and fuzzes consist mainly of catechin and catechin-derived polymers [31]. Since the majority of PAs in brown cotton fiber are gallocatechin and its polymers, brown cotton fiber may have a different PA biosynthesis pathway independent of seed coat and fuzz. Additionally, with flavan-3-ols trihydroxylated on B ring as main units, brown cotton fiber may represent a novel PA resource compared to Arabidopsis, grapevine and *Medicago truncatula* which consist mainly of epicatechin and/or catechin [15]. Given the simplicity of cotton production, PAs from brown cotton fiber are also potential to be applied in food and medicine industry [12,13].

In higher plant, PAs include a large number of oligomers or polymers of flavan-3-ols. In addition to various degrees of polymerization, difference in monomeric composition is a key factor influencing the complexity of PA components. There are two major branching points in the PA pathway which lead to different PA monomers (Figure 1). Firstly, DFR converts dihydrokampferol to leucoparlegonidin finally leading to PA monomers with a single hydroxyl on B ring, while F3′5′H catalyzes the hydroxylation on C-3′ and/or C-5′ of B ring which results in PA monomers di- or trihydroylated on B ring [32]. Secondly, LAR directly converts leucoanthocyanidins to 2, 3-trans-flavan-3-ols (catechin and gallocatechin), while ANS catalyzes leucoanthocyanidins to form anthocyanindins and then 2, 3-cis-flavan-3-ols (epicatechin and epigallocatechin) with ANR activity (Figure 1). High percentages of flavan-3-ols trihydroxylated on B ring implied that dihydrokampferols in brown cotton fiber were primarily converted to dihydromyricetin instead of leucopelargonidin and therefore F3′5′H activity should be much higher than DFR. Likewise, LAR activity might be much higher than ANS in brown cotton fiber, for the free flavan-3-ols were mainly in 2, 3-trans form. Consistently, DGE analysis revealed that the expression levels of F3′5′H and LAR were dramatically higher than those of DFR and ANS, respectively. These results implied that the flavonoid profiles in cotton fiber were mainly regulated by the relative expression pattern of corresponding synthases at transcription level.

Flavonoids may play roles in many aspects of plant growth and development [33,34]. Several studies have suggested that pigment accumulation in cotton fiber may affect the fiber quality and yield [9,27,28]. Biochemical analyses indicated that the contents of PA and flavonoid precursors in brown cotton fiber were much higher than in white fibers [27]. Tan and coworkers showed that down-regulation of F3H and accumulation of flavonoid narigenin retarded the fiber development and reduced the final fiber quality and yield [27]. Our DGE analysis also showed that PA accumulation in developing cotton fiber might significantly affect several cellular components, KEGG pathways and other fiber-related physiological processes. However, the molecular basis of the negative influence of accumulation of pigment and other flavonoids on fiber quality and yield was largely unclear. Documenting the expression profiles of flavonoid synthases in white and brown fibers may facilitate to design transgenic strategy to engineer flavonoid pathway and to dissect the relationship between flavonoid accumulation and fiber development.

Conclusion

Transcriptome analysis revealed that a whole PA pathway from phenylalanine to flavan-3-ol was activated in cotton fiber by the brown fiber gene. LC-MS analyses demonstrated that most of free favan-3-ols in brown cotton fiber were in 2, 3-trans form, and the main PA units were favan-3-ols trihydroxylated on B ring. The PA monomeric composition was consistent with the expression profiles of PA synthase genes, and suggested that F3′5′H and LAR represented the major flow of the PA biosynthesis pathway in brown cotton fiber.

Author Contributions

Conceived and designed the experiments: YX. Performed the experiments: YX QY HD DY HL XL JZ. Analyzed the data: YX MZ ML. Contributed reagents/materials/analysis tools: LH MZ ML. Wrote the paper: YX YP.

References

1. Murthy MSS (2001) Never say dye: The story of coloured cotton. Resonance 6: 29–35.
2. Kimmel LB, Day MP (2001) New life for an old fiber: Attributes and advantages of naturally colored cotton. AATCC Review 1: 32–36.
3. Parmar MS, Chakraborty M (2001) Thermal and burning behavior of naturally colored cotton. Textile Research Journal 71: 1099–1102.
4. Hustvedt G, Crews PC (2005) The ultraviolet protection factor of naturally-pigmented cotton. The Journal of Cotton Science 9: 47–55.
5. Li Y-J, Zhang X-Y, Wang F-X, Yang C-L, Liu F, et al. (2013) A comparative proteomic analysis provides insights into pigment biosynthesis in brown color fiber. Journal of Proteomics 78: 374–388.
6. Feng H, Tian X, Liu Y, Li Y, Zhang X, et al. (2013) Analysis of flavonoids and the flavonoid structural genes in brown fiber of upland cotton. PLoS ONE 8: e58820.
7. Hua S, Yuan S, Shamsi IH, Zhao X, Zhang X, et al. (2009) A Comparison of three isolines of cotton differing in fiber color for yield, quality, and photosynthesis. Crop Science 49: 983.
8. Xiao Y-H, Zhang Z-S, Yin M-H, Luo M, Li X-B, et al. (2007) Cotton flavonoid structural genes related to the pigmentation in brown fibers. Biochemical and Biophysical Research Communications 358: 73–78.
9. Hua S, Wang X, Yuan S, Shao M, Zhao X, et al. (2007) Characterization of pigmentation and cellulose synthesis in colored cotton fibers. Crop Science 47: 1540–1546.
10. He F, Pan Q-H, Shi Y, Duan C-Q (2008) Biosynthesis and genetic regulation of proanthocyanidins in plants. Molecules 13: 2674–2703.
11. Winkel-Shirley B (2001) Flavonoid Biosynthesis. A Colorful model for genetics, biochemistry, cell biology, and biotechnology. Plant Physiol 126: 485–493.
12. Santos-Buelga C, Scalbert A (2000) Proanthocyanidins and tannin-like compounds – nature, occurrence, dietary intake and effects on nutrition and health. Journal of the Science of Food and Agriculture 80: 1094–1117.
13. Yokozawa T, Cho EJ, Park CH, Kim JH (2012) Protective effect of proanthocyanidin against diabetic oxidative stress. Evid Based Complement Alternat Med 2012: 623879.
14. Tanner GJ, Francki KT, Abrahams S, Watson JM, Larkin PJ, et al. (2003) Proanthocyanidin biosynthesis in plants. Journal of Biological Chemistry 278: 31647–31656.
15. Zhao J, Pang Y, Dixon RA (2010) The mysteries of proanthocyanidin transport and polymerization. Plant Physiol 153: 437–443.
16. Zhang Z-S, Xiao Y-H, Luo M, Li X-B, Luo X-Y, et al. (2005) Construction of a genetic linkage map and QTL analysis of fiber-related traits in upland cotton (Gossypium hirsutum L.). Euphytica 144: 91–99.
17. Ming L, Yue-Hua X, Lei H, Xiao-Ying L, De-Mou L, et al. (2003) Cloning and expression analysis of LIM-domain protein gene from cotton (*Gossypium hirsutrum* L.) Journal of Genetics and Genomics 30: 8.
18. 't Hoen PAC, Ariyurek Y, Thygesen HH, Vreugdenhil E, Vossen RHAM, et al. (2008) Deep sequencing-based expression analysis shows major advances in robustness, resolution and inter-lab portability over five microarray platforms. Nucleic Acids Research 36: e141.
19. Jiang H, Wu P, Zhang S, Song C, Chen Y, et al. (2012) Global analysis of gene expression profiles in developing physic nut (*Jatropha curcas* L.) seeds. PLoS ONE 7: e36522.
20. Wheeler DL, Church DM, Federhen S, Lash AE, Madden TL, et al. (2003) Database resources of the National Center for Biotechnology. Nucleic Acids Research 31: 28–33.
21. Paterson AH, Wendel JF, Gundlach H, Guo H, Jenkins J, et al. (2012) Repeated polyploidization of Gossypium genomes and the evolution of spinnable cotton fibres. Nature 492: 423–427.
22. Goodstein DM, Shu S, Howson R, Neupane R, Hayes RD, et al. (2012) Phytozome: a comparative platform for green plant genomics. Nucleic Acids Research 40: D1178-D1186.
23. Al-Ghazi Y, Bourot S, Arioli T, Dennis ES, Llewellyn DJ (2009) Transcript profiling during fiber development identifies pathways in secondary metabolism and cell wall structure that may contribute to cotton fiber quality. Plant Cell Physiol 50: 1364–1381.
24. Zhu Y-Q, Xu K-X, Luo B, Wang J-W, Chen X-Y (2003) An ATP-binding cassette transporter GhWBC1 from elongating cotton fibers. Plant Physiol 133: 580–588.
25. Hellström JK, Mattila PH (2008) HPLC determination of extractable and unextractable proanthocyanidins in plant materials. Journal of Agricultural and Food Chemistry 56: 7617–7624.
26. Gu L, Kelm M, Hammerstone JF, Beecher G, Cunningham D, et al. (2002) Fractionation of polymeric procyanidins from lowbush blueberry and quantification of procyanidins in selected foods with an optimized normal-phase HPLC−MS fluorescent detection method. Journal of Agricultural and Food Chemistry 50: 4852–4860.
27. Tan J, Tu L, Deng F, Hu H, Nie Y, et al. (2013) A genetic and metabolic analysis revealed that cotton fiber cell development was retarded by flavonoid naringenin. Plant Physiol 162: 86–95.
28. Efe L, Killi F, Mustafayev SA (2009) An evaluation of eco-friendly naturally coloured cottons regarding seed cotton yield, yield components and major lint quality traits under conditions of East Mediterranean region of Turkey. Pak J Biol Sci 12: 1346–1352.
29. Chaudhary B, Hovav R, Flagel L, Mittler R, Wendel J (2009) Parallel expression evolution of oxidative stress-related genes in fiber from wild and domesticated diploid and polyploid cotton (Gossypium). BMC Genomics 10: 378.
30. Shi Y-H, Zhu S-W, Mao X-Z, Feng J-X, Qin Y-M, et al. (2006) Transcriptome profiling, molecular biological, and physiological studies reveal a major role for ethylene in cotton fiber cell elongation. Plant Cell 18: 651–664.
31. Halloin JM (1982) Localization and changes in catechin and tannins during development and ripening of cottonseed. New Phytologist 90: 651–657.
32. Castellarin S, Gaspero G, Marconi R, Nonis A, Peterlunger E, et al. (2006) Colour variation in red grapevines (Vitis vinifera L.): genomic organisation, expression of flavonoid 3′-hydroxylase, flavonoid 3′,5′-hydroxylase genes and related metabolite profiling of red cyanidin−/blue delphinidin-based anthocyanins in berry skin. BMC Genomics 7: 1–17.
33. Woo H-H, Jeong B, Hawes M (2005) Flavonoids: from cell cycle regulation to biotechnology. Biotechnology Letters 27: 365–374.
34. Buer CS, Imin N, Djordjevic MA (2010) Flavonoids: new roles for old molecules. J Integr Plant Biol 52: 98–111.

15

Quantitative Phosphoproteomics Analysis of Nitric Oxide–Responsive Phosphoproteins in Cotton Leaf

Shuli Fan[1⁋], Yanyan Meng[2⁋], Meizhen Song[1], Chaoyou Pang[1], Hengling Wei[1], Ji Liu[1,3], Xianjin Zhan[2], Jiayang Lan[2], Changhui Feng[2], Shengxi Zhang[2*], Shuxun Yu[1*]

1 State Key Laboratory of Cotton Biology, Institute of Cotton Research, Chinese Academy of Agricultural Sciences, Anyang, Henan Province, China, **2** Key Laboratory of Cotton Biology and Breeding in the Middle Reaches of the Changjing River, Institute of Economic Crop, Hubei Academy of Agricultural Science, Wuhan, Hubei Province, China, **3** College of Agronomy, Northwest A & F University, Yangling, Shaanxi Province, China Crop, Hubei Academy of Agricultural Science, Wuhan, Hubei Province, China

Abstract

Knowledge of phosphorylation events and their regulation is crucial to understanding the functional biology of plant proteins, but very little is currently known about nitric oxide–responsive phosphorylation in plants. Here, we report the first large-scale, quantitative phosphoproteome analysis of cotton (*Gossypium hirsutum*) treated with sodium nitroprusside (nitric oxide donor) by utilizing the isobaric tag for relative and absolute quantitation (iTRAQ) method. A total of 1315 unique phosphopeptides, spanning 1528 non-redundant phosphorylation sites, were detected from 1020 cotton phosphoproteins. Among them, 183 phosphopeptides corresponding to 167 phosphoproteins were found to be differentially phosphorylated in response to sodium nitroprusside. Several of the phosphorylation sites that we identified, including RQxS, DSxE, TxxxxSP and SPxT, have not, to our knowledge, been reported to be protein kinase sites in other species. The phosphoproteins identified are involved in a wide range of cellular processes, including signal transduction, RNA metabolism, intracellular transport and so on. This study reveals unique features of the cotton phosphoproteome and provides new insight into the biochemical pathways that are regulated by nitric oxide.

Editor: Jens Schlossmann, Universität Regensburg, Germany

Funding: This research was sponsored by the National Basic Research Program of China (973 Program; 2010CB126006) and supported by the State Key Laboratory of Cotton Biology Open Fund (CB2013A01). The funders had no role in study design, data collection and analysis, decision to publish, or preparation of the manuscript.

Competing Interests: The authors have declared that no competing interests exist.

* E-mail: zhangsx01@126.com (SZ); yu@cricaas.com.cn (SY)

⁋ These authors contributed equally to this work.

Introduction

Nitric oxide (NO) is an important growth regulator that modulates diverse developmental processes in plants over the entire lifespan, including cell elongation and germination [1], root development [2], stomatal movement [3], pollen tube growth [4], and plant senescence [5]. NO is also involved in plant responses to abiotic stresses such as heavy metal exposure [6], low temperature [7], salt [8], and drought [9]. Additionally, NO has been demonstrated to play many roles in biotic interactions, such as fighting pathogen infections [10,11], deterring herbivore feeding [12] and building symbiotic interactions [13].

During the last decade, several medium- and large-scale transcriptomic analyses have identified hundreds of putative NO-regulated genes [14–19]. Moreover, eight families of transcription factor–binding sites have been identified in the promoter of NO-regulated genes [19]. Recent research has mainly focused on identifying NO target proteins, attempting to reveal NO involvement in plant cell biology. As such, more than 200 NO-responsive proteins have been identified in plants [20,21]. Unlike the genome, however, protein abundance, structure, stability, subcellular localization, and interactions with other biological macromolecules are in continuous flux and these changes are regulated by post-translational modifications such as phosphorylation, acetylation, glycosylation, and methylation.

Phosphorylation is the most common and the most important dynamic adjustment mechanism and is involved in nearly all cellular and extracellular processes including defense responses, signal transduction, cytoskeleton regulation, and apoptosis [22,23]. Given the importance of phosphorylation, there has been much interest in identifying new phosphorylated proteins and phosphorylation sites and in exploring the functional role of these phosphoproteins. Early studies looking at the effect of abscisic acid (ABA) on protein phosphorylation of rice (*Oryza sativa*) and *Arabidopsis thaliana* have identified 6 and 50 proteins, respectively, for which phosphorylation was found to be regulated by ABA [24,25]. Furthermore, examination of changes in the phosphoproteome of rape seeds during the filling stage identified 70 phosphoproteins and 16 non-redundant phosphoproteins, which were verified by mapping the phosphorylation sites [26]. Analysis of vacuolar and cell membranes of rice bud and root revealed 230 membrane and membrane-associated proteins, 20% of which are phosphorylated [27]. In addition, soybean (*Glycine max*) root hairs contain 1625 unique phosphopeptides, including 1659 non-redundant phosphorylation sites, which originate from 1126 phosphoproteins [28]. A recent study identified multiple components of ABA-responsive protein phosphorylation network by integrating genetics with phosphoproteomics [29]. Furthermore, Wang et al. have shown the role of the SnRK2 protein kinases in the ABA signaling pathway by using quantitative phosphoproteomics [30]. To date, only a few plant species have been annotated with protein phosphorylation data and these do not include the economically important plant, cotton (*Gossypium hirsutum*) [31]. We previously examined the proteome of cotton leaves in response to NO treatment [21]. To follow-up this research, we performed quantitative time-course measurements of the phosphoproteome of cotton leaves treated with sodium nitroprusside (SNP), a NO donor. This is the first study to show that NO exposure in leaf tissue alters the phosphorylation states of multiple proteins found in cotton. This information should accelerate research on NO metabolic regulation and will lay a novel, theoretical foundation for further related studies in cotton.

Materials and Methods

Plant materials

Seeds of *G. hirsutum* ecotype CCRI10 were cultured in a mix of sand and nutritional soil in a culture room under white fluorescent light (14 h light/10 h dark) with day/night temperatures of 30/22°C. Previously we evaluated the effects of nitric oxide on cotton seeds treated with 0, 0.05, 0.1 or 0.5 mM SNP for 24 h. The results showed that the effect of treatment with 0.05 mM SNP on plant growth was not severe, 0.5 mM SNP was lethal, and 0.1 mM SNP was significant but not toxic (Figure S1). Additionaly, we also found that the effect of treatment at 0.1 mM SNP for 6 h on cotton seedlings effected protein levels significantly [21]. Based on these results, proteins were extracted from the leaves treated with 0.1 mM SNP. Plants that were 30 days old after sowing were irrigated with 0.1 mM SNP (Sigma, USA) in distilled water for 0 h (control), 1 h, 3 h and 6 h, respectively during the light period. The experiment was performed in triplicate with 30 plants in each group. Fresh, fully expanded leaves were harvested after treatment and immediately frozen in liquid nitrogen and stored at −80°C.

Protein extraction

Leaf samples were ground in liquid nitrogen and lysed in buffer containing 4% SDS, 100 mM Tris-HCl, and 1 mM dithiothreitol (pH 7.6) at room temperature and briefly sonicated to reduce the viscosity of the lysate. The ratio of buffer to tissue was 5:1 (v/v). After 3 min incubation in boiling water, the suspensions were ultrasonicated (80 w, 10 s ultrasonic at a time, every 15 s, and 10 times) then incubated at 100°C for 3 min. The crude extract was clarified by centrifugation at 13,000×*g* at 25°C for 10 min [32]. The sample protein content was determined using the BCA Protein Assay Reagent (Promega, USA). The supernatants were stored at −80°C until use.

Protein digestion

Protein digestion was processed by using the method of filter-aided sample preparation (FASP) [32]. The protocol was as follows: For each sample, 300 µg of protein was diluted with 200 µL UA buffer (8 M urea with 0.1 M dithiothreitol in 0.15 M Tris-HCl pH 8.0) and centrifuged at 14,000×*g* at 20°C for 40 min. Another 200 µL of UA buffer was added and the samples were centrifuged again at 14,000×*g* for 20 min. Then 100 µL of 0.05 M iodoacetamide in UA buffer was added, and the samples were incubated at room temperature for 20 min in darkness. After 10 min centrifugation at 14,000×*g*, 100 µL UA buffer was added and centrifuged at 14,000×*g* for 15 min. This step was repeated twice. Subsequently, 100 µL ABC buffer (0.05 M NH_4HCO_3 in water) was added to the filters and the samples were centrifuged for 10 min at 14,000×*g*. This step was also repeated twice as reported previously [32]. Finally, 2 µg of trypsin (Promega, USA) in 40 µL ABC buffer was added to each filter, and incubated overnight at 37°C. The resulting peptides were collected by centrifugation at 14,000×*g* for 10 min and then the filters were rinsed with 40 µL 10× DS buffer (50 mM triethylammonium bicarbonate at pH 8.5) and centrifuged again [32]. The peptide content was estimated by absorbance at 280 nm using an extinction coefficient of 1.1 for a 0.1% (g/100 mL) solution that was calculated based on the frequency of tryptophan and tyrosine in vertebrate proteins [33].

Isobaric tag for relative and absolute quantitation (iTRAQ) labeling and phosphopeptide enrichment

Equal amounts of the trypsin-digested samples from the four groups were labeled as CK-114, 1 h-115, 3 h-116 and 6 h-117 utilizing the iTRAQ Multiplex (4-plex) kit (Applied Biosystems, USA); three biological replicates were performed. For the process of enriching phosphopeptides, each labeled peptide mixture was concentrated in a vacuum concentrator and resuspended in 500 µL buffer containing 2% glutamic acid, 65% acetonitrile, and 2% trifluoroacetic acid. TiO_2 beads (500 µg, GL Sciences, Japan) were added and the mixture was agitated for 40 min then centrifuged for 1 min at 5000×*g* [34]. The supernatant from the first centrifugation was mixed with another 500 µg of TiO_2 beads and agitated and centrifuged as before. The two bead collections were combined and washed three times with 50 µL of 30% acetonitrile, 3% TFA and then three times with 50 µL of 80% acetonitrile, 0.3% trifluoroacetic acid to remove the remaining non-adsorbed material. Finally, the phosphopeptides were eluted with 50 µL of elution buffer (40% acetonitrile, 15% NH_4OH) followed by lyophilization and mass spectrometry (MS) [35].

Nano-liquid chromatography

Each iTRAQ sample (5 µL) was mixed with 15 µL 0.1% trifluoroacetic acid (v/v), and subsequently a 5 µL of the mixture was loaded onto a C18-reversed phase column (15 cm long, 75 µm inner diameter, RP-C18 3 µm, packed in house) in buffer A (0.1% formic acid) and separated with a linear gradient of buffer B (80% acetonitrile and 0.1% formic acid) at a flow rate of 250 nL/min (controlled by IntelliFlow technology; Applied Flow Technology, USA) over 240 min on a Q Exactive MS (Thermo Finnigan,USA) equipped with Easy nLC (Thermo Fisher Scientific, USA). The peptides were eluted with a gradient of 0–55% buffer B from 0 to 220 min, 55–100% buffer B from 220 to 228 min, and 100% buffer B from 228 to 240 min.

Tandem MS analysis

We analyzed the peptides in positive ion mode and acquired the MS spectra using a data-dependent, top10 method, dynamically choosing the most abundant precursor ions from the survey scan (300–1800 m/z) for subsequent high-energy collisional dissociation (HCD) fragmentation [36]. Determination of the target value was based on predictive automatic gain control(pAGC). The dynamic exclusion duration was 25 s. Survey scans were acquired at a resolution of 70,000 at m/z 200 and resolution for HCD spectra was set to 17,500 at m/z 200. Normalized collision energy was 29 eV and the under fill ratio, which specifies the minimum percentage of the target value likely to be reached at maximum fill time, was defined as 0.1%. The instrument was operated with the peptide recognition mode enabled. Each iTRAQ experiment was analyzed at least three times.

Data analysis

Mascot 2.2 (Matrix Science) and Proteome Discoverer 1.3 software (Thermo Scientific, USA) were used for identification and quantitative analysis [37]. The reference sequences in the Cotton Genome Project (CGP) database derived from the *Gossypium raimondii* genome include 40,976 non-redundant protein-coding genes sequences and 39316 non-redundant peptide sequences (ftp://public.genomics.org.cn/BGI/cotton/Annotation/G.raimondii.pep.fasta.gz, updated to 07-23-2012, downloaded on 12-05-2012) [38]. A q-value is the minimal false discovery rate at which the identification is considered correct. The dataset was filtered to give a < 1% false discovery rate (q-Value ≤ 0.009, Table S1) at peptide level using the target decoy method [39]. The phosphorylated sites on the identified peptides were assigned again using the PhosphoRS algorithm, which calculated the possibility of the phosphorylated site from the spectra matched to the identified peptides. According to the instruction of the software, for each phosphorylation site(s) on all phosphopeptides, PhosphoRS probabilities were set above 75%, indicating that the site is truly phosphorylated, and PhosphoRS scores were set above 50, indicating a good peptide spectral match [40,41]. Proteome Discoverer 1.3 software was used to extract the peak intensity within 20 ppm of each expected iTRAQ reporter ion from each analyzed fragmentation spectrum. Only spectra in which all the expected iTRAQ reporter ions (four for HCD in this work) were detected were used for quantification.

The intensity of the reporter ions was used for phosphopeptide quantification. We normalized the phosphopeptide ratios by dividing by the median ratio of all peptides identified. As for the quantitative analysis, the $\log_2$ fold-change values (Treatment/Control) for each time point were calculated for each phosphopeptide. Only phosphopeptides detected in at least two out of the three biological replicates were used for assessment of significant change. In some cases, the quantitative values of certain time points were not available due to missing phosphopeptide identification in one particular sample or the intensity values failed to pass the cutoff. The t-test was employed to identify significant changes between the control and treatment sample among the three biological replicates. The phosphopeptides that passed t-test with p-value<0.05 were considered to be significantly regulated. We also included the cutoff for the $\log_2$ fold change values, in which the phosphorylation changes were considered highly significant if the $\log_2$ value ≥ 0.58 or ≤−0.58 (increasing or decreasing 1.5 fold in phosphorylation activity).

Protein annotation and classification

To identify the corresponding protein for each phosphopeptide, each phosphopeptide was searched against the CGP peptide database [38]. To overcome the challenge that multiple proteins may share the same peptide, the annotations were confirmed by comparison to the annotation of the top score protein hits from an in-house BLAST search against the CGP peptide database.

Identified phosphoproteins were classified based on: (a) the biological processes of each gene product according to annotations in the CGP database and (b) the gene product subcellular localization predicted using the SherLoc2 web server applying the defaulting settings (http://abi.inf.uni-tuebingen.de/Services/SherLoc2) [21].

Results and Discussion

Phosphopeptide identification

We profiled the phosphoproteome of the cotton leaf after exposure to SNP in order to gain insight into plant cellular responses to NO. Approximately 2291 phosphopeptides were identified across the cotton database. Subsequently, we removed redundant and invalid peptides and additionally set the PhosphoRS probabilities above 75% and PhosphoRS scores above 50, according to standard practices, to further select for unique phosphopeptides. Following this selection process, we identified 1315 unique phosphopeptides, collectively containing 1528 non-redundant phosphorylation sites (Table S1). Of these non-redundant phosphorylation sites, 1371 (89.7%) were found to be phosphorylated at serine residues, 145 (9.5%) at threonine, and 12 (0.8%) at tyrosine (Figure S2A). Of the 1315 unique phosphopeptides, 1110 were singly phosphorylated, 198 were doubly phosphorylated, 6 were phosphorylated at three sites, and in only one case the phosphopeptide was phosphorylated at four sites (Figure S2B). The distribution of phosphor-Ser (pS) was found to be consistent with that of rice (89.5%) [22] and Glycine max (89.3%) [28]. Moreover, the distribution of phosphor-Thr (pT) was similar to Arabidopsis (9.9%) [22]. However, the abundance of tyrosine phosphorylation in cotton (0.8%, this study) was slightly lower than in Arabidopsis (2.4%) [22], Medicago (1.3%) [42] and rice (1.6%) [22]. On the other hand, Laurent et al. reported a proportion of pY in soybean of only 0.48% [28]. Additionally, within the same species (e.g. Arabidopsis), the distribution of the phosphor-amino acids was not always consistent [22,43]. These differences may be attributed to differences in methodology (e.g. phosphopeptide enrichment and/or LC-MS) or in the biological system, where each cell type, tissue and organism has a unique phosphoproteome profile.

Phosphorylation site motifs

To determine the potential consensus sequences for cotton, the motifs of the 1528 non-redundant cotton phosphorylation sites were extracted using the Motif-X program (version v1.2 10.05.06; http://motif-x.med.harvard.edu/motif-x.html) with default parameters [44,45]. Because *Gossypium raimondii* Protein Fasta data has a size surpassing the upload restrictions to 10 MB, the database included in 20,000 sequences selected randomly from *Gossypium raimondii* Protein Fasta data was used as the background in Motif-X. Only confidently identified phosphorylation sites with Ascore >19 were used in analysis. All 25 of the phosphorylation site motifs identified in cotton are presented in Table 1. Serine was phosphorylated in 20 of the motifs, threonine in 4, and tyrosine in 1. Table 1 also lists the phosphorylation site motifs that contained at least one fixed position aside from the central phosphorylated residue.

With the purpose of investigating the specificity of the identified motifs to cotton, these motifs were compared with literatures [46,47] as well as in the PhosPhAt 4.0 database of Arabidopsis phosphorylation site motifs (http://phosphat.mpimp-golm.mpg.de/phosphat.html) [48] and Human protein reference database (HPRD) [49]. All the phosphorylation motifs from Arabidopsis and cotton are shown in Table S2A and the information of cotton peptide sequence can be found in Table S2B. Among the 25 motifs identified in cotton, 4 of the motifs (RQxS, DSxE, TxxxxSP and SPxT) were neither present in the Arabidopsis database nor present in the Human protein reference database (HPRD) (Table 1).

Phosphoprotein identification

To identify the corresponding proteins for each phosphopeptide, the phosphopeptides in our dataset were individually searched against the CGP peptide database. As expected, the longer peptides tended to identify single proteins, but the shorter peptides tended to be shared by multiple proteins. To overcome this potential biasing aspect, the annotations of all of phosphopep-

Table 1. Cotton phosphorylation site motifs detected in our study compared with motifs identified in Arabidopsis and PhosphoMotif Finder.

No.	Cotton*	Arabidopsis	PhosphoMotif Finder Database
1	......**S**PR....		
2	....R**S**.S....		
3	..SP..**S**......		
4	......**S**...SP.		
5	......**S**P.R...		
6	...RS.**S**......		
7	......**S**PK....		
8	......**S**.SP...		
9	R..S..**S**......		
10	..S...**S**P.....		
11	.....G**S**P.....		
12	...RQ.**S**......	N	N
13	......**S**DDE...		
14	.L.R..**S**......		
15	...SP**S**......		
16	......**S**D.E...		
17	...SP.**S**......		
18	......**S**R..S..		
19	......**S**D.D...		
20	.....D**S**.E....	N	N
21	.....R**T**.S....		
22	......**T**....SP	N	N
23	...SP**T**......		
24	...SP.**T**......	N	N
25	....P.**Y**......		

*The phosphorylated residues are indicated in bold and underlined, and (.) indicates any amino acid. N indicates the motif was not found.

tides were confirmed by comparing them to the annotation of the top score protein hits in CGP peptide database. From 1528 unique phosphopeptides, 1020 unique phosphoproteins were identified and all of sequence information could be found in Table S3. Passing through several steps of protein digestion, phosphopeptide enrichment, fractionation and mass spectrometry analysis, it is possible that for a single protein, two or more phosphopeptide is retained and can be detected in the LC/MS analysis. As much, our results indicated that, among the 1020 phosphoproteins, 81% were represented by a single phosphopeptide, 13.2% by two phosphopeptides, 3.7% by three phosphopeptides, and 2.1% by more than three phosphopeptides (Figure S3).

To obtain a general overview of the roles of phosphorylation in cotton, we next analyzed the biological functions/pathways and cellular localization of the 1020 phosphoproteins. The proteins were classified into 30 different categories based on their predicted functions, covering a wide range of pathways. The largest functional groups were ion/protein transport–related proteins, various enzymes (including kinases, transferases, and hydrolases), RNA splicing/processing, transcription, and sensory and signal transduction, accounting for 9.3, 7.7, 7.1, 6.4 and 5.3% of the 1020 proteins, respectively (Figure 1A). No function/pathway could be assigned to 32.0% of the phosphoproteins identified. For the subcellular localization analysis, these 1020 phosphoproteins were located in 12 different cellular compartments, including primarily the nucleus (22.5%), membranes (15.0%), cytoplasm (13.7%) and chloroplasts (4.4%). Exact subcellular localization information was not available for 35.0% of the proteins (Figure 1B).

Quantitative analysis of protein phosphorylation changes in response to NO

The fold change (treatment/control) for every phosphopeptide at each time point was calculated, and the amounts of 183 unique phosphopeptides derived from 167 phosphoproteins were found to change significantly (>1.5 fold change and P-value $\leq$ 0.05) after SNP exposure (Table S4). The total number of differentially expressed phosphopeptides was 95, 89, and 85 at 1, 3, and 6 h, respectively (Figure 2). A decreasing trend was observed for the number of upregulated phosphopeptides, which decreased from 53.7% in the 1 h group to 40.0% in the 6 h group. Conversely, the number of downregulated phosphopeptides increased gradually from 42.1% in the 1 h group to 52.8% in the 3 h group and 58.8% in the 6 h group. Quantitative analysis revealed that most of the phosphopeptide changes occurred rapidly, i.e., within an hour of first exposure to SNP, which is consistent with the initiation of signal transduction and is similar to other phosphoproteomics studies [28]. As such, we hypothesized that, for phosphoproteins regulated by NO, we should see a shift from the

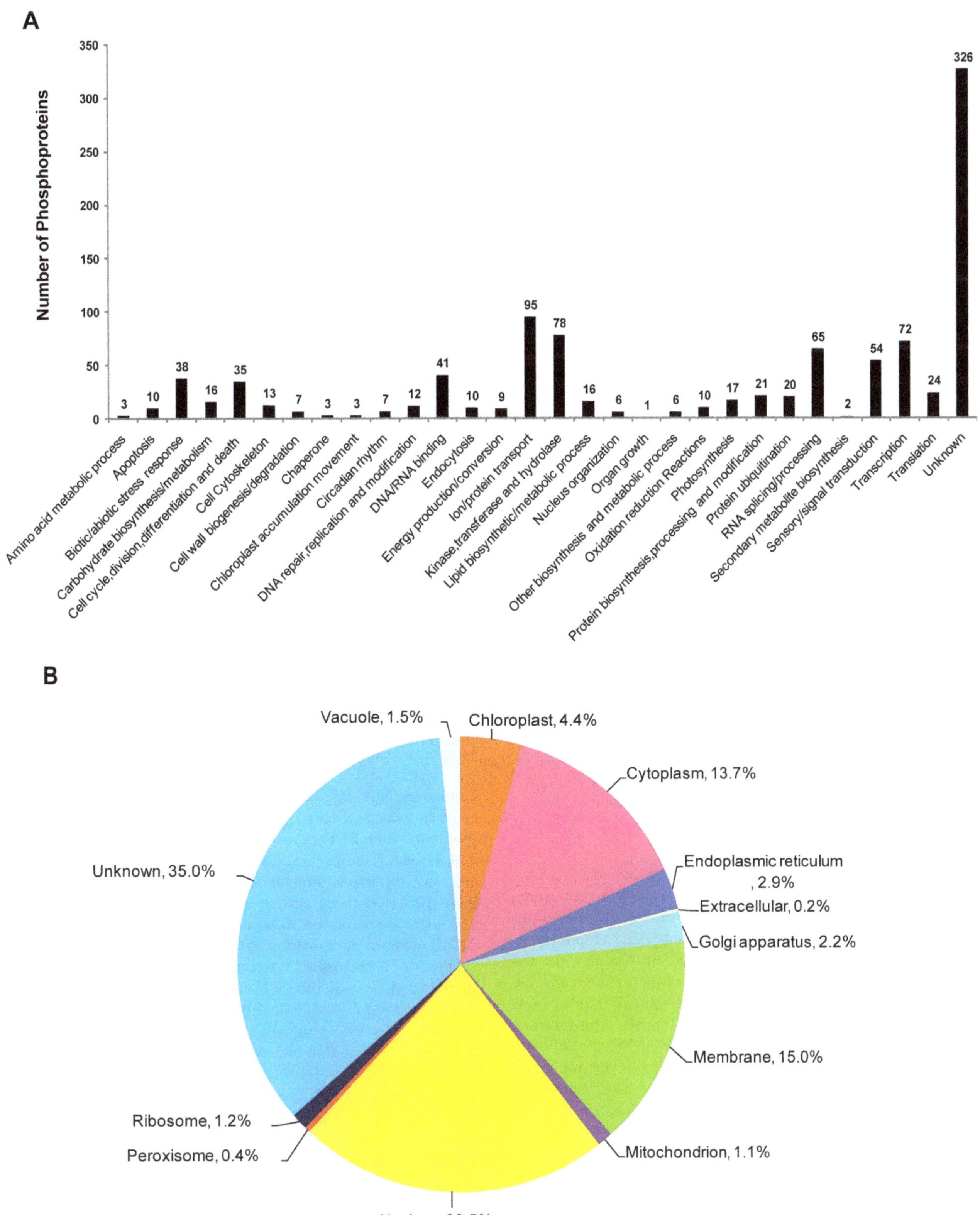

Figure 1. Functional classifications (A) and protein subcellular locations (B) of all 1020 phosphoproteins.

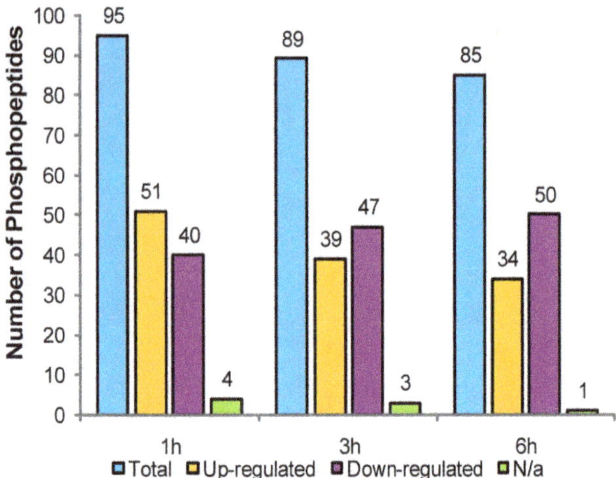

Figure 2. Analysis of the 183 phosphopeptides whose levels changed after exposure to SNP. N/a, not applicable, indicates that the phosphopeptides were not detected in this group and were only expressed in other treatment groups.

phosphorylated to the dephosphorylated forms as the treatment continued.

The 167 phosphoproteins whose phosphopeptides levels changed significantly in response to NO were classified into 23 functional categories based on their predicted biological function (Figure 3A). Except for the unknown proteins (25.7%), the largest functional groups were RNA splicing or processing (17.4%), ion or protein transport (7.8%), sensory or signal transduction (7.2%), and transcription (6.6%). Compared with the total dataset of all the non-redundant phosphoproteins identified, our results suggest that these functional categories are the most important in terms of response to NO. Correspondingly, the largest subcellular location groups were the nucleus (35.3%), unknown (26.3%), the cytoplasm (15.6%), and membranes (9.6%) (Figure 3B).

Ion/Protein transport proteins

Many (95) of the cotton phosphoproteins that we identified are involved in ion/protein trafficking processes, such as potassium, ammonium, sucrose, and peptide transport, or are transporter proteins themselves (i.e., aquaporin and ABC transporter). ATPases are post-translationally regulated by complex mechanisms, including phosphorylation [50,51]. In our present study, the abundance of phosphopeptides corresponding to 16 ATPases was found to be altered, including 2 V-type and 9 plasma membrane-type phosphopeptides. These results suggest that NO regulation of transporters by phosphorylation may play a role in quick and reversible activating/inactivating transport channels, leading to rapid changes in water potential and turgor pressure that are necessary for plant growth and stimulus response.

Miscellaneous enzymes

As shown in Figure 1A, a considerable number (78) of the cotton phosphoproteins identified were classified as "miscellaneous enzymes", including kinases, transferases, and hydrolases. However, only 4 of these proteins, all kinases, appeared in our quantitative phosphoproteomic data (Figure 3A). We hypothesize that changes in the abundance of these proteins might reflect a coordinated activity of one or more kinase cascades involved in the NO-response process. How these proteins fit into the context of NO signaling and their relationships with other well-known receptor and protein kinases need to be further explored in future studies.

Proteins involved in DNA/RNA binding and transcription

Among the 72 transcription-related phosphoproteins identified, the content levels of only 11 significantly changed in response to SNP (Figure 3A), including transcription initiation/elongation factors, transcriptional co-repressors, and some sequence-specific transcription factors. DNA/RNA-binding proteins have been identified in several phosphoproteomic studies [28,42]. Similarly, we found 41 phosphoproteins in cotton that appear to be involved in a variety of DNA/RNA processes (Figure 1A). In addition, we found 7 differentially expressed phosphoproteins involved in DNA/RNA binding, including RNA-binding motif protein (Cotton_D_gene_10033131) expressed predominantly in spematocytes [52], HMG1/2-like protein (Cotton_D_gene_10035211) containing 1HMG box DNA-binding domain that belongs to the HMGB family which was regulated by darkness and circadian rhythm [53], uncharacterized RNA-binding protein (Cotton_D_gene_10011862) and a number of predicted proteins (Table S1). We speculate that the phosphorylation and the presence of differential phosphorylation of these proteins may be essential for the rapid responses to NO treatment.

Phosphorylation changes of protein in response to plant hormones

The 187 phosphopeptides, corresponding to the 167 identified NO-responsive phosphoproteins, were placed into appropriate signaling pathways to better understand which physiological processes in cotton are regulated by NO. The signaling pathways analysis was carried out using KEGG Automatic Annotation Server (http://www.genome.jp/tools/kaas/) with the default settings.

Ethylene-insensitive protein 2 (EIN2) acts downstream of ethylene receptors and upstream of ethylene-regulated transcription factors and is involved in regulating early leaf senescence caused by NO deficiency [54,55]. We found that the levels of phosphorylated EIN2 (Cotton_D_gene_10027821) increased gradually after SNP exposure and were significantly higher at 6 h. Levels of another negative regulator in the ethylene response pathway, the serine/threonine-protein kinase CTR1 (Cotton_D_gene_10011591), were also increased in response to NO as early as 1 h after treatment began. In our previous study, we found that NO is involved in regulating ACS [21], a synthase catalyzing the synthesis of a precursor (1-Aminocyclopropane-1-carboxylic acid) in the ethylene biosynthesis pathway. Here, we found that NO appears to regulate the phosphorylation of EIN2 and CTR1, central factors in the signaling pathways regulated by ethylene. Thus, we postulate that NO regulates not only the biosynthesis of ethylene but also the ethylene-signaling pathway by modulating phosphorylation of EIN2 and CTR1.

In our study, NO also appeared to affect the signal transduction pathways of other plant hormones, such as cytokinine and ABA (Figure S4). Inhibitors of NO-synthase have been shown to inhibit expression from the cytokinine-responsive *ARR5* promoter [56]. The two-component response regulator ARR2 is also a transcriptional activator involved in cytokinine signaling and is phosphorylated in response to cytokinine [57,58]. In the current paper, we found that levels of phospho-ARR2 (Cotton_D_gene_10029097) were markedly reduced in the 1 and 3 h treatment groups. This decrease is likely related to the NO-regulated cytokinine signaling pathway. Protein phosphatase 2C (PP2C) 37 is a major negative regulator of ABA responses during seed germination and cold acclimation [59,60]. We also found that the levels of phosphor-

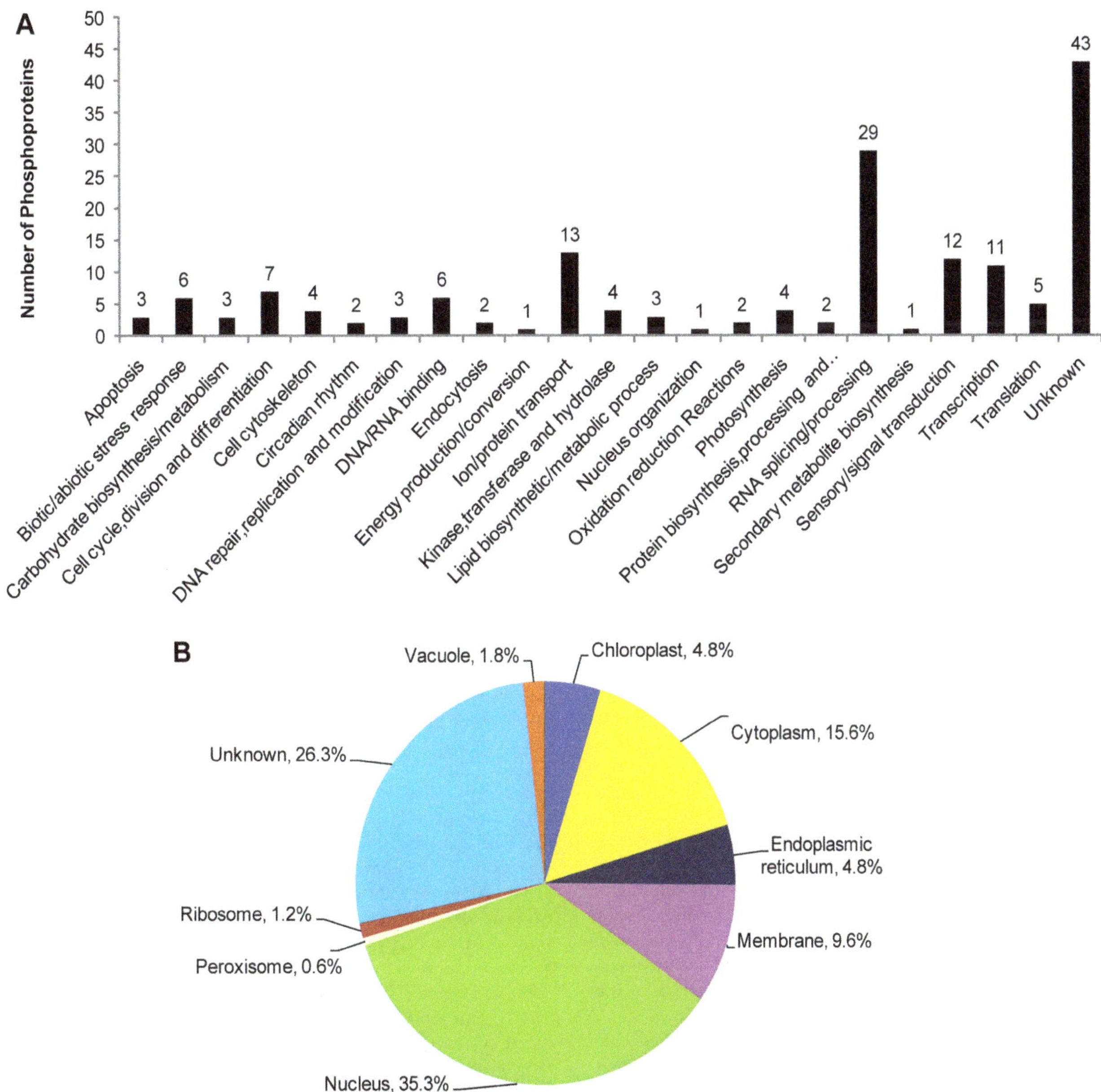

Figure 3. Functional classifications (A) and protein subcellular locations (B) of the 167 phosphoproteins.

ylated PP2C 37 (Cotton_D_gene_10021530) were increased at both the 1 and 3 h time points. Moreover, NO mediates the induction of ABA biosynthesis involved in oxidative stress tolerance in maize (*Zea mays*) [61,62]. Additionally, PP2C inhibits SnRK2, a vital kinase in the regulation of ABA-induced gene expression, thus inhibiting activation of the ABA pathway [25]. Given this extensive cross-talk between NO and ABA, it is plausible that the apparent dephosphorylation of PP2C 37 that we observed may play a role in ABA signal transduction regulated by NO.

Many components of the auxin efflux (but not influx) system are activated by phosphorylation [28,63]. In this study, the phosphopeptide levels of one auxin-related protein were significantly altered in response to NO. Furthermore, we found that a phosphoprotein required for auxin influx-facilitator (AUX1) polar trafficking (Cotton_D_gene_10018241) was significantly increased 1 h after SNP treatment. Thus, it appears that increases in phosphorylation of both efflux and influx carriers facilitate auxin transport in cotton.

Phosphoproteins involved in RNA metabolism

Our study also identified numerous phosphoproteins involved in RNA metabolism, including components of the spliceosome, mRNA surveillance, and RNA transport proteins. Specifically, 21 phosphoproteins, corresponding to 13 KEGG orthology entries involved in spliceosome activity were detected (Figure S5), 8 of which were SR proteins (serine/arginine-rich splicing factors; Cotton_D_gene_10019600, Cotton_D_gene_10022365, Cotton_D_gene_10039654, Cotton_D_gene_10032493, Cotton_D_gene_10026606, Cotton_D_gene_10024595, Cotton_D_gene_10024595, and Cotton_D_gene_10030715). SR proteins belong to a highly conserved family of splicing regulators involved in constitutive and alternative splicing in response to stress and hormones [64] and have been shown to be phosphorylated in both animals and plants [65–67]. Moreover, kinase-protein interactions related to pre-RNA splicing are modulated by the phosphorylation of SR proteins [68]. The 8 SR proteins we detected were extensively phosphorylated at serine residues in their serine/arginine-rich domains (Table S1), similar to the phosphorylation of SR proteins in Arabidopsis [69].

The phosphorylation of some helicases in humans and Arabidopsis has been shown to regulate pre-mRNA splicing [69–71]. Our analysis identified two ATP-dependent RNA helicases (Cotton_D_gene_10003086, Prp22; Cotton_D_gene_10004269, Prp2), suggesting that phosphorylation of the cotton homologs may also function in splicing. Small nuclear RNA conformation and the phosphorylation of related proteins strongly affect 3′-splice site recognition and catalytic activation of the spliceosome [72,73]. We found that the levels of phosphorylated forms of two small nuclear ribonucleoproteins (Cotton_D_gene_10032473, Prp3; and Cotton_D_gene_10038196, Prp4) and one small nuclear ribonucleoprotein helicase (Cotton_D_gene_10020278, Bn2) increased in response to NO within 3 h of exposure to SNP. In addition, the phosphorylated form of a pre-mRNA processing factor (Cotton_D_gene_10025929, FBP11), the nuclear cap-binding protein (Cotton_D_gene_10014749, CBP80/20), and a nuclear apoptotic chromatin-condensation inducer (Cotton_D_gene_10036545, ACINUS) also increased 1 h and 3 h after SNP treatment began. The levels of phosphorylated U2-associated protein (Cotton_D_gene_10039568, SR140) and RNA-binding protein (Cotton_D_gene_10022997, HnRNPs) also decreased to varying degrees. We also found three splicing factors in this pathway (Cotton_D_gene_10040894, SF36; Cotton_D_gene_10001970, U2AF; Cotton_D_gene_10027627, U2AF) that are involved in phosphorylation in response to NO, indicating a direct link between NO signaling and spliceosome regulation through phosphorylation and dephosphorylation of related proteins. Except for a special pattern that sharp increase followed by sharp decrease (Cotton_D_gene_10001970), the changes in the other 20 spliceosome-related phosphoproteins could be categorized as two distinct patterns–a gradual decline in phosphorylation followed by a sharp increase or a stable decline from 1 to 6 h (Figure 4).

The mRNA surveillance pathway is a quality-control mechanism that detects and degrades abnormal mRNAs. Upf3, together with Upf1 and Upf2, detects the presence of premature termination codon and instigates decapping and rapid exonucleolytic digestion of the mRNA [74]. The dephosphorylation of hUPF1 by PP2A, contribute to the remodeling of the mRNA surveillance complex [75]. We found that Upf3 (Cotton_D_gene_10004090, Upf3) levels were higher at 3 h in the treated samples than at 1 or 6 h, and the levels of the phosphorylated forms of three PP2A subunits (Cotton_D_gene_10023447, PP2A; Cotton_D_gene_10028017, PP2A; and Cotton_D_gene_10028691, PP2A) also changed over time. In addition, levels of the phosphorylated RNA-binding motif protein (Cotton_D_gene_10033131, CPSF6/7), heterogeneous nuclear ribonucleoprotein (Cotton_D_gene_10035148, Musashi), cleavage and polyadenylation specificity factor (Cotton_D_gene_10004099, CPSF4) and serine/threonine-protein kinase (Cotton_D_gene_10001769, SMG1) also changed significantly in response to NO (Figure 5, Figure S6). The changes in these proteins are consistent with the known critical role of the mRNA surveillance pathway in cellular homeostasis.

In addition to the apoptotic chromatin condensation inducer, nuclear cap-binding protein, and Upf3, all mentioned above, we identified another six phosphoproteins involved in RNA transport (Figure S7), including four translation initiation factors (Cotton_D_gene_10020489, eIF3; Cotton_D_gene_10004109, eIF3; Cotton_D_gene_10012222, eIF4G; and Cotton_D_gene_10010902, eIF4A), one ribonuclease (Cotton_D_gene_10011365, RNaseZ), and one unknown protein (Cotton_D_gene_10022069, Nup133). Eukaryotic initiation factor 3 (eIF3) is an essential, highly conserved multiprotein complex that is a key component in the recruitment and assembly of the translation initiation machinery [76]. eIF4F is a protein complex that mediates recruitment of ribosomes to mRNA [77,78]. The activity of both proteins is regulated by phosphorylation and in turn influences translation, apoptosis, and growth [78,79]. Analysis of the cotton eIFs revealed that the phosphorylated peptides of all four proteins increased to varying degrees in response to SNP. The pattern for all the RNA transport phosphopeptides showed a rapid increase followed by a steady downward trend (Figure 6).

RNA splicing/processing was the largest group of proteins whose phosphorylation level was significantly affected by NO (Figure 3A). Correspondingly, we found the numerous proteins in the splicesome, mRNA surveillance and RNA transport pathways that were altered noticeably in regards to their phosphorylation status as regulated by NO. These results suggest that the plant mRNA splicing machinery is a major target of phosphorylation or dephosphorylation induced by NO.

Other phosphorylated proteins

Other phosphoproteins, whose peptide levels changed noticeably in response to NO in the current paper, were found to be involved in protein processing and/or to be related to plant-pathogen interactions [21,80]. In protein processing, the binding protein, Bip, in the endoplasmic reticulum can recognize improperly folded or assembled and facilitates their refold and reassembly [81,82]. It has been reported that the transcription of the gene encoding Bip is induced by SNP [21]. Similarly, in the current paper we found that phosphopeptides of Bip (Cotton_D_gene_10017746) were increased significantly after treatment with SNP in the 6 h group. In terms of plant-pathogen interaction proteins, we found that the phosphorylated forms of three heat shock proteins were also dramatically increased by SNP (Cotton_D_gene_10018942, Cotton_D_gene_10030775 and Cotton_D_gene_10025013). A mitogen-activated protein kinase kinase (Cotton_D_gene_10039089, MKK1/2), a calcium-dependent protein kinase (Cotton_D_gene_10038462, CDPK), and a respiratory burst oxidase (Cotton_D_gene_10034071, Rboh) also showed significant upregulation in phosphorylation in varying degrees. On the other hand, an unknown protein (Cotton_D_gene_10020033, CaMCML) showed significant decreases in the phospho-forms (Figure 7, Figure S8). These results indicate that NO regulates plant disease resistance by modulating phosphorylation of the relevant proteins.

It is noteworthy that some key proteins involved in circadian rhythms were also revealed in the phosphoproteome of SNP-treated cotton. The phosphorylated forms of protein LHY (Cotton_D_gene_10035743) and the transcription factor HY5 (Cotton_D_gene_10010118), both important components in clock function, were significantly reduced following treatment with SNP (Figure S9). To our knowledge, this is the first report of a relationship between NO and plant circadian rhythms, especially in terms of phosphorylation regulation. The levels of several other phosphopeptides/phosphoproteins not previously implicated in NO response were significantly also altered by SNP treatment, including translocase of chloroplast 159 (Cotton_D_gene_10024721), stem-specific protein TSJT1 (Cotton_D_gene_10003896), and villin-2 (Cotton_D_gene_10038481). Further exploration of these

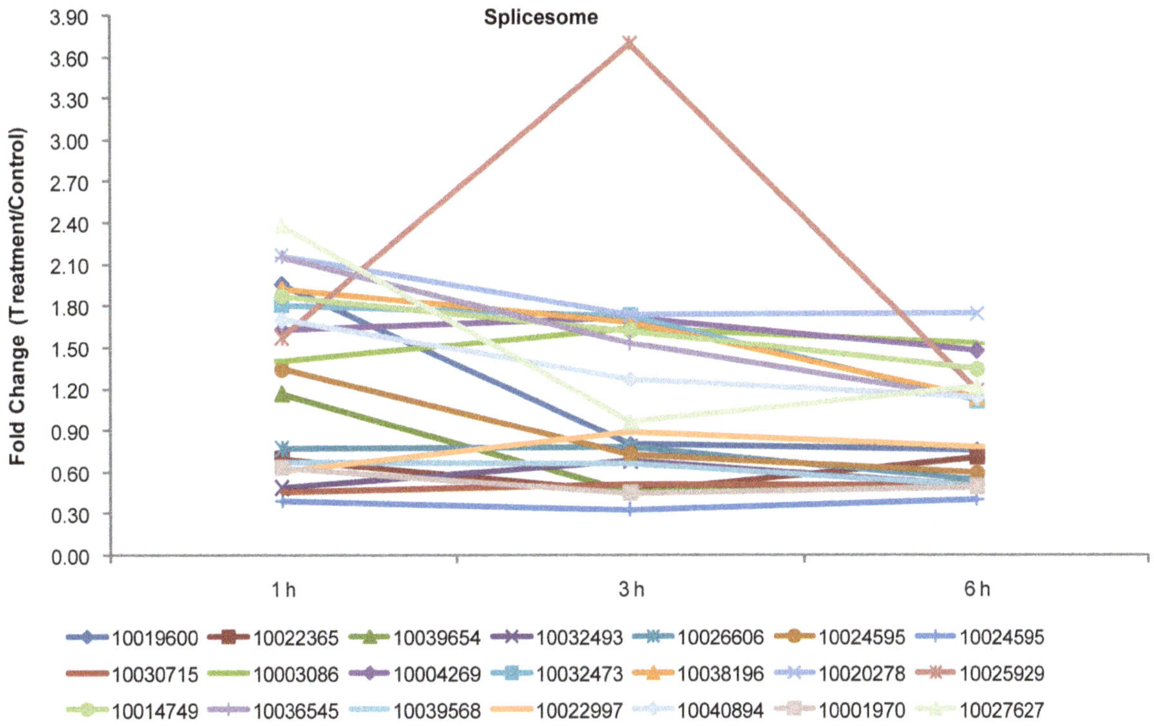

Figure 4. Fold changes of the phosphopeptides involved in spliceosome significantly altered following SNP. Protein identification numbers refer to the CGP database. To make the figure easier to read, the protein ID is represented only by the numbers in the key (i.e., remove "Cotton_D_gene_").

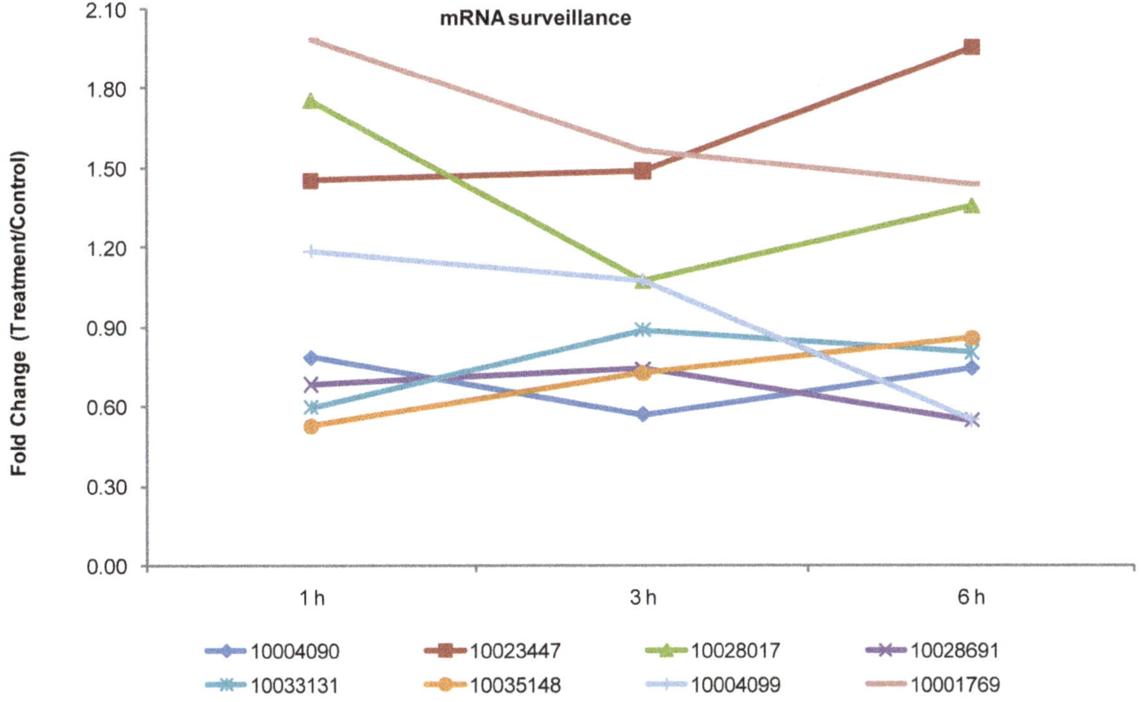

Figure 5. Fold changes of the phosphopeptides involved in mRNA surveillance significantly altered following SNP. Protein identification numbers refer to the CGP database. To make the figure easier to read, the protein ID is represented only by the numbers in the key (i.e., remove "Cotton_D_gene_").

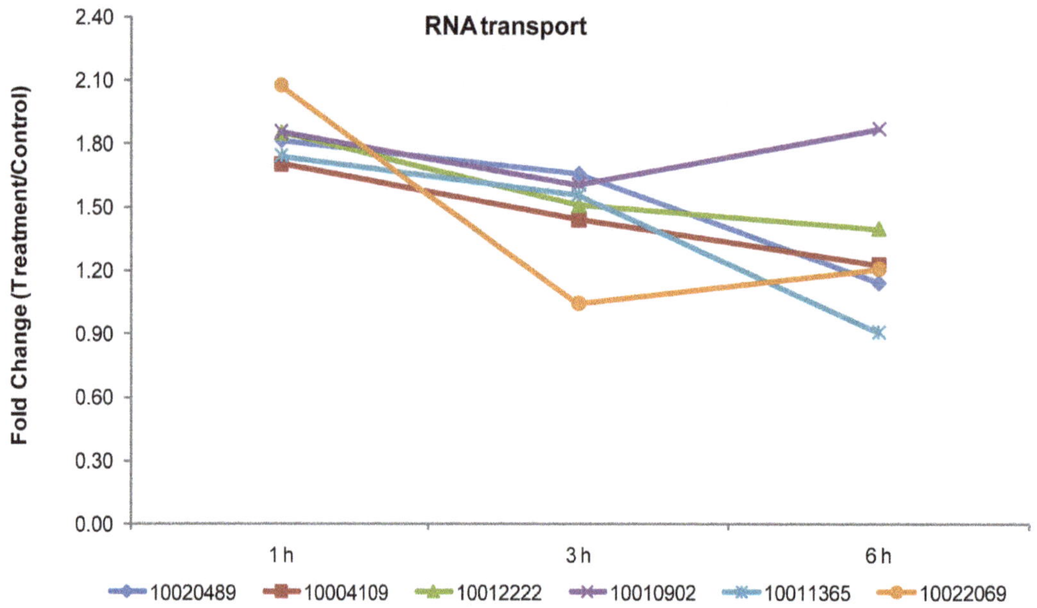

Figure 6. Fold changes of the phosphopeptides involved in RNA transport significantly altered following SNP. Protein identification numbers refer to the CGP database. To make the figure easier to read, the protein ID is represented only by the numbers in the key (i.e., remove "Cotton_D_gene_").

phosphoprotein targets may provide insight into previously unknown effectors of the NO signaling pathway.

Conclusions

This study represents one of the first large-scale phosphoproteomic analyses of the NO response in cotton, providing both a global and comparative analysis of protein phosphorylation regulated by NO and further insight into the dynamics of individual phosphorylation sites. Our results reveal phosphorylation site motifs and proteins not previously observed in plants, which may provide insight into NO signaling in cotton as well as plants in general. In addition, the novel plant phosphorylation site motifs could lead to future discovery of novel kinase-substrate interactions. We have submitted the peptide and protein sequences to the plant protein phosphorylation database (P3DB,

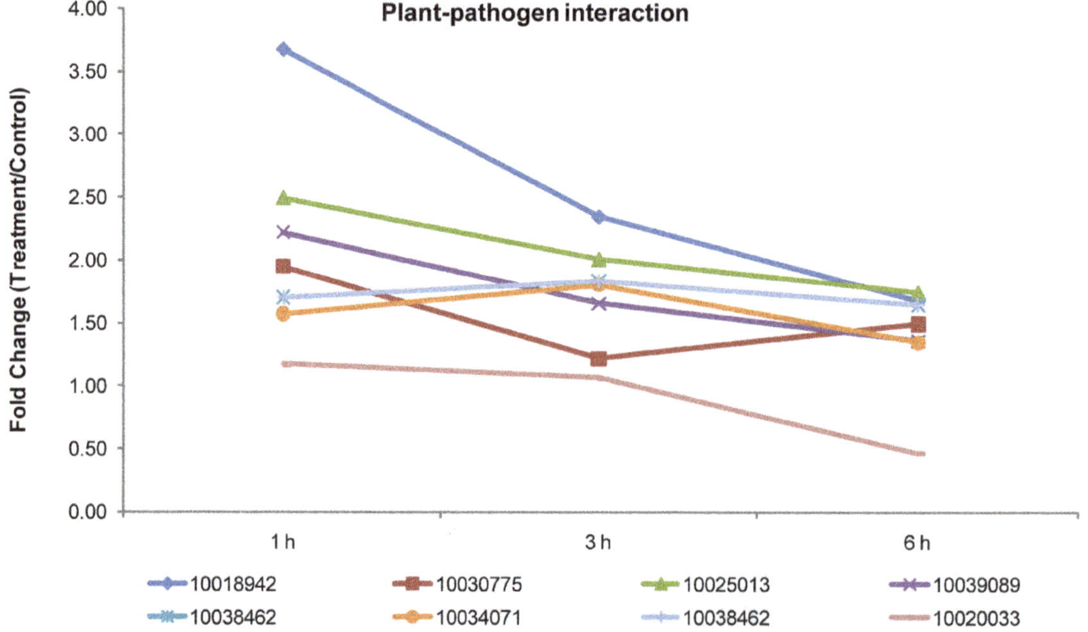

Figure 7. Fold changes of the phosphopeptides involved in plant-pathogen interaction significantly altered following SNP. Protein identification numbers refer to the CGP database. To make the figure easier to read, the protein ID is represented only by the numbers in the key (i.e., remove "Cotton_D_gene_").

http://p3db.org/download.php) [31,83] to help develop future, large-scale cotton proteomic datasets. In this perspective, a better understanding of phosphorylation events may eventually lead to the development of a comprehensive model for the mechanisms of action of NO. Functional analysis of these protein phosphorylation sites by site-directed mutagenesis or reverse genetic approaches will be necessary to further investigate their role in NO signaling.

Supporting Information

Figure S1 The effect of NO on cotton growth. Cotton seeds were treated with 0, 0.05, 0.1, or 0.5 mM SNP for 24 h. Changes in the morphology (A) and chlorophyll content (B) following the treatment of different concentrations of SNP.

Figure S2 The distribution of phosphorylation sites. A. Distribution of phosphorylation on serine, threonine, and tyrosine was assessed for all non-redundant localized phosphorylation sites. B. Distribution of single- and multi-phosphorylated peptides showing that the majority of phosphopeptides have only one phosphorylation site.

Figure S3 The number of phosphopeptides per protein. The distribution of the number of phosphopeptides per protein is shown for all 1315 unique phosphopeptides.

Figure S4 The plant hormone signal-transduction pathway derived from KEGG. Proteins indicated in red were found to be significantly upregulated or downregulated by SNP in our analysis.

Figure S5 The spliceosome pathway derived from KEGG. Proteins indicated in red were found to be significantly upregulated or downregulated by SNP in our analysis.

Figure S6 The mRNA surveillance pathway derived from KEGG. Proteins indicated red were found to be significantly upregulated or downregulated by SNP in our analysis.

Figure S7 The RNA transport pathway derived from KEGG. Proteins indicated red were found to be significantly upregulated or downregulated by SNP in our analysis.

Figure S8 The plant-pathogen interaction pathway derived from KEGG. Proteins indicated red were found to be significantly upregulated or downregulated by SNP in our analysis.

Figure S9 The circadian rhythm pathway derived from KEGG. Proteins indicated red were found to be significantly upregulated or downregulated by SNP in our analysis.

Table S1 Phosphopeptide identifications in cotton leaves.

Table S2 Phosphorylation motifs identified from cotton (A) and raw peptide data used to extract the given motif of cotton (B).

Table S3 Amino acid sequences of 1020 phosphoproteins.

Table S4 Quantitative analysis identified significant changes in the content levels of 183 phosphopeptides.

Acknowledgments

We would like to thank Shanghai Applied Protein Technology Co. Ltd. for providing technical support.

Author Contributions

Conceived and designed the experiments: SF SY YM. Performed the experiments: YM J. Liu. Analyzed the data: YM HW CP MS. Contributed reagents/materials/analysis tools: SF SZ XZ J. Lan CF. Wrote the paper: YM.

References

1. Beligni MV, Lamattina L (2000) Nitric oxide stimulates seed germination and de-etiolation, and inhibits hypocotyl elongation, three light-inducible responses in plants. Planta 210: 215–221.
2. Pagnussat GC, Simontacchi M, Puntarulo S, Lamattina L (2002) Nitric oxide is required for root organogenesis. Plant Physiol 129: 954–956.
3. Garcia-Mata C, Lamattina L (2007) Abscisic acid (ABA) inhibits light-induced stomatal opening through calcium-and nitric oxide-mediated signaling pathways. Nitric Oxide 17: 143–151.
4. Prado AM, Porterfield DM, Feijó JA (2004) Nitric oxide is involved in growth regulation and re-orientation of pollen tubes. Development 131: 2707–2714.
5. Mishina TE, Lamb C, Zeier J (2007) Expression of a nitric oxide degrading enzyme induces a senescence programme in Arabidopsis. Plant Cell Environ 30: 39–52.
6. Yang LM, Tian DG, Todd CD, Luo YM, Hu XY (2013) Comparative proteome analyses reveal that nitric oxide is an important signal molecule in the response of rice to aluminum toxicity. J Proteome Res 12: 1316–1330.
7. Zhao MG, Chen L, Zhang LL, Zhang WH (2009) Nitric reductase-dependent nitric oxide production is involved in cold acclimation and freezing tolerance in Arabidopsis. Plant Physiol 151: 755–767.
8. Zhao MG, Tian QY, Zhang WH (2007) Nitric oxide synthase-dependent nitric oxide production is associated with salt tolerance in Arabidopsis. Plant Physiol 144: 206–217.
9. Lamattina L, Carcía-Mata C (2001) Nitric oxide induces stomatal closure and enhances the adaptive plant responses against drought stress. Plant Physiol 126: 1196–1204.
10. Delledonne M, Xia Y, Dixon RA, Lamb C (1998) Nitric oxide functions as a signal in plant disease resistance. Nature 394: 585–588.
11. Hong JK, Yun BW, Kang JG, Raja MU, Kwon E, et al. (2008) Nitric oxide function and signaling in plant disease resistance. J Exp Bot 59: 147–154.
12. Wunsche H, Baldwin IT, Wu J (2011) S-Nitrosoglutathione reductase (GSNOR) mediates the biosynthesis of jasmonic acid and ethylene induced by feeding of the insect herbivore Manduca sexta and is important for jasmonate-elicited responses in Nicotiana attenuata. J Exp Bot 62: 4605–4616.
13. del Giudice J, Cam Y, Damiani I, Fung-Chat F, Meilhoc E, et al. (2011) Nitric oxide is required for an optimal establishment of the Medicago truncatula-Sinorhizobium meliloti symbiosis. New Phytol 191: 405–417.
14. Huang X, von Rad U, Durner J (2002) Nitric oxide induces transcriptional activation of the nitric oxide-tolerant alternative oxidase in Arabidopsis suspension cells. Planta 215: 914–923.
15. Parani M, Rudrabhatla S, Myers R, Weirich H, Smith B, et al. (2004) Microarray analysis of nitric oxide responsive transcripts in Arabidopsis. Plant Biotechnol J 2: 359–366.
16. Zeidler D, Zähringer U, Gerber I, Dubery I, Hartung T, et al. (2004) Innate immunity in Arabidopsis thaliana: lipopolysaccharides activate nitric oxide synthase (NOS) and induce defense genes. P Natl Acad Sci USA 101: 15811–15816.
17. Ahlfors R, Brosché M, Kollist H, Kangasjärvi J (2009) Nitric oxide modulates ozone-induced cell death, hormone biosynthesis and gene expression in Arabidopsis thaliana. Plant J 58: 1–12.
18. Ferrarini A, De Stefano M, Baudouin E, Pucciariello C, Polverari A, et al. (2008) Expression of Medicago truncatula genes responsive to nitric oxide in pathogenic and symbiotic conditions. Mol plant Microbe In 21: 781–790.
19. Palmieri MC, Sell S, Huang X, Scherf M, Werner T, et al. (2008) Nitric oxide-responsive genes and promoters in Arabidopsis thaliana: a bioinformatics approach. J Exp Bot 59: 177–186.

20. Tanou G, Job C, Belghazi M, Molassiotis A, Diamantidis G, et al. (2010) Proteomic signatures uncover hydrogen peroxide and nitric oxide cross-talk signaling network in citrus plants. J Proteome Res 9: 5994–6006.
21. Meng Y, Liu F, Pang C, Fan S, Song M, et al. (2011) Label-free quantitative proteomics analysis of cotton leaf response to nitric oxide. J Proteome Res 10: 5416–5432.
22. Nakagami H, Sugiyama N, Mochida K, Daudi A, Yoshida Y, et al. (2010) Large-scale comparative phosphoproteomics identifies conserved phosphorylation sites in plants. Plant Physiol 153: 1161–1174.
23. Benschop JJ, Mohammed S, O'Flaherty M, Heck AJR, Slijper M, et al. (2007) Quantitative phosphoproteomics of early elicitor signaling in Arabidopsis. Mol Cell Proteomics 6: 1198–1214.
24. He H, Li J (2008) Proteomic analysis of phosphoproteins regulated by abscisic acid in rice leaves. Biochem Bioph Res Co 371: 883–888.
25. Kline KG, Barrett-Wilt GA, Sussman MR (2010) In planta changes in protein phosphorylation induced by the plant hormone abscisic acid. P Natl Acad Sci USA 107: 15986–15991.
26. Agrawal GK, Thelen JJ (2006) Large scale identification and quantitative profiling of phosphoproteins expressed during seed filling in oilseed rape. Mol Cell Proteomics 5: 2044–2059.
27. Whiteman S, Nühse TS, Ashford DA, Sanders D, Maathuis FJ (2008) A proteomic and phosphoproteomic analysis of Oryza sativa plasma membrane and vacuolar membrane. Plant J 56: 146–156.
28. Nguyen THN, Brechenmacher L, Aldrich JT, Clauss TR, Gritsenko MA, et al. (2012) Quantitative phosphoproteomic analysis of soybean root hairs inoculated with Bradyrhizobium japonicum. Mol Cell Proteomics 11: 1140–1155.
29. Umezawa T, Sugiyama N, Takahashi F, Anderson JC, Ishihama Yasushi, et al. (2013) Genetics and phosphoproteomics reveal a protein phosphorylation net work in the abscisic acid signaling pathway in Arabidopsis thaliana. Sci Signal 6: (270):rs8. doi: 10.1126/scisignal.2003509.
30. Wang P, Xue L, Batelli G, Lee S, Hou YJ, et al. (2013) Quantitative phsophoproteomics identifies SnRK2 protein kinase substrates and reveals the effectors of abscisic acid action. Proc Natl Acad Sci U S A 110: 11205–11210.
31. Yao Q, Bollinger C, Gao J, Xu D, Thelen JJ (2012) P(3)DB: An Integrated Database for Plant Protein Phosphorylation. Front Plant Sci 3: 1–8.
32. Wiśniewski JR (2009) Universal sample preparation method for proteome analysis. Nat Methods 6: 359–362.
33. Zhuang Y, Ma F, Li LJ, Xu X, Li Y (2003) Comparative analysis of amino acid usage and protein length distribution between alternatively and non-alternatively spliced genes across six eukaryotic genomes. Mol Biol Evol 20: 1978–1985.
34. Unwin RD, Griffiths JR, Whetton AD (2010) Simultaneous analysis of relative protein expression levels across multiple samples using iTRAQ isobaric tags with 2D nano LC-MS/MS. Nat Protoc 5: 1574–1582.
35. Larsen MR, Thingholm TE, Jensen ON, Roepstorff P, Jørgensen TJ (2005) Highly selective enrichment of phosphorylated peptides from peptide mixtures using titanium dioxide microcolumns. Mol Cell Proteomics 4: 873–886.
36. Michalski A, Damoc E, Hauschild JP, Lange O, Wieghaus A, et al. (2011) Mass spectrometry-based proteomics using Q Exactive, a high-performance benchtop quadrupole Orbitrap mass spectrometer. Mol Cell Proteomics doi: 10.1074/mcp.M111.011015.
37. Sandberg A, Lindell G, Källström B, Branca RM, Danielsson KG, et al (2012) Tumor proteomics by multivariate analysis on individual pathway data for characterization of vulvar cancer phenotypes. Mol Cell Proteomics doi: 10.1074/mcp.M112.016998.
38. Wang K, Wang Z, Li F, Ye W, Wang J, et al (2012) The draft genome of a diploid cotton Gossypium raimondii. Nat Genet 44: 1098–1103.
39. Kim S, Gupta N, Pevzner PA (2008) Spectral probabilities and generating functions of tandem mass spectra: a strike against decoy databases. J Proteome Res 7: 3354–3363.
40. Olsen JV, Blagoev B, Gnad F, Macek B, Kumar C, et al. (2006) Global, in vivo, and site-specific phosphorylation dynamics in signaling networks. Cell 127: 635–648.
41. Beausoleil SA, Villn J, Gerber SA, Rush J, Gygi SP (2006) A probability-based approach for high-throughput protein phosphorylation analysis and site localization. Nat Biotechnol 24: 1285–1292.
42. Grimsrud PA, den Os Dse, Wenger CD, Swaney DL, Schwartz D, et al. (2010) Large-scale phosphoprotein analysis in Medicago truncatula roots provides insight into in vivo kinase activity in legumes. Plant Physiol 152: 19–28.
43. Sugiyama N, Nakagami H, Mochida K, Daudi A, Tomita M, et al. (2008) Large-scale phosphorylation mapping reveals the extent of tyrosine phosphorylation in Arabidopsis. Mol Syst Biol doi: 10.1038/msb.2008.32.
44. Chou MF, Schwartz D (2011) Biological sequence motif discovery using motif-x. Curr Protoc Bioinformatics 13.5. 1-.5. 24.
45. Chou MF, Schwartz D (2011) Using the scan-x Web site to predict protein post-translational modifications. Curr Protoc Bioinformatics 13.6. 1-.6. 8.
46. Reiland S, Messerli G, Baerenfaller K, Gerrits B, Endler A, et al. (2009) Large-scale Arabidopsis phosphoproteome profiling reveals novel chloroplast kinase substrates and phosphorylation networks. Plant Physiol 150: 889–903.
47. Wang X, Bian YY, Cheng K, Gu LF, Ye ML, et al. (2013) A large-scale protein phosphorylation analysis reveals novel phosphorylation motifs and phosphoregulatory networks in Arabidopsis. J Proteomics 78: 486–498.
48. Durek P, Schmidt R, Heazlewood JL, Jones A, MacLean D, et al. (2010) PhosPhAt: the Arabidopsis thaliana phosphorylation site database. An update. Nucleic Acids Res 38: D828–D834.
49. Keshava Prasad TS, Goel R, Kandasamy K, Keerthikumar S, Kumar S, et al. (2009) Human protein reference database-2009 update. Nucleic Acids Res 37: D767–D772.
50. Palmgren MG (2001) Plant plasma membrane H+-ATPases: powerhouses for nutrient uptake. Annu Rev Plant Biol 52: 817–845.
51. Lan P, Li W, Wen TN, Shiau JY, Wu YC, et al. (2010) iTRAQ protein profile analysis of Arabidopsis roots reveals new aspects critical for iron homeostasis. Plant Physiol 155: 821–834.
52. Elliott DJ, Venables JP, Newton CS, Lawson D, Boyle S, et al. (2000) An evolutionarily conserved germ cell-specific hnRNP is encoded by a retrotransposed gene. Hum Mol Genet 9: 2117–2124.
53. Zheng CC, Bui AQ, O'Neill SD (1993) Abundance of an mRNA encoding a high mobility group DNA-binding protein is regulated by light and an endogenous rhythm. Plant Mol Biol 23: 813–823.
54. Alonso JM, Hirayama T, Roman G, Nourizadeh S, Ecker JR (1999) EIN2, a bifunctional transducer of ethylene and stress responses in Arabidopsis. Science 284: 2148–2152.
55. Niu Y, Guo F (2012) Nitric oxide regulates dark-induced leaf senescence through ein2 in Arabidopsis. J Integr Plant Biol 54: 516–525.
56. Romanov GA, Lomin SN, Rakova NY, Heyl A, Schmülling T (2008) Does NO play a role in cytokinin signal transduction? FEBS Lett 582: 874–880.
57. Hass C, Lohrmann J, Albrecht V, Sweere U, Hummel F, et al. (2004) The response regulator 2 mediates ethylene signalling and hormone signal integration in Arabidopsis. EMBO J 23: 3290–3302.
58. Kim HJ, Ryu H, Hong SH, Woo HR, Lim PO, et al. (2006) Cytokinin-mediated control of leaf longevity by AHK3 through phosphorylation of ARR2 in Arabidopsis. P Natl Acad Sci USA 103: 814–819.
59. Sheen J (1998) Mutational analysis of protein phosphatase 2C involved in abscisic acid signal transduction in higher plants. P Natl Acad Sci USA 95: 975–980.
60. Kuhn JM, Boisson-Dernier A, Dizon MB, Maktabi MH, Schroeder JI (2006) The protein phosphatase AtPP2CA negatively regulates abscisic acid signal transduction in Arabidopsis, and effects of abh1 on AtPP2CA mRNA. Plant Physiol 140: 127–139.
61. Zhang A, Zhang J, Zhang J, Ye N, Zhang H, et al. (2011) Nitric oxide mediates brassinosteroid-induced ABA biosynthesis involved in oxidative stress tolerance in maize leaves. Plant and Cell Physiol 52: 181–192.
62. Ma F, Lu R, Liu H, Shi B, Zhang J, et al. (2012) Nitric oxide-activated calcium/calmodulin-dependent protein kinase regulates the abscisic acid-induced antioxidant defence in maize. J Exp Bot 63: 4835–4847.
63. Delbarre A, Muller P, Guern J (1998) Short-lived and phosphorylated proteins contribute to carrier-mediated efflux, but not to influx, of auxin in suspension-cultured tobacco cells. Plant Physiol 116: 833–844.
64. Palusa SG, Ali GS, Reddy AS (2007) Alternative splicing of pre-mRNAs of Arabidopsis serine/arginine-rich proteins: regulation by hormones and stresses. Plant J 49: 1091–1107.
65. Lorkovic ZJ, Lopato S, Pexa M, Lehner R, Barta A (2004) Interactions of Arabidopsis RS domain containing cyclophilins with SR proteins and U1 and U11 small nuclear ribonucleoprotein-specific proteins suggest their involvement in pre-mRNA splicing. J Biol Chem 279: 33890–33898.
66. Lopato S, Mayeda A, Krainer AR, Barta A (1996) Pre-mRNA splicing in plants: characterization of Ser/Arg splicing factors. P Natl Acad Sci USA 93: 3074–3079.
67. Lopato S, Kalyna M, Dorner S, Kobayashi R, Krainer AR, et al. (1999) atSRp30, one of two SF2/ASF-like proteins from Arabidopsis thaliana, regulates splicing of specific plant genes. Gene Dev 13: 987–1001.
68. Reddy AS (2004) Plant serine/arginine-rich proteins and their role in pre-mRNA splicing. Trends Plant Sci 9: 541–547.
69. de la Fuente van Bentem S, Anrather D, Roitinger E, Djamei A, Hufnagl T, et al. (2006) Phosphoproteomics reveals extensive in vivo phosphorylation of Arabidopsis proteins involved in RNA metabolism. Nucleic Acids Res 34: 3267–3278.
70. Tanner NK, Linder P (2001) DExD/H box RNA helicases: from generic motors to specific dissociation functions. Mol Cell 8: 251–262.
71. Arenas J, Abelson J (1991) Requirement of the RNA helicase-like protein PRP22 for release of messenger RNA from spliceosomes. Nature 347: 487–493.
72. Makarov EM, Makarova OV, Urlaub H, Gentzel M, Will CL, et al. (2002) Small nuclear ribonucleoprotein remodeling during catalytic activation of the spliceosome. Science 298: 2205–2208.
73. Zhang Y, Madl T, Bagdiul I, Kern T, Kang HS, et al. (2013) Structure, phosphorylation and U2AF65 binding of the N-terminal domain of splicing factor 1 during 3′-splice site recognition. Nucleic Acids Res 41: 1343–1354.
74. Kunz JB, Neu-Yilik G, Hentze MW, Kulozik AE, Gehring NH (2006) Functions of hUpf3a and hUpf3b in nonsense-mediated mRNA decay and translation. RNA 12: 1015–1022.
75. Ohnishi T, Yamashita A, Kashima I, Schell T, Anders KR, et al. (2003) Phosphorylation of hUPF1 induces formation of mRNA surveillance complexes containing hSMG-5 and hSMG-7. Mol Cell 12: 1187–1200.
76. Sun C, Todorovic A, Querol-Audí J, Bai Y, Villa N, et al. (2011) Functional reconstitution of human eukaryotic translation initiation factor 3 (eIF3). P Natl Acad Sci USA 108: 20473–20478.
77. Hinnebusch AG (2006) eIF3: a versatile scaffold for translation initiation complexes. Trends Biochem Sci 31: 553–562.

78. Gingras AC, Raught B, Sonenberg N (1999) eIF4 initiation factors: effectors of mRNA recruitment to ribosomes and regulators of translation. Annu Rev Biochem 68: 913–963.
79. Farley AR, Powell DW, Weaver CM, Jennings JL, Link AJ (2011) Assessing the components of the eIF3 complex and their phosphorylation status. J Proteome Res 10: 1481–1494.
80. Asai S, Yoshioka H (2009) Nitric oxide as a partner of reactive oxygen species participates in disease resistance to necrotrophic pathogen Botrytis cinerea in Nicotiana benthamiana. Mol Plant Microbe In 22: 619–629.
81. Rothman JE (1989) Polypeptide chain binding proteins: catalysts of protein folding and related processes in cells. Cell 59: 591–601.
82. Koizumi N (1996) Isolation and responses to stress of a gene that encodes a luminal binding protein in Arabidopsis thaliana. Plant Cell Physiol 37: 862–865.
83. Gao J, Agrawal GK, Thelen JJ, Xu D (2009) P3DB: a plant protein phosphorylation database. Nucleic Acids Res 37: D960–D962.

16

Distribution and Differentiation of Wild, Feral, and Cultivated Populations of Perennial Upland Cotton (*Gossypium hirsutum* L.) in Mesoamerica and the Caribbean

Geo Coppens d'Eeckenbrugge[1]*, Jean-Marc Lacape[2]

1 CIRAD, UMR 5175 CEFE, Campus du CNRS, Montpellier, France, **2** CIRAD, UMR AGAP, Montpellier, France

Abstract

Perennial forms of *Gossypium hirsutum* are classified under seven races. Five Mesoamerican races would have been derived from the wild race 'yucatanense' from northern Yucatán. 'Marie-Galante', the main race in the Caribbean, would have developed from introgression with *G. barbadense*. The racial status of coastal populations from the Caribbean has not been clearly defined. We combined Ecological Niche Modeling with an analysis of SSR marker diversity, to elucidate the relationships among cultivated, feral and wild populations of perennial cottons. Out of 954 records of occurrence in Mesoamerica and the Caribbean, 630 were classified into four categories cultivated, feral (disturbed and secondary habitats), wild/feral (protected habitats), and truly wild cotton (TWC) populations. The widely distributed three first categories cannot be differentiated on ecological grounds, indicating they mostly belong to the domesticated pool. In contrast, TWC are restricted to the driest and hottest littoral habitats, in northern Yucatán and in the Caribbean (from Venezuela to Florida), as confirmed by their climatic envelope in the factorial analysis. Extrapolating this TWC climatic model to South America and the Pacific Ocean points towards places where other wild representatives of tetraploid *Gossypium* species have been encountered. The genetic analysis sample comprised 42 TWC accessions from 12 sites and 68 feral accessions from 18 sites; at nine sites, wild and feral accessions were collected in close vicinity. Principal coordinate analysis, neighbor joining, and STRUCTURE consistently showed a primary divergence between TWC and feral cottons, and a secondary divergence separating 'Marie-Galante' from all other feral accessions. This strong genetic structure contrasts strikingly with the absence of geographic differentiation. Our results show that TWC populations of Mesoamerica and the Caribbean constitute a homogenous gene pool. Furthermore, the relatively low genetic divergence between the Mesoamerican and Caribbean domesticated pools supports the hypothesis of domestication of *G. hirsutum* in northern Yucatán.

Editor: Xianlong Zhang, National Key Laboratory of Crop Genetic Improvement, China

Funding: The authors have no support or funding to report.

* Email: geo.coppens@cirad.fr

Introduction

Cotton (*Gossypium* spp.) is unique among crop plants in that four species have been independently domesticated in four different regions of the world: two tetraploids, *G. hirsutum* L. in Mesoamerica, *G. barbadense* L. in South America, and two diploids, *G. herbaceum* L. in Arabia and Syria and *G. arboreum* L. in the Indus Valley of India and Pakistan [1]. In the process, they were transformed from photoperiod-sensitive perennial sprawling or upright shrubs into short, compact, annualized day-length-neutral plants; and their small impermeable seeds sparsely covered by coarse, poorly differentiated hairs became larger and covered with abundant and long, white lint. Simultaneously, their seeds lost their impermeability and dormancy. The wide diversity of cotton results from the successive waves of agronomic improvement and human-mediated germplasm diffusion [1,2].

Phylogenetic investigations in *Gossypium* distinguish 45 modern diploid species distributed among three major geographic lineages and eight genomes. The American tetraploid lineage originated within the last 1–2 million years from a single hybridization event between a maternal African A and an American D genome [1]. It diversified into five species, three wild endemic species, *G. darwinii* Watt native to the Galapagos, *G. tomentosum* Nutt. ex Seem. from the Hawaiian Islands, *G. mustelinum* Miers ex Watt restricted to Northeastern Brazil, and the two cultivated species *G. barbadense* and *G. hirsutum*. The latter provides over 90% of the world cotton, spreading North and South to subtropical and temperate latitudes well over 30° as an annual crop. Its indigenous (preindustrial) range encompasses most of Mesoamerica and the Caribbean, with two centers of morphological and genetic diversity, one in Southern Mexico-Guatemala, considered a primary center of diversity, and one in the Caribbean, where some introgression took place with *G. barbadense* [2–5].

In these two regions, *G. hirsutum* exhibits a diverse array of perennial forms, which Hutchinson [6] classified into seven geographical races. The primitive and highly variable race 'punctatum' is mostly found in Yucatán and round the coasts and islands of the Gulf of Mexico. Race 'latifolium' has a center of diversity in Guatemala and southern Mexico, but its range extends from most of Mexico to El Salvador and Nicaragua. Race 'Marie-Galante' is distinct both geographically and morphologically, with its pronounced apical dominance and tree-like habit. Its range includes the Antilles and Central America, South from El Salvador into northern to northeastern South America. Its origin and diffusion seems to be closely related to human migrations that would have resulted in the introduction of *G. barbadense* into Central America and the Antilles and its introgression with *G. hirsutum* in these areas [2,3,7,8]. Together, these three most widespread races, 'latifolium', 'punctatum' and 'Marie-Galante', encompass most of the morphological variation in *G. hirsutum*. The remaining four races present a more restricted geographic distribution, with race 'palmeri' in the Mexican states of Oaxaca and Guerrero, race 'morrilli' in the central Mexican plateau, race 'richmondi' along the Pacific side of the Isthmus of Tehuantepec, and race 'yucatanense' limited to the northern coast of Yucatán. The latter is known only as a small, highly branched, sprawling shrub forming a dominant constituent of undisturbed beach strand vegetation. Hutchinson [6] considered race 'yucatanense' an extreme case of feral populations derived from primitive 'punctatum' landraces.

The persistence of wild populations of *G. hirsutum* has been the subject of considerable debate. On one hand, most germplasm collections came from man-made habitats, such as field plots and house yards, or highly disturbed habitats, such as roadsides and secondary vegetation, indicating that spontaneous cotton plants were escapes from cultivation. Furthermore, morphological differentiation appears similar and parallel for both landraces and feral plants [6,9,10]. Testing materials from Yucatán, Hutchinson [6] observed no differences between progenies of 'punctatum' from plants cultivated in dooryards and plants established in natural vegetation. On the other hand, Sauer [11] observed that the northern Yucatán wild cottons are negatively associated with human settlements and form a dominant constituent of "a complex vegetation type occupying a coherent and extensive area with natural and edaphic and climatic boundaries." He maintained this interpretation in his study of the Cayman Islands shoreline vegetation [12].

In a study of the effects of domestication in *G. hirsutum*, Stephens [9] extended the question to the seemingly wild populations of race 'punctatum' observed on the dry leeward sides of some of the Greater Antilles, on Florida Cays [6,13], along the coasts of the Gulf of Mexico as well as in Venezuela. For a long time, he could not rule out the possibility that these forms are feral relics of pre-Columbian or early post-Columbian cultivation [9,14], even though they have retained their small impermeable seeds with an impressive capacity for long distance dispersal [14,15]. Only from 1967 did he abandon the views of Hutchinson et al. [16] and refer without restriction to coastal populations in the Caribbean and the Gulf of Mexico as wild [3,7]. In their extensive collecting travels, Ano et al. [17], Ano and Schwendiman [18], and Schwendiman et al. [19] went even further in underlining the similarity of these cotton populations with those of northern Yucatán shores, relating their distribution to sea currents, and classifying them in the same race 'yucatanense'.

Long-range seed dispersal also explains the presence of *G. hirsutum* in the Pacific Ocean. Fryxell and Moran [20] described a truly wild small 'punctatum' population in Socorro, an island of the Revillagigedo archipelago, some 600 km West of Mexico. Similar wild forms have diffused to even more distant Pacific islands (Tahiti, Marquesas, Samoa, Fiji, and Wake islands) [20–22]. Indeed, Fryxell [23] suggested a close relationship between the evolutionary history of the tetraploid cotton species and their particular adaptation to strand habitats along marine beaches, underlining the importance of oceanic seed diffusion and citing a dozen cases of such populations, eight of which concerned *G. hirsutum*. He presented a hypothesis relating this coastal adaptation and capacity for diffusion via ocean currents to the significant mobility of shorelines during the Pleistocene. In 1979, Fryxell further developed his views in his monograph on the Malvaceae [24], adding to his arguments those of Sauer [11]. Since then, the question of the natural dispersion of *G. hirsutum* has been further complicated by the recent description of wild populations of *G. hirsutum* in Paraguay [25], confirming an intuition of Stephens [7].

Despite its importance for cotton genetics and breeding, the question of truly wild cottons has spawned relatively few genetic studies. In their RFLP study, Brubaker and Wendel [8] observed three groups: (1) races 'yucatanense' and 'punctatum', (2) races 'latifolium' and 'palmeri', and (3) race 'Marie-Galante'. They refuted Hutchinson's views on the regressive status of race 'yucatanense', and proposed a model where "the morphological intergradation, geographical proximity, and genetic similarity of race 'yucatanense' to inland 'punctatum' populations – of Yucatán – reflects a relationship between the first domesticated form of *G. hirsutum* and its wild progenitor." Thus, the initial stages of cotton (*G. hirsutum*) domestication would have taken place in northern Yucatán and the human-mediated transfer of the first 'punctatum' cottons out of the species' natural range would have triggered the process of concomitant differentiation into new and improved races, agronomic developments, and long range germplasm diffusion. This process would explain the current distribution of *G. hirsutum* diversity. The SSR study of Lacape et al. [26] supported the racial classification [6], and the interracial relations appeared consistent with the model of progressive domestication, diffusion and differentiation proposed by Brubaker and Wendel [8], except for the geographically more distant 'Marie-Galante', which appeared closely related to 'punctatum'. Their three 'yucatanense' accessions from Guadeloupe (as classified by Ano et al. [27]) exhibited a high number of unique alleles. Similarly, in the study of Liu et al. [28], the unique representative of race 'yucatanense', from Yucatán, appeared highly divergent from the other accessions.

The views of Brubaker and Wendel [8], which explain the pre-Columbian *G. hirsutum* diversity by successive waves of diffusion of genetic and agronomic developments, from northern Yucatán to inland Yucatán (race 'punctatum'), then to southern Mexico and Guatemala (race 'latifolium'), and finally to all Mesoamerica and the Caribbean, imply an early cotton domestication. This is consistent with the contributions of historical linguistics and archaeology. Thus, words for cotton can be reconstructed in Proto-Otomangue, a language that was spoken in Central Mexico at least 6500 BP [29,30]. According to Smith and Stephens [31], the oldest remains of Mesoamerican cotton, found in the Tehuacán Valley and dated 5500 to 4300 BP, represent fully domesticated introductions, being comparable in form and size to the landraces currently existing in the same area.

The domestication and diffusion scenario proposed by Brubaker and Wendel [8] for *G. hirsutum* has been generally accepted and it

Table 1. Repartition among races of *G. hirsutum* of the 110 selected accessions per country of origin and type/race (Nota: 9 locations had both 'truly wild' (TWC) and feral specimens).

	Feral						Total feral	Total wild	Total
Country race:	**MG**	**MO**	**PA**	**RI**	**PU**	**unk**		**TWC**	
Antigua & Barbuda								1	1
Aruba	1						1		1
Bahamas	1						1		1
Barbados	1						1		1
Bonaire	2						2	2	4
BWI_Grand Cayman	1						1		1
Colombia	2						2		2
Costa Rica	1						1		1
Cuba	1						1		1
Curaçao	4						4	4	8
Dominican Rep	4						4	4*	8
Dominica	1						1		1
French Guiana	1						1		1
Guadeloupe	6						6	4	10
Guam Pacific					1		1		1
Guatemala	1						1		1
Haïti	1						1		1
Jamaica	3						3	1	4
Maldives					1		1		1
Martinique	1						1		1
Mexico		1	1	1	9		12	8[Y]	20
Nicaragua	1						1		1
Puerto Rico	8						8	7[†]	15
Saint-Kitts & Nevis	2						2	3	5
Saint-Vincent & Grenadines	1						1		1
Samoa						1[¶]	1		1
Trinidad & Tobago	2						2		2
USA-Florida								3	3
Venezuela	7						7	5	12
Total[‡]	**53**	**1**	**1**	**1**	**11**	**1**	**68**	**42**	**110**

Races are referred as MG for 'Marie-Galante', MO for 'morrilli', PA for 'palmeri', PU for 'punctatum', RI for 'richmondi', unk for 'unknown'. TWC collectively refers to truly wild cotton populations including those of race 'yucatanense' from Mexico (see text).
* = also referred-to as *G. ekmanianum*,
[Y] = 8 accessions from Mexico include 7 accessions from Yucatán and one from Socorro Islands,
[†] = local name "algodon brujo",
[‡] = one additional accession from Australia (FM966) as modern cultivated (total = 111),
[¶] = accession initially of «unknown» race (further assigned as race 'punctatum').

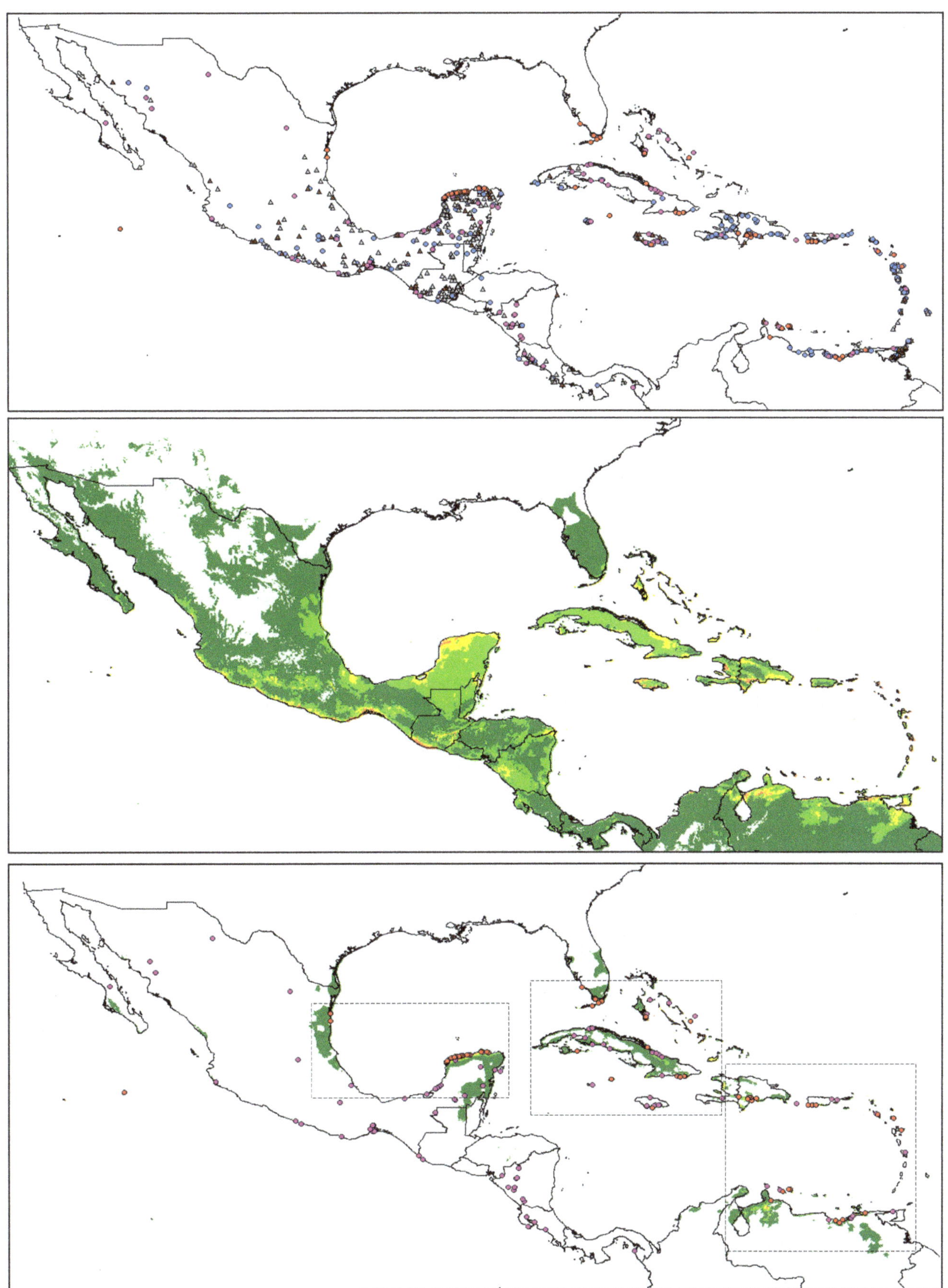

Figure 1. Distribution and climate model of perennial forms of *G. hirsutum* in Mesoamerica and the Caribbean. A. Distribution of 954 categorized datapoints for perennial forms of *G. hirsutum* in Mesoamerica and the Caribbean 'truly wild' (TWC) specimens/populations are represented by red dots, 'wild/feral' by purple dots, 'feral' (disturbed habitats) by blue dots, 'cultivated' by brown triangles, and unclassified plants by grey triangles. **B.** Climate model for distribution of both cultivated and spontaneous *G. hirsutum* in Mesoamerica and the Caribbean (complete set as presented in Figure 1A). Climate suitability is indicated by background color from unfavorable (no color) to marginal (dark green) or increasingly favorable (light green and warmer colors). **C.** Localization of the populations from categories TWC and 'wild/feral' and climate model for TWC populations. Red dots represent the datapoints used for the distribution model (TWC populations), whereas purple dots represent 'wild/feral' populations of uncertain status (truly or secondarily wild). Climate suitability is indicated as in Fig. 1B. Three dotted frames refer to map limits as magnified in Figure 2.

is found practically unchanged in the most recent syntheses [1,32]. The fact that it is based on only one wild population (from northern Yucatán) has not been challenged, and alternative scenarios have been overlooked. Nevertheless, as stated by Sauer [11], "if lint bearing cottons were naturally present in the New World as sea dispersed pioneers, they were not likely to be confined to Yucatán… The lint may have been widely gathered and perhaps traded long before regular cultivation began; the process of domestication may have been diffuse in space and time, involving wild cottons from Caribbean and Pacific coasts, as well as Yucatán." Indeed, if *G. hirsutum* is a perennial whose regressive forms thrive in disturbed human habitats and xerophytic secondary vegetation, domestication was not necessarily a linear process, moving from littoral strands to the agricultural field through the dooryard. Wendel et al. [4] questioned "whether *G. hirsutum* achieved widespread distribution and regional differentiation as a wild plant prior to domestication, or if it was widely distributed as a perennial semi-domestic by the pre-Columbian people from a much smaller native range". Casas et al. [33] have described how Mesoamerican societies have improved more than 200 plant species, through management practices that integrate cultivated areas, agroforestry systems and gathering from the wild, with or without conscious selection. As documented from many studies of cactus fruit species [34–36], the result of this *in situ* domestication process is a mosaic of habitats and useful plant populations with particular morphological, genetic, and even reproductive characteristics, according to management intensity. Similar management practices may have been used for cotton. Stephens [9] cites several accounts, from the first voyage of Columbus to much later periods in colonial times, mentioning the simultaneous exploitation of cultivated, feral and wild cottons, according to the quality objective.

If we recognize that Caribbean wild cotton populations may have been involved in the domestication process, we must also question the distinction between a primary centre of diversity in Mesoamerica and a secondary centre in the Caribbean. The strong dominance of race 'Marie-Galante' in the latter region, as well as in southern Central America and northern South America, poses the question of its origin and even of its possible separate domestication [7,32].

The question of the natural distribution of *G. hirsutum* is not only crucial for understanding the biogeography of tetraploid cottons, and their evolution and diffusion under domestication, but also for the continuation of the domestication process. Further improvement of the crop requires both a better exploitation of the available germplasm and better genetic tools to manipulate important economic traits [32]. For example, studies on the effects of domestication on such essential traits as fiber development [37] and the corresponding genetic transformations, with an altered expression of about 25% of the genes at transcriptome level [38], depend on the comparison of well-defined and representative samples of wild and domesticated germplasm.

We present here a double approach to investigate this question, combining Ecoclimatic Niche Modeling (ENM) and neutral genetic markers to assess whether coastal cotton populations are "truly wild," and investigate their relationship with inland perennial cottons. ENM methods derive an envelope for the environmental requirements of a taxon from a set of its occurrence localities. They have provided a powerful tool for investigating the ecology and distribution of both plant and animal species. An ENM study on *G. hirsutum* was recently published by Wegier et al. [39] aiming to understand not only the distribution of "wild" cotton populations from Mexico, but also the spatial organization of genetic diversity and potential gene flow from genetically modified cultivars using molecular markers. However, as compared to the wild cotton studies cited above, Wegier et al. [39] used much more permissive criteria to distinguish feral and wild cottons. In our approach, we have used ENM to document the relationship between perennial cotton domestication and distribution in Mesoamerica, the Caribbean and the Gulf of Mexico, by mapping and comparing the potential tropical/subtropical distributions of domesticated *G. hirsutum* populations, feral cotton (escaped from cultivation), and presumably or truly wild populations of races 'yucatanense' and 'punctatum'. Potential distribution of *G. hirsutum* was predicted for modern climatic conditions as well as for climatic parameters modeled for the Last Glacial Maximum (LGM). The underlying idea is that the original distribution of wild cotton in the early Holocene was necessarily related to its distribution during the Pleistocene, following an approach validated by several studies [40,41]. The identification of potential climatic refuges for the species should help in distinguishing natural and human factors in its dispersal.

As the ENM study confirmed the particular ecology and "truly wild" status of a number of coastal cotton populations, SSR neutral genetic markers were then used to characterize them and investigate their relationship with neighboring feral cottons.

Materials and Methods

Climatic modeling and analysis

Our ENM study focused on the centers of diversity of *G. hirsutum*, i.e., Mexico, Central America, and the Eastern Caribbean (from the coasts of Venezuela to Florida through the Antilles). From now on, we will collectively refer to this region as Mesoamerica and the Caribbean. Geographical and ecological information was extracted from the CIRAD cotton germplasm database and records [17–19,27,42,43] and related collecting reports by French and US scientists (collections in the 80s under the aegis of the former IBPGR), the scientific literature on wild cotton, regional floras, herbarium-label and germplasm-passport data obtained from the Global Biodiversity Information Facility (GBIF) portal, the Mexican Red Mundial de Información sobre Biodiversidad (REMIB), and relevant Mexican administrative documents. All geographic coordinates have been assigned or verified against associated geographic information with gazetteers (mostly Google Earth and Geonames). Incomplete or imprecise records were discarded, as were redundant data (dataset available upon request).

The information associated with the collections/observations was also used to classify cotton occurrences according to their status on a wild to cultivated scale, using four categories: 'cultivated' (fields and dooryards), 'feral' (plants found in disturbed habitats, such as roadsides and secondary vegetation), 'wild/feral' (plants found in preserved habitats and/or forming persistent populations), and 'truly wild' (populations described as such by experts, based on ecological and morphological grounds). This categorization is partially analogous to that used by Stephens [9], whose "wild forms" would include both our 'truly wild' and 'wild/feral' categories, whereas Stephens' "semiferal" and "commensal/cultivated" forms correspond to our 'feral' category and cultivated categories, respectively. The objective was also analogous: Stephens tested his categories on domestication traits (fiber and seeds) while we aimed at testing them on eco-climatic grounds.

For each occurrence record, 19 bioclimatic variables were extracted from WorldClim, a package consisting of global surfaces of climate, with a 2′30″ grid resolution (corresponding roughly to 4.4×4.6 km) [44]. These variables are: 1) annual mean temperature; 2) mean diurnal range (mean of monthly (max temp - min temp); 3) isothermality (Bio2/Bio7); 4) temperature seasonality; 5) maximal temperature of warmest month; 6) minimal temperature of coldest month; 7) temperature annual range; 8) mean temperature of wettest quarter; 9) mean temperature of driest quarter; 10) mean temperature of warmest quarter; 11) mean temperature of coldest quarter; 12) annual precipitation; 13) precipitation of wettest month; 14) precipitation of driest month; 15) precipitation seasonality; 16) precipitation of wettest quarter; 17) precipitation of driest quarter; 18) precipitation of warmest quarter; and 19) precipitation of coldest quarter.

For ENM, we chose the widely used Maxent machine learning method. It estimates the probability distribution of maximum entropy (i.e. closest to uniform) subject to the constraint that the expected value of each environmental variable (or its transform and/or interactions) under this estimated distribution matches its empirical average [45]. Maxent was run twice, firstly on the whole dataset, and secondly only on points in the 'truly wild' category. A logistic threshold value equivalent to the 10 percentile training presence was retained to separate climatically favorable areas from marginally fit areas. Maxent output provides measures of the contribution of each bioclimatic variable (percent contribution and permutation importance) and proposes a jackknife test to quantify the contribution of each variable from the gain when it is used in isolation and the gain loss when it is omitted from the model. However, the strong correlations among bioclimatic variables do not allow an easy interpretation of their relative importance. Therefore, we performed a principal component analysis (PCA) to characterize and compare the climatic envelopes of our categories of *G. hirsutum* observations, discarding those variables whose contribution appeared marginal. The factors with an eigenvalue above 1 were retained and a normalized varimax rotation was applied to maximize the sum of the variances of the squared loadings, simplifying the interpretation of the results. The different categories of populations were then plotted on the principal components plane to visualize and compare their ecoclimatic range.

To predict the potential distribution of *G. hirsutum* at LGM, the MIROC climatic model [46] derived from the PMIP2 database Paleoclimate Modelling Intercomparison Project Phase II for 21,000 BP was downloaded from the Worldclim website (http://www.worldclim.org/) and used on a dataset restricted to the 'truly wild' category.

Genetic analyses

The panel of accessions of perennial *G. hirsutum* cotton populations used for SSR genotyping comprised 110 feral and wild accessions supplemented by a modern cultivar, 'FM966' (Table 1). One hundred and eight accessions originated from the CIRAD seed bank, and three from USDA. Twenty-nine countries/provinces of Mesoamerica and the Caribbean were represented (Table 1). Particular attention was paid to geographic locations where both truly wild and feral populations could be identified in close proximity (such as for the populations of Pointe des Châteaux in Guadeloupe), or slightly more distant (such as for the populations from Yucatán sea-shores versus inland). Such sites with both truly wild and nearby feral specimens were identified in nine cases (Mexico/Yucatán, Jamaica, Dominican Republic, Puerto Rico, St Kitts & Nevis, Guadeloupe, Venezuela, Bonaire, and Curaçao). Three localities were represented only by wild specimens, Florida (one feral specimen discarded due to missing data), Antigua, and Socorro Islands of Mexico; and 18 additional localities were only represented by feral populations. A few additional locations where truly wild cotton (further abbreviated as TWC) populations had been reported (visible as red dots in Figure 1) could not be included in the genetic study due to lack of plant material, such as in Cuba, the western coast of the Gulf of Mexico (Tamaulipas), Bahamas and Grand Cayman. Detailed geographic information of the 110 accessions is available in Table S1; Figure S1 presents their localizations on the sites with TWC populations.

Five seeds per accession were sown in small pots in the greenhouse in Montpellier and DNA was extracted from pooled samples (1–3 different plants) of young leaves using the MATAB protocol [47]. Thirty-seven SSR markers were selected for genotyping based on previous experience [26], in order to optimize information and quality. They were mostly derived from non-coding genomic DNA sequences (majority from series 'BNL' and 'CIR'), preferably to the more frequent EST-derived SSRs, with presumptive neutrality (no evidence of having been targeted during domestication). They had shown in previous experiments the amplification of a single PCR product in tetraploid cotton, thus avoiding the ambiguity generated by homoeolog loci. SSRs were genotyped in multiplex panels of 8 SSRs (four dyes and two SSRs per dye). Simultaneous PCR amplifications in a final volume of 10 µl contained 5 ng of genomic DNA, 200 µM of each dNTP, 0.5 mM MgCl2, 1 U Taq polymerase, 0.08 µM of M13-tailed 'F' primer, 0.1 µM of both the 'R' primer and of an M13 oligonucleotide tailed with the ad hoc fluorochrome. PCR reactions were performed on an Eppendorf microcycler (Eppendorf, Madison, WI)) using the following profile, a hot start of 94°C for 5 min, 35 cycles of 30 sec at 94°C, 1 min at 55°C and 1 min at 72°C, and a final extension step of 30 min at 72°C. PCR products were pooled with 10 µl of GeneScan 600-LIZ size standard. PCR products were denaturated and size fractionated using capillary electrophoresis on an ABI 3500 Genetic Analyzer (Applied Biosystems). Subsequently, GeneMapper 4.1 (Applied Biosystems) software was used for allele size estimation.

Twenty-six SSRs showing strict and unambiguous bi-allelic patterns (coded as homozygote when a single peak/allele and heterozygote with 2 peaks/alleles) were selected. The 26 SSRs were mapped on 18 of the 26 chromosomes (Table S2). Expected heterozygosity at each locus was calculated as $He = 1 - \Sigma pi^2$ where pi is the frequency of the ith allele.

The data matrix of bi-allelic codings for the 26 SSRs and 111 genotypes was imported into the DARWin5 software [48] to calculate genetic dissimilarities. Bootstrap dissimilarity matrices were calculated by drawing 10 000 entries. A Principal Coordinate Analysis based on the similarity matrix was conducted also with DARWin package. In complement to this factorial analysis, unweighted trees without topological constraints were constructed using a neighbor joining (NJ) approach [49] to represent individual relations. Lastly, the methods implemented in the STRUCTURE software [50] were used to infer population clusters and estimate admixture (quantitative clustering). The number of clusters, K, was chosen based on 20 independent runs for K values ranging between 1 and 5 with a burn-in length of 500,000 followed by 750,000 MCMC iterations. The ΔK method [51] was then applied using Structure Harvester [52], and estimated membership for each genotype, in each cluster, was read from the STRUCTURE output.

Results and Discussion

Dataset composition and distribution for climatic modeling

A total of 954 datapoints were gathered, of which 630 could be ascribed to our four categories (Table 2). Figure 1A shows no clear differences in the distributions of the different categories, except for 'truly wild' cotton (TWC) populations, which only occur along the coasts of the Eastern Caribbean and the Gulf of Mexico. The sample is well balanced between Central America and Mesoamerica, on one hand, and the islands and shores from the Eastern Caribbean to Florida on the other hand. Feral and wild specimens are better represented than cultivated germplasm, which can be explained by a collecting bias of botanists, most often interested by spontaneous plants, and germplasm collectors, motivated by the rusticity expected from primitive and spontaneous materials. The poor representation of 'cultivated' cotton also reflects the decline of its cultivation in Mexico [10] and in the Caribbean [18].

Among the 544 datapoints from Central and Mesoamerica, few have been assigned to a geographical race: 2 for race 'morrilli' (state of Guerrero), 8 for 'palmeri' (Guerrero), 5 for 'richmondi' (Oaxaca), 41 for 'punctatum' (Yucatán peninsula and Socorro Island), and 32 for 'yucatanense' (state of Yucatán). Albeit poor, this information is consistent with their original description by Hutchinson [6] and, with the exception of 'yucatanense', all races are found in both 'cultivated' and 'feral' categories, illustrating the absence of morphological differentiation between cotton landraces and feral cottons within a same region, as reported by several collectors [9,10,24]. 'Punctatum' is the only race with important spontaneous populations classified as 'wild/feral', one in the state of Yucatán, around Celestún, and several ones on the southern coast of Campeche state, between Champotón and Isla del Carmen. The only 'truly wild' Mexican population of race 'punctatum' is the one described by Fryxell and Moran [20] in the Socorro Island (Revillagigedo archipelago).

For the Eastern Caribbean (410 accessions from Venezuela to Florida), most observations were from breeders, so the racial composition is much better documented. It shows a strong dominance of race 'Marie-Galante' (278 acc.). The only other identified race is 'punctatum', ascribed to the TWC category (64 datapoints) or, exceptionally, to the 'wild/feral' category (one datapoint). In our dataset, these TWC are classified as 'punctatum', following the early views of Hutchinson [13], author of the original classification, although the same materials collected by Ano et al. [27] and Schwendiman et al. [19] were later reclassified under race 'yucatanense'.

Ecoclimatic niche models for cultivated, feral, and wild *G. hirsutum*

Figure 1B presents the potential distribution extrapolated by the Maxent software for the whole dataset. Along the coasts of Mexico, climatically favorable lowland areas correspond to those identified by Wegier et al. [39], i.e., the Yucatán peninsula, the regions of Veracruz and Tamaulipas along the western shores of the Gulf of Mexico, and the tropical Pacific coast. The latter area appears particularly favorable. The state of Tabasco (southern shores of the Gulf of Mexico) is better represented than in the study of Wegier et al. [39]. Other favorable areas are found much further inland.

Given the relative over-representation of wild and feral materials in our sample, Figure 1B gives a likely picture of the Mesoamerican distribution of perennial *G. hirsutum* for the last three millennia at least, i.e. a period of very active agricultural development, during which modern climatic conditions were already established [53]. The distribution of favorable areas corresponds quite well with those areas where several of Hutchinson's geographic races were developed: Yucatán to Mexican shores of the Gulf of Mexico for race 'punctatum', Yucatán to Guatemala for race 'latifolium', Pacific regions and the southern side of the isthmus of Tehuantepec for races 'palmeri' and 'morrilli', and even regions of the central Mexican plateau for race 'richmondi'. In Central America, the pre-Columbian distribution of *G. hirsutum* appears related to the diffusion of race 'Marie-Galante', as the favorable areas close to the Guatemalan-Salvadoran border and in western Nicaragua show good correspondence with the distribution of this race, presented by Stephens [7]. As suggested by this author, these races probably differentiated under relative geographical, ecological and cultural isolation, the latter term covering "the combined effects of human selection, migration and diffusion."

Figure 1C presents the geographical distribution of 'wild/feral' and TWC populations, together with a distribution model based only on 'truly wild' populations (100 datapoints). The areas suitable for TWC populations (Figure 1C) cover a very small part of the favorable areas for the whole sample (Figure 1B). They are mostly found in three sub-regions: (i) Gulf of Mexico and northern Yucatán, (ii) Florida and western Greater Antilles and (iii) Venezuela and eastern Caribbean, as detailed in Figures 2A, B and C, respectively.

Table 2. Dataset composition and distribution among domestication status categories of perennial *G. hirsutum* as defined for the present study.

	Total	Uncategorized	cultivated	feral	wild/feral	truly wild
Meso- & Central America	544	308	61	80	59	36
Eastern Caribbean to Florida	410	16	96	188	46	64
Total	954	324	157	268	105	100

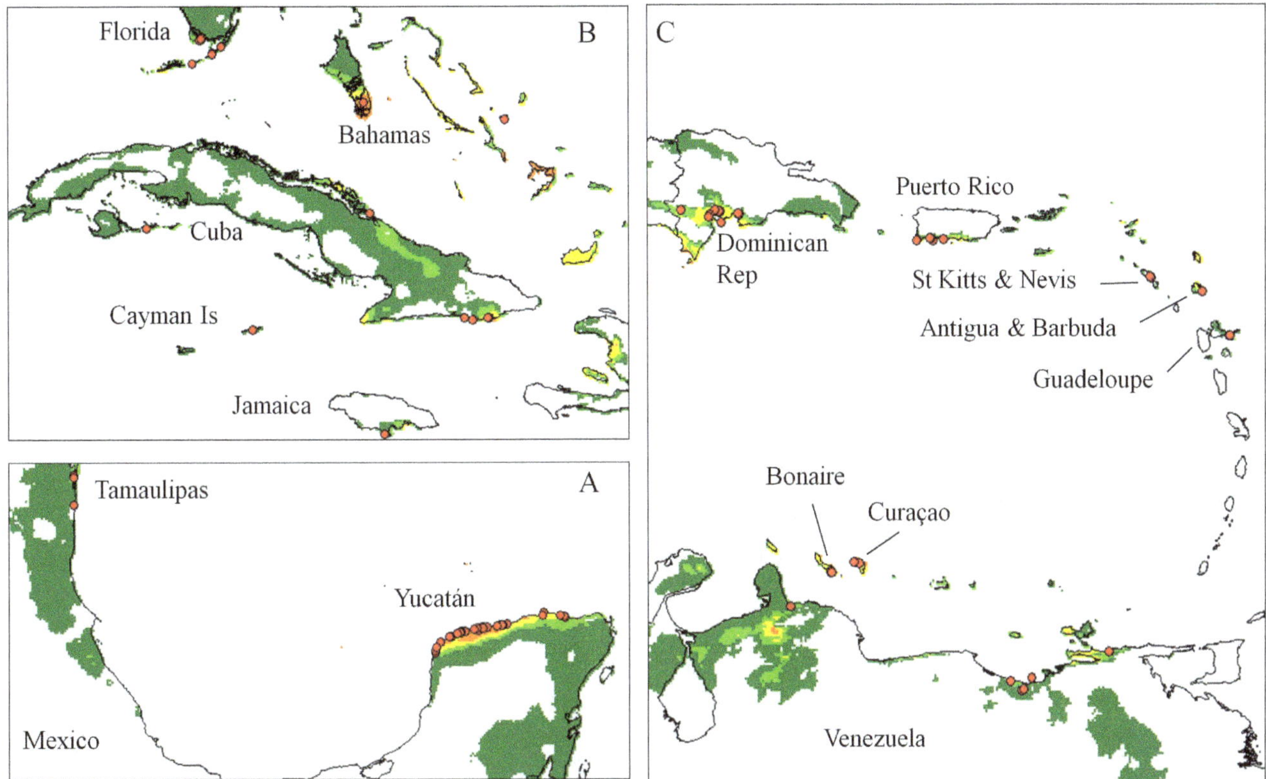

Figure 2. Localization of the truly wild cotton (TWC) populations and corresponding climate model. Map frames indicated as rectangles in Figure 1C. Climate suitability as indicated in Fig. 1. **A.** Gulf of Mexico. **B.** Florida and western Greater Antilles. **C.** Venezuela and eastern Caribbean.

The great majority of 'wild/feral' populations (purple dots on Figure 1C) fall in areas that are marginal (dark green areas) or unsuitable for TWC populations, validating our a priori categorization.

The 'yucatanense' population along the northern coast of Yucatán (Figure 2A), certainly constitutes the most extensive TWC population [43,54]. Our model confirms that its distribution is clearly limited by ecological parameters, as stated by Sauer [11]. Within this well-delimited area, a few specimens classified as 'wild/feral' are very probably incompletely documented representatives of race 'yucatanense'. Extensive spontaneous populations also exist on the western coast of the Yucatán peninsula, but we have found no indications that these are 'truly wild'. On the contrary, the model indicates that they have developed under climatic conditions that are not even marginally fit for TWC populations. West of the Gulf of Mexico, along the coast of Tamaulipas, 'truly wild' *G. hirsutum* was observed by Lukefahr cited in Stephens [7]. However, favorable areas are small and sparse in this region, and we could trace only three specimens whose labels mention that they were parts of natural coastal vegetation. Confirming the statement of Stephens [7], no population that could be classified as TWC has been documented for the Pacific coast of Mexico, where a very few small coastal areas appear climatically marginal for sustaining such populations. Thus, while the model confirms highly favorable climatic conditions in the Revillagigedo Islands, it gives no clear indication about areas where wild *G. hirsutum* could have developed on the western coast of Mexico before diffusing to islands in the Pacific Ocean.

In northern South-America and the southern Caribbean (Figure 2C), TWC populations are scattered along the coasts of Venezuela, between the Gulf of Venezuela (Saco de Maracaibo; state of Falcón) and the North of the state of Sucre, and on the shores of many islands along these coasts: Curaçao, Bonaire, Isla de Piritú. We have found only ambiguous information for the Chacachacaré Island. Mentions of colonial cotton plantation cast doubt on the only report of wild cotton populations in this area by Stephens [9]. On the other hand, the surroundings of Chacachacaré village in the Island of Margarita offer excellent conditions for TWC populations, suggesting that the homonymy of these neighbor sites may have created confusion. To the West, the shores of Colombia only offer marginal conditions for TWC (Figure 1C), which explains why Stephens [9] was not successful in his search for wild cotton in this area. To the Northeast of Venezuela, there seems to be another gap in the natural distribution of *G. hirsutum*, as no TWC populations have been identified in Trinidad and Tobago or in the southern half of the Lesser Antilles (Figure 2C), which is consistent with the descriptions of Hutchinson [13,55]. In the northern Lesser Antilles (Figure 2C), only three TWC populations have been described, in Guadeloupe [27], in Antigua and in Saint Kitts [9,18,19], and the model confirms favorable climatic conditions at these sites.

In the Greater Antilles (Figure 2B and 2C), the modeled distribution also agrees well with the wealth of previous reports of TWC populations of race 'punctatum', indicating favorable climatic conditions for the "algodón brujo" of southern Puerto Rico [9,13,19], for the populations around the Yaquí Valley of the Dominican Republic [9,19,56], in Haiti [13], Jamaica [19,57,58] and in the Cayman Islands [12,59,60]. In southern Cuba, similar populations exist around Guantánamo (specimen labels refer to the morphological type described by Britton in 1908 [57]). Further North, the modeled distribution is consistent with the observations of TWC in Florida [19,23] and in the Bahamas [61,62]. For Bermudas, much further North, the model indicates unfavorable conditions for TWC, which is consistent with the statement by Britton [63] about the absence of native cotton in these islands.

Climatic requirements of cultivated, feral and wild populations of perennial cotton

Seven variables were discarded for PCA on climatic variables, because of their poor specific contribution to the Maxent model obtained from the whole sample: isothermality (Bio3), maximal temperature of the warmest period (Bio5), precipitation of the wettest and driest periods (Bio13 and Bio14), precipitation seasonality (Bio15), and precipitation of the driest and warmest quarters (Bio17 and Bio 18).

The analysis on the remaining twelve variables produced three factors with an eigenvalue superior to 1 (Table 3). The first one is strongly associated with mean temperatures at all periods of the year (Bio1, and Bio8-11), with correlations between 0.82 and 0.95; the second one is associated with precipitation (Bio12, 16 and 19), with correlations between 0.80 and 0.95; and the third one is associated with variables related to latitude (Bio2-7: diurnal temperature range, temperature seasonality, minimal temperature of coldest period and temperature annual range). The third factor shows no clear differences among our categories, which is consistent with their similar latitudinal dispersion, from tropical Venezuela to subtropical northern Mexico and Florida. In contrast, the categories and origins present different patterns of dispersion in the plane formed by the two first principal component factors (Figure 3). On the continent (Central and Mesoamerica, Figure 3A), part of the observations come from cooler regions (along the x-axis of factor 1, to the left) or from wetter regions (along the y-axis of factor 2, upwards), while cotton-associated climates appear more uniform in the eastern Caribbean (Figure 3B). *G. hirsutum* was not observed in regions that are both cooler and wetter (upper left area in Figure 3), which gives the general shape of an inverted 'L' to the Mesoamerican dot cloud.

When considering domestication status, no clear distinction can be made between 'cultivated', 'feral', and 'wild/feral' materials (Figure 3C), as these categories share the same general inverted 'L' pattern of dispersion in the principal components plane. In contrast, TWC populations are clearly characterized by very uniform climatic conditions; thus the environment of both 'yucatanense' and truly wild 'punctatum' (Figure 3D) is clearly among the hottest and driest in our sample. The best represented geographical race, 'Marie-Galante', which is highly dominant throughout the Antilles, logically presents the same climatic dispersion as the general Caribbean sample, with occurrences under extremely arid conditions too (not shown). Indeed, several reports mention spontaneous 'Marie-Galante' populations in the vicinity of TWC populations, as in Puerto Rico [13], Saint Kitts [18], and Guadeloupe [27].

Potential distribution of native *G. hirsutum* in America and the Pacific

Both the ENM and factorial analyses clearly show that TWC populations of *G. hirsutum* present an exceptional combination of a narrow environmental niche and a highly geographically scattered distribution. Stephens [64] has related the capability for long distance dispersal of tetraploid cotton seeds to their buoyancy and tolerance to prolonged immersion in salt water. It is therefore interesting to extend the TWC climatic model derived from occurrences in Mesoamerica and the Caribbean to a larger area in South America and the Pacific. Figure 4 presents the results of this extrapolation in South America. Four areas offer favorable climatic conditions, two inland areas, Bolivia/Paraguay and Northeastern Brazil, and two coastal areas, Ecuador/Peru and Pacific islands. Strikingly, all of them are validated by the existence of wild populations of tetraploid cottons. The favorable area in Bolivia and Paraguay was suggested long ago by Stephens [7] and, indeed, a wild form of *G. hirsutum* has been reported there recently [25]. Its inland situation renews the question of tetraploid cotton dispersal, as it implies non-oceanic diffusion. A bird-related mechanism is the likely explanation [15]. The other potential inland area, in Northeastern Brazil, corresponds well to the distribution of *G. mustelinum*, a wild tetraploid endemic to the region [65–67]. The third area, in the arid coastal regions of southern Ecuador and northern Peru and in the Galapagos Islands, corresponds with the distribution of 2 other wild tetraploid *Gossypium species*: (i) the wild populations of *G. barbadense* (North and South of the Guayas estuary) and (ii) the wild tetraploid species *G. darwinii*, a close relative of *G. barbadense*, endemic to the Galapagos islands [68].

In the fourth favorable area (not shown), further west into the Pacific, Worldclim coverage is incomplete, particularly for small atolls, so all climatically suitable sites could not be detected. Among those cases where the extrapolation results can be compared to data from the literature, worth mentioning are the Hawaiian Islands (with marginal climatic conditions in leeward coastal areas of Honolulu, Lana'i, Kaua'i and Hawai'i), Wake Island, the Republic of Kiribati, Fiji, Samoa, and French Polynesia. Indeed, Hawaiian Islands are home of the endemic wild tetraploid *G. tomentosum*, while an unusual wild form of *G. hirsutum* is locally common in Wake Island [21,69]. The information available on the presence of wild cotton in Kiribati is less clear, with mentions of *G. tomentosum* [70–72], and/or another *Gossypium* species (probably *G. hirsutum*) [73]. Among the Pacific islands cited for wild populations of *G. hirsutum*, only Fiji and Samoa do not appear climatically fit for this species according to our extrapolation; however, this can be related to the rarity of *G. hirsutum* var. *taitense* Roberty in both archipelagos [74].

The excellent correspondence between areas potentially favorable to wild forms of *G. hirsutum* and the actual distributions of wild tetraploid species (*G. hirsutum* itself, *G. mustelinum*, *G. barbadense*, *G. darwinii* and *G. tomentosum*) provides a very interesting example of ecological niche conservatism in evolution [75]. In the present case, it constitutes a further confirmation that the model derived from our Caribbean and TWC population sample is accurate, and indicates that the main driver of tetraploid cotton radiation was geographic isolation, not environmental specialization.

Potential distribution of *Gossypium hirsutum* in Mesoamerica and the Caribbean at the Last Glacial Maximum

Figure 5 presents the potential distribution of 'truly wild' *G. hirsutum* for LGM climates, i.e. about 21,000BP. Sea level was ca. 125 m lower at that time, and rose markedly from 17,000 to 7,000 BP [76]. According to the MIROC model, most areas where 'truly wild' cotton populations are found under modern climates were only slightly less favorable at LGM. A few very small favorable areas, such as the one along the shores of Tamaulipas, were at best marginally fit for *G. hirsutum*. In contrast, three areas show a considerable extension at LGM, with many more favorable

Table 3. Principal component analysis (Varimax normalized rotation) on a set of bioclimatic variables retained for their contribution to the Maxent ecoclimatic model of distribution: factor loadings (values higher than 0.70 in bold characters).

Variable	Factor 1	Factor 2	Factor 3
1-Annual mean temperature	**0.93**	0.04	0.33
2- mean diurnal range	−0.06	0.04	**−0.82**
4- temperature seasonality	−0.01	−0.29	**−0.80**
6- minimal temperature of coldest month	0.54	0.12	**0.83**
7- temperature annual range	−0.07	−0.09	**−0.97**
8- mean temperature of wettest quarter	**0.90**	−0.18	−0.03
9- mean temperature of driest quarter	**0.82**	0.17	0.37
10- mean temperature of warmest quarter	**0.95**	−0.11	−0.15
11- mean temperature of coldest quarter	0.68	0.18	0.63
12- annual precipitation	−0.07	**0.95**	0.14
16- precipitation of wettest quarter	−0.00	**0.95**	−0.03
19- precipitation of coldest quarter	0.02	**0.80**	0.21
Proportion of total variance	0.34	0.22	0.31

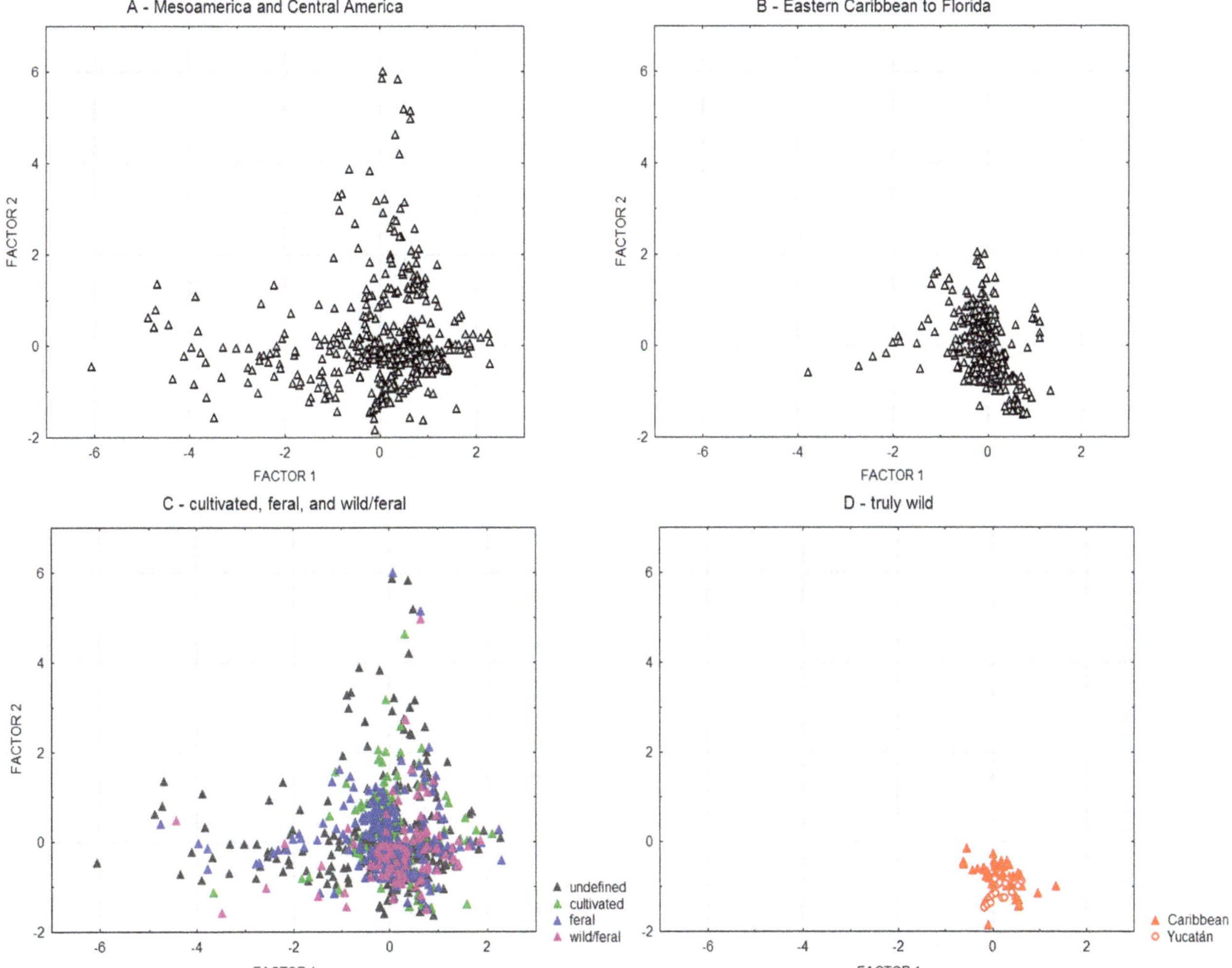

Figure 3. Principal component analysis of *G. hirsutum* climatic envelope. Climate variables are listed in Table 3. Comparison of different subsamples in Mesoamerica, the Eastern Caribbean and Florida, according to origin (A, B) and domestication status (C, D).

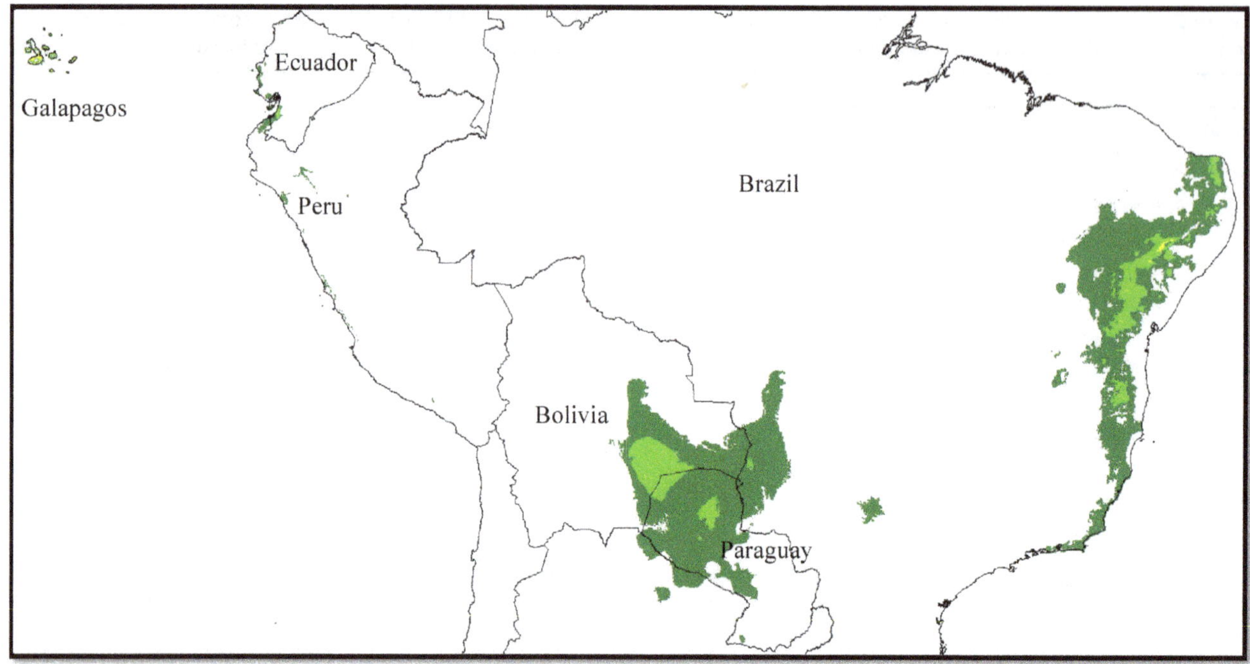

Figure 4. Potential distribution of truly wild *G. hirsutum* in South America. Distribution as extrapolated from the climate model presented in Figures 1C and 2.

emerged lands: (i) northern Yucatán, (ii) southern Florida, the Bahamas and Virgin Islands, and (iii) the western shore of Venezuela and a small area on the northeastern Colombian shores. On the whole, *G. hirsutum* distribution was probably much more extended in the Caribbean and in the Gulf of Mexico during late Pleistocene and early Holocene. The main picture is consistent with the hypothesis of Fryxell [24] that Pleistocene shoreline movements were decisive in the evolution and adaptation of tetraploid cottons.

Further south, in equatorial and southern America, LGM climatically favorable areas appear essentially similar to modern ones, except for the Brazilian Northeast, which was less favorable

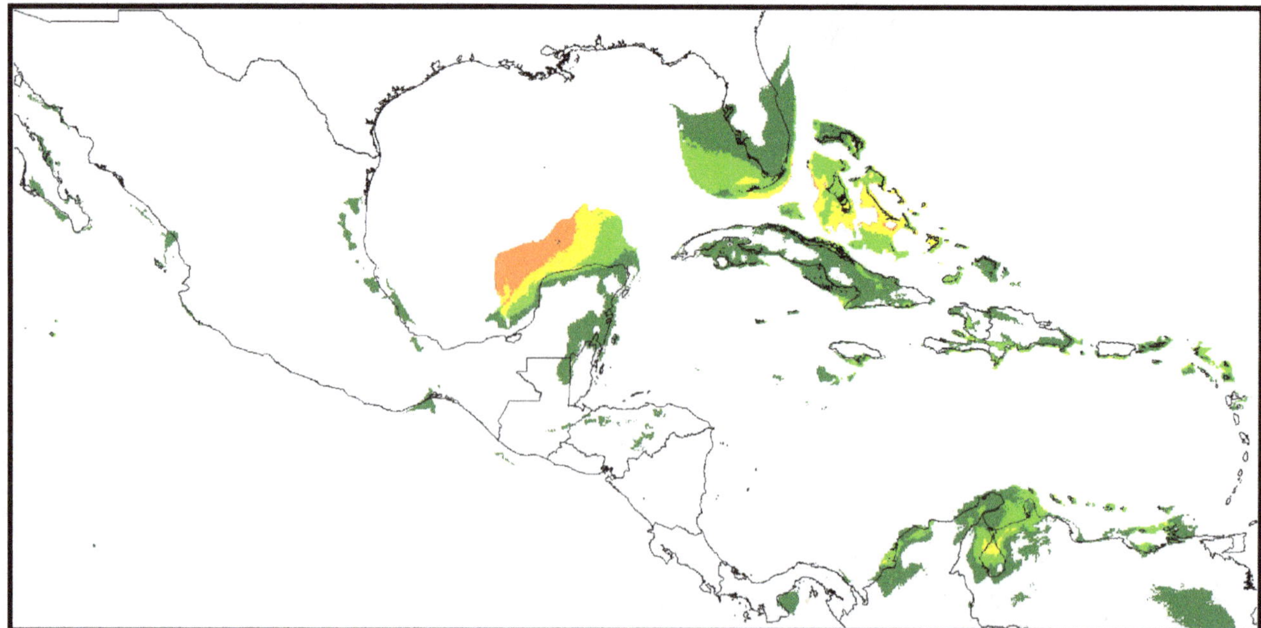

Figure 5. Potential distribution of *G. hirsutum* during the Last Glacial Maximum (21,000 BP). Potential distribution of *G. hirsutum* in the Caribbean and in the Gulf of Mexico, extrapolated according to the MIROC climatic model for LGM. (Note that sea level differences at LGM explain variation with modern sea shore delimitation).

(Figure S2) than in modern times (Figure 4). A few marginally favorable areas may have existed along the Mexican Pacific shores (Figure 5). In the Pacific Ocean, the situation appears similar to the modern one (not shown).

Genetic characterization of 'truly wild' cotton and their feral neighbors

The 42 TWC accessions from 11 different (2 for Mexico alone) countries (Figure 6, Table 1 and Table S1, Figure S1) ensure a good representation of the geographical range of truly wild *G. hirsutum* populations as described above; although several similar populations could not be sampled. It appeared very early in the analysis that these TWC populations, including the 'yucatanense' population of Yucatán as well as those from diverse places in the Caribbean, showed no racial or geographic differentiation, so we have pooled them in the following presentation. In our sample the 'feral' group was represented by 53 'Marie-Galante' accessions of northern South America and the Caribbean and by 15 other accessions ('punctatum' from Yucatán, other races from Mesoamerica, and un-ascribed material, Table 1).

SSR statistics are detailed in Table S2. In total, 204 alleles were coded over the 111 accessions and 26 SSR markers, ranging between 3 (HAU2861) and 19 (BNL3103) alleles per SSR. *He* values varied among markers, confirming previous results [26]. Unique alleles amounted to 37 in the feral 'Marie-Galante' group (53 accessions) and 43 in the TWC group (42 accessions). *He* shows only limited differences between the different races/categories (Table S3); globally it averages 24.2%, more than usually observed in cultivated cotton (between 5 and 15% under field conditions, but nil in the case of our cultivated control). *He* is slightly higher in wild accessions (28.2%) as compared to feral ones (22.0%). The genetic dissimilarity was also higher within the TWC group (D = 0.51) than within the feral group (D = 0.38) (Table S4).

Both distance-based methods implemented with DARwin, NJ classification (Figure S3) and principal coordinate analysis (Figure 7), separate TWC from feral accessions (first axis in the PCA, Figure 7, and basal branching in NJtree, Figure S3). Within the feral group the analyses further distinguished two subgroups. The first one includes 48 of the 53 accessions of race 'Marie-Galante' and the second one includes 17 accessions, 10 of race 'punctatum' from Mexico/Yucatán, 6 others (from races 'morrilli', 'palmeri' and 'richmondi' and 3 unassigned), as well as the modern cultivar. Thus, this clustering, which suffers only few exceptions, appears essentially to reflect domestication status (wild vs. feral), and secondarily race. In the nine locations that could be sampled for both feral and TWC accessions (Figure 6), the different analyses clearly indicated that wild/feral status was better than geographical distribution in determining genetic proximity among

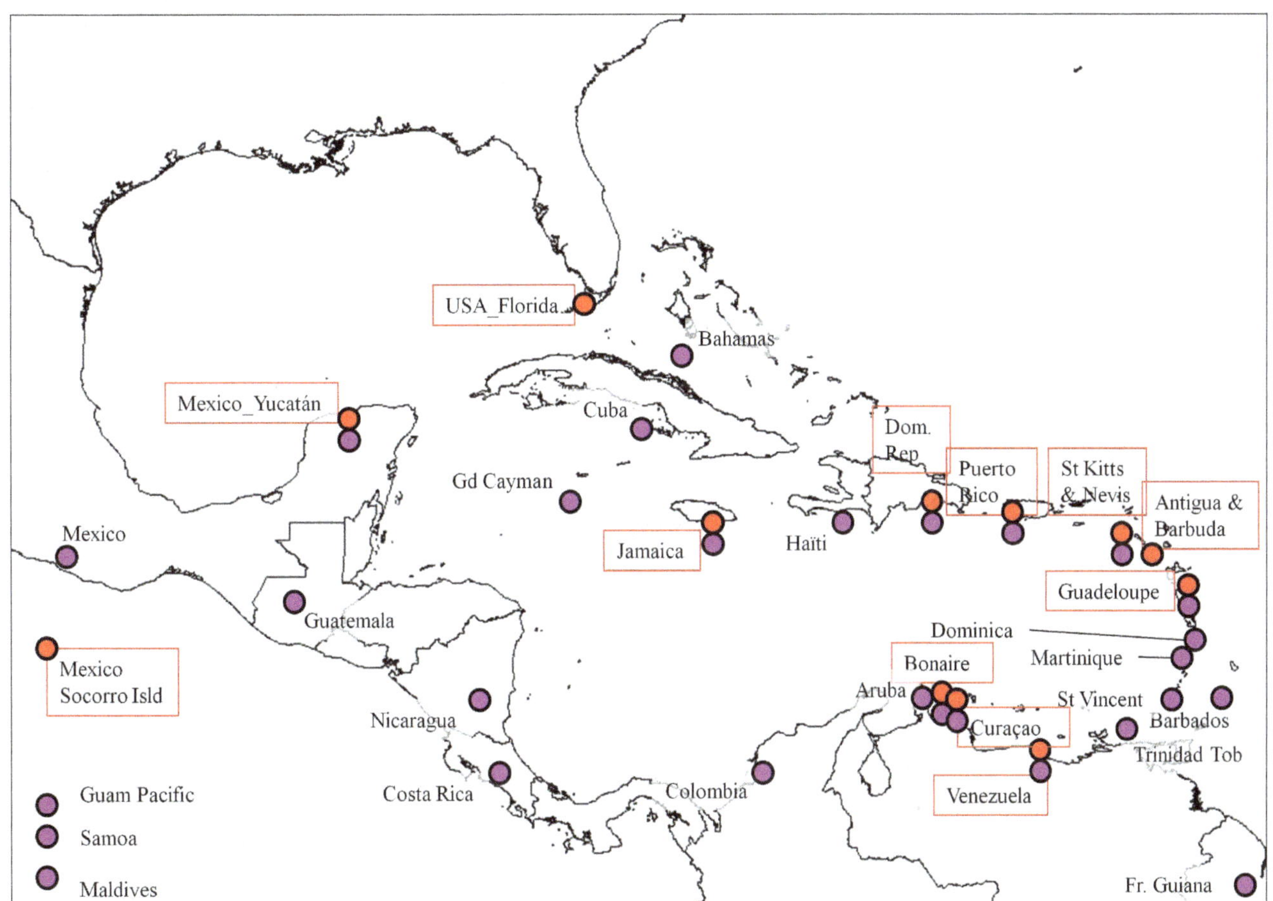

Figure 6. Distribution of the populations of perennial *G. hirsutum* sampled for the SSR-based genetic analysis. Samples include truly wild (TWC) and feral perennial populations in Mesoamerica and the Caribbean. TWC populations are shown as red dots and feral populations are shown as purple dots. Twelve locations where TWC were identified are labeled in red frame. All except USA/Florida, Antigua and Socorro Islands, are also represented by feral specimens, while 18 additional locations had only feral specimens. See Table S1 and Figures S1 for details and precise localizations of the accessions in the 11 sites with TWC populations (Socorro Islands not shown).

samples. For example, the TWC from the Atlantic shores of the Lesser Antilles (Guadeloupe, St Kitts and Antigua) were genetically much closer to TWC from northwestern Yucatán, distant by over 2,000 miles, than they were to the feral cottons of the same islands. The genetic relationship among feral cottons is not determined by geographical proximity either: for example, the six 'Marie-Galante' accessions from the island of Guadeloupe are not grouped in the same 'Marie Galante' branch of the dendrogram (Figure S3B).

The STRUCTURE analysis and ΔK method of Evanno [51] were fully consistent with the two previous ones, clustering the 111 accessions into either two or three clusters, both with high ΔK values (>1000) (Figure S4). Using K = 2 separated TWC from feral cottons (not shown). Using K = 3 further partitioned the feral group in two sub-groups. In Figure 8, we have organized our sample according to the same criteria inferred from both PCA and NJ analyses, but based on field observations: 'truly wild' vs. feral, and feral accessions assigned to 'Marie-Galante' vs. other feral accessions.

The 42 accessions from TWC populations form a fairly homogenous group (Figures 8 and S3) with an average 67% membership. Only few discrepancies were observed, whereby three accessions had very low (<5%) likelihood of membership to this cluster: W30 (acc. AS0340) from Venezuela, W102 (acc. BPS1240) and W103 (acc. BPS1247) from Puerto-Rico. These accessions were probably wrongly assigned due to an error in collection (although passport data are unambiguous) or a mixture at some stage of multiplication. For a few other TWC assignations, the possibility of *in situ* hybridization cannot be dismissed, as they show an important level of admixture (<50% membership to TWC): W105 (acc. INC035) from Socorro Island, W58 (acc. AS0653) from Yucatán, W86 (acc. BPS1157) from Bonaire, W148 (acc. BPS1239) from Puerto Rico, and W95 (acc. BPS1225) from Dominican Republic. For the latter, the collector mentioned a "different" phenotype with orange pollen and yellow petals [19].

The 53 accessions (22 countries) of race 'Marie-Galante' have an average membership of over 81%. This group encompasses the same geographical distribution as the TWC group except for Mexico (Figure 6). Four 'Marie-Galante' accessions present higher membership to the other feral group, probably because of wrong race assignation: W153 (acc. CR2000A) from Costa-Rica, W65 (acc. Texas184) from Guatemala, W150 (acc. BPS1243) from Puerto Rico, and W27 (acc. AS0335) from Venezuela. Two 'Marie-Galante' accessions, W98 (acc. BPS1230) from Puerto Rico and W59 (acc. AS0681) from St Kitts and Nevis, show high levels of admixture with the TWC cluster and both present unusually high rates of heterozygote SSR, of 64% and 58% respectively; they probably result from an hybridization. Of the two accessions sampled in Colombia, one (W32, acc. AS0435) presents 98% membership to the 'Marie Galante' cluster while the other one (W33, acc. AS0437) presents some admixture. It is noteworthy that the latter, W33, belongs to a series of 'Marie-Galante' from Northern Colombia, near Barraquilla, described by Ano and Schwendiman [42] as "híbrido nativo; offspring of ancient deliberate crossings between local spontaneous 'Marie-Galante' and commercial varieties of *G. hirsutum* or *G. barbadense*."

Lastly, the mostly 'punctatum' branch of feral cottons presents the lowest level of admixture (>97% membership). With 16 accessions, this group includes 12 'punctatum' accessions [9 from inland-Yucatán (as opposed to TWC from the northern shores of Yucatán, - 1 from Maldives in the Indian Ocean (W181, acc. KLM1872), - 2 from Pacific islands (W157, acc. TX-0997 from Guam, W108 acc. TX-1295 from Samoa)], one representative of Mexican races, 'morrilli', 'richmondi' and 'palmeri', and the modern cultivar (FM966) from Australia. The homogeneity of this group indicates that the genetic differentiation among Mexican races [26] is negligible as compared to their divergence from both 'Marie-Galante' and TWC populations.

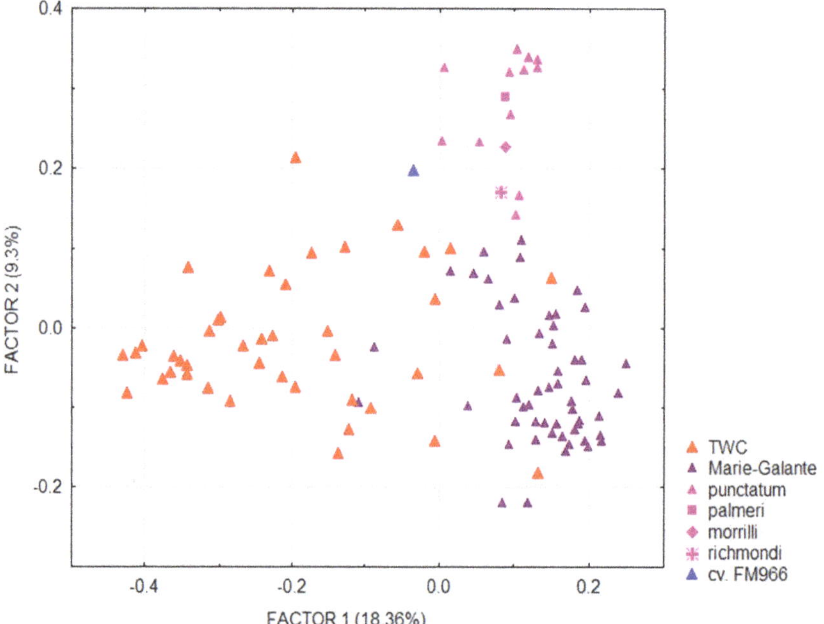

Figure 7. Principal coordinates analysis (PCA) on SSR data in truly wild (TWC) and feral *G. hirsutum*. PCA based on the similarity matrix for 26 SSR markers and 111 accessions represented according to their racial assignation. Factor 1 separates TWC from feral cottons and Factor 2 separates race 'Marie-Galante' from other feral cottons.

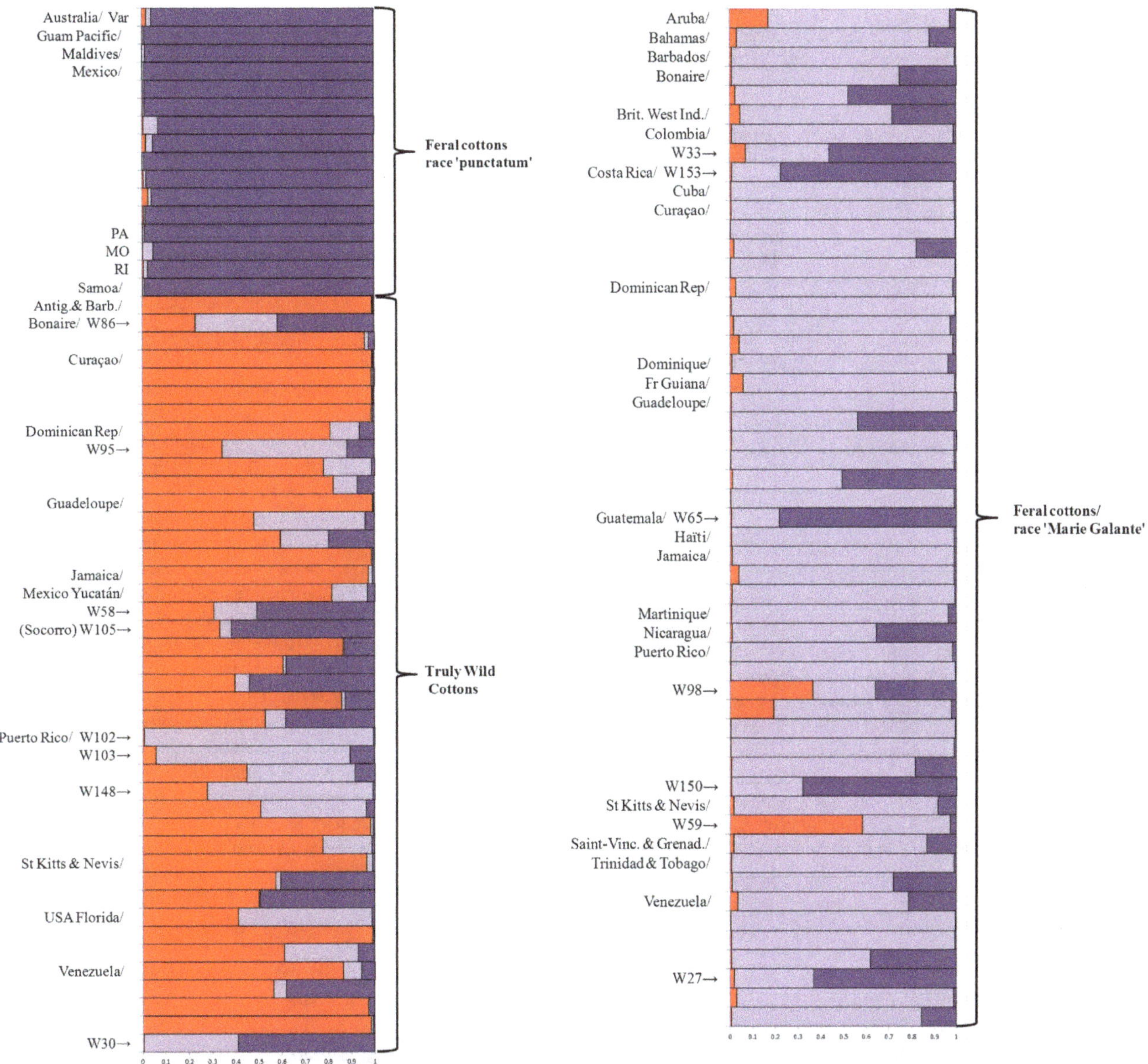

Figure 8. STRUCTURE plot of 111 perennial cottons of *Gossypium hirsutum* with K = 3 clusters. The y-axis shows the proportion membership to the cluster (three clusters depicted in light purple, deep purple and red). Each horizontal bar represents a single accession. The accessions are arranged according to their domestication status and, for feral accessions, their racial assignation, and then alphabetically per country of origin. Fourteen questionable cases (membership to cluster <33%) are indicated with their 'W' accession numbers as detailed in Table S1 (see also comments in the main text). Within cluster 'feral cotton/punctatum', MO, RI, PA, and Var refer to 'morrilli', 'richmondi', 'palmeri' and modern cultivar, respectively.

Distinctiveness of wild and feral populations

The genetic structure observed in a broad collection of cottons representing a vast region of Mesoamerica, Central America, the eastern Caribbean, and even Pacific islands (110 accessions, from 29 different countries or islands) demonstrates that the major driver organizing this collection is the status, feral or wild, of the cotton population, rather than any geographical factor. Thus, ENM and genetic analyses converge in discriminating TWC populations from feral populations, as assessed in our categorization exercise. We can conclude definitively that, not only do 'truly wild' populations of *G. hirsutum* still exist, but they are ecologically and genetically distinct, occupying a narrow and well defined habitat. Their genetic distinctiveness and homogeneity invalidate any racial or specific distinction among TWC populations of *G. hirsutum*, such as their classification into race 'yucatanense' in northern Yucatán and into race 'punctatum' in the Caribbean. *A fortiori*, our results do not support any particular status for the wild perennial cottons from the Dominican Republic, which were the most 'inland' collections among our TWC samples (see Figure S1). These wild cottons had been given racial status (*G hirsutum* race 'ekmanianum') or specific status, as *G. ekmanianum* Wittmack [56,77], and they had even been proposed as a new species by Wendel [78], and other authors, of genome AD_6 (other 5 tetraploid species being denoted as AD_1 to AD_5). Our results do not support such proposals as these specimens fall within the overall range of TWC accessions (Figure S3B). Instead, they unambiguously validate the opinion of Schwendiman and colleagues [19,27] who recognized their morphological unity, from Yucatán to Florida, the Antilles and Venezuela, grouping them under race 'yucatanense'.

The genetic and ecological divergence between race 'yucatanense' *sensu* Schwendiman and feral populations is clearly stronger than the splitting of the latter into two clusters corresponding to (i) races of pure *G. hirsutum* from Mesoamerica, and (ii) the Caribbean and Central American representatives of race 'Marie-Galante' resulting from an introgression with *G. barbadense*. This comparison indicates that domestication resulted in a major infraspecific division in *G. hirsutum*. In any case, the low level of admixture between neighboring TWC and feral populations shows the effects of surprisingly strong reproductive barriers and/or very strict ecological adaptation, resulting in very limited gene flow, despite their geographical proximity.

Distribution and domestication status in *Gossypium hirsutum*

The much stronger differentiation of TWC populations is reminiscent of the study of Stephens [9] who used a similar categorization approach to evaluate the effects of domestication on seed and fiber properties of perennial cottons, well before he formally admitted the existence of 'truly wild' populations of tetraploid cottons. As in our ecological and genetic analyses, his "wild" category was clearly the most distinct. Thus, there were highly significant differences in seed grade, seed index and lint index between the wild and feral categories, whereas differences among feral and cultivated categories were much less marked. The morphological, genetic, and ecological proximity between cultivated and feral cottons can be easily explained if they are closely related, i.e. if the latter are still part of the domesticated genepool. This is first suggested by the fact that feral plants show the same geographic patterns of morphological differentiation as cultivated materials [6,9,10]. Second, the correlation between the occurrence of feral populations and the cultivation of perennial cotton has been reported by most experts, including Ulloa et al. [10] who observed that feral populations are getting rarer as the cultivation of cotton declines in Mexico. This indicates that most feral populations depend on cultivation of ancient landraces for their perpetuation, following a sink-source dynamics model; in ecological terms, their realized niche is wider than their fundamental niche [79]. This double dependence on man, for their man-made habitat and for their reproduction from cultivated plants, contrasts with the long-term permanency of wild coastal populations of *G. hirsutum*. For example, the wild population of Portland Point in Jamaica was mentioned by Britton in 1908 [57], Schwendiman et al. in 1986 [19] and Stoddart and Fosberg in 1991 [58]. Such cases provide excellent illustrations of the fact that, in its original condition, *G. hirsutum* is a pioneer plant colonizing disturbed coastal habitats, but that "this (habitat) instability is in itself highly stable" and very ancient, so "that the pioneers are simultaneously old residents", as Fryxell [24] put it. The relationship between extreme aridity and the occurrence of wild cotton is obviously related to the fact that very few other plant species can compete under such conditions, suggesting that our TWC-specific climatic model is fairly representative of its realized niche. Indeed, as stated by Hutchinson [80], even the most mesophytic members of *Gossypium* are intolerant to competition, particularly at the seedling stage. Contrary to feral cotton, the realized niche of TWC populations is narrower than their fundamental niche.

Thus, the present study provides an opportunity to analyze the effect of domestication on the distribution of cultivated perennials, a rarely studied aspect of domestication. Miller and Knouft [81] have analyzed the case of the jocote or purple mombin (*Spondias purpurea* L.), a small fruit tree native from the dry forests of southern Mexico and Central America, and cultivated for its fruit and/or as a fence. They found that the climatic envelope of the wild populations is nested in that of the cultivated forms. In other words, domestication and cultivation mostly expanded the range of the species. Miller and Knouft [81] attributed this expansion to genetic adaptation, discarding the effect of tending cultivated trees, and, more surprisingly, neglecting the effect of the domestication syndrome itself. Indeed, the domesticated purple mombin produces mostly sterile fruits, so it is essentially reproduced from cuttings that grow much faster than seedlings [82], under much less intense competition.

As compared to purple mombin and the majority of perennial fruit crops, *G. hirsutum* differs in its relatively high level of autogamy and endogamy [83]. Domestication has considerably increased the diversity of the species [6,26] and apparently extended its ecoclimatic range (Figure 3C and D), well beyond the most peripheral and arid habitats of TWC populations. Thus the question remains fundamentally the same: have domestication and selection under cultivation widened the fundamental ecoclimatic envelope of perennial *G. hirsutum* through selection and genetic adaptation? As this envelope is common to feral and cultivated populations, and the feral populations depend on the permanent contribution of cultivated cotton, the most likely answer is that the much wider distribution of these two categories is essentially related to the reduction of competition in cultivated and neighboring disturbed habitats, not to a genetic effect. Furthermore, as the domestication syndrome involves seed characteristics (e.g. seed permeability, hardseededness, dormancy) that are essential for the survival of wild populations, most feral cottons are unable to re-colonize, and persist in, the original habitat of the species. Thus, the apparent paradox is that, although the geographical distribution of perennial *G. hirsutum* has been considerably widened by domestication and cultivation, its niche has been reduced by the loss of reproductive capacity in its natural habitat. In fact, there would have been a true paradox if cultivation, while reducing exposure to both extreme aridity and competition, had increased the competitive potential of *G. hirsutum* in secondary habitats.

Domestication of *Gossypium hirsutum*

Among the important reasons to study the natural distribution of perennial *G. hirsutum* in Mesoamerica and the Caribbean are the identification of potential areas for the early domestication processes and the comparative characterization of domesticated versus wild cottons. Our distribution maps do not contradict the hypothesis of Brubaker and Wendel [8] of an initial domestication of *G. hirsutum* in northern Yucatán, as this region, home of the most extensive wild populations, indeed corresponds to the largest favorable area (Figure 2A). This has been true not only for the last three millennia under modern climates [53] but very likely also for all the Holocene and even earlier, during the late Pleistocene (Figure 5). On the other hand, our maps also support the views of Sauer [11] on a more diffuse process in space and time, with early lint gathering and even trade preceding regular cultivation. Such a process is consistent with the descriptions of multiform exploitation of different cotton populations by Caribbean natives in early colonial chronicles [9], reminiscent of the domestication processes described by Casas et al. [33]. Clearly, *G. hirsutum* lends itself particularly well to such practices. It is naturally restricted to marginal habitats, where it does not suffer much from competition, but as a pioneer species it could have responded very fast and positively to disturbance by man. Its propensity to cross the boundaries between wild, disturbed and cultivated habitats is still obvious today. We can easily imagine how cotton may have invaded spontaneously the surroundings of fishing communities living close to a natural population. Some basic selection in this new habitat would have steadily brought some improvement, progressively providing the genetic basis for more intense management under managed cultivation, and thereby triggering the domestication process. Once the domestication syndrome was acquired, cultivated forms could not revert to the 'truly wild' condition, favoring spatial isolation between the two forms, and in turn further strengthening selection and domestication processes.

However, while the process described above may have taken place both in Mesoamerica and the Caribbean, our genetic data do not favor domestication in the latter area, as Caribbean feral populations appear more closely related to Mesoamerican cultivated and feral cottons than to local TWC populations. Thus, the most likely hypothesis remains that of Brubaker and Wendel [8], with a very early domestication of *G. hirsutum* in northern Yucatán, followed by its progressive diffusion and racial differentiation in all Mesoamerica, then Central America and northern South America. There, race 'Marie-Galante' would have developed through introgression with domesticated forms of *G. barbadense*, as hypothesized by Stephens [7], before reaching the Caribbean.

Conservation of the genetic diversity of *G. hirsutum* and potential interest of wild perennial cottons for breeding

Strategies for the conservation of cotton genetic resources must take into account the relationship between cultivated, feral and wild populations, and the risks of genetic erosion. In the case of the domesticated gene pool, Ulloa et al. [10] have underlined that in southern Mexico *G. hirsutum* perennial cottons survive only as curiosities in garden plots or dooryards, or as occasional feral plants; while attempts at commercial cotton production have been abandoned. In the case of wild cottons, their very ancient habitat is being increasingly threatened, as international tourism covets the same sea-and-sun ecoclimatic niche of dry tropical coasts [19].This point is important in considering the long term *in situ* conservation of perennial cotton *G. hirsutum* populations. Although not considered in this study, the cases of endangered wild *G. barbadense* populations of southern Ecuador/northern Peru, as well as of *G. mustelinum* from northeastern Brazil [84], are similar in ecology and climatic conditions. Only *G. darwinii* from Galapagos is not threatened [85]. The conservation and further plant exploration of wild cottons is important. As highlighted by the results of Liu et al. [28] and Lacape et al. [26], these cottons may have up to 70% unique alleles.

The ecological niche where these wild cotton populations are encountered in Mesoamerica clearly indicates that they represent a great reservoir for genes and alleles related to tolerance to abiotic stresses (water, high temperature or saline stresses). Even though these wild cottons are excellent sources for widening the genetic base for breeding because of their complete interfertility with modern cultivars of *G. hirsutum* [86], this type of material has so far been poorly characterized for its physiological and ecophysiological adaptive traits [87,88] and rarely exploited in breeding programs [89,90].

Lastly, a further understanding of the domestication process, through the comparison of the domesticated and wild pools of *G hirsutum*, for example at the transcriptome level [38], as well as for the identification of valuable phenotypic traits [91,92], can only benefit from an *ad hoc* categorization as attempted in the present study.

Conclusions

Ocean diffusion and ecological constraints, related to extreme aridity and low levels of competition, best explain the past and current distribution of truly wild populations of *G. hirsutum* restricted to littoral or related habitats, on the shores of the Caribbean Sea and the Gulf of Mexico from Venezuela to Florida, and even as far as Polynesian islands in the Pacific Ocean. The obvious niche conservatism expressed in the strong similarity of the natural habitats of all five allotetraploid species shows that their speciation was essentially driven by the geographic, rather than ecological, isolation of their highly scattered populations.

Our ecological and genetic data consistently support the hypothesis of Brubaker and Wendel [8], indicating that upland cotton domestication was very probably initiated in its largest native population, in northern Yucatán. Cultivated forms then diffused progressively to all the Mesoamerican cultural area, differentiating progressively into the five Mesoamerican races, following a process of geographical and cultural isolation [7]. The diffusion of both New World domesticated cottons, *G. hirsutum* and *G. barbadense*, would have allowed genetic introgression in southern Central America and/or northern South America, resulting in the development of race 'Marie-Galante'. The close genetic relatedness between 'Marie-Galante' and the Mesoamerican domesticated races shows that the introgression process was anterior to the diffusion of domesticated *G. hirsutum* to the Antilles.

Even where domesticated and TWC forms grow in close proximity, they hybridize only sporadically. As a result, the level of genetic divergence between them overwhelms differentiation among domesticated races and/or geographic regions.

Our understanding of plant evolution under domestication is more limited for perennial plants than for seed-propagated annual crops [93]. With their evolution from geographically limited wild populations and their concomitant diffusion and racial differentiation, allowing their establishment under warm temperate climates, the two cultivated tetraploid cottons present interesting parallels with the evolution and adaptation of maize in prehistoric and historic times. The persistence of truly wild populations of both species further increases their interest as unique models for understanding how the genomes of perennials respond to selection pressures operating on the relatively short time scale of the domestication process. The existence of three closely related wild species allows situating this process in the general context of the evolution of allotetraploid cottons from a unique hybridization event, 1–2 million years ago [94].

Wild forms of *G. hirsutum*, with seeds only sparsely covered with short fibers but with adaptation to extreme environmental conditions, contrast with cultivated cotton, with its highly valued long-fibered seeds but adaptation to less demanding ecologies. Owing to the advances of genomics and genome sequencing and the ability to scan the genomes of wild species for new and useful genes, we may now be in a position to unlock the genetic potential of the wild germplasm resources of crop plants [91,92], including cotton. The sequences of the two diploid species with genomes closest to the constitutive genomes of tetraploid cottons, the genome D of *G. raimondii* and the genome A of *G. arboreum*, have already been published [95,96]; and the sequencing of the AD genomes of *G. hirsutum* and *G. barbadense* is underway. It should be a relatively easy step now to systematically scan wild germplasm for useful genetic variants. However, this presupposes that *ex situ* collections are adequate, accessible and safe, and that *in situ* preservation efforts are effective in safeguarding material not yet in gene banks, and still evolving in the field. This work should facilitate the development not only of efficient strategies for exploiting cotton diversity for crop improvement, but also of strategies for its long-term conservation.

Supporting Information

Figure S1 Maps of the 11 locations/islands where truly wild cottons (red dots) were sampled for the SSR-based genetic analysis (Socorro Islands not shown). Purple dots represent sampling locations for feral populations in the vicinity. The comments attached to each map are extracted (and translated from French) from collecting reports as indicated.

Figure S2 Potential distribution of *G. hirsutum* during the Last Glacial Maximum (21,000 BP) in South America, extrapolated according to the MIROC climatic model.

Figure S3 Unrooted neighbor-joining tree based on dissimilarities between 110 perennial, and one cultivated, accessions of *Gossypium hirsutum* based for 26 SSR markers.

Figure S4 Prediction of the best value of K (ΔK method of Evanno et al., 2005), from K = 2 to 5 clusters, from the STRUCTURE analysis of 111 perennial cottons of *G. hirsutum*.

Table S1 Detailed records of the 111 accessions used in the SSR-based genetic analysis.

Table S2 List of 26 SSR markers, their chromosome localization on the consensus map (Blenda et al., 2012, *PlosONE* 0045739), and summary statistics (as calculated over 111 accessions): total number of alleles, *He* value as expected heterozygosity, numbers of alleles and unique alleles within groups (MG = race 'Marie-Galante', PU = race 'punctatum', TWC = truly wild cottons).

Table S3 Mean observed heterozygosity (in %) of 26 SSR markers as observed among 110 perennial accessions of *G. hirsutum* (cultivated variety FM966, as 111th accession was 100% homozygote).

Table S4 Mean dissimilarities within and between groups of 110 perennial accessions of *G. hirsutum*. MG, PU and TWC refer to 'Marie-Galante', 'punctatum' and truly wild cottons.

Acknowledgments

Authors are thankful to USDA and CIRAD seed banks curators, James Frelichowski and Dominique Dessauw, for providing seeds of accessions; to Manel Benhassen and Ronan Rivallan for their help in SSR genotyping; and to Luigi Guarino (Global Crop Diversity Trust) and Xavier Perrier (CIRAD) for their valuable comments and suggestions.

Author Contributions

Conceived and designed the experiments: GCE JML. Performed the experiments: GCE JML. Analyzed the data: GCE JML. Contributed to the writing of the manuscript: GCE JML.

References

1. Wendel JF, Brubaker CL, Seelenan T (2010) The origin and evolution of *Gossypium*. In: Stewart JM, Oosterhuis DM, Heitholt JJ, Mauney JR, editors. Physiology of cotton. Dordrecht: Spinger, Netherlands. pp. 1–18.
2. Brubaker CL, Bourland FM, Wendel JF (1999) The origin and domestication of cotton. In: Smith CW, Cothren JT, editors. Cotton, Origin, history, technology, and production: Wiley Series in crop science. pp. 3–31.
3. Stephens SG (1967) Evolution under domestication of the New world cottons (*Gossypium* spp.). Ciencia e cultura 19: 118–134.
4. Wendel JF, Brubaker CL, Percival AE (1992) Genetic diversity in *Gossypium hirsutum* and the origin of Upland cotton. Am J Bot 79: 1291–1310.
5. Wendel JF, Brubaker C, Alvarez I, Cronn RC, Stewart JM (2009) Evolution and natural history of the cotton genus. In: Paterson AH, editor. Plant genetics and genomics Crops and Models Volume 3 Genetics and genomics of cotton: Springer. pp. 3–22.
6. Hutchinson JB (1951) Intra-specific differentiation in *Gossypium hirsutum*. Heredity 5: 169–193.
7. Stephens SG (1973) Geographical distribution of cultivated cottons relative to probable centers of domestication in the New World. In: Adrian M, editor. Genes, Enzymes and Populations. pp. 239–254.
8. Brubaker CL, Wendel JF (1994) Reevaluating the origin of domesticated cotton (*Gossypium hirsutum*, Malvaceae) using nuclear restriction fragment length polymorphism (RFLP). Am J Bot 81: 1309–1326.
9. Stephens SG (1965) The effects of domestication on certain seed and fiber properties of perennial forms of cotton, *Gossypium hirsutum* L. The American Naturalist 94: 365–372.
10. Ulloa M, Stewart JM, Garcia EA, Godoy S, Gaytan A, et al. (2006) Cotton genetic resources in the Western states of Mexico: *in situ* conservation status and germplasm collection for *ex situ* preservation. Genet Resour Crop Evol 53: 653–668.
11. Sauer JD (1967) Geographic reconnaissance of seashore vegetation along the Mexican gulf coast; Mc Intire WG, editor. Baton Rouge. 59 p.
12. Sauer JD (1982) Cayman islands seashore vegetation. A study in comparative biogeography; Press UoC, editor. Berkeley. 161 p.
13. Hutchinson JB (1944) The cottons of Puerto-Rico. The Journal of Agriculture of the University of Puerto Rico 28: 35–42.
14. Stephens SG, Phillips LL (1966) Cotton collection in Colombia, Ecuador, N Peru and Surinam.
15. Stephens SG (1971) Some problems in interpreting transoceanic dispersal of the New world cotton. In: Riley CL, Kelley C, Pennington CW, Rands RL, editors. Man across the sea. University of Texas Press, Austin. pp. 401–415.
16. Hutchinson JB, Silow RA, Stephens SG (1947) The evolution of *Gossypium* and the differenciation of the cultivated cottons: Emprire Cotton Growing Corporation. 160 p.
17. Ano G, Fersing J, Lacape JM (1983) The cottons of Marie-Galante island. Cot Fib Trop 38: 206–208.
18. Ano G, Schwendiman J (1983) Multiphase collecting missions for cotton (I). Plant Genetic Resources Newsletter 54: 5p.
19. Schwendiman J, Percival AE, Belot J-L (1986) Cotton collecting on Carribean islands and South Florida. IBPGR: 4 p.
20. Fryxell PA, Moran R (1963) Neglected form of *Gossypium hirsutum* on Socorro island, Mexico. Empire Cotton Growing Review 40: 289–291.
21. Fosberg FR (1959) Vegetation and flora of Wake Island. Atoll Research Bulletin No67: 20p.
22. Stephens SG (1963) Polynesian cottons. Annals of the Missouri Botanical garden 50: 1–22.
23. Fryxell PA (1965) Stages in the evolution of *Gossypium*. Adv Fontiers Plant Sci 10: 31–56.
24. Fryxell PA (1979) The Natural History of the Cotton Tribe (Malvaceae, Tribe Gossypieae): Texas A&M University Press, College Station, TX. 245 p.
25. Mc Stewart J, Bertoni S (2008). The wild *Gossypium hirsutum* of Paraguay. Belwide Cotton Research Conferences. Nashville (Tennessee). pp. 833.
26. Lacape J-M, Dessauw D, Rajab M, Noyer JL, Hau B (2007) Microsatellite diversity in tetraploid *Gossypium* gene pool: assembling a highly informative genotyping set of cotton SSRs. Mol Breed 19: 45–58.
27. Ano G, Schwendiman J, Fersing J, Lacape J-M (1982) Les cotonniers primitifs de *G. hirsutum* race *yucatanense* de la Pointe des Châteaux en Guadeloupe et l'origine possible des cotonniers tétraploides du Nouveau Monde. Cot Fib Trop: 327–332.
28. Liu S, Cantrell RG, Mc Carty JCJ, Stewart JM (2000) Simple sequence repeat-based assessment of genetic diversity in cotton race stock accessions. Crop Sc 40: 1459–1469.
29. Rensch CR (1976) Comparative Otomanguean Phonology. Indiana University Publications (Language science monographs. Vol. 14). Bloomington.

30. Kaufman S (1990) Early Otomanguean homeland and cultures: Some premature hypotheses. University of Pittsburgh Working Papers in Linguistics 1. pp. 91–136.
31. Smith CE, Stephens SG (1971) Critical identification of Mexican archaeological cotton remains. Economic botany 25: 160–168.
32. Lubbers EL, Chee PW (2009) The worldwide gene pool of *G. hirsutum* and its improvement. In: Paterson AH, editor. Genetics and genomics of cotton: Springer. pp. 23–52.
33. Casas A, Otero-Arnaiz A, Perez-Negron E, Valiente-Banuet A (2007) In situ management and domestication of plants in Mesoamerica. Ann Bot 100: 1101–1115.
34. Ortiz F, Stoner KE, Perrez-Negron E, Casas A (2010) Pollination biology of *Myrtillocactus schenckii* (Cactaceae) in wild and managed populations of the Tehuacan Valley, Mexico. J Arid Env 74: 897–904.
35. Parra F, Casas A, Penaloza-Ramirez JM, Cortez-Palomec AC, Rocha-Ramirez V, et al. (2010) Evolution under domestication: ongoing artificial selection and divergence of wild and managed *Stenocereus pruinosus* (Cactaceae) populations in the Tehuacan Valley. Ann Bot 106: 483–496.
36. Guillen S, Terrazas T, la Barrera E, Casas A (2011) Germination differentiation patterns of wild and domesticated columnar cacti in a gradient of artificial selection intensity. Genet Resour Crop Evol 58: 409–423.
37. Butterworth KM, Adams DC, Horner HT, Wendel JF (2009) Initiation and early development of fiber in wild and cultivated cotton. Int J Plant Sci 170: 564–574.
38. Rapp RA, Haigler CH, Flagel L, Hovav RH, Udall JA, et al. (2010) Gene expression in developing fibers of Upland cotton (*Gossypium hirsutum* L.) was massively altered by domestication. BMC Biol 8: 139.
39. Wegier A, Pinero-Nelson A, Alarcon J, Galvez-Mariscal A, Alvarez-Buylla ER, et al. (2011) Recent long-distance transgene flow into wild populations conforms to historical patterns of gene flow in cotton (*Gossypium hirsutum*) at its centre of origin. Mol Ecol 20: 4182–4194.
40. Waltari E, Hijmans RJ, Peterson AT, Nyari AS, Perkins SL, et al. (2007) Locating pleistocene refugia: comparing phylogeographic and ecological niche model predictions. PLoS ONE 2: e563.
41. Martínez-Meyer E, Peterson AT (2006) Conservatism of ecological niche characteristics in North American plant species over the Pleistocene-to-Recent transition. Journal of Biogeography 33: 1779–1789.
42. Ano G, Schwendiman J (1981) Rapport de mission en Guyane française - Venezuela - Colombie sur la préservation des ressources génétiques du cotonnier. FAO-IBPGR. 44 p.
43. Ano G, Schwendiman J (1982) Rapport de mission au Mexique sur la préservation des ressources génétiques du cotonnier. FAO-IBPGR. 33 p.
44. Hijmans RJ, Cameron SE, Parra JL, Jones PG, Jarvis A (2005) Very high resolution interpolated climate surfaces for global land areas. Int J Climat 25: 1965–1978.
45. Phillips SJ, Anderson RP, Schapire RE (2006) Maximum entropy modeling of species geographic distributions. Ecol Modelling 190: 231–259.
46. Chan WL, Abe-Ouchi A, Oghaito R (2011) Simulating the mid-Pliocene climate with the MIROC general circulation model: experimental design and initial results. Geosci Model Dev 4.
47. Risterucci A-M, Grivet L, N'Goran JAK, Pieretti I, Flament MH, et al. (2000) A high density linkage map of *Theobroma cacao* L. Theor Appl Genet 101: 948–955.
48. Perrier X, Jaquemoud Collet JP (2006) DARwin software. Available: http://darwin.cirad.fr/darwin.
49. Saitou N, Nei M (1987) The neighbor-joining method: a new method for reconstructing phylogenetic trees. Mol Biol Evol 4: 406–425.
50. Pritchard JK, Stephens M, Donnelly P (2000) Inference of population structure using multilocus genotype data. Genetics 155: 945–959.
51. Evanno G, Regnau S, Goudet J (2005) Detecting the number of clusters of individuals using the software STRUCTURE: a simulation study. Mol Ecol 14: 2611–2620.
52. Earl DA, von Holdt BM (2012) STRUCTURE HARVESTER: a website and program for visualizing STRUCTURE output and implementing the Evanno method. Conservation Genet Resour 4: 359–361.
53. Metcalfe SE (2006) Late Quaternary Environments of the Northern Deserts and Central Transvolcanic Belt of Mexico Annals Missouri Botanical Garden 93: 258–273.
54. Percival AE, Kohel RJ (1990) Distribution, collection, and evaluation of *Gossypium*. Adv Agron 44: 225–256.
55. Hutchinson JB (1943) Notes on the native cottons of Trinidad. Tropical Agriculture 20: 235–238.
56. Krapovickas A, Seijo G (2008) *Gossypium ekmanianum* (Malvaceae), algodon silvestre de la Republica Dominicana. Bonplandia 17: 55–63.
57. Coville FV, Britton NL, Cook OF (1908) Wild Jamaica cotton. Science 24: 664–666.
58. Stoddart DR, Fosberg FR (1991) Plants of the Jamaican cays. Atoll Research Bulletin 352: 24 p.
59. Proctor GR (1980) Checklist of the plants of little Cayman. Atoll Research Bulletin 241: 71–80.
60. Stoddart DR (1980) Vegetation of little Cayman. Atoll Research Bulletin 241: 53–70.
61. Britton NL, Millspaugh CF (1920) The Bahama flora; Britton NL, editor. New York.
62. Nickrent DL, Eshbaugh WH, Wilson TK (2008) The vascular flora of Andros island, Bahamas.
63. Britton NL (1918) Flora of Bermuda. New York.
64. Stephens SG (1966) The potentiality for long range oceanic dispersal of cotton seeds. The American Naturalist 100: 199–210.
65. Pickersgill B, Barrett SCH (1975) Wild cotton in Northeast Brazil. Biotropica: 42–54.
66. Barroso PAV, Hoffmann LV, Freitas RBd, Araujo Batista CEd, Alves MF, et al. (2010) *In situ* conservation and genetic diversity of three populations of *Gossypium mustelinum* Miers ex Watt. Genet Resour Crop Evol 57: 343–349.
67. de Menezes IPP, Gaiotto FA, Hoffmann LV, Ciampi AY, Barroso PAV (2014) Genetic diversity and structure of natural populations of *Gossypium mustelinum*, a wild relative of cotton, in the basin of the De Contas River in Bahia, Brazil. Genetica 142: 99–108.
68. Wendel JF, Percy RG (1990) Allozyme diversity and introgression in the Galapagos Islands endemic *Gossypium darwinii* and its relationship to continental *G barbadense*. Bioch Syst Ecol: 517–528.
69. Fosberg FR, Sachet MH (1969) Wake island vegetation and flora. Atoll Research Bulletin 123.
70. Degener O, Degener I (1959) Canton island, South Pacific. Atoll Research Bulletin 64: 24 p.
71. Fosberg FR, Stoddart DR (1994) Flora of the Phoenix islands, Central Pacific. Atoll Research Bulletin 393: 61 p.
72. Bakineti T, Seluka S (1999) Country report for the Pacific sub-regional workshop on forest and tree genetic resources. 12–16 April 1999. In: SPRIG A, FAO, SPC, SPREP, editor. State of forest and tree genetic resources in the Pacific islands, and sub-regional action plan for their conservation and sustainable use. Apia, Samoa.
73. Fosberg FR, Sachet MH (1987) Flora of the Gilbert islands, Kiribati. Atoll Research Bulletin 295: 34 p.
74. Whistler A (2011) The rare plants of Samoa; Atherthon J, Duffy L, editors. Apia: Biodiversity Conservation. Lessons learned Technical Series. 212 p.
75. Peterson AT, Soberon J, Sanchez-Cordero V (1999) Conservatism of ecological niches in evolutionary time. Science 285: 1265–1267.
76. Fleming K, Johnston P, Zwartz D, Yokohama Y, Lambeck K, et al. (1998) Refining the eustatic sea-level curve since the Last Glacial Maximum using far- and intermediate-field sites. Earth and Planetary Science Letters 163: 327–342.
77. Grover C, Grupp KK, Wanzek RJ, Wendel JF (2012) Assessing the monophyly of polyploid *Gossypium* species. Plant Syst Evol 298: 1177–1183.
78. Wendel JF (2012) Genes, Jeans and Genomes: Exploring the Mysteries of Polyploidy in Cotton, in International Cotton Genome Initiative (ICGI) Research Conference, Raleigh (NC). Available: http://www.cottoninc.com/fiber/Agricultural-Research/Agricultural-Meetings-Conferences/ICGI-Presentations/p01-Wendel-Jonathan/01-Wendel-Jonathan.pdf. Accessed 2014 Jun 25.
79. Pulliam HR (2000) On the relationship between niche and distribution. Ecology Letters 3: 349–361.
80. Hutchinson JB (1954) New evidence on the origin of the old world cottons. Heredity 8: 225–241.
81. Miller AJ, Knouft JH (2006) GIS-based characterization of the geographic distributions of wild and cultivated populations of the Mesoamerican fruit tree *Spondias purpurea* (Anacardiaceae). Am J Bot 93: 1757–1767.
82. Morton J (1987) Purple Mombin. In: Morton JF, editor. Fruits of warm climates. Miami, FL. pp. 242–245.
83. Alves MF, Pereira FRA, de Andrade AM, de Menezes IPP, Hoffmann LV, et al. (2009) Molecular polymorphic markers between moco and herbaceous cotton. Rev Cienc Agron 40: 406–411.
84. Alves MF, Barroso PAV, Ciampi AY, Hoffmann LV, Azevedo VCR, et al. (2013) Diversity and genetic structure among subpopulations of *Gossypium mustelinum* (Mavaceae). Genetics and Molecular Research 12: 597–609.
85. Darwin-Foundation. *Gossypium darwinii*. Galapagos Species Checklist. Available: http://wwwdarwinfoundationorg/datazone/checklists/573/. Accessed 2014 Jun 25.
86. Stewart JM (1994). Potential for crop improvement with exotic germplasm and genetic engineering. In: Constable GA, Forrester NW, editors. Challenging the future, Proc World Cotton Res Conf -1. Brisbane (Aus). pp. 313–327.
87. Quisenberry JE, Jordan WR, Roark B, Fryrear DW (1981) Exotic cottons as genetic sources for drought resistance. Crop Sc 21: 889–896.
88. Wu T, Weaver DB, Locy RD, McElroy S, van Santen E (2014) Identification of vegetative heat-tolerant upland cotton (*Gossypium hirsutum* L.) germplasm utilizing chlorophyll fluorescence measurement during heat stress. Plant Breeding 133: 250–255.
89. Mc Carty JCJ, Jenkins JN, Wu J (2004) Primitive accession derived germplasm by cultivar crosses as sources for cotton improvement: II. Genetic effects and genotypic values. Crop Sc 44: 1231–1235.
90. Mc Carty JC, Wu J, Jenkins JN (2006) Genetic diversity for agronomic and fiber traits in day-neutral accessions derived from primitive cotton germplasm. Euphytica 148: 283–293.
91. Tanksley SD, Mc Couch S (1997) Seed banks and molecular maps: unlocking genetic potential from the wild. Science 277: 1063–1066.
92. McCouch S, Baute GJ, Bradeen J, Bramel P, Bretting PK, et al. (2013) Agriculture: Feeding the future. Nature 499: 23–24.
93. Miller AJ, Gross BL (2011) From forest to field: Perennial fruit crop domestication. Am J Bot 98: 1389–1414.

94. Wendel JF, Cronn RC (2003) Polyploidy and the evolutionary history of cotton. Adv Agron 78: 140–186.
95. Paterson AH, Wendel JF, Gundlach H, Guo H, Jenkins J, et al. (2012) Repeated polyploidization of *Gossypium* genomes and the evolution of spinnable cotton fibres. Nature 492: 423–427.
96. Li F, Fan G, Wang K, Sun F, Yuan Y, et al. (2014) Genome sequence of the cultivated cotton *Gossypium arboreum*. Nat Genet 46: 567–572.

17

Comparative Transcriptome Analysis of Short Fiber Mutants Ligon-Lintless 1 And 2 Reveals Common Mechanisms Pertinent to Fiber Elongation in Cotton (*Gossypium hirsutum* L.)

Matthew K. Gilbert[1], Hee Jin Kim[1], Yuhong Tang[2], Marina Naoumkina[1], David D. Fang[1]*

1 Cotton Fiber Bioscience Research Unit, USDA-ARS, Southern Regional Research Center, New Orleans, Louisiana, United States of America, **2** The Samuel Roberts Noble Foundation, Genomics Core Facility, Ardmore, Oklahoma, United States of America

Abstract

Understanding the molecular processes affecting cotton (*Gossypium hirsutum*) fiber development is important for developing tools aimed at improving fiber quality. Short fiber cotton mutants Ligon-lintless 1 (Li_1) and Ligon-lintless 2 (Li_2) are naturally occurring, monogenic mutations residing on different chromosomes. Both mutations cause early cessation in fiber elongation. These two mutants serve as excellent model systems to elucidate molecular mechanisms relevant to fiber length development. Previous studies of these mutants using transcriptome analysis by our laboratory and others had been limited by the fact that very large numbers of genes showed altered expression patterns in the mutants, making a targeted analysis difficult or impossible. In this research, a comparative microarray analysis was conducted using these two short fiber mutants and their near isogenic wild type (WT) grown under both field and greenhouse environments in order to identify key genes or metabolic pathways common to fiber elongation. Analyses of three transcriptome profiles obtained from different growth conditions and mutant types showed that most differentially expressed genes (DEGs) were affected by growth conditions. Under field conditions, short fiber mutants commanded higher expression of genes related to energy production, manifested by the increasing of mitochondrial electron transport activity or responding to reactive oxygen species when compared to the WT. Eighty-eight DEGs were identified to have altered expression patterns common to both short fiber mutants regardless of growth conditions. Enrichment, pathway and expression analyses suggested that these 88 genes were likely involved in fiber elongation without being affected by growth conditions.

Editor: Jinfa Zhang, New Mexico State University, United States of America

Funding: The research was funded by the USDA-ARS CRIS project 6435-21000-017D. The funders had no role in study design, data collection and analysis, decision to publish, or preparation of the manuscript.

Competing Interests: The authors declare no conflict of interests.

* E-mail: david.fang@ars.usda.gov

Introduction

Cotton fibers are single-celled trichomes that initiate from the ovule epidermal cells on or about the day of anthesis (DOA) [1]. Approximately 25% of the ovule epidermal cells differentiate into fiber cells during the initiation stage of cotton fiber development and subsequently undergo a period of rapid elongation known as the elongation stage [2,3]. The rate of fiber elongation peaks at approximately 6 to 12 days post-anthesis (DPA) and nears cessation around 22 DPA [4]. During peak elongation fiber cells can increase in length at rates of 2 mm/day or more depending on environmental factors and genotypes [5–7]. Beginning at 12–16 DPA and overlapping with the elongation phase is the secondary cell wall (SCW) biosynthesis stage. During this stage cellulose is synthesized and deposited between the primary cell wall and the plasmalemma [8,9]. Elongation and SCW biosynthesis continue until the fibers reach full length [25–35 mm in Upland cotton (*Gossypium hirsutum* L.) cultivars] [10], after which the cotton bolls open and the fibers desiccate under exposure to the environment. The environmental and genetic factors that influence the timing of these processes are shown to influence the development of desirable fiber traits such as lint yield and fiber quality [5,11–13].

Understanding the molecular mechanisms of fiber development is essential for cotton researchers to devise strategies for developing cotton lines with superior fiber quality. Furthermore, it is important to identify key genes that could be genetically engineered to improve fiber properties. Toward these goals, scientists have been using fiber mutants to study the molecular and genetic mechanisms of fiber development [14–17]. Among them, two short fiber mutants Ligon-lintless 1 and Ligon-lintless 2 (Li_1 and Li_2, respectively) were extensively studied by our group [18–20] and others [14,21,22] in order to develop a comprehensive understanding of the molecular and metabolomic mechanisms related to cotton fiber length development. In a near-isogenic state with the cotton cultivars Texas Marker-1 (TM-1) or DP5690, both the Li_1 and Li_2 mutants have seed fibers that are extremely short (<6 mm) as compared to wild type (WT) fibers that are typically longer than 25 mm in length [18,19,23–25]. As a monogenic dominant trait, the short fiber phenotypes of Li_1 and Li_2 are similar (Fig. 1A). However, unlike the Li_2 mutant, which appears

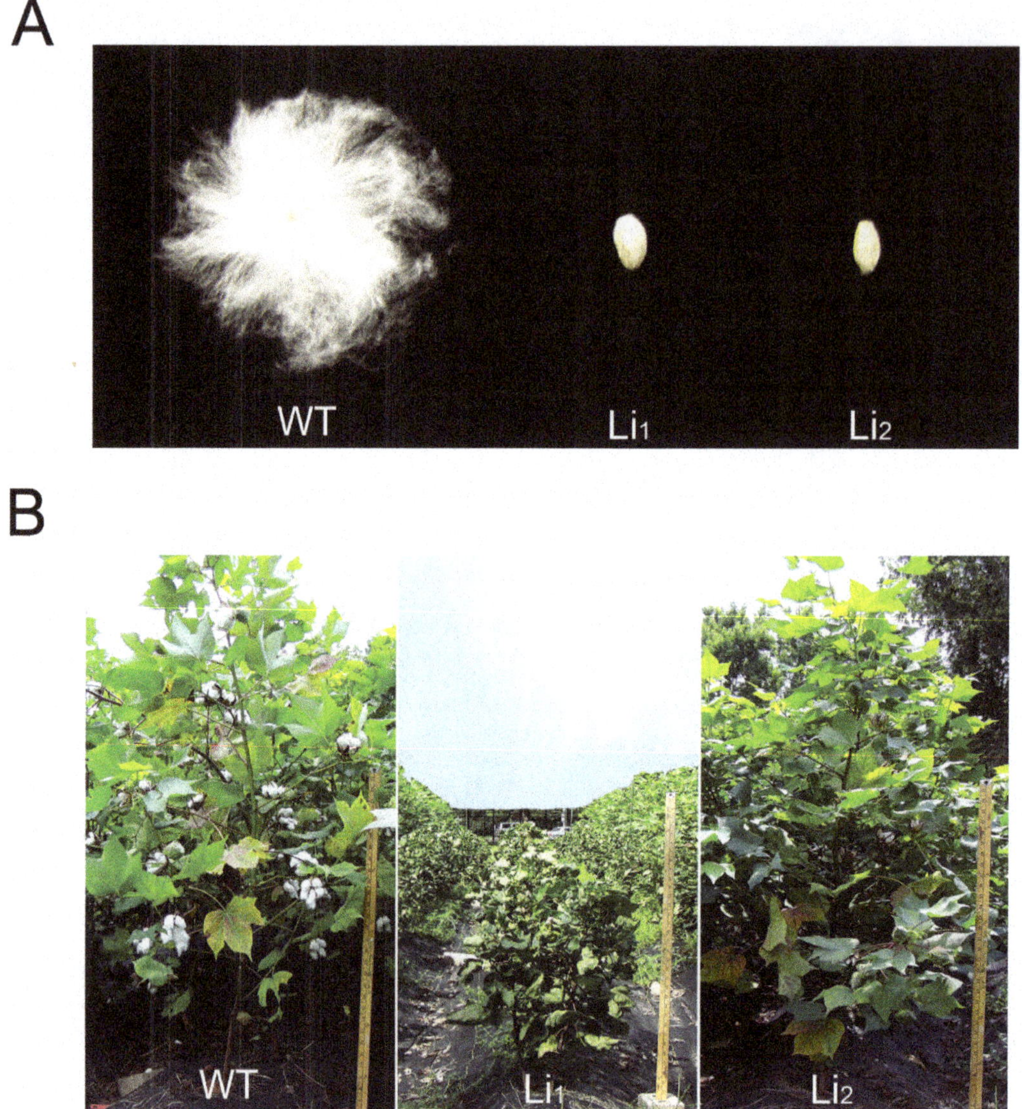

Figure 1. Cotton seed fibers (A) and plants (B) of wildtype DP5690 (WT), Li_1 mutant and Li_2 mutant. Plants were grown in the USDA-ARS Southern Regional Research Center field in New Orleans, LA.

morphologically similar to the WT plants with the exception of short seed fibers, the Li_1 mutant exhibits pleiotropy in the form of severely stunted and deformed plants in both the homozygous dominant and heterozygous state (Fig. 1B) [18,23]. Although the Li_1 and Li_2 mutants have similar phenotype in cotton fiber, these two genes reside on different chromosomes with Li_1 on chromosome (Chr.) 22 and Li_2 on Chr.18 [18,19,26,27]. These two mutants, when taken in combination, provide an excellent experimental system to find both common and mutant locus-specific mechanisms related to fiber elongation.

Analyzing the microarray or RNA-seq data collected to date for the Li_1 or Li_2 mutant is limited in one aspect by the fact that very large number of genes showed altered expression patterns between a mutant and its WT near isogenic line (NIL). For example, previous microarray data obtained from fibers of Li_1 or Li_2 showed approximately 1,500 to 2,500 differentially expressed genes (DEGs) including many genes that may be affected or regulated by environments [18,19,21]. It is difficult to decipher which of the genes are truly vital and common to fiber elongation-related processes, and which are due to different environmental, genetic and physiological cues if using only one mutant in an experiment as reported in all the previous studies. In recognition of this issue, we conducted the present experiment with two mutant lines and two growth conditions in order to identify the genes that are differentially regulated in both mutants, with the goal of identifying the common molecular mechanisms involved in cotton fiber length development. First we took advantages of our unique NILs of the Li_1 and Li_2 mutants. As reported earlier [18,19], both Li_1 and Li_2 NILs were developed using the Upland cotton cultivar DP5690 as the recurrent parent. The Li_1, Li_2 and WT DP5690 are mutually near isogenic. Second, we conducted experiments in both field and greenhouse for Li_2 mutant, allowing the identification of genes impacted by environmental conditions and response to stress in this mutant. Third, we did a comparative analysis of transcriptome profiles between Li_1, Li_2 and WT to identify genes that had altered expression patterns in a short fiber mutant,

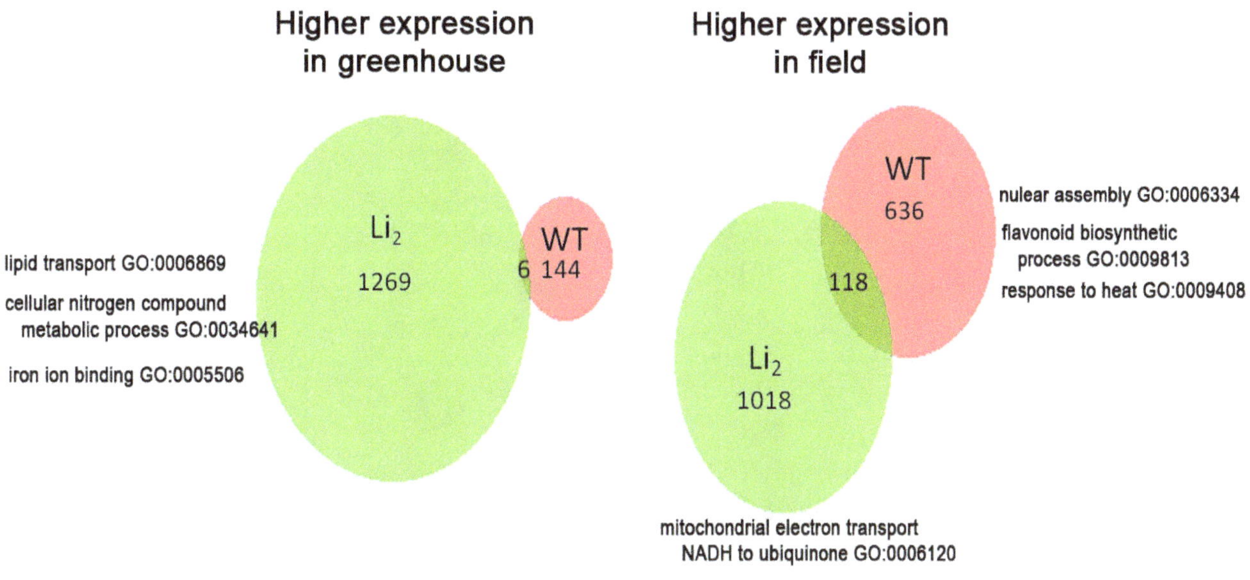

Figure 2. Probes showing altered regulation between greenhouse and field grown cotton. Each number represents the number of probes that showed altered expression between field and greenhouse conditions. Samples shown are 12 days post anthesis fibers. The included Gene Ontology labels are selected representative categories identified by Gene Ontology Enrichement Analysis conducted on AgriGO.

regardless of the growth conditions (field or greenhouse) or the nature of the mutation (*Li₁* or *Li₂*). Our major objective was to identify common genes (or molecular mechanisms) that were essential to the fiber elongation regardless of environment or a specific mutation. Herein we report our findings.

Materials and Methods

Plant Materials

The *Li₁*, *Li₂* and WT DP5690 used in the present study were mutual NILs. Development of these NILs was described in our earlier reports [18,19]. For the greenhouse-grown *Li₂* plants utilized in this study, growth and sample conditions were described in Hinchliffe et al. (2011) [19], and the growing period was between October, 2009 and March, 2010. Each plant was grown in an 18.9 L pot. A commercial service provider periodically sprayed pesticides to control insects or diseases. Automatic drip irrigation was used throughout the growing season. For field grown plants, a total of 200 *Li₁*, 100 *Li₂*, and 100 WT DP5690 plants were grown in a field at the USDA-ARS Southern Regional Research Center, New Orleans, LA in the summer of 2012. The distance between two plants within a plot was 30 cm. The plot

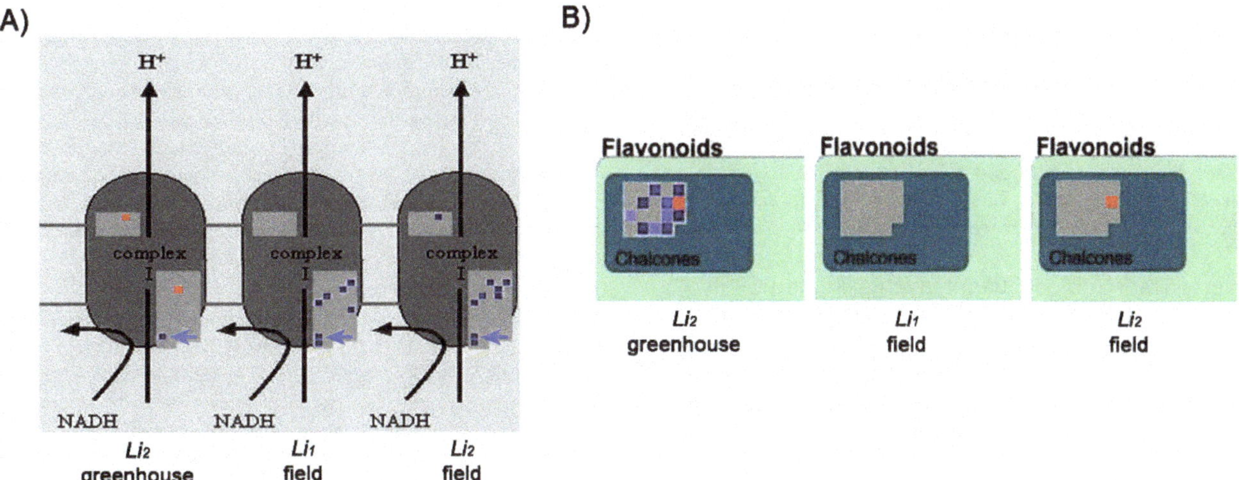

Figure 3. MapMan software illustrates different stress response of the short fiber mutants. Probes showing differential regulation were analyzed by MapMan software. The identification of processes affected differently in the wildtype and mutant lines was done manually. A) The Li_1 and Li_2 mutants showed increased (blue) expression of NADH dehydrogenase in field conditions as compared with greenhouse. Blue arrow indicating the common probe GhiAffix45916.1. B) Li_2 in greenhouse conditions exhibited increase chalcone synthase-related expression, which were not replicated in field conditions.

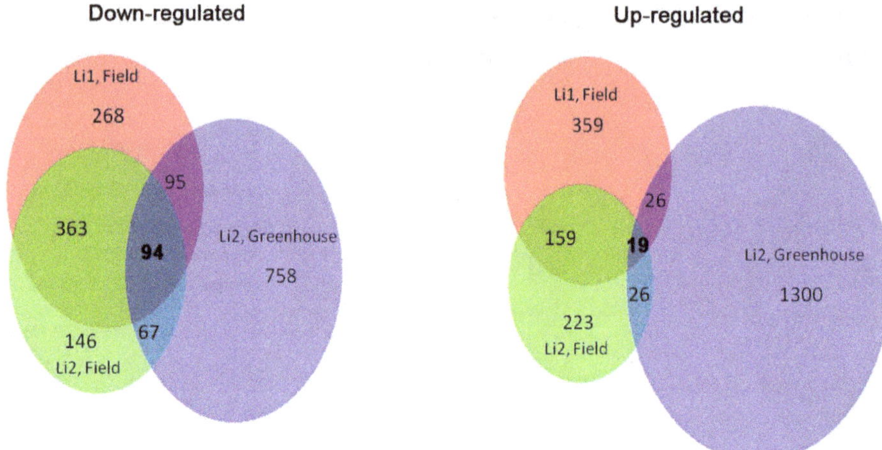

Figure 4. Venn Diagrams illustrating the similarities and differences between the experimental groups. Each number represents the number of probes showing different expression levels (>2 fold difference) between the mutants and the wildtype near-isogenic line.

distance was 45 cm. The soil type in this field was aquent dredged over alluvium in an elevated location to provide adequate drainage. Throughout the growing season, no pest control spray was applied. Supplemental sprinkle irrigation was provided when needed. Flowers were tagged and cotton boll sample collections were made before 10:00 a.m. and immediately placed on ice. To minimize environmental effects, boll samples were not collected from plants on the perimeter of the field and samples were only collected when 15–30 bolls were available for analysis. All samples of the same developmental stage were tagged and collected on the same day. Bolls were randomly separated into 3 replicates with 5–10 bolls per replicate. Bolls were then dissected, frozen in liquid nitrogen and stored at −80°C until further processing.

RNA Isolation from Cotton Fibers

RNA was isolated as previously described [19]. To separate the fibers from the ovules the samples were shaken vigorously enough to break fibers without damaging the ovules. Isolation of RNA was conducted using the Sigma Spectrum Plant Total RNA Kit (Sigma-Aldrich, St. Louis, MO) with on-column DNaseI digestion and used according to the manufacturer's instructions. RNA quantity was determined by using a Nanodrop 2000 spectrophotometer (NanoDrop Technologies Inc., Wilmington, DE). RNA integrity number (RIN) was determined for each sample using an Agilent Bioanalyzer 2100 and the RNA 6000 Nano Kit Chip (Agilent Technologies Inc., Santa Clara, CA). Only samples with RIN values of 7.0 or higher were used for expression analysis.

Reverse Transcription Quantitative Real-time PCR (RT-qPCR)

The experimental procedures and data analysis related to RT-qPCR were performed according to the Minimum Information for Publication of Quantitative Real-Time PCR Experiments (MIQE) guidelines [28]. We used RNA samples from 12 DPA fibers in two biological replicates for cDNA synthesis and in two technical replicates for qPCR. The cDNA synthesis reactions were performed using the iScript cDNA Synthesis Kit (Bio-Rad Laboratories, Hercules, CA) according to the manufacturer's instructions with 1 µg of total RNA per reaction used as template. The RT-qPCR reactions were performed with iTaq SYBR Green Supermix (Bio-Rad Laboratories) in a Bio-Rad CFX96 real time PCR detection system. The detail description of amplification parameters and calculation reported before [19]. Normalization of RT-qPCR data was performed by geometric averaging three internal control genes, including 18 S rRNA, ubiquitin-conjugating protein, and α-tubulin 4 [29]. The primer sequences of the 14 probe sets and the three internal control genes are listed in Table S1.

Microarray Hybridizations and Data Analysis

The microarray technology used for this study was the commercially available Affymetrix GeneChip Cotton Genome Microarray (Affymetrix Inc., Santa Clara CA), comprised of 239,777 probes representing 21,854 cotton transcripts from a variety of expressed sequence tag (EST) databases. The source material for the EST data was derived from *G. arboreum*, *G. barbadense*, *G. hirsutum*, and *G. raimondii*. Labeling of the RNA was conducted using the Affymetrix GeneChip 3′ IVT Express Kit and cotton genome hybridizations were conducted according to the manufacturer's protocols. Our earlier studies [18–20] indicated that a significant difference in both transcript profiles and fiber length measurement was observed at 12 DPA (peak elongation) between the Li_1, or Li_2 and its WT NIL. Thus in this experiment, we used RNA samples from 12 DPA fibers (in two biological replicates) for microarray analysis. Probes sets demonstrating a two-fold or greater difference in expression levels between experimental samples were considered differentially regulated. Data normalization and the determination of statistically relevant deviations in expression patterns was performed as described [30].

Gene Annotation Analysis

Microarray data obtained from greenhouse-grown Li_2 plants, and field-grown Li_1 and Li_2 plants were first subjected to Venn analysis utilizing BioVenn [31] to determine which probes demonstrate consistent expression profiles between experimental sets. To assist in the identification of biological processes represented in the data, Gene-Ontology Enrichment Analysis (GOEA) was performed using the agriGO Singular Enrichment Analysis (SEA) [32] by comparing to the *Gossypium raimondii* reference genome sequence [33]. The statistical test method used was the Fisher's Exact test (significance level 0.05). Annotation of the probes was accomplished with Blast2Go [34]. For pathway analysis, MapMan software [35] was used to identify and illustrate pathways of interest using the January 12, 2013 *Gossypium hirsutum*

Table 1. Annotation of common DEGs identified from both Li_1 and Li_2 mutants regardless of growth conditions.

No.	Probes	Homologous Gene	G raimondii locus
1	Ghi.7820.1.S1_s_at	18S ribosomal RNA (mitochondria)	Gorai.013G213000
2	GraAffx.32667.2.A1_s_at	18S rRNA (mitochondrial)	Gorai.013G213000
3	Ghi.581.1.S1_at	2-oxoglutarate (2OG) and Fe(II)-dependent oxygenase	Gorai.002G036400
4	GhiAffx.6286.1.S1_at	4-coumarate:CoA ligase (4CL)	Gorai.009G005900
5	Gra.1375.1.A1_at	actin 5 (ACT5)	Gorai.013G022400
6	GhiAffx.33535.1.S1_at	actin depolymerizing factor 5 (ADF5)	Gorai.008G035300
7	Ghi.1209.1.S1_at	ARM repeat superfamily protein	Gorai.010G025700
8	Ghi.8448.1.S1_x_at	beta-tubulin 1 (BTub1)	Gorai.004G211800
9	GraAffx.8388.1.S1_s_at	cellulose synthase-like protein	Gorai.009G066500
10	Ghi.3452.1.S1_s_at	cellulose synthase-like protein	Gorai.009G222300
11	GhiAffx.23257.1.S1_s_at	cellulose synthase-like protein	Gorai.009G066500
12	Ghi.2235.1.A1_s_at	chaperonin-60kD, ch60	Gorai.007G151700
13	Ghi.1908.1.S1_s_at	cofactor assembly	Gorai.007G003100
14	Ghi.8534.1.A1_s_at	cyclin-U2-1	Gorai.004G164400
15	Ghi.2840.3.S1_s_at	cysteine proteinase	Gorai.013G224100
16	GhiAffx.39816.1.S1_s_at	cytidine deaminase 1	Gorai.013G228000
17	GhiAffx.39795.1.S1_at	EF hand calcium-binding protein family	Gorai.008G075000
18	Ghi.68.1.A1_s_at	fasciclin-like arabinogalactan protein 1	Gorai.008G155400
19	Ghi.5801.1.A1_at	GDSL esterase/lipase	Gorai.011G103900
20	Ghi.5444.1.S1_at	gibberellin 20-oxidase	Gorai.004G149700
21	GhiAffx.11707.1.A1_at	gland development related protein 23-like	Gorai.004G208300
22	GhiAffx.8010.1.S1_at	glycolipid transfer protein	Gorai.005G138500
23	GhiAffx.19944.1.S1_at	glycoprotein membrane precursor GPI-anchored	Gorai.005G041200
24	Ghi.5081.1.S1_s_at	glyoxal oxidase-related protein	Gorai.002G125000
25	Ghi.632.1.S1_at	GroES-like zinc-binding alcohol dehydrogenase family protein	Gorai.002G066100
26	Gra.2833.1.S1_at	homeodomain-leucine zipper protein 56 (HDL56)	Gorai.003G041500
27	Ghi.5889.1.A1_x_at	HXXXD-type acyl-transferase family protein	Gorai.012G006700
28	Ghi.5889.2.S1_s_at	HXXXD-type acyl-transferase family protein	Gorai.012G006600
29	Ghi.7819.1.A1_at	hydrolase, alpha/beta fold family protein	Gorai.011G272800
30	Ghi.5515.1.A1_s_at	iron-binding protein (Fer1)	Gorai.006G184700
31	GraAffx.1241.1.S1_s_at	leucine-rich receptor-like protein kinase (LRPKm1)	Gorai.009G166500
32	Ghi.6301.1.S1_s_at	lung seven transmembrane receptor family protein	Gorai.007G019200
33	Ghi.6548.1.S1_s_at	MAP kinase-like protein	Gorai.002G096100
34	GhiAffx.29423.1.S1_s_at	MAR-binding protein	Gorai.004G245100
35	Ghi.5146.1.A1_x_at	NADP-dependent malic enzyme	Gorai.009G048600
36	GhiAffx.6438.1.S1_at	nodulin family protein	Gorai.007G034700
37	GhiAffx.21685.1.S1_at	nuclear transport factor 2 (NTF2)	Gorai.013G010700
38	Ghi.7430.2.S1_s_at	octicosapeptide/Phox/Bem1p (PB1) domain-containing protein/tetratricopeptide repeat (TPR)-containing protein	Gorai.013G214100
39	Ghi.1352.1.S1_s_at	O-fucosyltransferase family protein isoform 1	Gorai.003G038000
40	Ghi.10656.1.S1_s_at	photosystem I subunit PsaD (PSAD)	Gorai.005G042000
41	GhiAffx.51155.1.S1_s_at	PIP protein (PIP2;7)	Gorai.011G098100
42	Ghi.5186.1.A1_at	plant invertase/pectin methylesterase inhibitor superfamily	Gorai.001G018200
43	GhiAffx.33585.1.S1_at	plant invertase/pectin methylesterase inhibitor superfamily	Gorai.002G031100
44	Ghi.8118.1.S1_at	putative carboxyl-terminal proteinase	Gorai.005G180300
45	GraAffx.34131.2.S1_x_at	pyridoxal phosphate phosphatase (PHOSPHO2)	Gorai.011G067300
46	Ghi.4013.2.S1_at	root iron transporter protein IRT1	Gorai.011G049700
47	GhiAffx.22857.1.A1_at	rps16 (chloroplast)	Gorai.001G180700
48	GhiAffx.43008.1.S1_at	SAUR family protein	Gorai.001G017600

Table 1. Cont.

No.	Probes	Homologous Gene	G raimondii locus
49	GhiAffx.24789.1.S1_at	SAUR family protein (SAUR54)	Gorai.N011800.1
50	Ghi.5484.1.S1_s_at	SKU5-like 5 protein	Gorai.009G189900
51	Ghi.978.1.S1_at	tetratricopeptide repeat-like superfamily protein isoform 1	Gorai.013G142800
52	GhiAffx.44664.1.S1_at	thiosulfate sulfurtransferase	Gorai.007G049000
53	GhiAffx.53295.1.A1_at	UDP-glucuronosyl and UDP-glucosyl transferase	Gorai.008G273200
54	Ghi.9654.1.S1_s_at	UDP-glycosyltransferase UGT73C14	Gorai.008G273200
55	Ghi.3235.1.A1_at	UDP-glycosyltransferase UGT73C14	Gorai.009G411800
56	Ghi.10822.1.S1_at	xyloglucan endotransglucosylase/hydrolase (XTH2)	Gorai.003G033600
57	Ghi.4013.1.A1_at	zinc transporter 10 precursor	Gorai.011G049700
58	Ghi.10311.1.S1_s_at	SNARE protein Syntaxin 1 and related proteins	Gorai.006G148600
59	Ghi.4533.1.A1_x_at	SNARE protein Syntaxin 1 and related proteins	Gorai.006G148600
60	GhiAffx.7289.1.S1_at	SNARE protein Syntaxin 1 and related proteins	Gorai.006G148600
61	Ghi.7279.1.S1_at	ABC transporter	Gorai.009G022400
62	GhiAffx.58403.1.S1_at	G1/S-specific Cyclin D	Gorai.005G185600
63	GhiAffx.23478.1.S1_at	NTKL-BINDING PROTEIN 1	Gorai.001G120300
64	GhiAffx.4465.1.S1_s_at	F-BOX/LEUCINE RICH REPEAT PROTEIN	Gorai.009G049800
65	GhiAffx.44162.1.S1_s_at	Extracellular protein with conserved cysteines	Gorai.007G359700
66	Ghi.8451.1.S1_s_at	integral to membrane (GO:0016021)	Gorai.001G148000

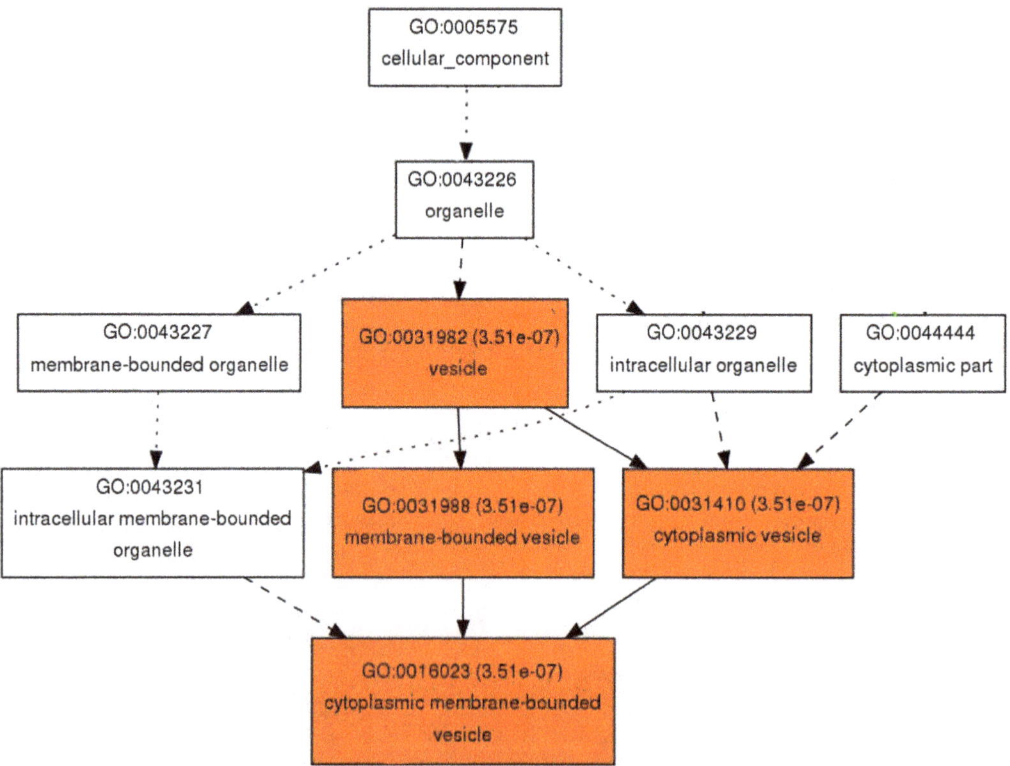

Figure 5. Identification of a potential cellular component in ***Li_1*** **and** ***Li_2*** **mutants.** AgirGo SEA analysis identified that nine of the DEGs presented in Table 1 were involved in vesicle transportation. The significant terms (adjusted $P \leq 0.05$) were shown in red color boxes, whereas non-significant terms were shown as white boxes. Solid, dashed, and dotted lines represented two, one and zero enriched terms at both ends connected by the line, respectively.

Table 2. Common DEGs involved in vesicle (GO:0031982) in both Li_1 and Li_2 mutants regardless of growth conditions (P value, 1.7e-11; FDR, 3.1e-10).

DEGs	Annotation
Ghi.68.1.A1_s_at	fasciclin-like arabinogalactan protein 1
GhiAffx.19944.1.S1_at	glycoprotein membrane precursor GPI-anchored
GraAffx.1241.1.S1_s_at	leucine-rich receptor-like protein kinase
Ghi.5186.1.A1_at	plant invertase/pectin methylesterase inhibitor superfamily
Ghi.7819.1.A1_at	hydrolase, alpha/beta fold family protein
Ghi.8118.1.S1_at	putative carboxyl-terminal proteinase
Ghi.7402.1.S1_at	Protein of unknown function, DUF642
Ghi.8451.1.S1_s_at	integral to membrane
Ghi.2840.3.S1_s_at	cysteine proteinase

mapping file. Co-expression analysis was conducted utilizing ATTEDII version 7.1 [36] with a mutual rank value of <200.

Results and Discussion

Effects of Growing Conditions (Field and Greenhouse) on Gene Expression

The data obtained in these experiments allowed for an analysis of the environmental effects on transcriptome data, i.e.; a field *vs* greenhouse comparison. This was useful for our purpose as it allowed the identification of transcripts affected by variable environmental conditions, which could then be excluded from consideration of strictly fiber-related transcripts. Although it is known that many fiber-related genes are environmentally impacted [5], the exclusion of these genes permitted a more targeted analysis of the genes that are essential to fiber elongation. It also allowed us to investigate how the mutant line differed from its WT in responding to environmental stressors, providing insight into the interaction between stress response and fiber elongation. Microarray analysis was conducted on WT and Li_2 12 DPA bolls collected from cotton plants grown in both field and greenhouse conditions. This comparison identified 150 probes in the WT that were expressed higher in the greenhouse than in a field, and 754 probes that were expressed higher when grown in field conditions (numbers are sums shown in orange ovals of Fig. 2). Gene enrichment analysis of the probes higher in the field showed that genes related to nucleosome assembly (GO:0006334), flavonoid biosynthetic process (GO:0009813), and response to heat (GO:0009408) were enriched in the field conditions (Fig. 2 and Table S2). Probes showing higher expression patterns in the greenhouse in WT were not enriched in any particular GO category to a statistically significant degree due to smaller number of probes, however did consist of ethylene and ap2 erf domain-containing transcription factors, expansin, and probes related to NADH dehydrogenase. The results of a similar analysis conducted in the Li_2 mutant differed dramatically from what was observed in the WT, with 1,275 probes showing higher expression in the greenhouse and 1,136 probes showing higher expression in the field (numbers are sums shown in green ovals of Fig. 2). The probes showing higher expression in Li_2 in the greenhouse were enriched in lipid transport (GO:0006869), cellular nitrogen compound metabolic process (GO:0034641), and iron ion binding proteins (GO: 0005506), whereas, probes showing higher expression in the field were enriched in mitochondrial electron transport (GO:0006120) and response to reactive oxygen species (GO: 000302) (Fig. 2 and Table S2). Only 124 probes showed the same expression profile in WT and Li_2 in both field and greenhouse conditions. This difference between WT and Li_2 in response to the environmental conditions is profound, as very few probes showed similar expression profiles between the two genotypes, (i.e., only 6 common probes were higher in the green house and only 118 were higher in the field).

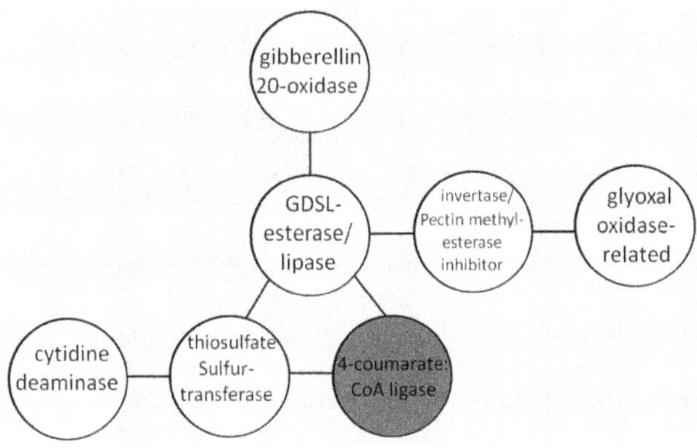

Figure 6. Co-Expression analysis. The ATTED-II database of genes co-expressed in Arabidopsis was utilized to identify potential interactions among the list of 66 targeted genes (see text). Circle with gray background indicates gene that is up-regulated, circles with white background indicate down regulated genes. A line connecting two genes means a mutual rank value of <200.

Gene enrichment analysis for Li_2 specific greenhouse/field condition response identified two categories of probes that are of particular interest; probes related to mitochondrial electron transport (GO:0006120) and response to reactive oxygen species (ROS) (GO:0000302) were both enriched in Li_2 in the field relative to their expression in the greenhouse. Probes for genes in these categories were not enriched in WT. It has been suggested in previous studies that the inability of certain *Gossypium* species and *G. hirsutum* mutants to produce long fibers may be due to the inability to modulate ROS homeostasis [19,37,38]. The mitchondrial electron transport gene NADH dehydrogenase (GhiAffx.21609.1.S1_at, GhiAffx.53261.1.A1_at, GhiAffx.45916.1.A1_s_at) and NADH plastiquinone reductase (GhiAffx.4260.1.S1_at, GhiAffx.61308.1.S1_at, GhiAffx.9732.1.A1_at) were among those highly expressed in Li_2 in the

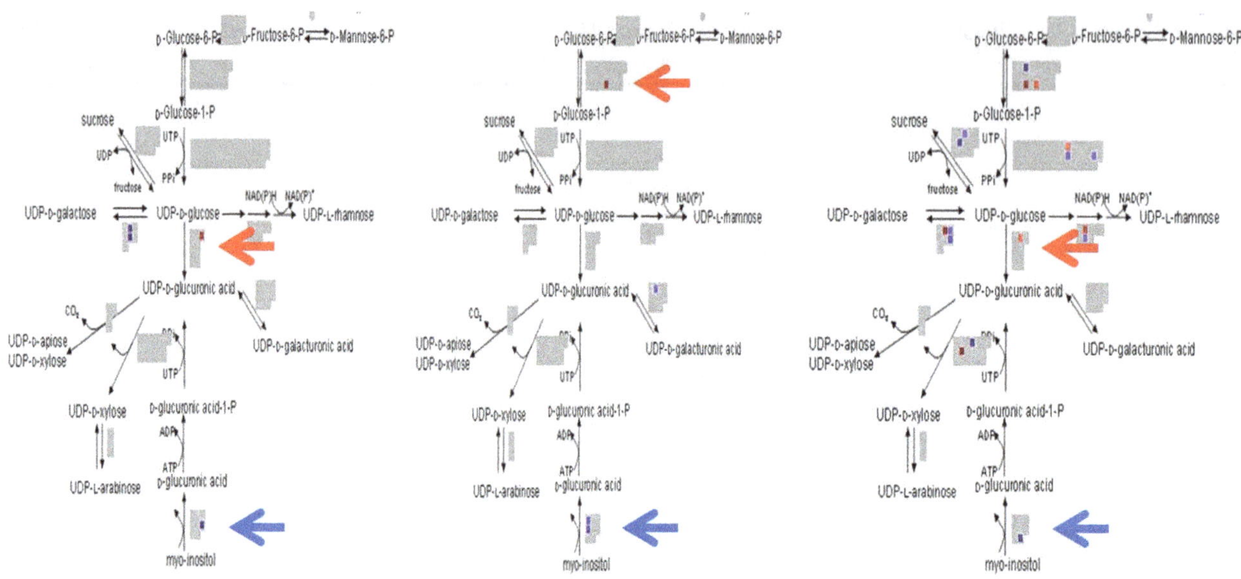

Figure 7. MapMan software illustrates the cell wall precursor pathway and effects of the Li_1 and Li_2 mutations. Enzymes involved in the synthesis of UDP-D-glucaronic acid are altered in both Li_1 and Li_2 mutants. Blue indicates up-regulated probes and red indicates down regulation. The arrows point to steps of the pathway that are discussed in more detail in the text.

field condition. Because the field conditions presented both abiotic and biotic stresses that were absent or minimized in greenhouse conditions, it was likely that the field grown Li_2 plants needed to divert limited cellular resources to manage these additional stresses, leading to higher ROS accumulation and an even higher expression of ROS homestasis genes.

Expression levels of mitochondrial-related genes were studied across all three microarray data sets. In field conditions Li_1 had 7 probes (Ghi.7225.1.s1_s_at, Ghiaffx.53261.1.a1_at, Ghi.7032.2.a1_at, Ghi.7032.2.s1_s_at, Ghiaffx.21609.1.s1_at, Ghiaffx.18012.1.s1_at, Ghiaffx.45916.1.a1_s_at), and Li_2 had 9 probes (Ghi.7032.2.s1_s_at, Ghi.7032.2.a1_at, Ghiaffx.3647.1.-s1_at, Ghiaffx.5964.1.s1_s_at, Ghi.7225.1.s1_s_at, Ghiaffx.53261.1.a1_at, Ghiaffx.18012.1.s1_at, Ghiaffx.45916.1.a1_s_at, Ghi.648.1.a1_at) up-regulated that had high sequence identity to NADH dehydrogenases, whereas in the greenhouse Li_2 had only one probe up-regulated, Ghiaffx.45916.1.a1_s_at (blue squares in Fig. 3A). The 2 probes down-regulated (red squares in Fig. 3A) consisted of Gra.1550.1, which has sequence identity to a Choline transporter-related transcript (AT4G38640) and Ghi.648.1.a1_at, which has sequence similarity to an NADH dehydrogenase. The probe that was up-regulated in all 3 data sets in the Li mutants is homologous to the probe GhiAffx.45916.1, a subunit of NADP dehydrogenase (NAD2) (shown by a blue arrow in Fig. 3A). These results further demonstrated the potential relevance of ROS and stress response in fiber developing processes.

Chalcone synthase is the upstream enzyme in the flavanoids synthesis pathway responsible for production of secondary metabolites (flavonols, proanthocyanins and anthocyanin) that are often produced in response to stresses [39,40]. Here, the chalcone synthase-related probes in WT plants had significantly higher expression in field conditions than in the greenhouse as indicated by a significant enrichment in flavonoid biosynthetic processes (GO:0009813) (Table S2). Additionally, pathway analysis utilizing MapMan software indicated that under greenhouse condition, Li_2 had higher expression levels of chalcone synthase than WT (Fig. 3B). Thus, all of the plants in the field (WT, Li_1 and Li_2), and the Li_2 plants in the greenhouse had high levels of genes related to flavonol production relative to the WT plants in the greenhouse. This further supports the hypothesis that the Li_2 mutant, even in greenhouse conditions, were in a stressed state. Ghi.6103.1 that codes for *chalcone synthase 3* (GhCHS3) (Fig. 3B, red squares) was the only identified flavonoid probe that decreased in Li_2 in both field and greenhouse conditions. The remaining probes that increased under field conditions demonstrate varying degrees of homology to Transparent Testa 4 (TT4) (8 probes) and TT5 (2 probes), both naringenin-chalcone synthases.

In brief, it was likely that field-grown plants were under more stressful conditions than greenhouse grown plants. Short fiber mutants (more specifically Li_2 mutant) commanded higher expression of genes related to energy production and transport such as mitochondrial electron transport or responding to ROS in order to fight against the stresses than the WT.

Identification and Annotation of DEGs Common to both Short Fiber Mutants Regardless of Growth Conditions

Although the Li_1 and Li_2 mutations are caused by different genes located on different chromosomes, both result in a similar short fiber phenotype. Thus it is possible that probes which are similarly altered in expression pattern between a mutant and WT in both field and greenhouse conditions are highly likely to be fiber elongation-related or specific genes. In all three data sets, 113 probes are commonly affected in the mutants in comparison with WT with 94 down-regulated and 19 up-regulated (Fig. 4). Of these 113 probes that showed altered regulation, 25 also showed to be differentially regulated between field and greenhouse, implying their altered regulation could be an environmentally controlled factor and were excluded from further consideration. The remaining 88 probes are likely the genes specific to fiber elongation. Among them, 66 genes were annotated by blastn or blastx and were used for further analyses (Table 1 for annotated genes and Table S3 for full list).

Among the fiber elongation specific genes, actin depolymerizing factor (ADF5) (GhiAffx.33535.1) was decreased in all conditions analyzed. ADF family proteins have previously been shown to affect the 3-dimensional structure of actin filaments [41] and to alter the disassociation rates of actin subunits [42]. The down-regulation of GhADF1 in transgenic cotton show altered fiber length and strength [43] and GhADF5 has been shown to localize to elongating cells of the root stem in *A. thaliana* [44]. As shown in Table 1, actin (Gra.1375.1.A1_at) and tubulin (Ghi.8448.1.S1_-x_at) were commonly down-regulated in short fiber mutants. It was proposed that actin and microtubule play important roles in fiber elongation [45]. Functional analysis showed that actin was indeed required for fiber elongation [46].

Two probes (Ghi.3235.1 and Ghi.9654.1.) showing altered regulation in all conditions are different regions of the same gene, UGT73C14, which codes for an UDP glycosyltransferase that glycosylates ABA *in vivo* and *in vitro* [47]. One probe (GhiAff-x.53295.1.A1_at) also codes for UDP-glucuronosyl and UDP-glucosyl transferase, but has not been characterized in the context of elongation and warrant further investigation. Ghi.6548.1.S1 codes for a MAP kinase-like protein that was previously identified as preferentially expressed from elongating fibers in *G. hirsutum* by subtractive PCR but remains otherwise uncharacterized [48]. Two probes, GhiAffx.43008.1 and GhiAffx.24789.1 code for SAUR (Small Auxin Up RNA) family genes, which comprise the largest family of auxin-responsive genes. However, some members have also reportedly been associated with cell expansion [49]. This is potentially evidence of a substantial and specific link between auxin regulation and fiber elongation.

Two probes that decreased in all conditions analyzed code for either plant invertase or pectin methylesterase (PME) inhibitors (GhiAffx.33585.1 and Ghi.5186.1). Invertases catalyze the hydrolysis of sucrose, and PMEs inhibit the enzyme that catalyzes the de-methylesterification of pectins, a process important in initiation and for the polysaccharides role in the primary cell wall of cotton fibers [50,51]. Li_1 and Li_2 have previously demonstrated a decrease in de-esterified pectin localized to the primary cell wall during elongation stages as indicated by antibody staining [52]. Previous studies have also demonstrated that there are at least 5 PMEs showing fluctuating expression profiles throughout fiber initiation, elongation, and transition to secondary cell wall synthesis, indicating that different PME regulated different fiber development processes [53]. It was further demonstrated that the timing of individual PME gene expression differed between the longer fibered Pima (*G. barbadense*) and Upland (*G. hirsutum*) species in such a manner that suggested the timing of PME activity affected the onset of the transition stage from elongation to secondary cell wall synthesis, in turn affecting length of the fiber [53]. Since it is suggested that PME activity generally corresponds to longer fibers [54–56], it is possible that the increased expression of PME inhibitor(s) identified in our study play a role in inhibiting mid or late elongation stage PMEs, thus low levels of inhibitor would induce early transition to secondary cell wall synthesis. This would effectively decrease the elongation period. In support of this, the PME inhibitor probed by Ghi.5186.1 also exhibited decrease expression ratios of 0.42 in Li_1 and 0.58 in Li_2 at 3 DPA (data not shown).

Gene Ontology Analysis of the DEGs Common to both Short Fiber Mutants

To identify the potential biological processes governing differential expressions of genes that affect fiber length development of the two short fiber mutants, we analyzed the identified 88 DEGs using agriGo SEA analyses. The GO enrichment analysis classified 9 DEGs as a group that is involved in vesicle transportation (GO:0031982) in cotton fiber cells (Fig. 5). The vesicle plays an important role of carrying membrane components to the growing site of elongating cells [57,58]. Among the 9 vesicle-related DEGs (Table 2), fasciclin-like arabinogalactan protein (*GhFLA1*, Ghi.68.1.A1_s_at) was reported as an essential cotton fiber gene for fiber elongation and primary cell wall biosynthesis [59].

Co-expression Analysis of Annotated DEGs

Beginning with our list of genes showing altered regulation in both mutants in all three data sets and no variability between field and greenhouse (Table 1), we wanted to identify genes that were previously identified as being co-expressed, and further support their roles in a common pathway and their involvement in the phenotype of the mutant plants. Subjecting the aforementioned list to the ATTED-II Arabidopsis co-expression database revealed a putative gibberellic acid-regulated pathway (Fig. 6). Several GDSL esterase/lipase family genes have been found with giberellin-responsive elements (P-box and GARE motif) in their 5′ upstream regulatory regions [60]. These genes are characterized by the presence of a conserved motif and have been shown to be involved in a wide range of functions, including stress response and development [61,62], however to our knowledge this is the first report of a possible link with fiber development. The only gene up-regulated in the identified pathway, 4-coumarate-coenzyme A ligase, has been shown to play a role in lignin deposition in the plant cell wall of several species, which has exhibited a decrease in plant height when the gene was suppressed [63,64]. In cotton fibers, lignin is a structural component of the primary cell wall, and genes that are involved in lignin deposition are up regulated in parallel with fiber elongation [65]. Finally, an additional topic that is re-visited in this analysis is the role of ROS-related genes in fiber development identified by this and other studies. Experimental evidence has demonstrated that the membrane bound glyoxal oxidases, which decreases in the mutants examined here, are known to produce hydrogen peroxide [66,67], which is then used as a cofactor for lignin-synthesizing peroxidases [68]. Thus, the genes represented in Fig. 5 could help explain the apparent perturbation to ROS-related processes affected in the Li_1 and Li_2 mutants.

Cell-wall Precursors

Because our analysis was focused on cell wall development we utilized MapMan software to determine which steps in the synthesis of cell wall-related polysaccharides are affected in the mutants. Fig. 7 illustrates that the conversion of myo-inositol to D-glucaronic acid by myo-inositol oxygenase (MIOX), an early step in the synthesis UDP-D-glucaronic acid, was affected. Between the 3 experimental groups, there are 4 separate probes that were up-regulated (Fig. 7, blue arrows) (Li_1/field, GraAffx.9655.1; Li_2/Field, Ghi.8187.2, and GhiAffx.4265.1; Li_2/greenhouse, Gra.2699.2). tBLASTx analysis of these EST sequences revealed GraAffx.9655.1 shares highest sequence identity with AtMIOX4, where as the other three share highest identity with AtMIOX1. Likely related to this, MapMan software also revealed the alternative pathway for UDP-D-glucaronic acid synthesis contains probes for genes that were down-regulated (red dots). The conversion of UDP-D-glucose to UDP-D-glucaronic acid is mediated by UDP-glucose 6 dehydrogenase, which was down-regulated in Li_2/greenhouse (Gra.2095.2) and Li_1/field (Ghi.8750.1) (red arrows). This gene was unaffected in Li_2/field, however GhiAffx.64086.1, which codes for phosphofructokinase 3 and mediates the formation of D-Glucose-1-P, was down-regulated, likely indicating an alternative site of regulation (red arrow). The regulation of UDP-D-glucaronic acid is a vital step, as it serves as the common precursor for arabinose, xylose, galacturonic acid, and apiose residues in the cell wall. It has been reported in null Arabidopsis mutants that the limited supply of myo-inositol generally prevents the effective induction of this pathway as an alternative to UDP-glucose formation [69]. Thus it remains a strong possibility that the UDP-D-glucose levels and other polysaccharides necessary for cell wall development were perturbed and insufficient in the Li_1 and Li_2 mutant fiber tissues.

Verification of Microarray Results by RT-qPCR Analysis

To test the reliability of microarray data, RT-qPCR analysis was performed for a subset of 14 genes selected from Table S3 showing altered regulation in Li_1, Li_2 (field) and Li_2 (greenhouse) as compared to the wild type. Expression of selected genes was tested at 12 DPA of fiber development, the common time point between greenhouse and field experiments. Overall, the results of RT-qPCR analysis were consistent with results of microarray for 14 selected genes (Table S4).

Conclusions

The development of cotton fibers is a very complicated and poorly understood biological process, however understanding this process is vital for the targeting of genes to use in the creation of value-added crop. We have developed a genetic model system consisting of two short fiber cotton mutants, Li_1 and Li_2, which when combined together with their near-isogenic WT line allows for the study of genes and processes specific to fiber elongation. Here we analyzed multiple transcriptome profiles obtained from Li_1, Li_2 short fiber mutants and their WT grown under different environmental conditions. We classified the differentially expressed genes into two groups: one was mainly affected by environmental conditions, and the other was largely regulated by Li_1 and Li_2. Our results provide new insight to how environmental factors affect fiber elongation by transcriptional regulation. Further, the short list of 88 genes required for fiber elongation without being affected by environmental conditions would warrant further investigation in hope to identify targets for improving cotton fiber property.

Supporting Information

Table S1 Primer sequences used for RT-qPCR.

Table S2 Complete results of gene ontology analysis.

Table S3 Expression levels of probes showing altered regulation in Li_1, Li_2 (field) and Li_2 (greenhouse) as compared to the wild type (WT).

Table S4 Comparison between RT-qPCR and microarray under different growing environments.

Acknowledgments

We thank Dr. Doug Hinchliffe for performing Li_2 experiment in the greenhouse, Dr. Rickie Turley for making the near isogenic lines used in this research, and Stacy Allen at Noble Foundation for conducting the microarray experiments. Mention of trade names or commercial products in this article is solely for the purpose of providing specific information and does not imply recommendation or endorsement by the U. S. Department of Agriculture that is an equal opportunity provider and employer.

Author Contributions

Conceived and designed the experiments: DDF. Performed the experiments: MKG MN. Analyzed the data: MKG HJK YT MN. Contributed reagents/materials/analysis tools: DDF. Wrote the paper: MKG DDF.

References

1. Lee JJ, Woodward AW, Chen ZJ (2007) Gene expression changes and early events in cotton fibre development. Ann Bot 100: 1391–1401.
2. Basra AS, Malik CP (1984) Development of the cotton fiber. Int Rev Cytol 89: 65–113.
3. Tiwari SC, Wilkins TA (1995) Cotton (Gossypium hirsutum) seed trichomes expand via diffuse growing mechanism. Can J Bot 73: 746–757.
4. Meinert MC, Delmer DP (1977) Changes in biochemical composition of the cell wall of the cotton fiber during development. Plant Physiol 59: 1088–1097.
5. Hinchliffe DJ, Meredith WR, Delhom CD, Thibodeaux DP, Fang DD (2011) Elevated growing degree days influence transition stage timing during cotton (*Gossypium hirsutum* L.) fiber development and result in increased fiber strength. Crop Sci 51: 1683–1692.
6. Kim HJ, Triplett BA (2001) Cotton fiber growth *in planta* and *in vitro*. Models for plant cell elongation and cell wall biogenesis. Plant Physiol 127: 1361–1366.
7. Schubert AM, Benedict CR, Gates CE, Kohel RJ (1976) Growth and development of the lint fibers of Pima S-4 cotton. Crop Sci 16: 539–543.
8. Willison JH, Brown RM (1977) An examination of the developing cotton fiber: wall and plasmalemma. Protoplasma 92: 21–42.
9. Seagull RW (1986) Changes in microtubule organization and wall microfibril orientation during in vitro cotton fiber development: an immunofluorescent study. Can J Bot 64: 1373–1381.
10. Ruan YL (2005) Recent advances in understanding cotton fibre and seed development. Seed Sci Res 15: 269–280.
11. Davidonis GH, Johnson AS, Landivar JA, Fernandez CJ (2004) Cotton fiber quality is related to boll location and planting date. Agronomy J 96: 42–47.
12. Liakatas A, Roussopoulos D, Whittington WJ (1998) Controlled-temperature effects on cotton yield and fibre properties. J Agric Sci 130: 463–471.
13. Roussopoulos D, Liakatas A, Whittington WJ (1998) Controlled-temperature effects on cotton growth and development. J Agric Sci 130: 451–462.
14. Bolton JJ, Soliman KM, Wilkins TA, Jenkins JN (2009) Aberrant expression of critical genes during secondary cell wall biogenesis in a cotton mutant, Ligon lintless-1 (*Li-1*). Comp Funct Genomics Article ID 659301.
15. Cai C, Tong X, Liu F, Lv F, Wang H, et al. (2013) Discovery and identification of a novel Ligon lintless-like mutant (*Lix*) similar to the Ligon lintless (Li_1) in allotetraploid cotton. Theor Appl Genet 126: 963–970.
16. Lee JJ, Hassan OS, Gao W, Wei NE, Kohel RJ, et al. (2006) Developmental and gene expression analyses of a cotton naked seed mutant. Planta 223: 418–432.
17. Wu Y, Machado AC, White RG, Llewellyn DJ, Dennis ES (2006) Expression profiling identifies genes expressed early during lint fibre initiation in cotton. Plant Cell Physiol 47: 107–127.
18. Gilbert MK, Turley RB, Kim HJ, Li P, Thyssen G, et al. (2013) Transcript profiling by microarray and marker analysis of the short cotton (*Gossypium hirsutum* L.) fiber mutant Ligon lintless-1 (Li_1). BMC Genomics 14: 403.
19. Hinchliffe DJ, Turley RB, Naoumkina M, Kim HJ, Tang Y, et al. (2011) A combined functional and structural genomics approach identified an EST-SSR marker with complete linkage to the Ligon lintless-2 genetic locus in cotton (*Gossypium hirsutum* L.). BMC Genomics 12: 445.
20. Naoumkina M, Hinchliffe DJ, Turley RB, Bland JM, Fang DD (2013) Integrated metabolomics and genomics analysis provides new insights into the fiber elongation process in Ligon lintless-2 mutant cotton (*Gossypium hirsutum* L.). BMC Genomics 14: 155.
21. Liu K, Sun J, Yao L, Yuan Y (2012) Transcriptome analysis reveals critical genes and key pathways for early cotton fiber elongation in Ligon lintless-1 mutant. Genomics 100: 42–50.
22. Ding M, Jiang Y, Cao Y, Lin L, He S, et al. (2014) Gene expression profile analysis of Ligon lintless-1 (Li1) mutant reveals important genes and pathways in cotton leaf and fiber development. Gene 535: 273–285.
23. Kohel RJ (1972) Linkage tests in Upland cotton, *Gossypium hirsutum* L. II. Crop Sci 12: 66–69.
24. Narbuth EV, Kohel RJ (1990) Inheritance and linkage analysis of a new fiber mutant in cotton. J Hered 81: 131–133.
25. Kohel RJ, Narbuth EV, Benedict CR (1992) Fiber development of Ligon lintless-2 mutant of cotton. Crop Sci 32: 733–735.
26. Rong J, Pierce GJ, Waghmare VN, Rogers CJ, Desai A, et al. (2005) Genetic mapping and comparative analysis of seven mutants related to seed fiber development in cotton. Theor Appl Genet 111: 1137–1146.
27. Karaca M, Saha S, Jenkins JN, Zipf A, Kohel R, et al. (2002) Simple sequence repeat (SSR) markers linked to the Ligon lintless (Li_1) mutant in cotton. J Hered 93: 221–224.

28. Bustin SA, Benes V, Garson JA, Hellemans J, Huggett J, et al. (2009) The MIQE guidelines: minimum information for publication of quantitative real-time PCR experiments. Clinical Chem 55: 611–622.
29. Vandesompele J, De Preter K, Pattyn F, Poppe B, Van Roy N, et al. (2002) Accurate normalization of real-time quantitative RT-PCR data by geometric averaging of multiple internal control genes. Genome Biology 3: RESEARCH0034.
30. Benedito VA, Torres-Jerez I, Murray JD, Andriankaja A, Allen S, et al. (2008) A gene expression atlas of the model legume Medicago truncatula. Plant J 55: 504–513.
31. Hulsen T, de Vlieg J, Alkema W (2008) BioVenn - a web application for the comparison and visualization of biological lists using area-proportional Venn diagrams. BMC Genomics 9: 488.
32. Du Z, Zhou X, Ling Y, Zhang Z, Su Z (2010) agriGO: a GO analysis toolkit for the agricultural community. Nucleic Acids Res 38: W64–70.
33. Paterson AH, Wendel JF, Gundlach H, Guo H, Jenkins J, et al. (2012) Repeated polyploidization of *Gossypium* genomes and the evolution of spinnable cotton fibres. Nature 492: 423–427.
34. Conesa A, Gotz S, Garcia-Gomez JM, Terol J, Talon M, et al. (2005) Blast2GO: a universal tool for annotation, visualization and analysis in functional genomics research. Bioinformatics 21: 3674–3676.
35. Thimm O, Blasing O, Gibon Y, Nagel A, Meyer S, et al. (2004) MAPMAN: a user-driven tool to display genomics data sets onto diagrams of metabolic pathways and other biological processes. Plant J 37: 914–939.
36. Obayashi T, Nishida K, Kasahara K, Kinoshita K (2011) ATTED-II Updates: Condition-Specific Gene Coexpression to Extend Coexpression Analyses and Applications to a Broad Range of Flowering Plants. Plant Cell Physiol 52: 213–219.
37. Chaudhary B, Hovav R, Flagel L, Mittler R, Wendel J (2009) Parallel expression evolution of oxidative stress-related genes in fiber from wild and domesticated diploid and polyploid cotton (Gossypium). BMC Genomics 10: 378.
38. Chaudhary B, Hovav R, Rapp R, Verma N, Udall JA, et al. (2008) Global analysis of gene expression in cotton fibers from wild and domesticated Gossypium barbadense. Evolution & Development 10: 567–582.
39. Holton TA, Cornish EC (1995) Genetics and Biochemistry of Anthocyanin Biosynthesis. Plant Cell 7: 1071–1083.
40. Hernández I, Alegre L, Van Breusegem F, Munné-Bosch S (2009) How relevant are flavonoids as antioxidants in plants? Trends Plant Sci 14: 125–132.
41. McGough A, Pope B, Chiu W, Weeds A (1997) Cofilin changes the twist of F-actin: implications for actin filament dynamics and cellular function. J Cell Biol 138: 771–781.
42. Carlier MF, Pantaloni D (1997) Control of actin dynamics in cell motility. J Mol Biol 269: 459–467.
43. Wang HY, Wang J, Gao P, Jiao GL, Zhao PM, et al. (2009) Down-regulation of GhADF1 gene expression affects cotton fibre properties. Plant Biotech J 7: 13–23.
44. Dong CH, Xia GX, Hong Y, Ramachandran S, Kost B, et al. (2001) ADF proteins are involved in the control of flowering and regulate F-actin organization, cell expansion, and organ growth in Arabidopsis. Plant Cell 13: 1333–1346.
45. Qin YM, Zhu YX (2011) How cotton fibers elongate: a tale of linear cell-growth mode. Curr Opin Plant Biol 14: 106–111.
46. Li XB, Fan XP, Wang XL, Cai L, Yang WC (2005) The cotton ACTIN1 gene is functionally expressed in fibers and participates in fiber elongation. Plant Cell 17: 859–875.
47. Gilbert MK, Bland JM, Shockey JM, Cao H, Hinchliffe DJ, et al. (2013) A transcript profiling approach reveals an abscisic acid-specific glycosyltransferase (UGT73C14) induced in developing fiber of Ligon lintless-2 mutant of cotton (*Gossypium hirsutum* L.). PLoS ONE 8: e75268.
48. Ji SJ, Lu YC, Feng JX, Wei G, Li J, et al. (2003) Isolation and analyses of genes preferentially expressed during early cotton fiber development by subtractive PCR and cDNA array. Nucleic Acids Res 31: 2534–2543.
49. Spartz AK, Lee SH, Wenger JP, Gonzalez N, Itoh H, et al. (2012) The SAUR19 subfamily of SMALL AUXIN UP RNA genes promote cell expansion. Plant J 70: 978–990.
50. Vaughn KC, Turley RB (1999) The primary walls of cotton fibers contain an ensheathing pectin layer. Protoplasma 209: 226–237.
51. Bowling AJ, Vaughn KC, Turley RB (2011) Polysaccharide and glycoprotein distribution in the epidermis of cotton ovules during early fiber initiation and growth. Protoplasma 248: 579–590.
52. Turley RB, Vaughn KC (2007) De-esterified pectins of the cell walls of cotton fibers: a study of fiber mutants. In: McMichaels B, editor; 2007 September 10–14, 2007; Lubbock, TX.
53. Liu Q, Talbot M, Llewellyn DJ (2013) Pectin methylesterase and pectin remodelling differ in the fibre walls of two gossypium species with very different fibre properties. PLoS ONE 8: e65131.
54. Siedlecka A, Wiklund S, Peronne MA, Micheli F, Lesniewska J, et al. (2008) Pectin methyl esterase inhibits intrusive and symplastic cell growth in developing wood cells of Populus. Plant Physiol 146: 554–565.
55. Wen F, Zhu Y, Hawes MC (1999) Effect of pectin methylesterase gene expression on pea root development. Plant Cell 11: 1129–1140.
56. Pilling J, Willmitzer L, Fisahn J (2000) Expression of a Petunia inflata pectin methyl esterase in Solanum tuberosum L. enhances stem elongation and modifies cation distribution. Planta 210: 391–399.
57. Augustine RC, Vidali L, Kleinman KP, Bezanilla M (2008) Actin depolymerizing factor is essential for viability in plants, and its phosphoregulation is important for tip growth. Plant J 54: 863–875.
58. Vidali L, Rounds CM, Hepler PK, Bezanilla M (2009) Lifeact-mEGFP reveals a dynamic apical F-actin network in tip growing plant cells. PLoS ONE 4: e5744.
59. Huang GQ, Gong SY, Xu WL, Li W, Li P, et al. (2013) A fasciclin-like arabinogalactan protein, GhFLA1, is involved in fiber initiation and elongation of cotton. Plant Physiol 161: 1278–1290.
60. Jiang Y, Chen R, Dong J, Xu Z, Gao X (2012) Analysis of GDSL lipase (GLIP) family genes in rice (Oryza sativa). Plant omics 5: 351–358.
61. Hong JK, Choi HW, Hwang IS, Kim DS, Kim NH, et al. (2008) Function of a novel GDSL-type pepper lipase gene, CaGLIP1, in disease susceptibility and abiotic stress tolerance. Planta 227: 539–558.
62. Kiba T, Naitou T, Koizumi N, Yamashino T, Mizuno T, et al. (2005) Combinatorial microarray analysis revealing Arabidopsis genes implicated in cytokinin responses through the His→Asp phosphorelay circuitry. Plant Cell Physiol 46: 339–355.
63. Wagner A, Donaldson L, Kim H, Phillips L, Flint H, et al. (2009) Suppression of 4-coumarate-CoA ligase in the coniferous gymnosperm Pinus radiata. Plant Physiol 149: 370–383.
64. Voelker SL, Lachenbruch B, Meinzer FC, Jourdes M, Ki C, et al. (2010) Antisense down-regulation of 4CL expression alters lignification, tree growth, and saccharification potential of field-grown poplar. Plant Physiol 154: 874–886.
65. Shi H, Liu Z, Zhu L, Zhang C, Chen Y, et al. (2012) Overexpression of cotton (Gossypium hirsutum) dirigent1 gene enhances lignification that blocks the spread of Verticillium dahliae. Acta Biochimica et Biophysica Sinica 44: 555–564.
66. Kersten PJ (1990) Glyoxal oxidase of Phanerochaete chrysosporium: its characterization and activation by lignin peroxidase. PNAS 87: 2936–2940.
67. Kersten PJ, Cullen D (1993) Cloning and characterization of cDNA encoding glyoxal oxidase, a H2O2-producing enzyme from the lignin-degrading basidiomycete Phanerochaete chrysosporium. PNAS 90: 7411–7413.
68. Pomar F, Caballero N, Pedreo MA, Ros Barceló A (2002) H_2O_2 generation during the auto-oxidation of coniferyl alcohol drives the oxidase activity of a highly conserved class III peroxidase involved in lignin biosynthesis. FEBS Letters 529: 198–202.
69. Reboul R, Geserick C, Pabst M, Frey B, Wittmann D, et al. (2011) Down-regulation of UDP-glucuronic acid biosynthesis leads to swollen plant cell walls and severe developmental defects associated with changes in pectic polysaccharides. J Biol Chem 286: 39982–39992.

18

Distribution and Frequency of *kdr* Mutations within *Anopheles gambiae* s.l. Populations and First Report of the *Ace.1*G119S Mutation in *Anopheles arabiensis* from Burkina Faso (West Africa)

Roch K. Dabiré[1]*, Moussa Namountougou[1], Abdoulaye Diabaté[1], Dieudonné D. Soma[1], Joseph Bado[1], Hyacinthe K. Toé[1,2], Chris Bass[3], Patrice Combary[4]

1 IRSS (Institut de Recherche en Sciences de la Santé), Centre Muraz, Bobo-Dioulasso, Burkina Faso, **2** Department of Vector Biology, Liverpool School of Tropical Medicine, Liverpool, United Kingdom, **3** Biological Chemistry and Crop Protection Rothamsted Research, Harpenden, United Kingdom, **4** National Malaria Control Programme, Ministry of Health, Ouagadougou, Burkina Faso

Abstract

An entomological survey was carried out at 15 sites dispersed throughout the three eco-climatic regions of Burkina Faso (West Africa) in order to assess the current distribution and frequency of mutations that confer resistance to insecticides in *An. gambiae* s.l. populations in the country. Both knockdown (*kdr*) resistance mutation variants (L1014F and L1014S), that confer resistance to pyrethroid insecticides, were identified concomitant with the *ace*-1 G119S mutation confirming the presence of multiple resistance mechanisms in the *An. gambiae* complex in Burkina Faso. Compared to the last survey, the frequency of the L1014F *kdr* mutation appears to have remained largely stable and relatively high in all species. In contrast, the distribution and frequency of the L1014S mutation has increased significantly in *An. gambiae* s.l. across much of the country. Furthermore we report, for the first time, the identification of the *ace.1* G116S mutation in *An. arabiensis* populations collected at 8 sites. This mutation, which confers resistance to organophosphate and carbamate insecticides, has been reported previously only in the *An. gambiae* S and M molecular forms. This finding is significant as organophosphates and carbamates are used in indoor residual sprays (IRS) to control malaria vectors as complementary strategies to the use of pyrethroid impregnated bednets. The occurrence of the three target-site resistance mutations in both *An. gambiae* molecular forms and now *An. arabiensis* has significant implications for the control of malaria vector populations in Burkina Faso and for resistance management strategies based on the rotation of insecticides with different modes of action.

Editor: Basil Brooke, National Institute for Communicable Diseases/NHLS, South Africa

Funding: This work was supported by the National Malaria Control Program (NMCP) of Burkina Faso. The funders had no role in study design, data collection and analysis, decision to publish, or preparation of the manuscript.

Competing Interests: The authors have declared that no competing interests exist.

* Email: dabire_roch@hotmail.com

Introduction

The pyrethroid class of insecticides have become a mainstay for vector control since the ban of DDT due to off-target toxicity and the development of resistance. They have been most widely used to treat bed nets (ITNs) dedicated to personal and community protection [1,2,3]. Unfortunately, knock down resistance (*kdr*) to pyrethroids, which also confers cross-resistance to DDT, was first reported in *Anopheles gambiae* populations from Côte d'Ivoire [4]. Resistance likely resulted from the earlier intensive use of DDT and selection from pyrethroid use in crop protection particularly in cotton areas [5,6]. *kdr* was initially shown to result from a point mutation (L1014F) in the pyrethroid target protein the voltage-gated sodium channel [7]. Based on a simple PCR diagnostic developed in the first report of the *kdr* mutation [7] several studies have been carried out on the distribution and the frequency of this mechanism throughout Africa. Initial studies showed that L1014F *kdr* was most widely distributed in West African *An. gambiae s.l.* populations [6,8,9]. This mutation was observed initially in the S molecular form of *An. gambiae* s.s. reaching high frequency but was not found either in sympatric mosquitoes of the M molecular form or *An. arabiensis* populations [5]. This provided further evidence of reproductive barrier between the M and S molecular forms [10,11] and the two molecular forms of *An. gambiae* s.s. were recently confirmed as two distinct species termed *Anopheles coluzzii* for the M form and *Anopheles gambiae* for the S form [12]. However, a few years after the initial finding of the *kdr* mutation in the S molecular form, this mutation was also reported in the M form from the littoral of Benin and Côte d'Ivoire [13]. In-depth investigations carried out later in these geographic regions confirmed that this phenomenon was frequently observed in littoral but was rare inland [11]. DNA sequencing of these mosquitoes suggested that the mutation emerged in the M form by genetic introgression from the S form [14,15]. In contrast, the emergence of the Leu-Phe *kdr* mutation within *Anopheles arabiensis* resulted from a *de novo* mutation event [15]. An extensive monitoring program in Burkina Faso has revealed that the L1014F *kdr* mutation initially detected in low frequency in the *An. gambiae* M molecular form and *An. arabiensis* [11,15] has spread throughout the country and is observed in mosquito populations at relatively high frequency [16,17]. Recently the L1014S *kdr*, which initially predominated in East Africa [18,19], was reported in West Africa, first in Benin and then Burkina Faso within *An. arabiensis* populations [20,21]. More recently this mutation was reported in a small number of individuals of the M and S forms of *An. gambiae* in Burkina Faso [22]. Taken together these results provide fundamental insight into the evolutionary processes underlying resistance in *Anopheles gambiae s.l.* Furthermore from an applied perspective, the emergence of resistance has significant implications for vector control programmes, especially those focused on the use of ITNs/Long-Lasting Insecticidal Nets (LLINs) or indoor residual sprayings (IRS). Although LLINs had shown good control of certain pyrethroid resistant populations [23] reduced efficacy of treated nets against *An. gambiae* populations with *kdr* resistance has since been reported [24].

Other insecticides belonging to the organophosphate (OP) and carbamate (CM) classes have been investigated to be used in mosaic, or in combination, with pyrethroids for bednet impregnation [25]. In addition to the use of LLINs, bendiocarb was recently used in IRS applications in West Africa through the President's Malaria Initiative (PMI) roadmap [26]. Initially described in *Culex* populations from Côte-d'Ivoire [27] reduced susceptibility to OPs and CMs was observed in *An. gambiae* populations in the North of Côte d'Ivoire and related to the domestic use of insecticide [28]. *An. gambiae* populations from Benin with resistance to the CM bendiocarb were reported after just three year of IRS use [29]. A common mechanism of resistance to OP and CM insecticides results from a single point mutation (termed *ace-1*^{R})in the target protein the acetylcholinesterase enzyme [30]. This mutation results in a glycine to serine replacement at amino acid position 119 and can be detected by a simple PCR-Restriction Fragment Length Polymorphism (RFLP) diagnostic [31]. This approach has been used to examine the frequency and distribution of this mutation in Burkina Faso where it was found predominately in the *An. gambiae* S form and in low frequency in the M form [9,16,32]. A recent study suggested that the mutation had introgressed from one form to the other but the precise origin of the introgression could not be determined due to the small sample size [33]. Since then, extensive country-wide surveys were performed in Burkina Faso from 2008 to 2010 and no case of *An. arabiensis* carrying this mutation was reported, although sample sizes for this species were sometimes small [16,17].

However insecticide resistance may also occur by other physiological mechanisms such as metabolic detoxification through increased enzyme activities (monooxygenases, esterases, or glutathione S- transferases) [34,35].

Burkina Faso is composed of three agro-climatic areas which exhibit different patterns of insecticide use especially in relation to crop protection. The present study provides an update on the distribution and the prevalence of the *kdr* L1014 and L1014S and*ace-1*^{R} mutations in *An. gambiae* s.l. populations throughout the 13 health regions dispersed across these different agro-climatic areas. We report here, for the first time, the occurrence of the *ace-1*^{R} mutation at remarkably high frequencies in *An. arabiensis*.

Materials and Methods

Study sites

Burkina Faso covers three ecological zones, the Sudan savannah zone in the south and west where rainfall is relatively heaviest (5–6 months), the arid savannah zone (Sudan-sahelian) which extends throughout much of the central part of the country and the aridland (Sahel) in the north. The northern part of the country has a dry season of 6–8 months. The varied ecological conditions are reflected in the different agricultural systems practiced throughout the country, from arable to pastoral lands. The western region constitutes the main cotton belt extending to the south where some new cotton areas have been cultivated since 1996. All ecological zones support the existence of *Anopheles* species that vector malaria and the disease is widespread throughout the country. Larvae were sampled from 15 sites dispersed throughout the three ecological zones (Table 1). The GPS coordinates were incorporated in Table 1.

Mosquito sampling

Larvae of *An. gambiae* s.l. were collected from at least 10 breeding sites dispersed throughout each sampling site mainly comprising pools of standing water and other small water collections. Larvae were pooled to constitute a colony, which was reared in the insectary to adulthood. A sample of 100 adult females were randomly sorted, killed and kept on silica gel in 1.5-ml tubes and stored at −20°C prior to PCR analysis. Anopheline species were identified morphologically using the standard identification keys of Gillies and Cootzee [36].

PCR analyses

An average of 30 mosquitoes was sampled per site by PCR analysis. Genomic DNA was extracted from single specimens and used as template for PCR to determine the species within the *An. gambiae* complex using the protocol SINE 200 of Santalomazza *et al.* [37] that allows the concomitant identification of *An. gambiae* M and S (respectively known as *Anopheles coluzzii* and *Anopheles gambiae*) and *An. arabiensis*. The same individuals were then tested for both the L1014F and L1014S *kdr* mutations using the protocols of Martinez-Torres *et al.* [7] (using specific primers Agd1, Agd2, Agd3 and Agd4) and Ranson *et al.* [18] (using Agd1, Agd2, Agd4 and Agd5) respectively:

- Agd1: 5′-ATAGATTCCCCGACCATG-3′;
- Agd2: 5′-AGACAAGGATGATGAACC-3′;
- Agd3: 5′-AATTTGCATTACTTACGACA-3′;
- Agd4: 5′-CTGTAGTGATAGGAAATTTA-5′;
- Agd5: 5′-TTTGCATTACTTACGACTG-3′.

The *ace-1*^{R} mutation was detected from the same samples by PCR according to the protocol of Weill *et al.* [31] using specific primers *Ex3AGdir* (GATCGTGGACACCGTGTTCG) and *Ex3-AGrev* (AGGATGGCCCGCTGGAACAG). Then the PCR products were digested using *Alu 1* enzyme at 37°C for 3 hours.

Statistical analysis

Data were compared between ecological zones and pooled for each species to compare the genotypes frequency between *An. gambiae* species by Chi^2 tests. The genotypic frequencies of L1014F and L1014S and *ace-1*^{R} in mosquito populations were compared to Hardy-Weinberg expectations using the exact test procedures implemented in GenePOP (ver.3.4) software [38].

Table 1. Distribution of *Anopheles gambiae s.l.* from 15 sites in Burkina Faso.

Study sites	Geographic references	Social environment	Climatic areas	Agricultural practices	Date of collection	*An. gambiae* s.l.	*An. gambiae*		*An. coluzzii*		*An. arabiensis*	
						N	n1	%	n2	%	n3	%
Gaoua	10°40′N; 3°15′W	sub-urban	Sudanian	cereals, cotton, old area	30/10/2012	43	39	90,69	1	32,33	3	6,98
Banfora	10°40′N; 3°15′W	sub-urban	Sudanian	cereals, cotton, old area	09/07/2012	30	24	80,00	6	20,00	0	0
Sindou	10°40′N; 3°15′W	rural	Sudanian	cotton, old area	01/10/2012	35	24	68,57	6	17,14	5	14,29
Orodara	10°40′N; 3°15′W	sub-urban	Sudanian	fruits, cotton, old area	23/19/2012	28	23	82,14	4	14,29	1	3,57
Dioulassoba	10°40′N; 3°15′W	traditional-urban	Sudanian	swamp	23/11/2012	29	4	13,79	5	17,24	20	68,97
Soumousso	10°40′N; 3°15′W	rural	Sudanian	cotton, old area	30/12/2012	30	20	66,67	3	10,00	7	23,33
Boromo	10°40′N; 3°15′W	sub-urban	Sudan-sahelian	cotton, old area	08/10/2012	33	16	48,48	0	0	17	51,52
Dédougou	10°40′N; 3°15′W	sub-urban	Sudan-sahelian	cotton, old area	06/10/2012	30	12	40,00	2	6,67	16	53,33
Koudougou	10°40′N; 3°15′W	urban	Sudan-sahelian	cotton, since 1996	07/11/2012	37	19	51,35	5	13,51	13	35,14
Nanoro	10°40′N; 3°15′W	rural	Sudan-sahelian	cereals	09/07/2012	32	4	12,50	24	75,00	4	12,50
Koupela	10°40′N; 3°15′W	sub-urban	Sudan-sahelian	cotton since 1996	06/10/2012	30	14	46,67	8	26,67	8	26,67
Fada	10°40′N; 3°15′W	sub-urban	Sudan-sahelian	cotton since 1996	25/08/2012	60	19	31,67	27	45,00	14	23,33
Kaya	10°40′N; 3°15′W	sub-urban	Sahelian	cereals, vegetables	03/10/2012	32	15	46,88	5	15,63	12	37,50
Ouahigouya	10°40′N; 3°15′W	sub-urban	Sahelian	cereals, vegetables	08/10/2012	31	20	64,52	10	32,26	1	3,23
Dori	10°40′N; 3°15′W	sub-urban	Sahelian	cereals, vegetables	01/10/2012	33	12	36,36	5	15,15	16	48,48

N: number total of mosquitoes.
n1: number of *An. gambiae.*
n2: number of *An. coluzzii.*
n3: number of *An. arabiensis.*

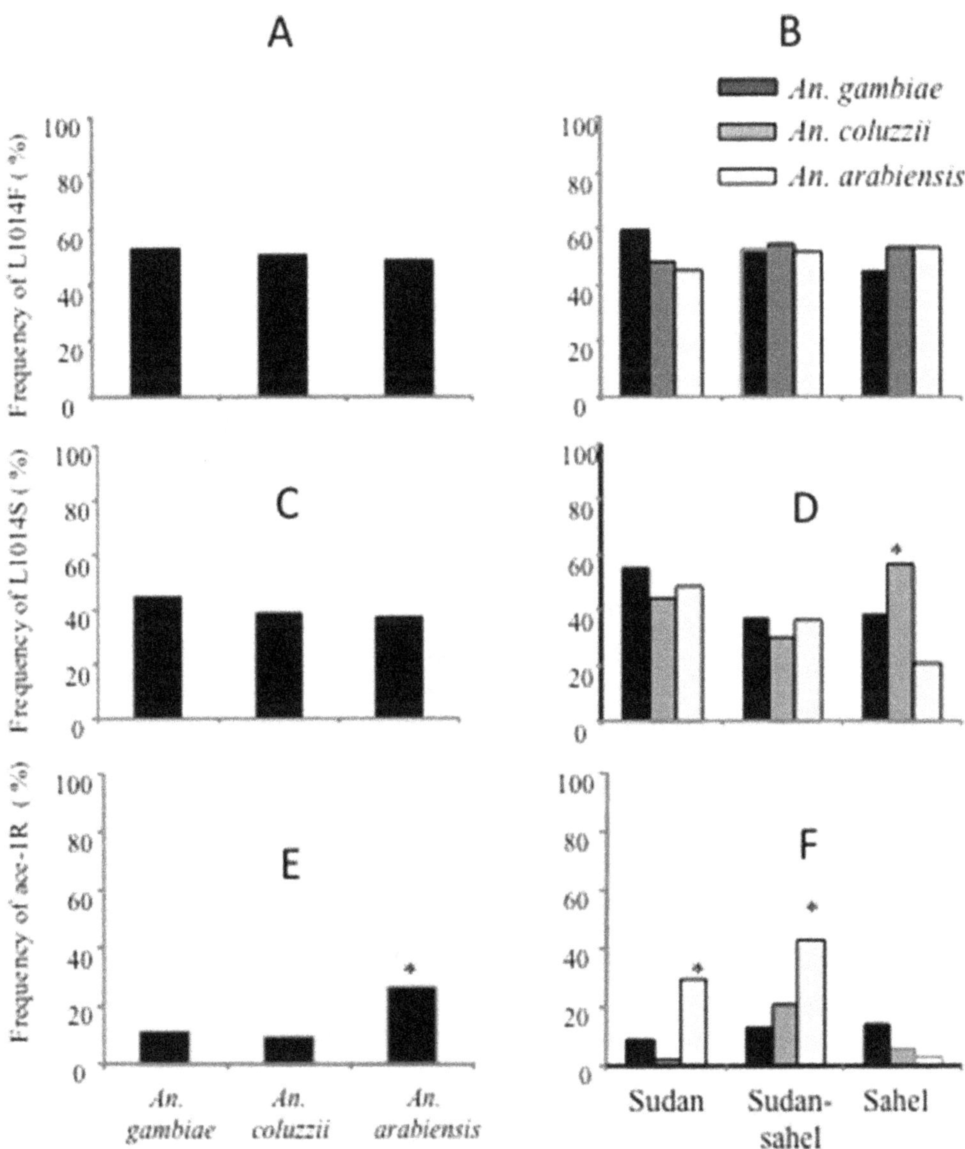

Figure 1. Comparison of allele frequencies of 1014F, 1014S and *ace-1^R* mutations within *Anopheles gambiae, An. coluzzii* and *An. arabiensis* populations from 15 sites dispersed across the 3 agro-ecological regions of Burkina Faso.

Ethical issues

Ethical approval was not required in this study.

This study was not carried out on private land. For each, no permission was required our study does not degrade the environment. No permission was required for these locations/ activities as the field activities did not involve damaged of protected species. We did not use any vertebrate during this study.

Results

Out of 516 mosquitoes analysed in PCR, 513 successfully scored (less than 5% failure rate). Overall species composition of the collected mosquitoes comprised a higher proportion of *An. gambiae* (51.7%) than *An. coluzzii* (21.6%) and *An. arabiensis* (26.7%) (Table 1). The species repartition across the three ecological regions revealed that *An. gambiae* was the predominant species in all regions including, in the Sahel where it comprised more than 49% of the *An. gambiae s.l.* population. *Anopheles arabiensis* was the second most predominant vector found in samples collected from the three regions. Somewhat *An. coluzzii* was found at a relatively low proportion of less than 15%. The central areas were characterised by an overlapped repartition of the three species 38.4%, 27.81% and 33.75% for *An. gambiae, An. coluzzii* and *An. arabiensis* respectively and proportions did not differ significantly ($\chi^2 = 1.95$, df = 1, $P > 0.05$). In the Sahel region, *An. gambiae* also predominated (49.75%) and the proportions of the two other species did not differ significantly at 21.01% and 29.74% for *An. coluzzii* and *An. arabiensis* respectively ($\chi^2 = 4.88$, df = 1, $P > 0.05$).

The overall frequency of the L1014F mutation averaged 50% and did not significantly differ between species (Figure 1A) whatever the ecological zone (Figure 1B) ($\chi^2 = 0.14$, df = 1, $P > 0.05$) even though the highest values were observed in the sudan zone (Figure 2). However some deviation from Hardy-Weinberg expectations was observed within the *An. arabiensis* populations in Dedougou and Dori and within *An. coluzzii* populations in Fada,

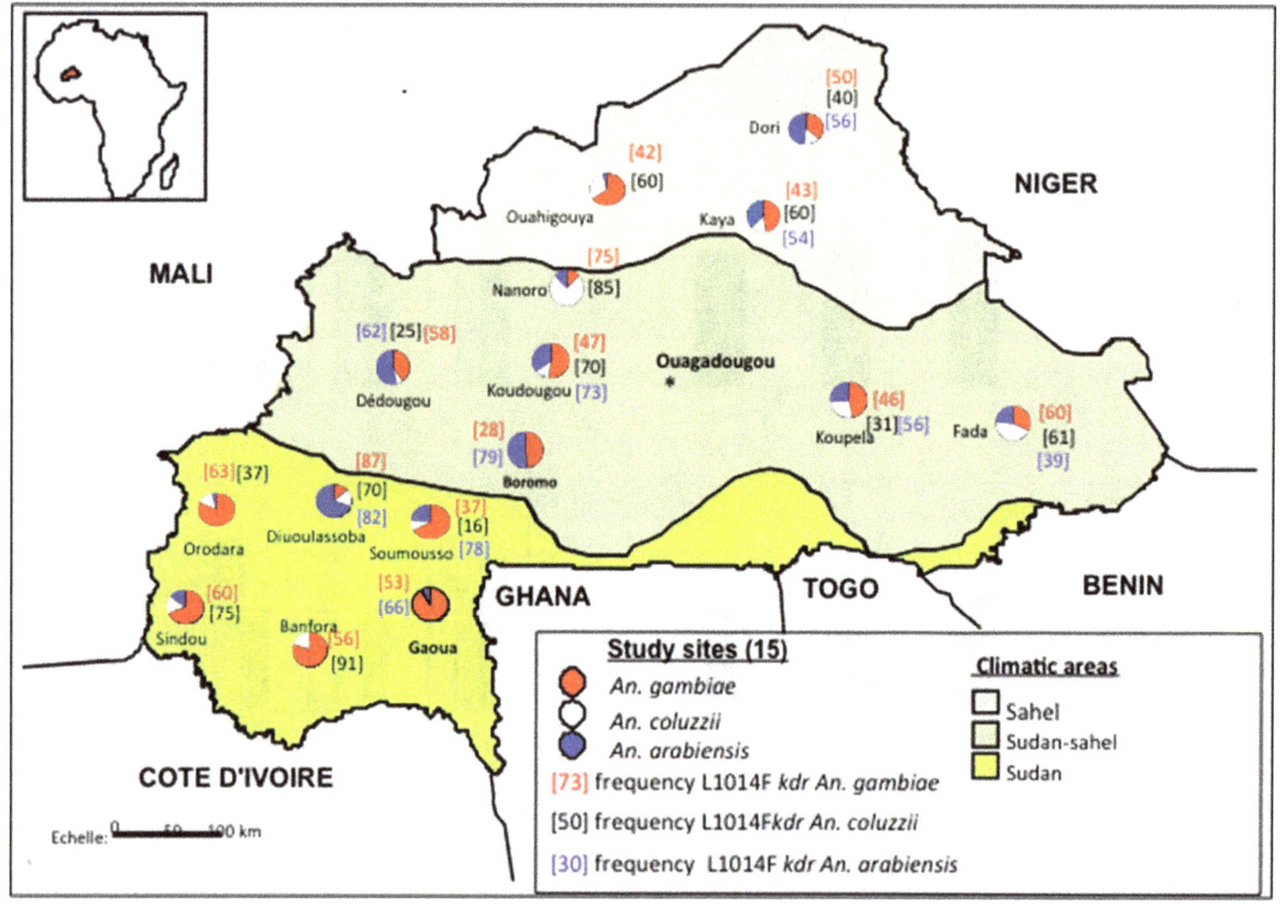

Figure 2. Distribution the 1014F *kdr* allele frequency from 15 sites dispersed across Burkina Faso.

Kaya, Ouahigouya and Dori with an excess of resistant homozygous alleles (Table 2). The same patterns were found in seven sites for *An. gambiae* (Gaoua, Banfora, Sindou in the West, Dedougou, Koudougou and Koupela in the central region and Ouahigouya in the Sahel) ($P<0.05$).

The overall allele frequency of the L1014S *kdr* mutation (Figure 3) was relatively higher in *An. gambiae* (48%) followed by *An. coluzzii* (38%) and *An. arabiensis* populations (37%) with no significant difference between the last two ($\chi^2=3.24$, df = 1, $P>0.05$) (Figure 1C). Comparing between ecological regions, L1014S *kdr* frequency did not differ significantly between species, except in the Sahel where it was significantly higher in *An. coluzzii* than *An. arabiensis* ($\chi^2=10.21$, df = 1, $P<0.001$) and *An. gambiae* ($P<0.04$) (Figure 1D). The observed genotypic frequencies were not significantly different from Hardy-Weinberg expectations at the 95% confidence level (Table 2) in populations from any site except in the *An. gambiae* populations from Orodara, Soumousso, Koupela, Fada, and in the *An. arabiensis* populations from Dioulassoba and Kaya where a heterozygous deficit was observed ($P=0.005$) and *An. gambiae* populations in two sites (Dedougou and Kaya) where an excess of heterozygotes was observed ($P<0.05$).

The *ace-1*^{R} mutation (Figure 4) was recorded in all the 15 sites under study with a wider distribution within the *An. gambiae* populations (Table 3). The overall allele frequency of *ace-1*^{R} was significantly higher in *An. arabiensis* (0.26) than in *An. gambiae* (0.11) ($\chi^2=14.4$; df = 1, $P=0.001$) and *An. coluzzii* (0.09) ($\chi^2=11.77$, df = 1, $P=0.006$) (Figure 1E) with no significant difference between the last two ($\chi^2=0.37$, df = 1, $P=0.54$). Compared between zones, the *ace-1*^{R} allele frequency in *An. arabiensis* was higher than that of *An. coluzzii* ($\chi^2=8.15$, df = 1, $P=0.004$) and *An. gambiae* ($\chi^2=9.79$, df = 1, $P<0.001$) in the Sudan and Sudan-sahelian savannah (with respectively $\chi^2=6.89$, df = 1, $P<0.008$ and $\chi^2=17.34$, df = 1, $P<0.0003$) (Fig. 1F). In the Sahel no significant difference was observed between the three species ($\chi^2=0.89$–0.021, df = 1, $P>0.05$). The observed genotypic frequencies were significantly different from Hardy-Weinberg expectations at the 95% confidence level (Table 3) in *An. gambiae* population from Orodara, Soumousso, Koudougou, Fada, Ouahigouya, Dori and Dioulassoba, Koudougou and Kaya for *An. arabiensis* where a heterozygote deficit was observed ($P=0.005$). Furthermore, the percentage of homozygous resistant individuals was significantly higher in *An. arabiensis* (25%) than in *An. gambiae* (6.25%). No homozygous resistant individual was recorded in *An. coluzzii* from any site.

Discussion

This study provides current information on the distribution of three members of the *Anopheles gambiae* complex across Benin and the frequency and distribution of three important target-site resistance mechanisms in these populations. In regards to the distribution of *An. gambiae* species throughout the country, the most significant finding is that *An. arabiensis* appears to be spreading in the Sudan whereas in the past it comprised only

Table 2. Allelic and genotypic frequencies at the *kdr* 1014F and 1014S locus in *An. gambiae s.l* populations.

Species	Sites	N	Genotypes						Genotypes				
			1014L 1014L	1014L 1014F	1014F 1014F	f(L1014F)	[95%CI]	p(HW)	1014L 1014L	1014L 1014F	f(L1014F)	[95%CI]	p(HW)
An. arabiensis	Gaoua	5	1	0	2	0.66	[8.5–9.82]	-	0	2	0.66	[8.5–9.82]	0.2000
	Banfora	0	0	0	0	-	-	-	0	0	0.9	-	-
	Sindou	10	5	0	0	0	-	-	1	4	0	[7.38–9.18]	-
	Orodara	1	1	0	0	0	-	0.4678	0	0	0.45	-	-
	Dioulassoba	30	1	5	14	0.82	[3.13–4.71]	0.2308	2	8	0.42	[2.34–3.38]	0.0003
	Soumousso	11	1	1	5	0.78	[5.74–7.3]	0.0956	2	2	0.37	[4.37–5.21]	0.2914
	Boromo	17	2	3	12	0.79	[3.42–5.00]	0.000	6	3	0.28	[2.31–3.35]	0.3405
	Dédougou	23	6	0	10	0.62	[3.23–4.47]	0.1652	5	2	0	-	0.3213
	Koudougou	13	2	3	8	0.73	[3.9–5.36]	-	0	0	0.5	[6.41–7.41]	-
	Nanoro	6	4	0	0	0	-	0.4406	0	2	0.5	[4.39–5.39]	0.0857
	Koupela	13	2	3	3	0.56	[4.61–5.73]	0.2970	2	3	0.53	[4.39–5.39]	0.1795
	Fada	25	6	5	3	0.39	[2.87–3.65]	0.0933	8	3	0.57	[3.42–4.48]	0.9035
	Kaya	17	4	3	5	0.54	[3.61–4.69]	-	1	4	0.37	[3.07–3.81]	0.0061
	Ouahigouya	1	0	1	0	0.5	[3.32–4.32]	0.0031	0	0	0	[18.5–20.5]	-
	Dori	22	6	2	8	0.56	[3.1–4.22]	-	4	2	0.26	[2.32–2.84]	0.2260
An. coluzzii	Gaoua	1	1	0	0	0	[18.5–20.5]	-	0	0	0	[18.5–20.5]	-
	Banfora	7	0	1	5	0.91	[6.69–8.51]	-	0	1	0.16	[3.04–3.36]	0.0909
	Sindou	12	1	1	4	0.75	[6.15–/.65]	0.2727	1	5	0.91	[6.69–8.51]	-
	Orodara	5	2	1	1	0.37	5.58–6.32]	0.4286	1	0	0.12	[3.27–3.51]	-
	Dioulassoba	9	1	1	3	0.7	[6.61–8.01]	0.3333	1	3	0.7	[6.61–8.01]	0.3333
	Soumousso	4	2	1	0	0.16	[4.36–4.68]	-	0	1	0.33	[6.16–6.82]	0.2000
	Boromo	0	0	0	0	-	-	-	0	0	-	-	-
	Dédougou	3	1	1	0	0.25	[6.67–7.17]	-	0	1	0.5	[9.28–10.28)	0.6190
	Koudougou	7	0	3	2	0.7	[6.6–8.01]	1	2	0	0.2	[3.72–4.12]	-
	Nanoro	39	1	5	18	0.85	[2.82–4.52]	0.3983	12	3	0.37	[2.06–2.8]	0.3333
	Koupela	9	3	5	0	0.31	[3.54–4.16]	1	1	0	0.06	[1.64–1.76]	0.7446
	Fada	46	7	7	13	0.61	[2.33–3.55]	0.0186	17	2	0.38	[1.94–2.7]	0.0817
	Kaya	8	2	0	3	0.6	[6.17–7.37]	0.0476	2	1	0.4	[5.13–5.93]	0.3333
	Ouahigouya	17	4	0	6	0.6	[4.19–5.39]	0.0017	2	5	0.6	[4.19–5.39]	1
	Dori	9	3	0	2	0.4	[5.13–5.93]	0.0476	1	3	0.7	[6.61–8.01]	-
An. gambiae	Gaoua	74	14	8	17	0.53	[3.75–2.81]	0.0002	0	35	0.92	[2.12–3.96]	1
	Banfora	29	7	7	10	0.56	2.43–3.55]	0.0434	3	2	0.14	[1.36–1.64]	0.1518
	Sindou	46	8	3	13	0.6	[2.49–3.69]	0.0003	5	17	0.81	[2.78–4.4]	0.0611

Table 2. Cont.

Species	Sites	N	Genotypes						Genotypes				
			1014L 1014L	1014L 1014F	1014F 1014F	f(L1014F)	[95%CI]	p(HW)	1014L 1014L	1014L 1014F	f(L1014F)	[95%CI]	p(HW)
	Orodara	33	5	7	11	0.63	[2.6–3.86]	0.0904	1	9	0.41	[2.2–3.02]	0.0420
	Dioulassoba	8	0	1	3	0.87	[8.239.97]	-	2	2	0.75	[7.71–9.21]	0.3257
	Soumousso	29	8	9	3	0.37	[2.29–3.63]	0.5690	5	4	0.32	[2.16–2.8]	0.0000
	Boromo	25	8	7	1	0.28	[2.31–2.87]	0.7912	4	5	0.43	[2.78–3.64]	0.1201
	Dédougou	19	5	0	7	0.58	[3.72–4.88]	0.0004	7	0	0.29	[2.75–3.33]	0.0150
	Koudougou	26	9	2	8	0.47	[2.61–3.55]	0.0005	4	3	0.26	[2.03–2.55]	1
	Nanoro	5	1	0	3	0.75	[7.71–9.21]	0.1429	0	1	0.25	[4.64–5.14]	0.1429
	Koupela	24	7	1	6	0.46	[3.08–4.00]	0.0013	4	6	0.57	[3.37–4.51]	0.0003
	Fada	30	3	9	7	0.6	[2.87–4.07]	0.6254	5	6	0.44	[2.54–3.42]	0.0473
	Kaya	19	5	7	3	0.43	[2.88–3.74]	0.5785	3	1	0.16	[1.86–2.18]	0.0000
	Ouahigouya	30	10	3	7	0.42	[2.4–3.25]	0.0020	2	8	0.45	[2.48–3.38]	0.0632
	Dori	18	4	4	4	0.5	[3.49–4.49]	0.2300	1	5	0.55	[4.03–5.13]	0.0520

N: number of mosquitoes.
f(1014F): frequency of the kdr W resistant allele.
f(1014S): frequency of the kdr E resistant allele.
p(HW): probability of the exact test for goodness of fit to Hardy Weinberg equilibrium.
-: not determined.

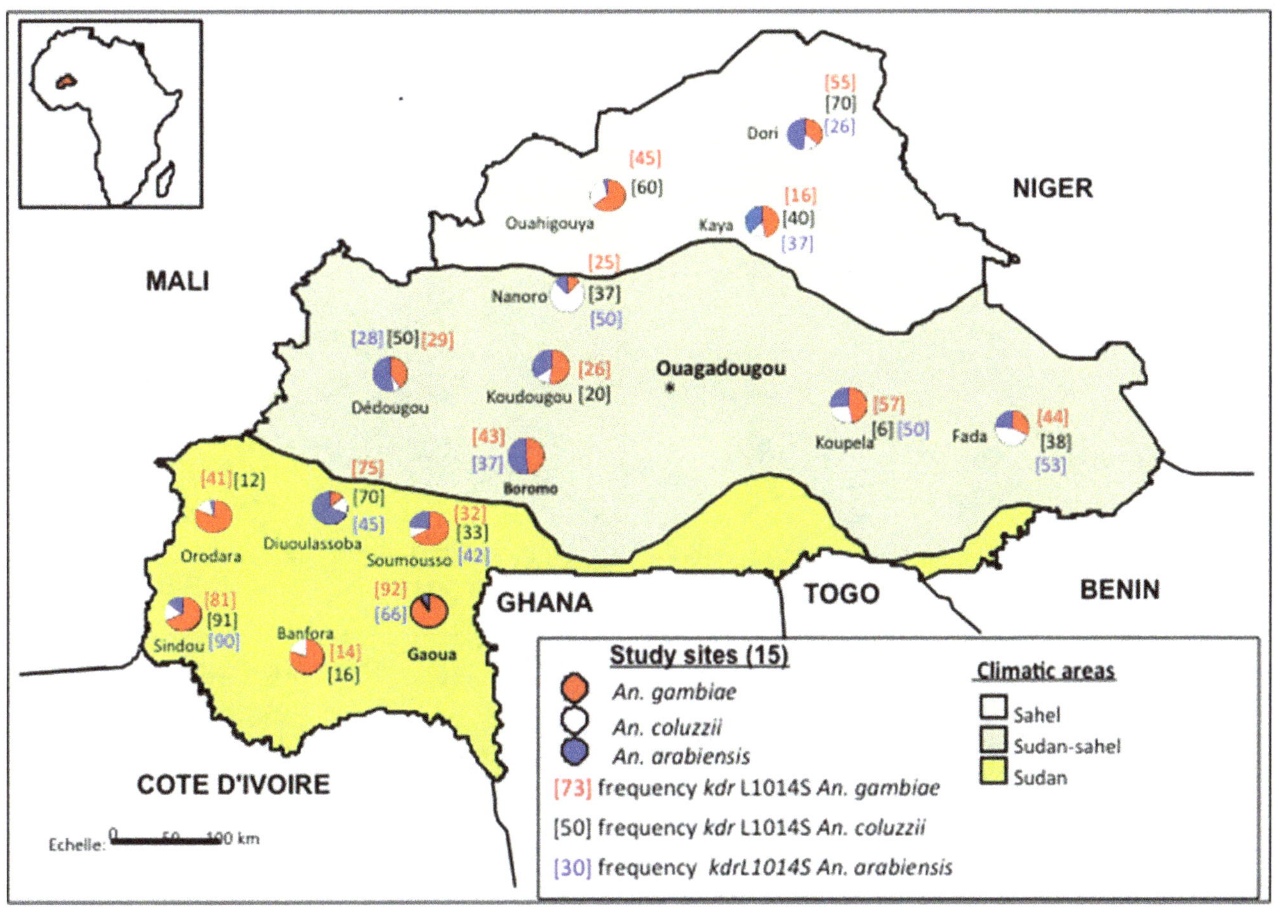

Figure 3. Distribution the 1014S *kdr* allele frequency from 15 sites dispersed across Burkina Faso.

around 5% of the *An. gambiae* complex species [6]. Furthermore, this species is now present in Sindou at 14.29% (nearest the frontier of Cote-d'Ivoire) where it was absent a decade ago [9]. The reason for this is not clear but could be related to climatic changes, such as irregularities in rainfall observed in the boundaries of the Sudan region that may make the landscape more favourable to the establishment of this species.

Across sampling covering 15 sites we identified the L1014F and L1014S *kdr* mutations concomitant with the *ace*-1 G119S mutation confirming the presence of multiple resistance mechanisms in the *An. gambiae* complex in Burkina Faso [16,17]. The distribution and the prevalence of the L1014F *kdr* mutation in *An. gambiae* species including *An. gambiae, An. coluzzii* and *An. arabiensis,* has been well documented in Burkina Faso for over a decade [9,16]. Many studies reported this mutation at high frequency within *An. gambiae* and *An. coluzzii* populations especially in *An. gambiae* populations from the Sudan area where mutation frequency was approaching fixation [9,15,16]. Over recent years the frequency of this mutation has increased within both *An. coluzzii* and *An. arabiensis.* In this study although the L1014F mutation remains widespread in all three ecological regions and is present at relatively high frequency within the three species (averaging 50%), the frequencies reported in this current study were lower in the Sudan ecological regions (West and South West covering the old cotton belt) than those from previous studies [9,16,22]. For the other climatic zones i.e. central and northern regions the allele frequencies of L1014F varied within the three species with particularly high frequencies in *An. arabiensis.* The reason(s) for the reduction of L1014F frequency in *An. gambiae* populations in the Sudan area is not known, however, a similar trend was recently observed in the Western region of Burkina Faso where transgenic and biological control practices have been implemented for crop protection of cotton over the last four years (a long side conventional crop protection approaches) (Namountougou, unpublished). These alternative cotton-growing practices would be expected to reduce the quantity and frequency of insecticide use in agriculture and this may in turn reduce the selection pressure experienced by local mosquito populations. The analysis of observed genotypic frequencies revealed a heterozygote deficit for the L1014F mutation in the three species of *An. gambiae* s.l. from many sites especially in the Sahel for *An. coluzzii* and *An. arabiensis* and in the Sudan and Sudan-Sahel for *An. gambiae* which deviated significantly from Hardy-Weinberg expectations. This finding is not surprising as the same patterns were observed in the West (Orodara and Soumousso) four years ago [9] in combination with a novel mutation, N1575Y, in the voltage-gated sodium channel, recently reported in *An. gambiae* s.l. populations in Soumousso [39].

The L1014S *kdr* mutation was recently recorded at highest frequency in *An. arabiensis* populations in the centre on the country [21] and in Bobo-Dioulasso at frequencies averaging 38% [40]. Previous studies have recorded only a few individuals of *An. gambiae* and *An. coluzzii* from the Centre-East part of the country [17] carrying this mutation in the heterozygous form. The present

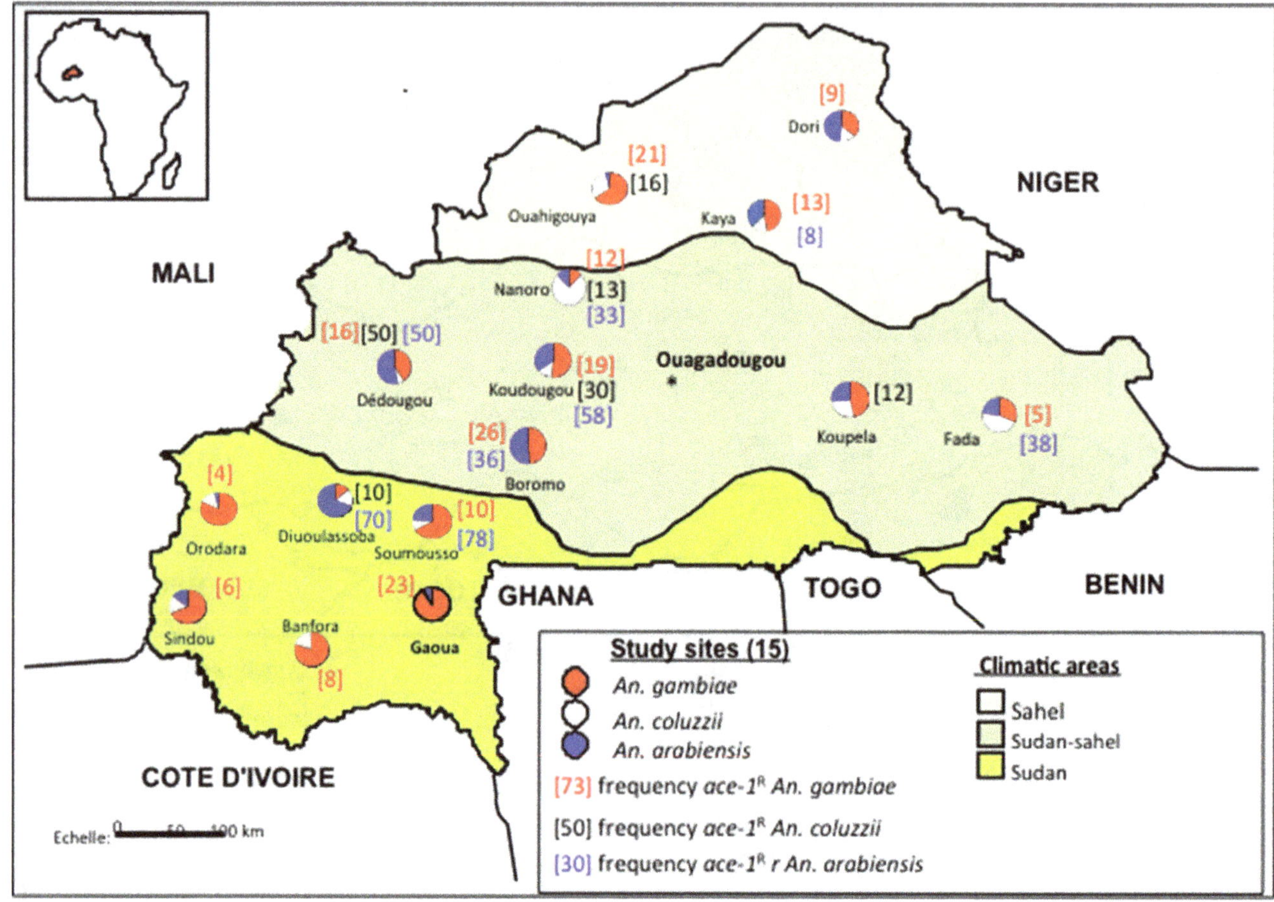

Figure 4. Distribution the *ace-1*^{R} allele frequency from 15 sites dispersed across Burkina Faso.

study reveals that this mutation has since spread across the whole country and is now observed at relatively high and similar frequencies (40%) between the three species. The comparison of the observed genotypic frequencies of this mutation with that expected for Hardy-Weinberg equilibrium indicated, depending on the site, a deficit or excess of heterozygotes, mainly for *An. gambiae* populations. The occurrence of the L1014F *kdr* mutation in *An. coluzzii* had been suggested to have occurred by introgression from *An. gambiae* and via a *de novo* mutation event in *An. arabiensis* [15], however, the origin of the L1014S mutation in *An. gambiae*, *An. coluzzii* and *An. arabiensis* species in West Africa is not so clearly understood. The proximity of Burkina Faso from the Benin frontier where the L1014S mutation was first reported in *An. arabiensis* populations [20] suggests that it arrived in Burkina Faso via migration of *An. arabiensis* carrying the mutation from Benin, however, the origin of this mutation in *An. gambiae* and *An. coluzzii* populations in Burkina Faso remains to be elucidated.

In this study we report, for the first time, the presence of the *ace.1* G119S mutation in *An. arabiensis* populations from eight sites: Dioulassoba, Soumousso in the West, Boromo, Dédougou, Koudougou, Nanoro and Fada in the Centre-North and East and Kaya in the North. In these sites *An. arabiensis* was observed as the second major vector after *An. gambiae* except at Fada and Nanoro where the proportion of *An. arabiensis* was lower than that of *An. coluzzii*. To confirm this finding, we repeated the PCR amplification of *ace.1*^{R} for our *An. arabiensis* specimens and used, as a control, 30 specimens of *An. Arabiensis* which we had confirmed in a previous study do not have this mutation. No false positives were observed in these samples suggesting our data is robust. The *ace.1*^{R} allele was observed in this study in *An. arabiensis* at varying frequency reaching a maximum value of 78% in populations from Dioulassoba and the lowest value in Kaya at 8%. Except for samples from Soumousso and Nanoro where the sample size was not sufficient ($n<10$) to compare genotype frequencies, deviations from Hardy-Weinberg equilibrium were observed at three sites (Dioulassoba, Koudougou and Kaya) as a result of a high heterozygote deficit. The same pattern was observed in *An. gambiae* from Orodara, Soumousso, Koudougou, Fada, Ouahigouya and Dori. The deficit of heterozygous genotypes observed in Orodara and Soumousso is not new as Dabiré *et al.* [41] reported similar results from the these areas from which the duplicated allele (*ace.1*^{D}) was reported by Djogbenou *et al.* [33]. It is possible that this duplicated allele *ace.1*^{D} is also present within *An. arabiensis* especially in Dioulassoba where the proportion of homozygous mutants was atypically high (60%). The high frequency of this mutation in Dioulassoba populations is intriguing as recent studies failed to find any L1014F *kdr* or *ace-1*^{R} in *An. arabiensis* population from this site [40,42]. As for the L1014S mutation, additional sequence analysis of the region flanking the *ace.1* locus are necessary to confirm whether the *ace.1* mutation in *An. arabiensis* has evolved along the same pathway as *kdr* e.g. as a *de novo* mutation or introgression from *An. gambiae* or *An. coluzzii*. Unfortunately our PCR data is not backed up by

Table 3. Allelic and genotypic frequencies at the ace-1 locus in *An. gambiae s.l* populations from 15 sites in Burkina Faso.

Species	Sites	N	Genotypes			f(119S)	[95%CI]	p(HW)
			119G 119G	119G 119S	119S 119S			
An. arabiensis	Gaoua	3	3	0	0	0	-	-
	Banfora	0	0	0	0	-	-	-
	Sindou	5	5	5	0	0	-	-
	Orodara	1	1	1	0	0	-	-
	Dioulassoba	20	4	4	12	0.7	[2.95–7.13]	0.0264
	Soumousso	7	1	1	5	0.78	[5.74–7.57]	0.2308
	Boromo	15	5	9	1	0.36	[2.67–5.42]	0.9488
	Dédougou	14	4	6	4	0.5	[3.19–7.25]	0.0444
	Koudougou	12	5	0	7	0.58	[3.72–9.1]	0.0004
	Nanoro	3	2	0	1	0.33	[6.16–17.45]	0.2000
	Koupela	8	8	0	0	0	-	-
	Fada	13	4	8	1	0.38	[2.96–6.26]	0.9449
	Kaya	12	11	0	1	0.08	[1.52–2.27]	0.0435
	Ouahigouya	1	1	0	0	0	-	-
	Dori	14	14	0	0	0	-	-
An. coluzzii	Gaoua	1	1	0	0	0	-	-
	Banfora	6	6	0	0	0	-	-
	Sindou	6	6	0	0	0	-	-
	Orodara	4	4	0	0	0	-	-
	Dioulassoba	5	4	1	0	0.1	[2.67–4.71]	-
	Soumousso	3	3	0	0	0	-	-
	Boromo	0	0	0	0	-	-	-
	Dédougou	2	0	0	2	0.5	[9.28–34.65]	1
	Koudougou	5	2	3	0	0.3	[4.49–10.78]	1
	Nanoro	23	17	6	0	0.13	[1.34–2.04]	1
	Koupela	8	6	2	0	012	[2.28–3.9]	1
	Fada	27	27	0	0	0	-	-
	Kaya	5	5	0	0	0	-	-
	Ouahigouya	9	6	3	0	0.16	[2.64–4.39]	1
	Dori	5	5	0	0	0	-	-
An. gambiae	Gaoua	36	22	11	3	0.23	[1.33–2.2]	0.2811
	Banfora	24	20	4	0	0.08	[1.05–1.46]	1
	Sindou	24	21	3	0	0.06	[0.92–1.23]	1
	Orodara	23	22	0	1	0.04	[0.74–0.99]	0.0222

Table 3. Cont.

Species	Sites	N	Genotypes			f(119S)	[95%CI]	p(HW)
			119G 119G	119G 119S	119S 119S			
	Dioulassoba	4	4	0	0	0	-	-
	Soumousso	20	18	0	2	0.1	[1.29–1.88]	0.0021
	Boromo	15	9	4	2	0.26	[2.32–4.31]	0.2260
	Dédougou	12	8	4	0	0.16	[2.1–3.59]	1
	Koudougou	18	14	1	3	0.19	[1.82–3.07]	0.0029-
	Nanoro	4	3	1	0	0.12	[3.27–6.29]	-
	Koupela	12	12	0	0	0	-	-
	Fada	19	18	0	1	0.05	[0.96–1.27]	0.0270
	Kaya	15	11	4	0	0.13	[1.69–2.62]	1
	Ouahigouya	19	14	2	3	0.21	[1.85–3.16]	0.0096
	Dori	11	10	0	1	0.09	[1.68–2.59]	0.0476

N: number of mosquitoes.
f(119S): frequency of the 119S resistant ace.1 allele.
p(HW): probability of the exact test for goodness of fit to Hardy Weinberg equilibrium.
-: not determined.

insecticide susceptibility bioassays and so we cannot assess the correlations between *kdr* and *ace*-1 mutations and the phenotypic expression of resistance.

The emergence of the *ace-1*^{R} mutation in *An. gambiae* s.l. population from the cotton-growing areas may be linked to the agricultural use of OP and CM insecticides used for crop protection. Other sources of selection pressure outside the cotton belt include insecticide use for vegetable growing and domestic use of insecticide in public health. Bioassays performed in 2012 on *An. gambiae* populations from sites located in the cotton belt of the West of Burkina Faso revealed the development of resistance to CMs and OPs especially to benidocarb (Dabiré, unpublished) correlating with the prevalence and frequency of genetic resistance revealed in the present study. However, further bioassays on a wider scale are now required in order to understand the implications of the current status of the *ace-1*^{R} mutation for the efficacy of OP and CM insecticides in vector control in Burkina Faso. The information provided by such studies combined with the genetic data presented here is a prerequisite for the informed use of CM and OP based-combinations for bednet impregnation and/or indoor residual spraying.

Acknowledgments

This work was supported by the National Malaria Control Program (NMCP) of Burkina Faso.

Author Contributions

Conceived and designed the experiments: RKD AD PC. Performed the experiments: DDS JB HKT. Analyzed the data: RKD MN. Wrote the paper: RKD CB. Supervised field work: MN. Revised the manuscript: MN AD CB. Performed PCR analyses: DDS JB HKT. Assured the financial support of the study through the Ministry of Health: CB. Read and approved the final version of the manuscript: RKD MN AD DDSJB HKT CB PC.

References

1. Carnevale P, Robert V, Boudin C, Halna JM, Pazart L, et al. (1998) La lute contre le paludisme par des moustiquaires imprégnées de pyrthrinoides au Burkina Faso. Bull Soc Pathol Exot 81: 832–846.
2. D'Alessandro U, Olaleye BO, McGuire W, Langerock P, Bennett S, et al. (1995) Mortality and morbidity from malaria in Gambian children after introduction of an impregnated bednet programme. Lancet 345: 479–483.
3. Binka FN, Kubaje A, Adjuik M, Williams LA, Lengeler C, et al. (1996) Impact of permethrin impregnated bednets on child mortality in Kassena-Nankana district, Ghana: a randomized controlled trial. Trop Med Int Health 1: 147–154.
4. Elissa N, Mouchet J, Riviere F, Meunier JY, Yao K (1993) Resistance of Anopheles gambiae s.s. to pyrethroids in Cote d'Ivoire. Ann Soc Belg Med Trop 73: 291–294.
5. Chandre F, Darrier F, Manga L, Akogbeto M, Faye O, et al. (1999) Status of pyrethroid resistance in *Anopheles gambiae* sensu lato. Bull World Health Organ 77: 230–234.
6. Diabate A, Baldet T, Chandre F, Akogbeto M, Guiguemde TR, et al. (2002) The role of agricultural use of insecticides in resistance to pyrethroids in *Anopheles gambiae* s.l. in Burkina Faso. Am J Trop Med Hyg 67: 617–622.
7. Martinez-Torres D, Chandre F, Williamson MS, Darriet F, Berge JB, et al. (1998) Molecular characterization of pyrethroid knockdown resistance (*kdr*) in the major malaria vector *Anopheles gambiae s.s.* Insect Mol Biol 7: 179–184.
8. Awolola TS, Oyewole IO, Amajoh CN, Idowu ET, Ajayi MB, et al. (2005) Distribution of the molecular forms of *Anopheles gambiae* and pyrethroid knock down resistance gene in Nigeria. Acta Tropica 95: 204–209.
9. Dabire KR, Diabate A, Namountougou M, Toe KH, Ouari A, et al. (2009a) Distribution of pyrethroid and DDT resistance and the L1014F *kdr* mutation in *Anopheles gambiae* s.l. from Burkina Faso (West Africa). Trans R Soc Trop Med Hyg 103: 1113–1120.
10. Favia G, Lanfrancotti A, Spanos L, Siden Kiamos I, Louis C (2001) Molecular characterization of ribosomal DNA polymorphisms discriminating among chromosomal forms of *Anopheles gambiae* s.s. Insect Mol Biol 10: 19–23.
11. Diabate A, Baldet T, Chandre C, Dabire KR, Kengne P, et al. (2003) KDR mutation, a genetic marker to assess events of introgression between the molecular M and S forms of *Anopheles gambiae* (Diptera: Culicidae) in the tropical savannah area of West Africa. J Med Entomol 40: 195–198.
12. Coetzee M, Hunt R, Wilkerson R, Della Torre A, Coulibaly BM, et al. (2013) *Anopheles coluzzii* and *Anopheles amharicus*, new members of the *Anopheles gambiae* complex. Zootaxa 3619: 246–274.
13. Fanello C, Akogbeto M, della Torre A (2000) Distribution of the knock down resistance gene (kdr) in *Anopheles gambiae s.l.* from Benin. Trans R Soc Trop Med Hyg 94: 132.
14. Weill M, Chandre F, Brengues C, Manguin S, Akogbeto M, et al. (2000) The kdr mutation occurs in the Mopti form of *Anopheles gambiae s.s.* through introgression. Insect Molecular Biology 9: 451–455.
15. Diabate A, Brengues C, Baldet T, Dabire KR, Hougard JM, et al. (2004) The spread of the Leu-Phe *kdr* mutation through *Anopheles gambiae* complex in Burkina Faso: genetic introgression and de novo phenomena. Trop Med Int Health 9: 1267–1273.
16. Dabiré KR, Diabaté A, Namountougou M, Djogbenou L, Wondji C, et al. (2012a) Trends in Insecticide Resistance in Natural Populations of Malaria Vectors in Burkina Faso, West Africa: 10 Years' Surveys. In: Perveen F, editors. Insecticides - Pest Engineering. ISBN. InTech: 479–502.
17. Namountougou M, Diabate A, Etang J, Bass C, Sawadogo SP, et al. (2013) First report of the L1014S kdr mutation in wild populations of *Anopheles gambiae* M and S molecular forms in Burkina Faso (West Africa). Acta Tropica 125: 123–127.
18. Ranson H, Jensen B, Vulule JM, Wang X, Hemingway J, et al. (2000) Identification of a point mutation in the voltage-gated sodium channel gene of Kenyan *Anopheles gambiae* associated with resistance to DDT and pyrethroids. Insect Mol Biol 9: 491–497.
19. Verhaeghen K, Van Bortel W, Roelants P, Backeljau T, Coosemans M (2006) Detection of the East and West African kdr mutation in *Anopheles gambiae* and *Anopheles arabiensis* from Uganda using a new assay based on FRET/Melt Curve analysis. Malaria J 5: 16.
20. Djegbe I, Boussari O, Sidick A, Martin T, Ranson H, et al. (2011) Dynamics of insecticide resistance in malaria vectors in Benin: first evidence of the presence of L1014S *kdr* mutation in *Anopheles gambiae* from West Africa. Malaria J 10: 261.
21. Badolo A, Traore A, Jones CM, Sanou A, Flood L, et al. (2012) Three years of insecticide resistance monitoring in *Anopheles gambiae* in Burkina Faso: resistance on the rise? Malaria J 11: 232.
22. Namountougou M, Simard F, Baldet T, Diabate A, Ouedraogo JB, et al. (2012) Multiple insecticide resistance in *Anopheles gambiae s.l.* populations from Burkina Faso, West Africa. PLoS One 7: e48412.
23. Henry MC, Assi SB, Rogier C, Dossou-Yovo J, Chandre F, et al. (2005) Protective efficacy of lambda-cyhalothrin treated nets in *Anopheles gambiae* pyrethroid resistance areas of Cote d'Ivoire. Am J Trop Med Hyg 73: 859–864.
24. N'Guessan R, Corbel V, Akogbeto M, Rowland M (2007) Reduced efficacy of insecticide-treated nets and indoor residual spraying for malaria control in pyrethroid resistance area, Benin. Emerg Infect Dis 13: 199–206.
25. Guillet P, N'Guessan R, Darriet F, Traore-Lamizana M, Chandre F, et al. (2001) Combined pyrethroid and carbamate 'two-in-one' treated mosquito nets: field efficacy against pyrethroid-resistant *Anopheles gambiae* and *Culex quinquefasciatus*. Med Vet Entomol 15: 105–112.
26. Ossè R, Aikpon R, Padonou GG, Oussou O, Yadouleton A, et al. (2012) Evaluation of the efficacy of bendiocarb in indoor residual spraying against pyrethroid resistant malaria vectors in Benin: results of the third campaign. Parasit Vectors 5: 163.
27. Chandre F, Darriet F, Doannio JM, Riviere F, Pasteur N, et al. (1997) Distribution of organophosphate and carbamate resistance in *Culex pipiens quinquefasciatus* (Diptera: Culicidae) in West Africa. J Med Entomol 34: 664–671.
28. N'Guessan R, Darriet F, Guillet P, Carnevale P, Traore-Lamizana M, et al. (2003) Resistance to carbosulfan in *Anopheles gambiae* from Ivory Coast, based on reduced sensitivity of acetylcholinesterase. Med Vet Entomol 17: 19–25.
29. Aikpon R, Agossa F, Osse R, Oussou O, Aizoun N, et al. (2013) Bendiocarb resistance in *Anopheles gambiae s.l.* populations from Atacora department in Benin, West Africa: a threat for malaria vector control. Parasit vectors 6: 192.
30. Weill M, Lutfalla G, Mogensen K, Chandre F, Berthomieu A, et al. (2003) Comparative genomics: Insecticide resistance in mosquito vectors. Nature 423: 136–137.
31. Weill M, Malcolm C, Chandre F, Mogensen K, Berthomieu A, et al. (2004) The unique mutation in *ace*-1 giving high insecticide resistance is easily detectable in mosquito vectors. Insect Mol Biol 13: 1–7.
32. Djogbenou L, Dabire R, Diabate A, Kengne P, Akogbeto M, et al. (2008) Identification and geographic distribution of the *ACE*-1R mutation in the malaria vector *Anopheles gambiae* in south-western Burkina Faso, West Africa. Am J Trop Med Hyg 78: 298–302.
33. Djogbenou L, Chandre F, Berthomieu A, Dabire R, Koffi A, et al. (2008) Evidence of introgression of the *ace*-1(R) mutation and of the *ace*-1 duplication in West African *Anopheles gambiae s.s.* PLoS ONE 3: e2172.
34. Scott JG (1996) Cytochrome P450 monooxygenase-mediated resistance to insecticides. J Pest Sci 21: 241–245.

35. Hemingway J, Karunaratne SH (1998) Mosquito carboxylesterases: a review of the molecular biology and biochemistry of a major insecticide resistance mechanism. Med Vet Entomol 12: 1–12.
36. Gillies MT, Coetzee M (1987) A supplement to the Anophelinae of Africa south of the Sahara. Pub. South Afr. Inst Med Res 55: 143.
37. Santolamazza F, Calzetta M, Etang J, Barrese E, Dia I, et al. (2008) Distribution of knock-down resistance mutations in *Anopheles gambiae* molecular forms in west and west-central Africa. Malar J 7: 74.
38. Raymond M, Rousset F (1995) GENEPOP Version 1.2 A population genetics software for exact tests and ecumenicism. J Hered: 248–249.
39. Jones CM, Toe HK, Sanou A, Namountougou M, Hughes A, et al. (2012a) Additional selection for insecticide resistance in urban malaria vectors: DDT resistance in *Anopheles arabiensis* from Bobo-Dioulasso, Burkina Faso. PLoS One 7: e45995.
40. Jones CM, Liyanapathirana M, Agossa FR, Weetman D, Ranson H, et al. (2012) Footprints of positive selection associated with a mutation (N1575Y) in the voltage-gated sodium channel of *Anopheles gambiae*. Proc Nati Acad Sci U S A 109: 6614–6619.
41. Dabire KR, Diabate A, Namountougou M, Djogbenou L, Kengne P, et al. (2009b) Distribution of insensitive acetylcholinesterase (*ace*-1R) in *Anopheles gambiae s.l.* populations from Burkina Faso (West Africa). Trop Med Int Health 14: 396–403.
42. Dabire RK, Namountougou M, Sawadogo SP, Yaro LB, Toe HK, et al. (2012) Population dynamics of *Anopheles gambiae s.l.* in Bobo-Dioulasso city: bionomics, infection rate and susceptibility to insecticides. Parasit vectors 5: 127.

19

Genome-Wide Transcriptome Profiling Revealed Cotton Fuzz Fiber Development Having a Similar Molecular Model as *Arabidopsis* Trichome

Qun Wan, Hua Zhang, Wenxue Ye, Huaitong Wu, Tianzhen Zhang*

National Key Laboratory of Crop Genetics and Germplasm Enhancement, Cotton Research Institute, Nanjing Agricultural University, Nanjing, China

Abstract

The cotton fiber, as a single-celled trichome, is a biological model system for studying cell differentiation and elongation. However, the complexity of gene expression and regulation in the fiber complicates genetic research. In this study, we investigated the genome-wide transcriptome profiling in Texas Marker-1 (TM-1) and five naked seed or fuzzless mutants (three dominant and two recessive) during the fuzz initial development stage. More than three million clean tags were generated from each sample representing the expression data for 27,325 genes, which account for 72.8% of the annotated *Gossypium raimondii* primary transcript genes. Thousands of differentially expressed genes (DEGs) were identified between TM-1 and the mutants. Based on functional enrichment analysis, the DEGs downregulated in the mutants were enriched in protein synthesis-related genes and transcription factors, while DEGs upregulated in the mutants were enriched in DNA/chromatin structure-related genes and transcription factors. Pathway analysis showed that ATP synthesis, and sugar and lipid metabolism-related pathways play important roles in fuzz initial development. Also, we identified a large number of transcription factors such as MYB, bHLH, HB, WRKY, AP2/EREBP, bZIP and C2H2 zinc finger families that were differently expressed between TM-1 and the mutants, and were also related to trichome development in *Arabidopsis*.

Editor: Samuel P. Hazen, University of Massachusetts Amherst, United States of America

Funding: The State Basic Research Development Program of China (973 Program, 2011CB109300), the National High Technology Research and Development Program of China (863 Program) (2011AA10A102), and the Priority Academic Program Development of Jiangsu Higher Education Institutions. The funders had no role in study design, data collection and analysis, decision to publish, or preparation of the manuscript.

Competing Interests: The authors have declared that no competing interests exist.

* E-mail: cotton@njau.edu.cn

Introduction

Cotton (*Gossypium* spp.) is an important commercial crop and the largest source of natural textile fibers grown throughout the world. Cotton fibers used in textiles originate from the outer epidermal layer of the maturing seed, and are classified into two types: lint and fuzz. Initiation of lint fibers is a quasi-synchronous process that occurs in developing ovules during anthesis. The fuzz fibers initiate growth at 4 DPA (days post anthesis) and elongate to approximately 0.5 cm, much shorter than lint fibers [1].

Many genes from *Arabidopsis* have been identified that control the initiation and morphogenesis of trichomes, and most of them encode transcription factors including MYB (*GLABROUS1, TRIPTYCHON, CAPRICE, WEREWOLF*) [2–5], WD-40 type (*TRANSPARENT TESTA GLABRA1*) [6,7], bHLH (*GLABROUS3*) [8,9], HD-ZIP (*GLABROUS2*) [10] and a WRKY-related transcription factor (*TRANSPARENT TESTA GLABRA2*) [11]. *TRANSPARENT TESTA GLABRA 1* (*TTG1*) encodes a small protein with WD-repeats, although no WD-repeat protein has either enzymatic activity or a DNA binding domain [12]. The identification of TTG1 as a WD40 repeat-containing protein suggests that TTG1 regulates MYC transcription factors or pathways in which MYC factors are involved [7]. *TTG2* encodes a WRKY transcription factor and acts downstream of the trichome initiation genes, *TTG1* and *GL1* [11]. bHLH family members have a basic helix–loop–helix domain [13]. Mutant analyses have identified several plant bHLH proteins involved in anthocyanin biosynthesis, such as GL3, EGL3 and TT8 [8,9,14–16].

Cotton fibers share many similarities with *A. thaliana* leaves trichome development, and several studies have demonstrated a close relationship between these two types of cells using cotton fiber-related genes (Table S1). Six putative cotton MYB genes (*GhMYB1-6*) have been isolated, and these DNA-binding factors were shown to be involved in the differentiation and expansion of cotton seed trichomes [17]. *GhMYB109*, which encodes a R2R3 MYB transcription factor, was shown to be expressed specifically in fiber cell initials and expanding fibers [18]. Another R2R3 MYB gene, *GaMYB2*, which is homologous to *AtGL1*, was predominantly expressed early in cotton fibers and complemented *gl1* phenotypes in *Arabidopsis* [19]. Overexpressing *GhMYB2* or its downstream gene *GhRDL1* in *Arabidopsis* activates fiber-like hair production on 4–6% of the seed coats and has no obvious effect on trichome development in leaves or siliques [20]. In addition, overexpression of *GbMYB2* in *Arabidopsis* caused thicker leaf trichomes and longer roots to develop due to the activation of trichome development-related genes such as *GL2* [21]. *GhMYB25* encodes a homolog of *AmMIXTA/AmMYBML1* which involved in epidermal cell differentiation, is highly expressed in ovules, fiber cell initials and trichomes on leaf. Silencing of *GhMYB25* in cotton showed fiber and trichome development were suppressed, while overexpression of *GhMYB25* increased cotton fibre initiation and leaf trichome number [22–26]. *GhMYB25-like* had a similar expression pattern with *GhMYB25* which significantly higher expression during fiber cell initiation (−3∼3 DPA). Transgenic plants showed *GhMYB25-like* had significant regulatory roles in cotton fiber development. RNA interference suppression of *GhMYB25-like* resulted in cotton plants with fibreless seeds, but normal trichomes elsewhere implying *GhMYB25-like* playing a crucial role in the very early stages of fiber cell differentiation [26,27]. A cotton gene encoding an *Arabidopsis CPC* ortholog (*R3 MYB* gene) was identified and downregulated in fiber initials at 1 DPA [28]. In addition to the MYB genes, four putative homologues of *Arabidopsis TTG1* (*GhTTG1–GhTTG4*), have been isolated and were shown to form two groups, with *GhTTG1* and *GhTTG3* being closely related to each other, and *GhTTG2* and *GhTTG4* forming the second group, based on sequence comparisons of the four deduced proteins and *Arabidopsis TTG1* [29]. Three homeobox (HOX) genes, *GhHOX1*, *GhHOX2*, and *GhHOX3*, have been identified from cotton, showing 66%, 34%, and 37% protein sequence similarity to *Arabidopsis* GL2, respectively. *GhHOX1* was able to restore the glabrous phenotype of *gl2* mutant, indicating that this protein is a functional homologue of GL2 in controlling trichome development and may function in fiber development [30]. Two GL3-like bHLH cDNAs from cotton ovule, *GhDEL65* and *GhDEL61*, have been deposited in the Genbank [31,32]. It will be interesting to examine if they work like GL3 during cotton fiber development. Also, several ESTs (expressed sequence tags) from cotton have been published that share identity with *Arabidopsis* homologues in the NCBI database [33,34]. As many homologous genes have been isolated from cotton and shown to play similar roles in trichome initiation in *Arabidopsis*, the GL1-GL3/EGL3-TTG1 protein complex may control fiber formation in cotton [33].

Several "qualitative" mutants in fiber development have been reported. The best characterized of these are the naked seed loci, N_1N_1 and n_2n_2. The dominant naked seed mutant (NSM) N_1NSM is fuzzless but with a little lint on the seed [35]. The recessive naked seed mutant n_2NSM produces regular lint, but bears a naked seed phenotype with very little fuzz fibers present at the micropyle tips of the seed [36]. Fuzzless-lintless mutants (FLM) XZ142FLM, MD17FLM and SL1-7-1FLM are all completely without any fiber; SL1-7-1FLM possess the dominant naked seed gene N_1 [37,38], XZ142FLM possess recessive naked seed gene n_2 [39,40], while MD17FLM possess both N_1 and n_2 [37,41]. TM-1 with lint and fuzz fiber is upland genetic standard line, which widely used in research programs [42]. Although these six materials have different genetic background, critical genes or pathways can be identified by studying the common different expressed genes between WT and several same genotype mutants.

To gain a better understanding of gene regulation in the early stage of fuzz development, we present here the first genome-wide analysis of gene expression during cotton fuzz initial cell development using massively parallel deep-sequencing developed by Solexa/Illumina. As cell fate determination for fiber (lint and fuzz) must occur prior to the formation of fiber cell initials, we selected +1, +3 and +5 DPA ovules to analyze fuzz initial development. In this study, we annotated thousands of read signatures matching predicted genes, and quantified the transcript abundance in developing ovules and fibers. In addition, we have profiled gene expression in the mutants against fuzz-bearing ovules

Figure 1. Cotton fiber morphology in the wild-type line and the mutant lines. a: wild line: TM-1; b: recessive naked-seed mutant: n_2NSM; c: dominant naked-seed mutant: N_1NSM; d: fuzzless-lintless mutants: SL1-7-1FLM (with N_1 gene), XZ142FLM (with n_2 gene), MD17FLM (with N_1 and n_2). Matured seed were separated from the opened bolls on the cotton plant. Ginned seeds (right) and matured seeds (left) showed on linted-fuzzy and linted-fuzzless panel.

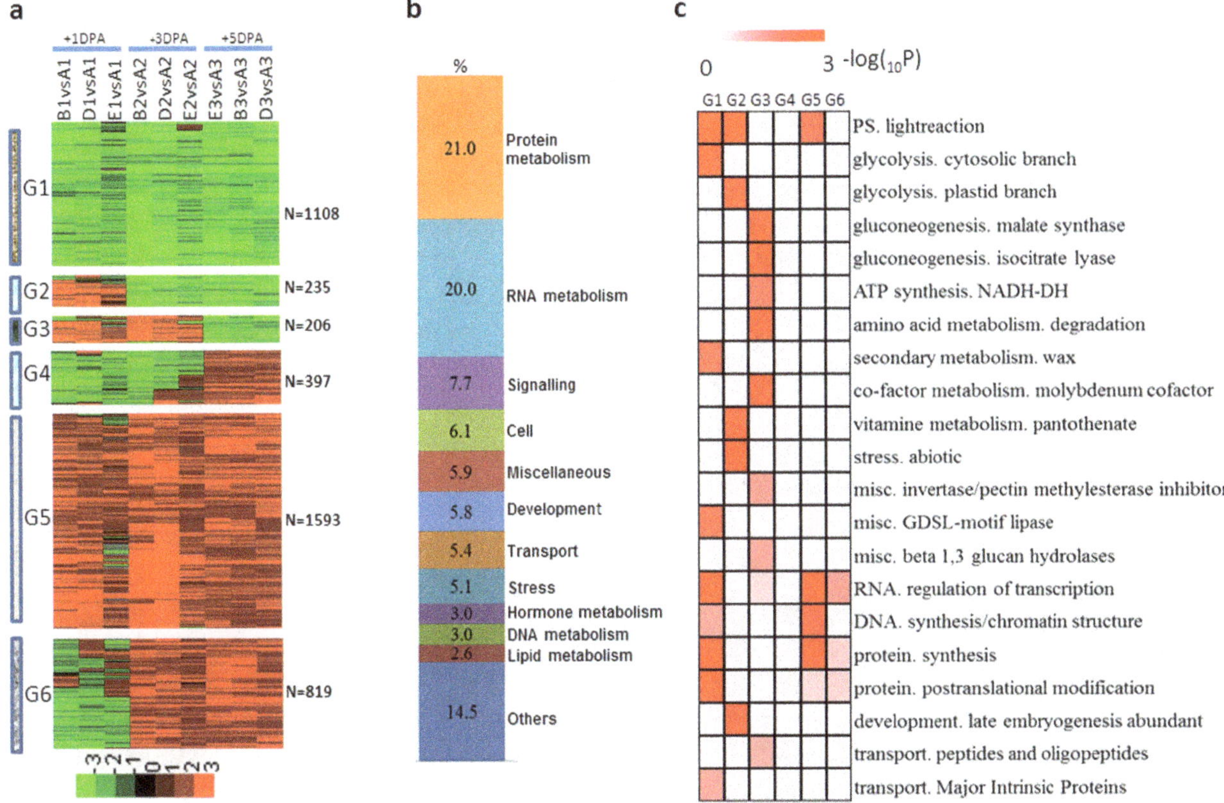

Figure 2. Dynamic progression of common differentially expressed genes in the dominant mutants. (a) Unsupervised hierarchical clustering of the 4,358 common DEGs in the dominant mutants. Common DEGs were clustered into six groups and the number of genes of each group was listed at right. Red region, genes upregulated in the mutants; green region, genes downregulated in the mutants. A, TM-1; B, SL1-7-1FLM; D, MD17FLM; E, N_1NSM; 1, +1 DPA; 2, +3 DPA; 3, +5 DPA. (b) Functional distribution of common DEGs in the dominant mutants. (c) Functional category enrichment of common differentially expressed genes in the dominant mutants.

(wild type, WT), and found large changes in gene expression in the mutants.

Materials and Methods

Plant Material Preparation and Total RNA Isolation

G. hirsutum cv. Texas Marker-1 (TM-1) and five naked-seed or fuzzless mutants (XZ142FLM, MD17FLM, SL1-7-1FLM, N_1NSM and n_2NSM) were used in this study (Figure 1). SL1-7-1FLM, MD17FLM and N_1NSM each possess the dominant naked seed gene N_1, while XZ142FLM and n_2NSM carry the recessive naked seed gene n_2.

Plants were grown at Jiangpu Breeding Station, Nanjing (JBS/NAU) in 2010. All lines were self-pollinated, and the progeny were tested to verify the initial pattern. Buds were tied up the day before anthesis to ensure self-pollination. Bolls were harvested at +1, +3 and +5 DPA. Ovules were excised carefully from bolls, frozen in liquid nitrogen immediately, and stored at −70°C. Total RNA was extracted using the CTAB method [43].

Sequencing and Digital Tag Profiling

Library construction, sequencing and raw data processing were performed commercially by BGI (Beijing Genomics Institute at Shenzhen, China) *via* the sequencing by synthesis (SBS) on Illumina HiSeq 2000 System as described previously [44]. Digital tag profiling was perfomed as descriped by Wang et al [44] and *Gossypium raimondii* primary transcript sequences (http://www.phytozome.net) was used as reference gene database.

Defining Differentially Expressed Genes and Cluster Analysis

Statistical analysis was performed to identify differentially expressed genes between the libraries using a rigorous algorithm described previously [45]. Gene expression was normalized to transcripts per million (TPM) clean tags. For gene expression variance, the statistical *t*-test was used to identify genes differently expressed between the libraries. *P* values were adjusted using the multiple testing procedures described by Benjamini and Yekutieli [46] for controlling the false discovery rate (FDR). In this study, we used a stringent value of FDR <0.001, and the absolute value of $|\log_2\text{Ratio}| \leq 1$ as the threshold to judge the significant difference of gene expression.

K means clustering was performed with the open-source program Cluster3.0 (http://bonsai.hgc.jp/~mdehoon/software/cluster/software.htm). The genes in each cluster were then classified into Mapman functional categories [47]. Functional categories of the MapMan annotation were tested for significance of expression change by applying a two-sided Wilcoxon rank test with a Benjamini Yekutieli correction for multiple tests. Pathway analysis was mainly based on the Kyoto Encyclopedia of Genes and Genomes (KEGG) database [48].

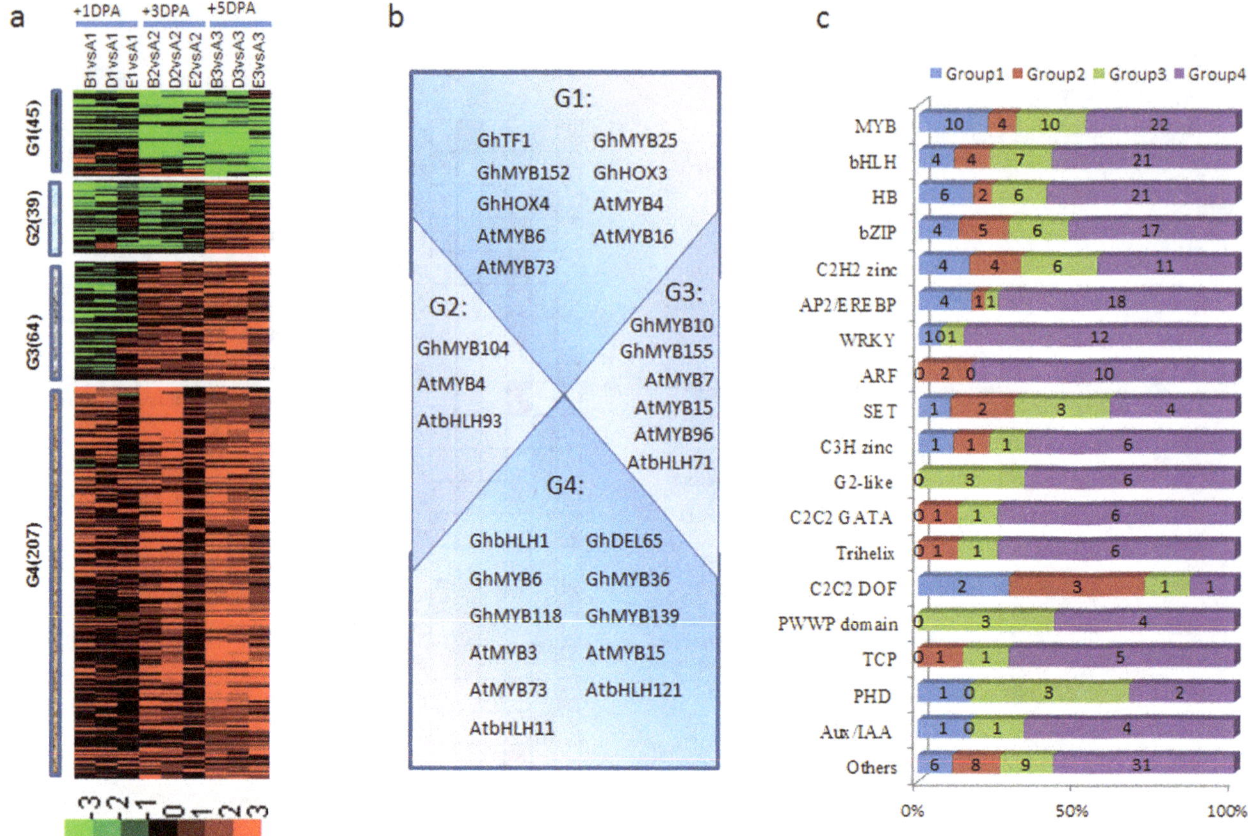

Figure 3. Dynamics of transcription factor expression profiles in various dominant mutants. (a) Unsupervised hierarchical clustering of 355 transcription factor genes included in the 4,358 common DEGs in the dominant mutants. Four groups were generated and the number of each group was in parentheses. Red region, genes upregulated in the mutants; green region, genes downregulated in the mutants. A, TM-1; B, SL1-7-1FLM; D, MD17FLM; E, N_1NSM; 1, +1 DPA; 2, +3 DPA; 3, +5 DPA. (b) Representative functions and genes showing expression gradients. (c) Distribution of transcription factor families among G1, G2, G3 and G4.

Quantitative Real Time RT-PCR (qPCR)

Verification of some differentially expressed genes (DEGs) was performed by real-time quantitative PCR (qPCR). The primers for the various genes were designed with Primer 3.0 (http://frodo.wi.mit.edu/cgi-bin/primer3/primer3), and synthesized commercially (Genscript, Nanjing, China); sequences are given in Table S10. Two microgramme total RNA was reversely transcripted using PrimeScript RT reagent Kit with gDNA Eraser (Perfect Real Time) (TaKaRa, Shiga, Japan). QPCR was performed using the LightCycler FastStart DNA Master SYBR Green I kit (Roche, Basel, Switzerland) in an ABI7500 Real-Time PCR detection system (Applied Biosystems, San Francisco, CA, USA). Each sample was PCR-amplified using 100ng cDNA template in triple reactions. The cotton *histone 3* gene [19] (ACC No. AF024716) was used as the positive control and amplified with the primer pair (F: 5′-GGTGGTGTGAAGAAGCCTCAT-3′, and R: 5′-AATTTCACGAACAAGCCTCTGGAA-3′). The amplification efficiency of each gene was calculated. The qRT-PCR cycles were as follows: (1) 95°C, 10 min; (2) 40 cycles of 95°C for 15 s, ~60°C (temperature varied for different primers, Table S10) for 30 s and 72°C for 30 s; (3) a melting curve analysis from 65 to 95°C (1 s hold per 0.2°C increase) to check the specificity of the amplified product. Relative expression levels were determined by the $2^{-\Delta Ct}$ method.

Results

Sequencing Data Analysis

To obtain a global view of transcription relevant to cotton fuzz development, we used the Illumina HiSeq 2000 System to perform high-throughput tag-sequencing (Tag-seq) analysis on poly(A)-enriched RNAs from eighteen cotton ovule libraries including the cultivar TM-1 and five mutants during the fuzz initiation stage (+1 DPA, +3 DPA and +5 DPA). The total number of tags per library ranged from 3.5 to 4.7 million, and the number of tags with distinct sequences ranged from 0.27 to 0.44 million (Table S2). After removal of low quality tags, we obtained a total of 3.4 to 4.5 million clean tags that corresponded to about 0.15 million distinct tags (Table S2). The distribution of total and distinct tag counts over different tag abundance categories showed very similar profiles for all libraries (Figure S1). Among the distinct tags, less than 5% had a copy number higher than 100, whereas 38% of the tags were present between 5 and 50 copies, and more than 57% of the transcripts had 2–5 copies.

As there was no allotetraploid cotton genome sequence available, clean tags were mapped to *G. raimondii* genome sequence (http://www.phytozome.net). Approximately 73%–82% of the distinct tags (83–87% of the total tags) could be mapped to the reference genome (Table S2). All clean tags were aligned to the reference *G. raimondii* primary transcript sequences. Approximately 26%–35% of the distinct tags could be uniquely mapped to the

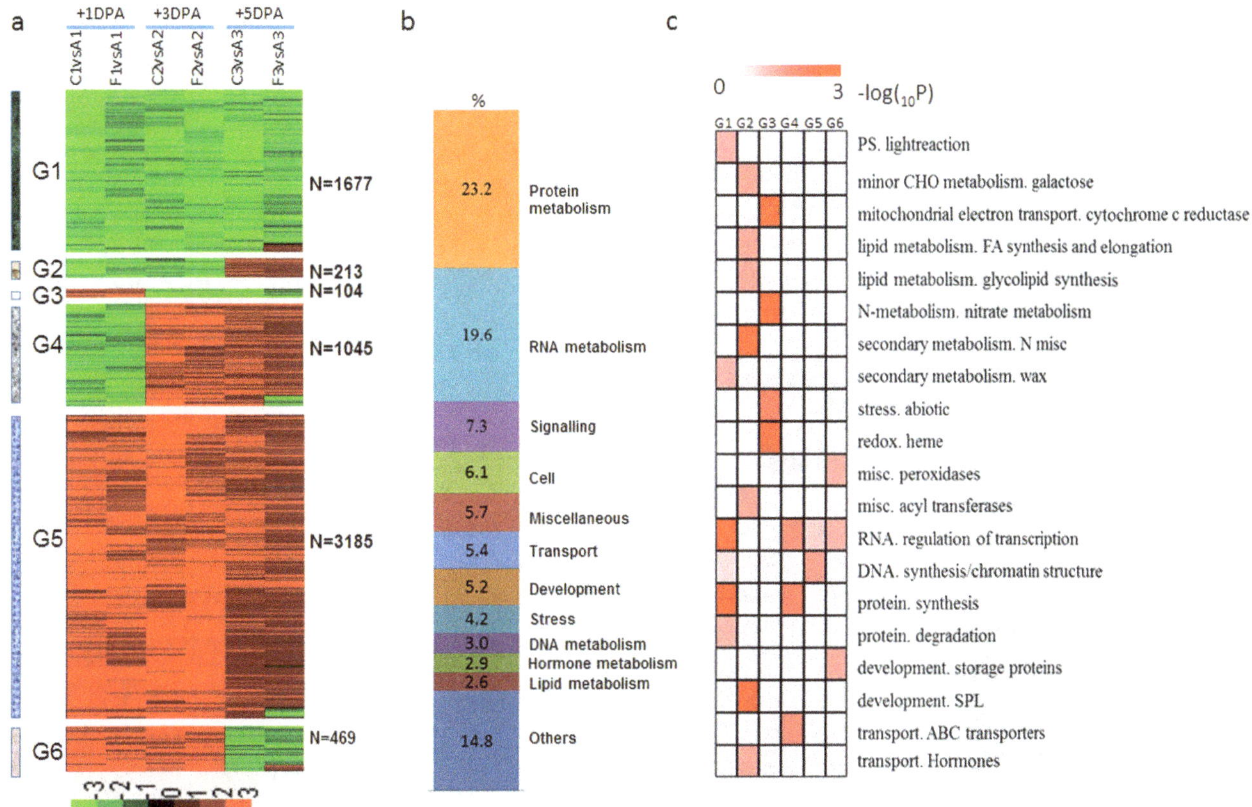

Figure 4. Dynamic progression of common differentially expressed genes in the recessive mutants. (a) Unsupervised hierarchical clustering of 6,693 common differentially expressed genes in the recessive mutants. Common DEGs were clustered into six groups and the number of genes of each group was listed at right. Red region, genes upregulated in the mutants; green region, genes downregulated in the mutants. A, TM-1; C, XZ142FLM; F, n_2NSM; 1, +1 DPA; 2, +3 DPA; 3, +5 DPA. (b) Functional category distribution of common DEGs in the recessive mutants. (c) Functional category enrichment of common differentially expressed genes in the recessive mutants.

reference sequence. The tags that mapped to the database generated 19,829–22,213 tag-mapped transcripts for the libraries (Table S2).

Common DEGs between Dominant Naked-seed Mutants and TM-1 during Fuzz Development

To understand the molecular mechanisms of the dominant fuzzy phynotype, 4,358 common DEGs differentially expressed between the mutants MD17FLM, SL1-7-1FLM, N_1NSM and the wild-type TM-1 were identified. Of these, 268 genes were up regulated and 557 genes down regulated at +1 DPA; 792 genes were up regulated and 699 genes down regulated at +3 DPA in the mutants; and 2,000 genes were up regulated with 957 genes down regulated at +5 DPA. Ten common differentially up regulated genes and 62 down regulated genes were identified at +1 DPA, +3 DPA and +5 DPA (Figure S2, Table S3).

We then used MapMan annotation to assign genes to functional categories and grouped the genes into six groups using the hierarchical clustering algorithm. Two main groups (Groups 1 and 5) accounted for ~62% of the DEGs at the three sampling times (Figure 2a). Excluding 866 genes belonging to the 'not assigned or unknown' categories, 3,492 genes had MapMan annotation assignments. Among these, 21.0% are related to protein metabolism, 20.0% to RNA metabolism, 7.7% to signaling, and the remaining genes to cell functions, development, transport, stress, hormone metabolism, DNA metabolism, or lipid metabolism (Figure 2b). To further explore this dataset, we tested for enrichment by MapMan functional category using Fisher's exact test ($P<0.01$, FDR = 5%). Most of the MapMan bins showed enrichment for particular groups of expressed genes (Figure 2c); for example, genes that encode enzymes for protein synthesis and transcription factors in Group 1, light reaction and abiotic stress in Group 2, ATP synthase, amino acid metabolism, glyoxylate cycle in Group 3, and transcription factors and DNA synthesis in Group 5.

The dynamics of transcription factor accumulation during fuzz initiation were particularly well resolved in our data. Of the 1,596 differentially expressed transcription factor genes that we detected in the fuzz initial stage, 355 were common to the dominant mutants (Figure 3). Most of these genes (207) were upregulated in the dominant mutants (Group4), including *GhbHLH1*, *GhDEL65*, *GhMYB6*, *GhMYB118*, *GhMYB139*, *AtMYB3*, *AtMYB73*, *AtbHLH121* and *AtbHLH11*. Only 45 transcription factors were downregulated (Group 1), including *GhTF1*, *GhMYB25*, *GhMYB152*, *GhHOX3*, *GhHOX4*, *AtMYB4*, *AtMYB6*, *AtMYB16* and *AtMYB73*. Thirty nine transcription factors showed downregulation at +1 DPA and +3 DPA, and upregulation at +5 DPA (Group 2); these included *GhMYB104*, *AtMYB4*, and *AtbHLH93*. An additional 64 transcription factors showed downregulation at +1 DPA, and upregulation at +3 DPA and +5 DPA (Group 3); these included *GhMYB10*, *GhMYB155*, *AtMYB7*, *AtMYB15*, *AtMYB96*, *AtbHLH71*(Figure 3a and 3b). We also identified family-specific expression trends (Figure 3c). Members of the C2C2 DOF

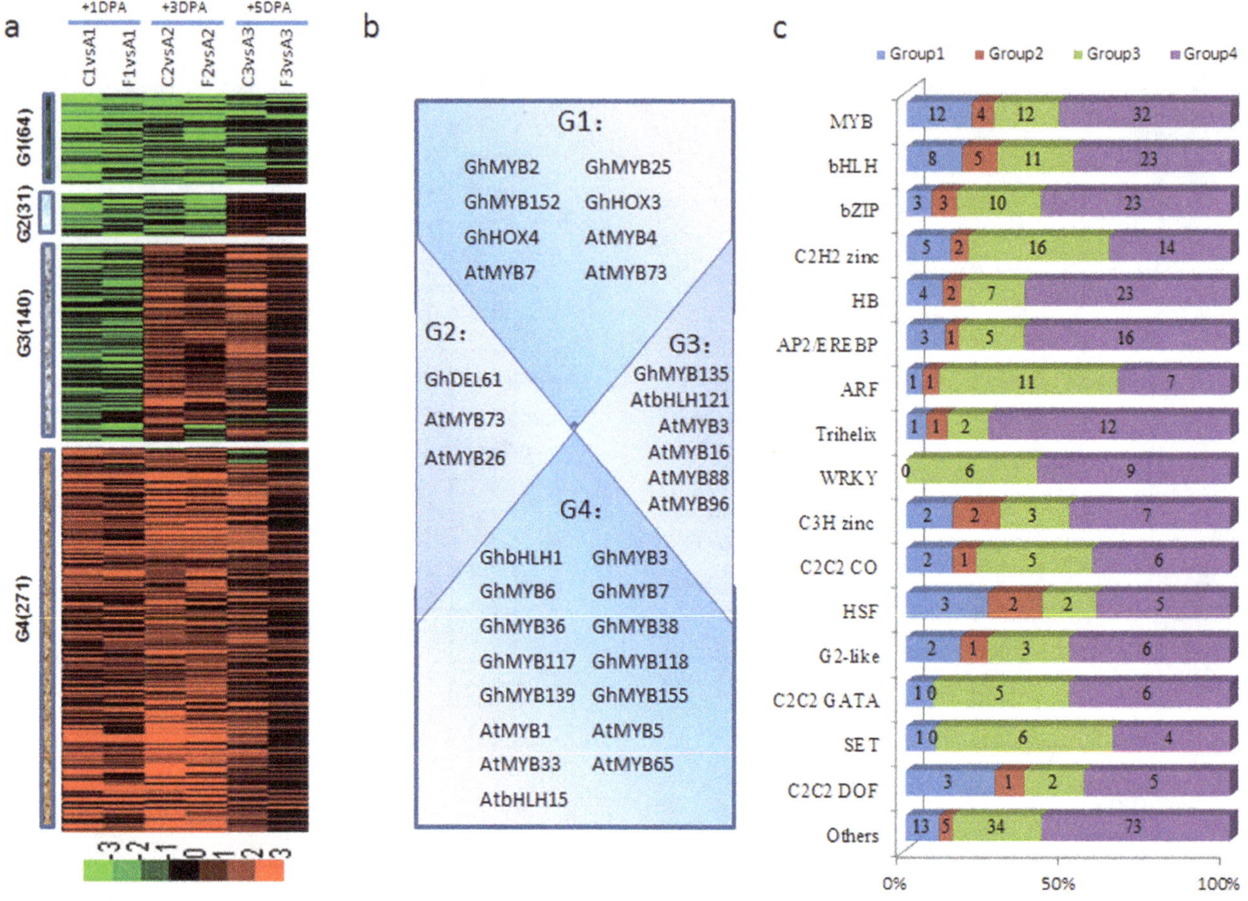

Figure 5. Dynamics of transcription factor accumulation profiles in various recessive mutants. (a) Unsupervised hierarchical clustering of 506 transcription factor genes included in the 6,693 common DEGs in the recessive mutants. Four groups were generated and the number of each group was in parentheses. Red region, genes upregulated in the mutants; green region, genes downregulated in the mutants. A, TM-1; C, XZ142FLM; F, n_2NSM; 1, +1 DPA; 2, +3 DPA; 3, +5 DPA. (b) Representative functions and genes showing expression gradients. (c) Distribution of transcription factor families among G1, G2, G3 and G4.

zinc-finger families of transcriptional regulators are highly expressed in TM-1; Aux/IAA, WRKY, AP2/EREBP and G2-like families are highly expressed in the dominant mutants.

To identify the metabolic pathways that are active during fuzz initiation, we mapped the 4,358 commonly expressed genes in the dominant mutants to the reference KEGG canonical pathways. In total, we assigned 1448 genes to KEGG pathways. Most of these mapped to ATP synthesis, or sugar and lipid metabolism-related metabolic pathways such as starch and sucrose metabolism (82 members), oxidative phosphorylation (30 members), galactose metabolism (22 members), glycolysis/gluconeogenesis (31 members), fatty acid degradation (20 members; Table S4). These annotations provide a valuable resource for investigating the processes, functions, and pathways specific to the initiation of fuzz development.

Common DEGs between Recessive Naked-seed Mutants and TM-1 during Fuzz Development

To understand the molecular mechanisms underlying the recessive naked seed phenotype, 6,693 DEGs were identified that are common to mutants XZ142FLM and n_2NSM compared with TM-1. Of these, 1,978 genes were up-regulated and 1,480 genes downregulated at +1 DPA; 2,971 genes were upregulated and 980 genes downregulated at +3 DPA; and 1,264 genes were upregulated and 666 genes downregulated at +5 DPA. There were 192 upregulated genes and 120 downregulated genes common to the differentially expressed genes at +1 DPA, +3 DPA and +5 DPA (Figure S2, Table S5).

We identified six groups using the hierarchical clustering algorithm. Group 1 and Group 5 accounted for ~73% of the differentially expressed genes at the three sampled times (approx. 4,852 genes; Figure 4a). Five-thousand four-hundred twenty-one genes had MapMan annotations, excluding 19.0% belonging to the 'not assigned or unknown' categories. Among the annotated genes, 23.2% are related to protein metabolism, 19.6% to RNA metabolism, 7.3% to signaling, and the rest to cell functions, transport, development, stress, DNA metabolism, hormone metabolism or lipid metabolism (Figure 4b). Most of the MapMan bins showed enrichment for particular groups (Figure 4c); for example, genes that encode enzymes for protein synthesis and transcription factors in Group 1, SPL protein in Group 2, abiotic stress in Group 3, ABC transport in Group 4, DNA synthesis in Group 5, and peroxidases and storage protein in Group 6.

We identified 506 transcription factors that were expressed in common in the two recessive mutants (Figure 5). Among these, 64 transcription factors (Group 1) including *GhMYB2*, *GhMYB25*, *GhMYB152*, *GhHOX3*, *GhHOX4* were downregulated in the recessive mutants, and 271 transcription factors (Group4) were

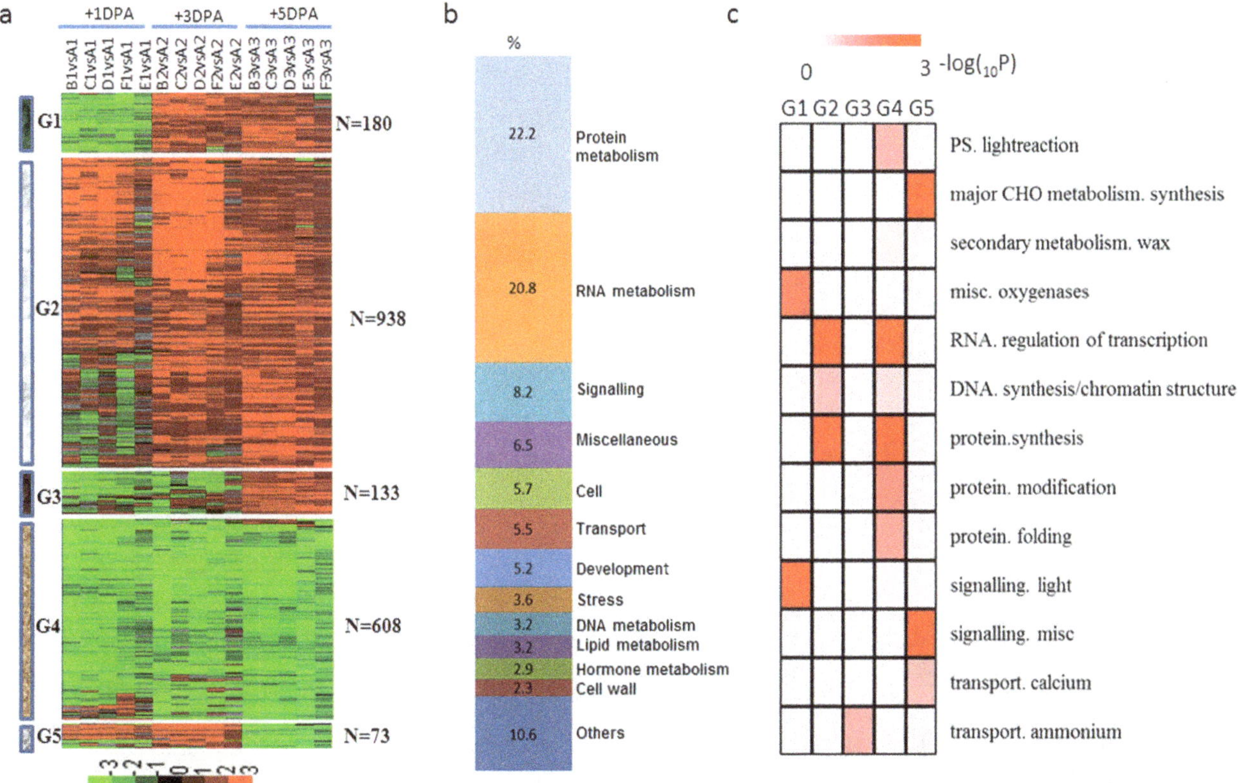

Figure 6. Dynamic progression of common differentially expressed genes in the dominant/recessive mutants. (a) Hierarchical clustering of the 1,932 common DEGs in five mutants. Common DEGs were clustered into five groups and the number of genes of each group was listed at right. Red region, genes upregulated in the mutants; green region, genes downregulated in the mutants. A, TM-1; B, SL1-7-1FLM; C, XZ142FLM; D, MD17FLM; E, N_1NSM; F, n_2NSM; 1, +1 DPA; 2, +3 DPA; 3, +5 DPA. (b) Functional distribution of common DEGs in the dominant/recessive mutants. (c) Functional category enrichment of common DEGs in the dominant/recessive mutants.

upregulated including *GhbHLH1*, *GhMYB3*, *GhMYB36*, *GhMYB7*, *GhMYB36*, *GhMYB38*, *GhMYB117*, *GhMYB118*, *GhMYB139*, *GhMYB155*. Thirty-one transcription factors (Group 2), such as *GhDEL61*, *GhGL2-like1* were downregulated at +1 DPA and +3 DPA, but were upregulated at +5 DPA, 140 transcription factors (Group 3), such as *GhMYB135* were downregulated at +1 DPA, were upregulated at +5 DPA (Figure 5a and 5b). Members of the HSF families of transcriptional regulators were highly expressed in TM-1; the MYB, WRKY, AP2/EREBP, bHLH, ARF and C2C2(Zn) GATA families were highly expressed in the recessive mutants (Figure 5c).

Of 6,693 common DEGs in the recessive mutants, 2,356 were assigned to KEGG pathways. The pathways with the most representation for the unique sequences were ATP synthesis, or sugar and lipid metabolism-related metabolic pathways including starch and sucrose metabolism (123 members), oxidative phosphorylation (44 members), glycolysis/gluconeogenesis (43 members), galactose metabolism (40 members) and fatty acid degradation (28 members) (Table S6).

Common DEGs between Naked-seed Mutants and TM-1

To uncover shared molecular mechanisms in the dominant and recessive fuzz-less mutants, we identified 1,932 DEGs that were common to the five mutants. Of these, 106 genes were upregulated and 314 downregulated at +1 DPA, 473 genes were upregulated and 215 downregulated at +3 DPA, and 737 genes were upregulated and 432 downregulated at +5 DPA (Figure S2, Table S7). Four were three upregulated genes and 29 downregulated genes common to the +1 DPA, +3 DPA and +5 DPA samples (Table S8).

The 1,932 common DEGs were classified into four groups by hierarchical clustering. Nine-hundred and thirty-eight genes (Group 2) were upregulated at +1 DPA, +3 DPA and +5 DPA; 608 genes (Group 4) were downregulated at +1 DPA, +3 DPA and +5 DPA; 180 genes (Group 1) were downregulated at +1 DPA, and upregulated at +3 DPA and +5 DPA; 133 genes (Group 3) were downregulated at +1 DPA and +3 DPA, and upregulated at +5 DPA; 73 genes (Group 5) were upregulated at +1 DPA and +3 DPA, and downregulated at +5 DPA (Figure 6a).

One-thousand six-hundred and two genes were annotated by MapMan, excluding 17% belonging to the 'not assigned or unknown' categories. Among these genes, 22.2% were related to protein metabolism, 20.8% were related to RNA metabolism, 8.2% were related to signaling, with the remaining genes were related to cell functions, transport, development, stress, DNA metabolism, lipid metabolism, hormone metabolism, and cell wall (Figure 6b). Genes that encode oxygenases and light signaling were enriched in Group 1, encode enzymes for protein synthesis and regulation of transcription in Group 2 and 4, ammonium transport in Group 3, and major CHO synthesis in Group 5 (Figure 6c).

We found 153 differentially expressed transcription factor genes in common between the wild-type TM-1 and the naked mutants (Figure 7). Fifteen transcription factors were downregulated in the naked mutants (Group 1), including *GhMYB25 and GhHOX3*. Ten

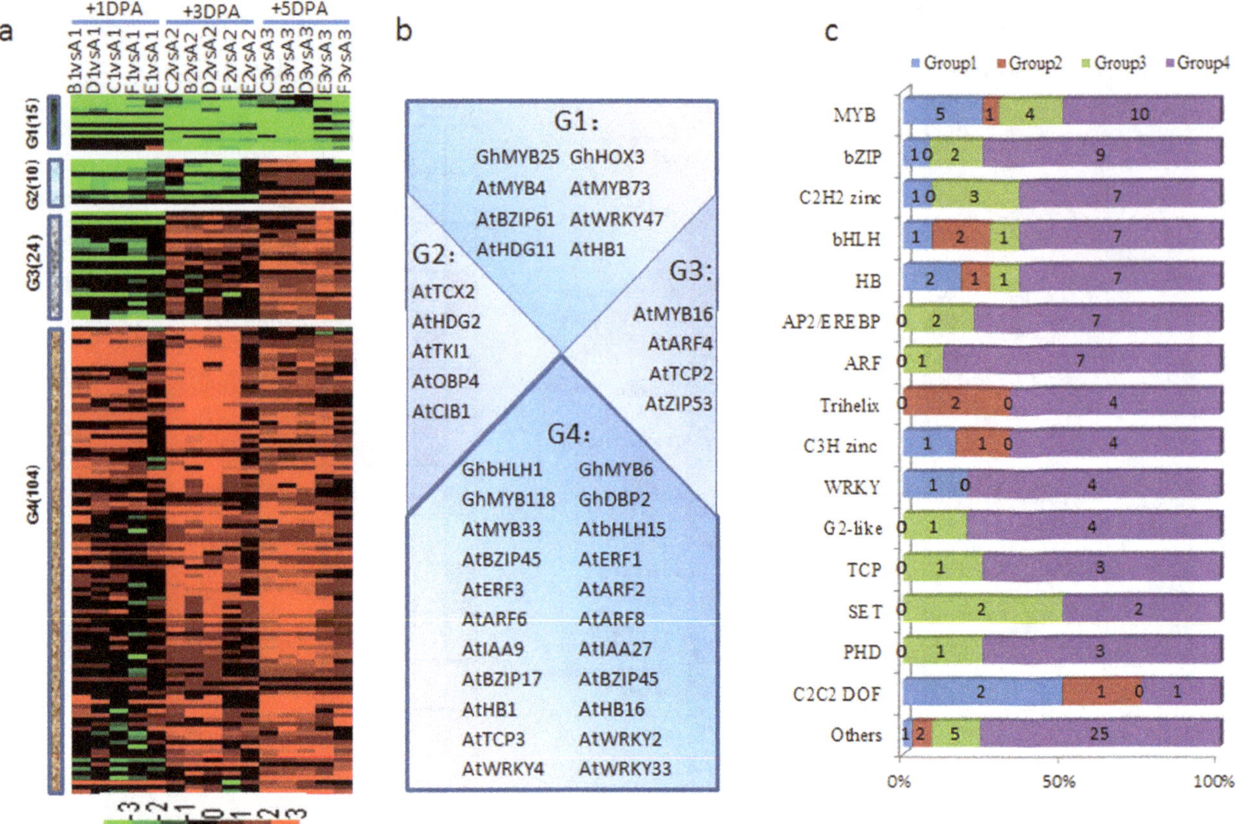

Figure 7. Dynamics of transcription factor accumulation profiles in various dominant/recessive mutants. (a) Unsupervised hierarchical clustering of 153 transcription factor genes included in the 1,932 common DEGs in the recessive mutants. Four groups were generated and the number of each group was in parentheses. Red region, genes upregulated in the mutants; green region, genes downregulated in the mutants. A, TM-1; B, SL1-7-1FLM; C, XZ142FLM; D, MD17FLM; E, N_1NSM; F, n_2NSM; 1, +1 DPA; 2, +3 DPA; 3, +5 DPA. (b) Representative functions and genes showing expression gradients. (c) Distribution of transcription factor families among G1, G2, G3 and G4.

transcription factors were downregulated at +1 DPA and +3 DPA, and upregulated at +5 DPA (Group 2), including *AtTCX2*, *AtHDG2*, *AtTKI1*, *AtOBP4* and *AtCIB1*. Fourty transcription factors were downregulated at +1 DPA, and upregulated at +3 DPA and +5 DPA (Group 3), including *AtMYB16*, *AtARF4*, *AtTCP2*, *AtZIP53*. Another 271 transcription factors showed upregulation in the naked mutants (Group 4), including *GhbHLH1*, *GhMYB6*, *GhMYB118* and *GhDBP2* (Figure 7a and 7b). Members of the most families of transcriptional regulators such as MYB, bHLH, bZIP, C2C2(Zn), HB, AP2/EREBP, ARF and WRKY families were highly expressed in the recessive mutants (Figure 7c).

To identify the differential metabolic pathways active in fuzz initiation, we mapped the 1,932 common DEGs in the dominant and recessive mutants in the KEGG database. In total, we assigned 620 genes to KEGG pathways. Similar to our earlier results, most of these genes were related to ATP synthesis, and sugar and lipid metabolism pathways. For example, 38 genes were annotated to starch and sucrose metabolism 13 genes were annotated to galactose metabolism, and 11 genes to oxidative phosphorylation (Table S9).

Validation of Differentially Expressed Genes by qPCR

To determine whether the digital gene expression results were reliable, we conducted qPCR analysis of the expression levels of 21 representative differentially expressed genes, most of them transcription factors. The qPCR results (Table S11) indicated that the expression levels estimated by DGE and qPCR were highly correlated ($r^2 = 0.72–0.93$). The qPCR validation results confirmed the accuracy and reliability of the expression levels determined by digital gene expression analysis, which means that we can make reasonable deductions from the functional enrichment analysis of the DEGs. The qPCR results for expression of transcription factors indicated that *GhMYB3* had a high level of expression in fuzzy ovules at +1 DPA and +3 DPA, *GhDEL61* had a low expression level at +1 DPA and +3 DPA, and *GhDEL65* had a low expression level at +3 DPA and +5 DPA. Additionally, *GhMYB25*, *GhHOX3* and *GhMYB2* had low levels of expression in the fuzzless mutants ovules (Figure 8).

Discussion

Choice of Materials is Important for the Study of Fuzz Initial Cell Development

Cotton lint fibers are extremely long, single epidermal cells that develop on the outer surface of ovules, reaching upwards of 5 cm in some species [50]. Fibers initiate between −1 DPA and +1 DPA, and the fiber initials begin to elongate rapidly immediately after fertilization, extending out from the surface of the seed coat epidermis. Zhang et al. [1] showed that fuzz initiation begins at +4 DPA by scanning electron microscopy (SEM) examination of TM-

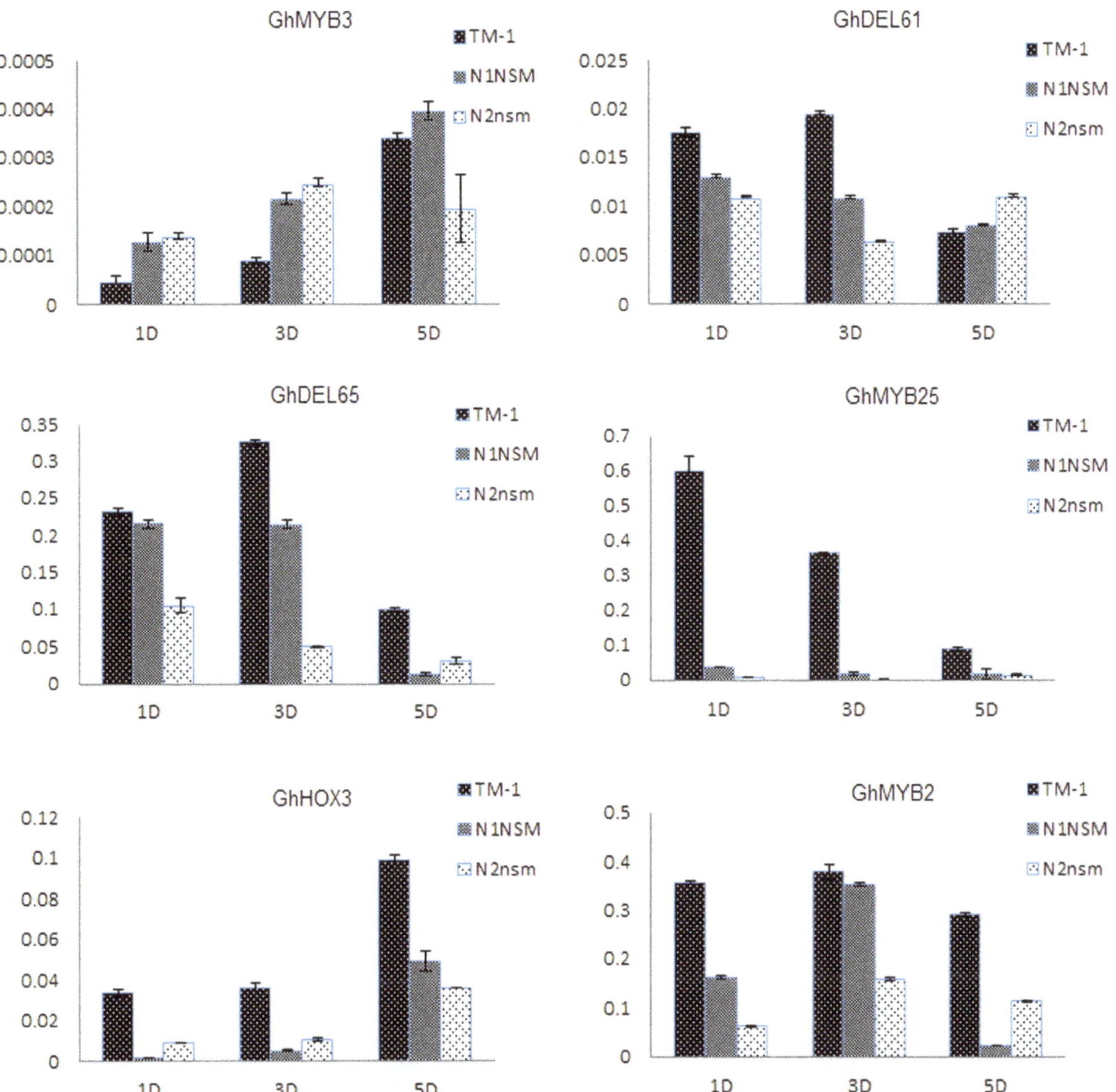

Figure 8. QRT-PCR confirmation for the six selected differentially-expressed transcription factor genes. The expression level of selected genes at +1 DPA, +3 DPA and +5 DPA in TM-1, N_1NSM and n_2NSM. Data shown are the means ± SD of three biological replicates.

1 ovules, although the shape of fuzz protrusions differed from that of lint fibers. In our study, +1, +3 and +5 DPA ovules were employed for fuzz initial development. Two types of fibers, the long lint fibers and the short fuzz fibers, probably share common developmental pathways at least in early differentiation. However, the fuzz fiber appears to be under separate genetic control, as a number of genetic loci specifying absence of fuzz fiber, but with normal lint, have been identified [37]. Lintless mutants, however, only occur in conjunction with lack of fuzz fiber, so are essentially fiberless [49]. Cotton fiber mutants are invaluable for the investigation of genes that control fiber development at the molecular level. The natural fiber mutants are well suited for genetic, physiological, and molecular characterization of the mutant phenotype, avoiding the complex and time-consuming progress of inducing, screening, and verifying fiber mutants. In this study, we selected five mutant lines, all of them fuzzless mutants that possessed different naked-seed genes. Thus, we can more clearly understand the mechanism of regulation of fuzz initial development by studying the five fuzzless mutants.

Many Specific Proteins that Relate to Fuzz Initial Development were Identified

A global analysis of the transcriptome will facilitate the characterization of gene expression and identification of regulatory mechanisms involved in fiber development [51,32]. In this study, we performed transcriptome profiling of fuzz-bearing and fuzzless ovules to identify genes that were differentially expressed during the fuzz initiation stage. Using a tag-based deep-sequencing approach [52], we could obtain a direct digital readout of cDNAs and achieve an essentially dynamic range of genes from the

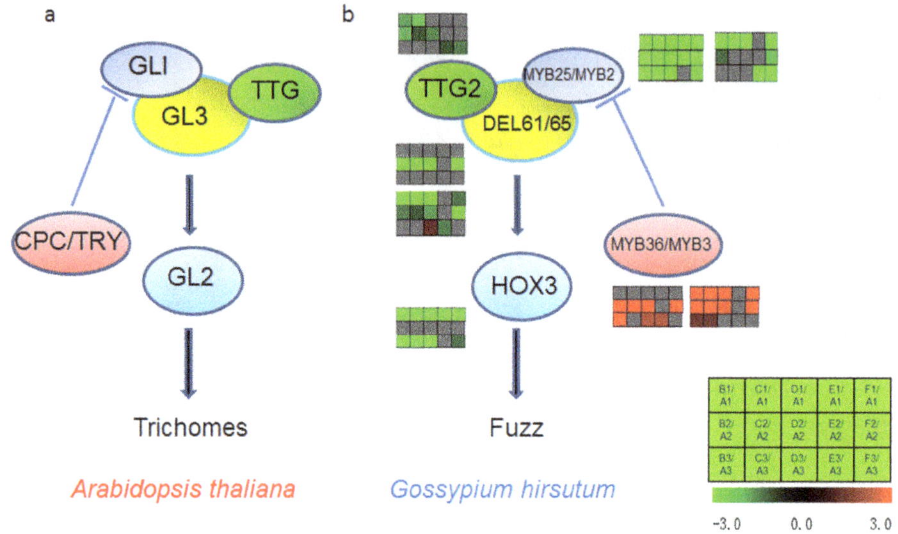

Figure 9. Model for the action of GL1-activating trichome development in *Arabidopsis thaliana* and fuzz development in *Gossypium hirsutum*. a: Model for the action of *GL1*-activating trichome development in *Arabidopsis thaliana*. b: Model for the action of *GL1*-activating fuzz development in *Gossypium hirsutum* A: TM-1, B: SL1-7-1FLM, C: XZ142FLM, D: MD17FLM, E: N_1NSM, F: n_2NSM, 1: +1 DPA, 2: +3 DPA, 3: +5 DPA. Light red/green bars indicate cotton fiber gene expression in the upper/lower group.

libraries. Thus, the present study represents the most comprehensive analyses of the cotton fuzz transcriptome. Approximately 19,829–22,213 tag-mapped reference genes were identified for each library. Unfortunately, sequencing of the upland cotton genome is incomplete, so there are still a large proportion of unique tags mismatched. These unmatched unique sequences probably represent novel genes to be identified in future studies.

From +1 to +5 DPA, the cotton fibers and ovules are in a state of rapid development. Jensen (1968) observed the ultrastructure and composition of the cotton zygote and described a dramatic series of alterations in cell structure including zygote size, endoplasmic reticulum, microtubes, mitochondria, ribosomes and plastids [53]. During early development, fiber cells produced from the surface of the ovules and elongate quickly. We found that DEGs between TM-1 and fuzzless mutants involved in protein metabolism, RNA metabolism, and signaling categories were enriched significantly. The large number of RNA metabolism-related genes is consistent with the sharp increase in the total number of ribosomes observed in the zygote [53].

Based on the large number of genes with fiber-specific expression, the molecular dissection of cotton fiber initiation has provided new insights [33,34]. Lee et al. identified more than 20 genes that were greatly enriched at the fiber-bearing (+3 DPA) stage in TM-1 as compared to the N_1NSM mutant [34,51]. Few studies have been performed to examine the initial pattern of cotton fuzz fiber development. In this study, we identified many DEGs between TM-1 and the fuzzless mutants. Protein synthesis-related genes had low levels of representation in both dominant and recessive mutants, while DNA and chromatin structure-related genes were highly represented. ATP synthesis, and sugar and lipid metabolism-related metabolic pathways play important roles in fuzz initial development. Recently, Du et al. [54] identified proteins related to fuzz fiber initiation in wild-type diploid cotton (*Gossypium arboreum* L.) and its fuzzless mutant by comparative proteomic analysis. They found 71 differentially expressed proteins between diploid Asiatic cotton DPL971 and the fuzzless mutant DPL972, mainly involved in cell response/signal transduction, redox homeostasis, protein metabolism, and energy/carbohydrate metabolism [51]. The differential expression of these proteins demonstrated that rapidly differentiating and expanding fuzz fiber cells experience active protein metabolism [55,56].

Fuzz Development May Share Similar Molecular Mechanisms with Leaf Trichome Development in *Arabidopsis*

Illuminating the functions of key regulators in fuzz development could help explain the reasons for the delayed developmental and elongation steps of fuzz fiber development. Through molecular improvement of key transcriptional factors in cultivated varieties, cotton could produce longer fuzz fibers and have higher yields. Cell fate determination is a critical step in the developmental processes of plants, and involves the participation of a large number of transcription factors. The pattern of trichome development has been studied in depth in the model plant *Arabidopsis* [57]. We found many common differentially expressed transcription factors in the dominant and recessive mutants. Most of these were in the MYB, bHLH, HB and WRKY gene families. *GhTTG2*, a putative homolog of *Arabidopsis TTG2*, were downregulated in lintless-fuzzless mutants at +1 DPA and +3 DPA, and also in ovules of lint-fuzzless mutants in +5 DPA. *GhMYB25* and *GhMYB2*, putative homologues of *Arabidopsis GL1*, showed low expression levels in fuzzless mutants. The bHLH domain of *DEL61* and *DEL65*, which share a high degree of sequence identity with *Arabidopsis* GL3 and EGL3, both had low expression levels in the fuzzless mutants. HOX3, a full-length coding sequence homolog of *AtGL2*, shares 72% identity with the homeobox conserved domain, and the expression level of HOX3 in the fuzzless mutants was extremely low compared to that in TM-1. *GhMYB3/GhMYB36* pertaining to the MYB family had the high expression level in fuzzy ovules in dominant/recessive mutants (Figure 9). A model of the MYB25/MYB2-DEL61/65-TTG2 protein complex was described as the initial pattern of cotton fuzz, similar to the model of trichomes and root hairs in *Arabidopsis* [33,53]. Future studies including analyses of protein function may shed light on the mechanism of cell initiation and formation of cotton fiber.

Supporting Information

Figure S1 Distribution of clean tag copy numbers for the 18 libraries.

Figure S2 Identity analysis of differentially expressed genes in various mutants.

Table S1 Genes involed in cotton fiber initial development and their *Arabidopsis* homologs.

Table S2 Raw sequence data for the 18 transcriptome libraries.

Table S3 List of 4,358 common DEGs between the mutants MD17FLM, SL1-7-1FLM, N1NSM and the wild-type TM-1.

Table S4 Pathway enrichment analysis for common differentially expressed genes in various dominant mutants.

Table S5 List of 6,693 common DEGs between the mutants XZ142FLM and n2NSM and the wild-type TM-1.

Table S6 Pathway enrichment analysis for common differentially expressed genes in various recessive mutants.

Table S7 List of 1,932 common DEGs between the mutants MD17FLM, SL1-7-1FLM, N1NSM, XZ142FLM, n2NSM and the wild-type TM-1.

Table S8 The common differentially expressed genes in various mutants and times.

Table S9 Pathway enrichment analysis for common differentially expressed genes in various dominant and recessive mutants.

Table S10 Primers used for qRT-PCR analysis.

Table S11 Correlation between qRT-PCR and DGE for 21 differentially expressed genes.

Acknowledgments

We thank Prof. Dr. Wangzhen Guo for his support and comments during the preparation of this manuscript.

Author Contributions

Conceived and designed the experiments: TZ. Performed the experiments: QW HZ WY HW. Analyzed the data: QW. Wrote the paper: QW TZ.

References

1. Zhang DY, Zhang TZ, Sang ZQ, Guo WZ (2007) Comparative development of lint and fuzz using different cotton fiber-specific developmental mutants in *Gossypium hirsutum*. J Integr Plant Bio 49 (7): 975–983
2. Larkin JC, Oppenheimer DG, Pollock S, Marks MD (1993) *Arabidopsis GLABROUS1* gene requires downstream sequences for function. Plant Cell 5: 1739–1748.
3. Schellmann S, Schnittger A, Kirik V, Wada T, Okada K, et al. (2002) *TRIPTYCHON* and *CAPRICE* mediate lateral inhibition during trichome and root hair patterning in *Arabidopsis*. EMBO J 21: 5036–5046.
4. Wada T, Tachibana T, Shimura Y, Okada K (1997) Epidermal cell differentiation in *Arabidopsis* determined by a Myb homolog, CPC. Science 277: 1113–1116.
5. Lee MM, Schiefelbein J (2001) Developmentally distinct MYB genes encode functionally equivalent proteins in *Arabidopsis*. Development 128: 1539–1546.
6. Larkin JC, Oppenheimer DG, Lloyd AM, Paparozzi ET, Marks MD (1994) Roles of the *GLABROUS1* and *TRANSPARENT TESTA GLABRA* genes in *Arabidopsis* trichome development. Plant Cell 6: 1065–1076.
7. Walker AR, Davison PA, Bolognesi-Winfield AC, James CM, Srinivasan N, et al. (1999) The *TRANSPARENT TESTA GLABRA1* locus, which regulates trichome differentiation and anthocyanin biosynthesis in *Arabidopsis*, encodes a WD40 repeat protein. Plant Cell 11: 1337–1349.
8. Payne CT, Zhang F, Lloyd AM (2000) *GL3* encodes a bHLH protein that regulates trichome development in *Arabidopsis* through interaction with GL1 and TTG1. Genetics 156: 1349–1362.
9. Zhang F, Gonzalez A, Zhao M, Payne CT, Lloyd A (2003) A network of redundant bHLH proteins functions in all TTG1-dependent pathways of *Arabidopsis*. Development 130: 4859–4869.
10. Szymanski DB, Jilk RA, Pollock SM, Marks MD (1998) Control of *GL2* expression in *Arabidopsis* leaves and trichomes. Development 125: 1161–1171.
11. Johnson CS, Kolevski B, Smyth DR (2002) *TRANSPARENT TESTA GLABRA2*, a trichome and seed coat development gene of *Arabidopsis*, encodes a WRKY transcription factor. Plant Cell 14: 1359–1375.
12. Neer EJ, Schmidt CJ, Nambudripad R, Smith TF (1994) The ancient regulatory protein family of WD-repeat proteins. Nature 371: 297–300.
13. Murre C, McCaw PS, Baltimore D (1989) A new DNA-binding and dimerization motif in immunoglobulin enhancer binding, daughterless, MYOD and MYC proteins. Cell 56: 777–783.
14. Bernhardt C, Lee MM, Gonzalez A, Zhang F, Lloyd A, et al. (2003) The *bHLH* genes *GLABRA3* (*GL3*) and *ENHANCER OF GLABRA3* (*EGL3*) specify epidermal cell fate in the *Arabidopsis* root. Development 130: 6431–6439.
15. Ramsay NA, Walker AR, Mooney M, Gray JC (2003) Two basic-helix-loop-helix genes (*MYC-146* and *GL3*) from *Arabidopsis* can activate anthocyanin biosynthesis in a white-flowered *Matthiola incana* mutant. Plant Mol Biol 52(3): 679–688.
16. Nesi N, Debeaujon I, Jond C, Pelletier G, Caboche M, et al. (2000) The *TT8* gene encodes a basic helix-loop-helix domain protein required for expression of *DFR* and *BAN* genes in *Arabidopsis* siliques. Plant Cell 12: 1863–1878.
17. Loguercio LL, Zhang JQ, Wilkins TA (1999) Differential regulation of six novel MYB-domain genes defines two distinct expression patterns in allotetraploid cotton (*Gossypium hirsutum* L.). Mol Genet Genomics 261: 660–671.
18. Suo J, Liang X, Pu L, Zhang YS, Xue YB (2003) Identification of *GhMYB109* encoding a R2R3 MYB transcription factor that expressed specifically in fiber initials and elongating fibers of cotton (*Gossypium hirsutum* L). Biochim Biophys Acta 1630: 25–34.
19. Wang S, Wang JW, Yu N, Li CH, Luo B, et al. (2004) Control of plant trichome development by a cotton fiber *MYB* gene. Plant Cell 16: 2323–2334.
20. Guan XY, Lee JJ, Pang MX, Shi XL, Stelly DM, et al. (2011) Activation of *Arabidopsis* seed hair development by cotton fiber-related genes. PLoS One 6(7): e21301
21. Huang YQ, Liu X, Tang KX, Zuo KJ (2013) Functional analysis of the seed coat-specific gene *GbMYB2* from cotton. Plant Physiol. Biochem.73: 16–22.
22. Noda K, Glover BJ, Linstead P, Martin C (1994) Flower colour intensity depends on specialized cell shape controlled by a Myb-related transcription factor. Nature 369: 661–664.
23. Perez-Rodriguez M, Jaffe FW, Butelli E, Glover BJ, Martin C (2005) Development of three different cell types is associated with the activity of a specific MYB transcription factor in the ventral petal of *Antirrhinum majus* flowers. Development 132: 359–370.
24. Wu Y, Llewellyn DJ, White R, Ruggiero K, Al-Ghazi Y, et al. (2007) Laser capture microdissection and cDNA microarrays used to generate gene expression profiles of the rapidly expanding fibre initial cells on the surface of cotton ovules. Planta 226: 1475–1490.
25. Machado A, Wu YR, Yang YM, Llewellyn DJ, Dennis ES (2009)The MYB transcription factor GhMYB25 regulates early fibre and trichome development. Plant J 59: 52–62
26. Wu YR, Machado AC, White RG, Llewellyn DJ, Dennis ES (2006) Expression profiling identifies genes expressed early during lint fibre initiation in cotton. Plant and Cell Physiology 47 (1): 107–127
27. Walford SA, Wu YR, Llewellyn DJ, Dennis ES (2011) GhMYB25-like: a key factor in early cotton fibre development. Plant J 65(5): 785–797

28. Taliercio EW, Boykin D (2007) Analysis of gene expression in cotton fiber initials. BMC Plant Biol 7: 22.
29. Humphries JA, Walker AR, Timmis JN, Orford SJ (2005) Two WD-repeat genes from cotton are functional homologues of the *Arabidposis thaliana TRANSPARENT TESTA GLABRA1* (*TTG1*) gene. Plant Mol Biol 57: 67–81.
30. Guan XX, Li QJ, Shan CM, Wang S, Mao YB, et al. (2008) The HD-Zip IV gene *GaHOX1* from cotton is a functional homologue of *Arabidopsis GLABRA2*. Physiol Plant 134: 174–182.
31. Mandaokar A, Kumar VD, Amway M, Browse J (2003) Microarray and differential display identify genes involved in jasmonate-dependent anther development. Plant Mol Biol 52: 775–786.
32. Shangguan XX, Xu B, Yu ZX, Wang LJ, Chen XY (2008) Promoter of a cotton fiber MYB gene functional in trichomes of *Arabidopsis* and glandular trichomes of tobacco. J Exp Bot 59(13): 3533–3542.
33. Guan XY, Yu N, Shangguan XX, Wang S, Lu S, et al. (2007) *Arabidopsis* trichome research sheds light on cotton fiber development mechanisms. Chinese Sci Bull 52: 1734–1741.
34. Lee JJ, Woodward AW, Chen ZJ (2007) Gene expression changes and early events in cotton fibre development. Ann Bot 100: 1391–1401.
35. Kearney TH, Harrison GJ (1927) The inheritance of smoothness seeds in cotton. J Agric Res 35: 193–217.
36. Turley JO, Benedict LI, Rolfe WH (1947) A recessive naked-seed character in upland cotton. J Hered 38: 313–320.
37. Turley RB, Kloth RH (2002) Identification of a third fuzzless seed locus in upland cotton (*Gossypium hirsutum* L.). J Hered 93(5): 359–364.
38. Turley RB, Kloth RH (2008) The inheritance model for the fiberless trait in upland cotton (*Gossypium hirsutum* L.) line SL1-7-1: variation on a theme. Euphytica 164: 123–132.
39. Zhang T, Pan J (1991) Genetic analysis of a fuzzless-lintless mutant in *Gossypium hirsutum*. J Agric Sci 7(3): 13–16.
40. Du XM, Pan JJ, Wang RH, Zhang TZ, Shi TZ (2001) Genetic analysis of presence and absence of lint and fuzz in cotton. Plant Breed 120: 519–522
41. Turley RB (2002) Registration of MD 17 fiberless upland cotton as a genetic stock. Crop Sci 42: 994–995.
42. Kohel RJ (1973) Genetic nomenclature in cotton. J Hered 64: 291–295.
43. Jiang JX, Zhang TZ (2003) Extraction of total RNA in cotton tissues with CTAB-acidic phenolic method. Cotton Sci 15: 166–167.
44. Wang QQ, Liu F, Chen XS, Ma XJ, Zeng HQ, et al. (2010) Transcriptome profiling of early developing cotton fiber by deep-sequencing reveals significantly differential expression of genes in a fuzzless/lintless mutant. Genomics 96(6): 369–376.
45. Audic S, Claverie JM (1997) The significance of digital gene expression profiles. Genome Res 7: 986–995.
46. Benjamini Y, Yekutieli D (2001) The control of the false discovery rate in multiple testing under dependency. Ann Stat 29: 1165–1188.
47. Thimm O, Blasing O, Gibon Y, Nagel A, Meyer S, et al. (2004) MapMan: a user-driven tool to display genomics data sets onto diagrams of metabolic pathways and other biological processes. Plant J 37: 914–939.
48. Kanehisa M, Araki M, Goto S, Hattori M, Hirakawa M, et al. (2008) KEGG for linking genomes to life and the environment. Nucleic Acids Res 36: D480–D484.
49. Hegedus Z, Zakrzewska A, Agoston VC, Ordas A, Racz P, et al. (2009) Deep sequencing of the zebrafish transcriptome response to mycobacterium infection. Mol Immunol 46: 2918–2930.
50. Stewart JM (1975) Fiber initiation on the cotton ovule (*Gossypium hirsutum*). Am J Bot 62: 723–730.
51. Lee JJ, Hassan OSS, Gao W, Wei NE, Kohel RJ, et al. (2006) Developmental and gene expression analyses of a cotton naked seed mutant. Planta 233: 418–432.
52. Morrissy AS, Morin RD, Delaney A, Zeng T, McDonald H, et al. (2010) Next-generation tag sequencing for cancer gene expression profiling. Genome Res 19: 1825–1835.
53. Jensen WA (1968) Cotton Embryogenesis: The Zygote. Planta (Berl.) 79: 346–366
54. Du SJ, Dong CJ, Zhang B, Lai TF, Du XM (2013) Comparative proteomic analysis reveals differentially expressed proteins correlated with fuzz fiber initiation in diploid cotton (*Gossypium arboreum* L.). J Proteomics 82: 113–129.
55. Schaller A (2004) A cut above the rest: the regulatory function of plant proteases. Planta 220: 183–197.
56. Hovav R, Udall JA, Hovav E, Rapp R, Flagel L, et al. (2008) A majority of cotton genes are expressed in single-celled fiber. Planta 227: 319–329.
57. Ramsay NA, Glover BJ (2005) MYB-bHLH-WD40 protein complex and the evolution of cellular diversity. Trends Plant Sci 10(2): 63–70.

PERMISSIONS

All chapters in this book were first published in PLOS ONE, by The Public Library of Science; hereby published with permission under the Creative Commons Attribution License or equivalent. Every chapter published in this book has been scrutinized by our experts. Their significance has been extensively debated. The topics covered herein carry significant findings which will fuel the growth of the discipline. They may even be implemented as practical applications or may be referred to as a beginning point for another development.

The contributors of this book come from diverse backgrounds, making this book a truly international effort. This book will bring forth new frontiers with its revolutionizing research information and detailed analysis of the nascent developments around the world.

We would like to thank all the contributing authors for lending their expertise to make the book truly unique. They have played a crucial role in the development of this book. Without their invaluable contributions this book wouldn't have been possible. They have made vital efforts to compile up to date information on the varied aspects of this subject to make this book a valuable addition to the collection of many professionals and students.

This book was conceptualized with the vision of imparting up-to-date information and advanced data in this field. To ensure the same, a matchless editorial board was set up. Every individual on the board went through rigorous rounds of assessment to prove their worth. After which they invested a large part of their time researching and compiling the most relevant data for our readers.

The editorial board has been involved in producing this book since its inception. They have spent rigorous hours researching and exploring the diverse topics which have resulted in the successful publishing of this book. They have passed on their knowledge of decades through this book. To expedite this challenging task, the publisher supported the team at every step. A small team of assistant editors was also appointed to further simplify the editing procedure and attain best results for the readers.

Apart from the editorial board, the designing team has also invested a significant amount of their time in understanding the subject and creating the most relevant covers. They scrutinized every image to scout for the most suitable representation of the subject and create an appropriate cover for the book.

The publishing team has been an ardent support to the editorial, designing and production team. Their endless efforts to recruit the best for this project, has resulted in the accomplishment of this book. They are a veteran in the field of academics and their pool of knowledge is as vast as their experience in printing. Their expertise and guidance has proved useful at every step. Their uncompromising quality standards have made this book an exceptional effort. Their encouragement from time to time has been an inspiration for everyone.

The publisher and the editorial board hope that this book will prove to be a valuable piece of knowledge for researchers, students, practitioners and scholars across the globe.

LIST OF CONTRIBUTORS

Mingwei Du, Yi Li, Xiaoli Tian, Liusheng Duan, Mingcai Zhang, Weiming Tan and Zhaohu Li
State Key Laboratory of Plant Physiology and Biochemistry, Engineering Research Center of Plant Growth Regulator, Ministry of Education, College of Agronomy and Biotechnology, China Agricultural University, Beijing, China

Dongyong Xu
Hebei Provincial Engineering Research Center of Cotton Seed, Hejian, Hebei, China

Wenfang Gong, Shoupu He, Jiahuan Tian, Junling Sun, Zhaoe Pan, Yinhua Jia and Xiongming Du
State Key Laboratory of Cotton Biology, Institute of Cotton Research, Chinese Academy of Agricultural Sciences, Anyang, China

Gaofei Sun
State Key Laboratory of Cotton Biology, Institute of Cotton Research, Chinese Academy of Agricultural Sciences, Anyang, China

Department of Computer Science and Information Engineering, Anyang Institute of Technology, Anyang, China

Wajad Nazeer
State Key Laboratory of Crop Genetics and Germpalsm Enhancement, MOE Hybrid Cotton R&D Engineering Research Center, Nanjing Agricultural University, Nanjing, Jiangsu Province, China

Cotton Research Station, Multan, Ayub Agricultural Research Institute, Faisalabad, Punjab, Pakistan

Baoliang Zhou
State Key Laboratory of Crop Genetics and Germpalsm Enhancement, MOE Hybrid Cotton R&D Engineering Research Center, Nanjing Agricultural University, Nanjing, Jiangsu Province, China

Abdul Latif Tipu, Saghir Ahmad and Khalid Mahmood
Cotton Research Station, Multan, Ayub Agricultural Research Institute, Faisalabad, Punjab, Pakistan

Abid Mahmood
Cotton Research Institute, Ayub Agricultural Research Institute, Faisalabad, Punjab, Pakistan

Guanze Liu
National Key Laboratory of Crop Genetic Improvement, Huazhong Agricultural University, Wuhan, Hubei, P. R. China

College of Tobacco Science, Yunnan Agricultural University, Kunming, Yunnan, P. R. China

Shuangxia Jin, Xuyan Liu, Longfu Zhu, Yichun Nie and Xianlong Zhang
National Key Laboratory of Crop Genetic Improvement, Huazhong Agricultural University, Wuhan, Hubei, P. R. China

Xuelin Li
Agricultural College, Henan University of Science and Technology, Luoyang, Henan, P. R. China

Xiuling Wang, Yan Yan, Yuzhen Li, Xiaoqian Chu, Changai Wu and Xingqi Guo
State Key Laboratory of Crop Biology, College of Life Sciences, Shandong Agricultural University, Taian, Shandong, PR China

Kenji Osabe, Jenny D. Clement, Frank Bedon, Filomena A. Pettolino, Lisa Ziolkowski, Danny J. Llewellyn, E. Jean Finnegan and Iain W. Wilson
CSIRO, Plant Industry, ACT, Australia

Natália A. Leite, Alberto S. Corrêa and Celso Omoto
Departamento de Entomologia e Acarologia, Escola Superior de Agricultura "Luiz de Queiroz", Universidade de São Paulo, Piracicaba, São Paulo, Brazil

Alessandro Alves-Pereira
Departamento de Genética, Escola Superior de Agricultura "Luiz de Queiroz", Universidade de São Paulo, Piracicaba, São Paulo, Brazil

Maria I. Zucchi
Agência Paulista de Tecnologia dos Agronegócios, Piracicaba, São Paulo, Brazil

Wu Zheng, Xueyan Zhang, Zuoren Yang, Fenglian Li, Lanling Duan, Chuanliang Liu, Lili Lu, Chaojun Zhang and Fuguang Li
State Key Laboratory of Cotton Biology, Institute of Cotton Research, Chinese Academy of Agricultural Sciences, Anyang, Henan, China

Jiahe Wu
State Key Laboratory of Cotton Biology, Institute of Cotton Research, Chinese Academy of Agricultural Sciences, Anyang, Henan, China
Institute of Microbiology, Chinese Academy of Sciences, Beijing, China

Xiaoyu Zhi, Yingchun Han, Shuchun Mao, Guoping Wang, Lu Feng, Beifang Yang, Zhengyi Fan, Wenli Du, Jianhua Lu and Yabing Li
Institute of Cotton Research of the Chinese Academy of Agricultural Sciences/State Key Laboratory of Cotton Biology, Anyang, 455000, Henan, China

Dhivyaa Rajasundaram and Joachim Selbig
Institute of Biochemistry and Biology, University of Potsdam, Potsdam-Golm, 14476, Germany

Max-Planck Institute of Molecular Plant Physiology, Potsdam-Golm, 14476, Germany

Jean-Luc Runavot and Frank Meulewaeter
Bayer CropScience NV-Innovation Center, Technologiepark 38, 9052 Gent, Belgium

Xiaoyuan Guo and William G. T. Willats
Department of Plant and Environmental Sciences, Faculty of Sciences, University of Copenhagen, Thorvaldsensvej, 40 1.1871, Fredriksberg C, Denmark

Jeffrey A. Fabrick
U.S. Department of Agriculture, Agricultural Research Service, U.S. Arid Land Agricultural Research Center, Maricopa, Arizona, United States of America

Jeyakumar Ponnuraj
National Institute of Plant Health Management, Rajendranagar, Hyderabad, Andhra Pradesh, India

Amar Singh and Raj K. Tanwar
National Centre for Integrated Pest Management, Indian Agricultural Research Institute, New Delhi, Delhi, India

Gopalan C. Unnithan, Alex J. Yelich, Xianchun Li, Yves Carriére and Bruce E. Tabashnik
Department of Entomology, University of Arizona, Tucson, Arizona, United States of America

Lei Fang, Ruiping Tian, Jiedan Chen, Sen Wang, Xinghe Li, Peng Wang and Tianzhen Zhang
National Key Laboratory of Crop Genetics and Germplasm Enhancement, Cotton Hybrid R & D Engineering Center (the Ministry of Education), Nanjing Agricultural University, Nanjing, China

Lei Zhang, Huijuan Ma, Tingting Chen, Jun Pen, Shuxun Yu and Xinhua Zhao
State Key Laboratory of Cotton Biology, Institute of Cotton Research of CAAS, Anyang, P. R. China

Yue-Hua Xiao, Qian Yan, Hui Ding, Ming Luo, Lei Hou, Mi Zhang, Dan Yao, Hou-Sheng Liu, Xin Li,Jia Zhao and Yan Pei
Biotechnology Research Center, Southwest University, Beibei, Chongqing, China

Shuli Fan, Meizhen Song, Chaoyou Pang, Hengling Wei and Shuxun Yu
State Key Laboratory of Cotton Biology, Institute of Cotton Research, Chinese Academy of Agricultural Sciences, Anyang, Henan Province, China

Yanyan Meng, Xianjin Zhan, Jiayang Lan, Changhui Feng and Shengxi Zhang
Key Laboratory of Cotton Biology and Breeding in the Middle Reaches of the Changjing River, Institute of EconCormopic, Hubei Academy of Agricultural Science, Wuhan, Hubei Province, China

Ji Liu
State Key Laboratory of Cotton Biology, Institute of Cotton Research, Chinese Academy of Agricultural Sciences, Anyang, Henan Province, China
College of Agronomy, Northwest A & F University, Yangling, Shaanxi Province, China Crop, Hubei Academy of Agricultural Science, Wuhan, Hubei Province, China

Geo Coppens d'Eeckenbrugge
CIRAD, UMR 5175 CEFE, Campus du CNRS, Montpellier, France

Jean-Marc Lacape
CIRAD, UMR AGAP, Montpellier, France

Matthew K. Gilbert, Hee Jin Kim, Marina Naoumkina and David D. Fang
Cotton Fiber Bioscience Research Unit, USDA-ARS, Southern Regional Research Center, New Orleans, Louisiana, United States of America

Yuhong Tang
The Samuel Roberts Noble Foundation, Genomics Core Facility, Ardmore, Oklahoma, United States of America

Roch K. Dabiré, Moussa Namountougou, Abdoulaye Diabaté, Dieudonné D. Soma and Joseph Bado
IRSS (Institut de Recherche en Sciences de la Santé), Centre Muraz, Bobo Dioulasso, Burkina Faso

Hyacinthe K. Toé
IRSS (Institut de Recherche en Sciences de la Santé), Centre Muraz, Bobo-Dioulasso, Burkina Faso
Department of Vector Biology, Liverpool School of Tropical Medicine, Liverpool, United Kingdom

Chris Bass
Biological Chemistry and Crop Protection Rothamsted Research, Harpenden, United Kingdom

Patrice Combary
National Malaria Control Programme, Ministry of Health, Ouagadougou, Burkina Faso

Qun Wan, Hua Zhang, Wenxue Ye, Huaitong Wu and Tianzhen Zhang
National Key Laboratory of Crop Genetics and Germplasm Enhancement, Cotton Research Institute, Nanjing Agricultural University, Nanjing, China

Index

www.ingramcontent.com/pod-product-compliance
Lightning Source LLC
Chambersburg PA
CBHW061254190326
41458CB00011B/3665

* 9 7 8 1 6 8 2 8 6 3 7 4 9 *